Encyclopedia of Cell Biology

Encyclopedia of Cell Biology

Edited by **Ralph Becker**

New York

Published by Callisto Reference,
106 Park Avenue, Suite 200,
New York, NY 10016, USA
www.callistoreference.com

Encyclopedia of Cell Biology
Edited by Ralph Becker

© 2015 Callisto Reference

International Standard Book Number: 978-1-63239-219-0 (Hardback)

This book contains information obtained from authentic and highly regarded sources. Copyright for all individual chapters remain with the respective authors as indicated. A wide variety of references are listed. Permission and sources are indicated; for detailed attributions, please refer to the permissions page. Reasonable efforts have been made to publish reliable data and information, but the authors, editors and publisher cannot assume any responsibility for the validity of all materials or the consequences of their use.

The publisher's policy is to use permanent paper from mills that operate a sustainable forestry policy. Furthermore, the publisher ensures that the text paper and cover boards used have met acceptable environmental accreditation standards.

Trademark Notice: Registered trademark of products or corporate names are used only for explanation and identification without intent to infringe.

Printed in the United States of America.

Contents

	Preface	IX
Section 1	Molecular and Cellular Regulatory Mechanisms	1
Chapter 1	Exploring Secrets of Nuclear Actin Involvement in the Regulation of Gene Transcription and Genome Organization Yong Zhong Xu, Cynthia Kanagaratham and Danuta Radzioch	3
Chapter 2	Signaling of Receptor Tyrosine Kinases in the Nucleus Sally-Anne Stephenson, Inga Mertens-Walker and Adrian Herington	33
Chapter 3	The Kinetochore and Mitosis: Focus on the Regulation and Correction Mechanisms of Chromosome-to-Microtubule Attachments Rita M. Reis and Hassan Bousbaa	57
Chapter 4	*Drosophila*: A Model System That Allows *in vivo* Manipulation and Study of Epithelial Cell Polarity Andrea Leibfried and Yohanns Bellaïche	77
Chapter 5	Development and Cell Polarity of the C. elegans Intestine Olaf Bossinger and Michael Hoffmann	97
Chapter 6	Intercellular Communication Nuri Faruk Aykan	123
Chapter 7	Molecular and Sub-Cellular Gametogenic Machinery of Stem and Germline Cells Across Metazoa Andrey I. Shukalyuk and Valeria V. Isaeva	139

Chapter 8	**G Protein-Coupled Receptors-Induced Activation of Extracellular Signal-Regulated Protein Kinase (ERK) and Sodium-Proton Exchanger Type 1 (NHE1)** Maria N. Garnovskaya	175
Section 2	**Cellular Basis of Disease and Therapy**	199
Chapter 9	**Biology of Cilia and Ciliopathies** David Alejandro Silva, Elizabeth Richey and Hongmin Qin	201
Chapter 10	**Adult Stem Cells in Tissue Homeostasis and Disease** Elena Lazzeri, Anna Peired, Lara Ballerini and Laura Lasagni	231
Chapter 11	**Claudins in Normal and Lung Cancer State** V. Morales-Tlalpan, C. Saldaña, P. García-Solís, H. L. Hernández-Montiel and H. Barajas-Medina	257
Chapter 12	**The Roles of ESCRT Proteins in Healthy Cells and in Disease** Jasmina Ilievska, Naomi E. Bishop, Sarah J. Annesley and Paul R. Fisher	275
Chapter 13	**Autologous Grafts of Mesenchymal Stem Cells – Between Dream and Reality** Frédéric Torossian, Aurelie Bisson, Laurent Drouot, Olivier Boyer and Marek Lamacz	303
Section 3	**New Methods in Cell Biology**	323
Chapter 14	**Salivary Glands: A Powerful Experimental System to Study Cell Biology in Live Animals by Intravital Microscopy** Monika Sramkova, Natalie Porat-Shliom, Andrius Masedunkas, Timothy Wigand, Panomwat Amornphimoltham and Roberto Weigert	325
Chapter 15	**Regeneration and Recycling of Supports for Biological Macromolecules Purification** Marcello Tagliavia and Aldo Nicosia	347
Chapter 16	**Evaluation of Mitochondrial DNA Dynamics Using Fluorescence Correlation Analysis** Yasutomo Nomura	369

Permissions

List of Contributors

Preface

This book provides an extensive analysis of the present knowledge regarding cell biology. The text consists of multiple chapters which can broadly be divided into categories such as regulatory mechanisms, cellular therapy and new methods in biology. However, due to the interdisciplinary approaches utilized by the authors this categorization is not rigid. The current frontiers explored within this book, sets the foundation for further research. Also, the views represented are visible in various areas of fundamental biology, biotechnology, biomedicine and other applications of the information regarding cell biology. This book will help beginners to gain interest and provide experts with new information in the field.

Significant researches are present in this book. Intensive efforts have been employed by authors to make this book an outstanding discourse. This book contains the enlightening chapters which have been written on the basis of significant researches done by the experts.

Finally, I would also like to thank all the members involved in this book for being a team and meeting all the deadlines for the submission of their respective works. I would also like to thank my friends and family for being supportive in my efforts.

Editor

Section 1

Molecular and Cellular Regulatory Mechanisms

Exploring Secrets of Nuclear Actin Involvement in the Regulation of Gene Transcription and Genome Organization

Yong Zhong Xu, Cynthia Kanagaratham and Danuta Radzioch
McGill University,
Canada

1. Introduction

Actin is one of the most abundant proteins in eukaryotic cells. It is a 43-kDa protein that was originally identified and purified from skeletal muscle. Once thought to be simply a component of muscle cells, actin has later been shown to be a highly conserved and ubiquitiously distributed protein in eukaryotic cells. It has been extensively studied as a cytoplasmic cytoskeletal protein that is involved in a wide range of cellular processes, including cell motility, growth and cytokinesis; endocytosis, exocytosis and secretion; signal transduction, synaptic transmission as well as intracellular trafficking (Ascough, 2004;Brakebusch and Fassler, 2003;Suetsugu and Takenawa, 2003). In the cytoplasm, actin exists in equilibrium between monomers (globular- or G-actin) and polymers (filamentous- or F-actin). The dynamics of actin, the coordinated assembly and disassembly of actin filaments in response to cellular and extracellular signaling, is critical for the diverse functions of actin and is tightly regulated by a plethora of actin-binding proteins (ABPs) in the cytoplasm (dos Remedios *et al.*, 2003). To date, over 70 distinct classes of ABPs have being identified and the inventory is still far from been completed (Pollard and Borisy, 2003).

While the cytoplasmic functions of actin are well established, the findings obtained from studies on nuclear actin have encountered consistent skepticism for many years. Presence of actin in the nucleus was considered to be cytoplasmic contamination from extraction or fixation procedures, or antibody cross-reactivity (Pederson and Aebi, 2002;Shumaker *et al.*, 2003). In addition, many known functions of actin in the cytoplasm are associated with the polymerization of actin into filaments, which can be detected by phalloidin staining. However, under normal conditions, nuclei cannot be stained by phalloidin. Nevertheless, in the past decade, there has been convincing data demonstrating that actin, actin-related proteins (Arps) as well as ABPs are not only present in the nucleus but also play important roles in diverse nuclear activities. Actin has been localized to specialized subnuclear compartments such as the nucleoli, splicing speckles and Cajal bodies (Fomproix and Percipalle, 2004;Gedge *et al.*, 2005;Saitoh *et al.*, 2004). In these subnuclear compartments, actin proves to be involved in almost all the processes associated with gene expression, from chromatin remodeling via transcription to ribonucleoprotein (RNP) assembly and maturation, as well as mRNA nuclear export (Blessing *et al.*, 2004;Chen and Shen, 2007;Olave *et al.*, 2002). Other nuclear processes in which actin is implicated, include assembly of the nuclear structure (Krauss et al., 2003;Krauss et al., 2002;Olave et al., 2002), genome organization, and regulation of transcription factor activity (Olave et al., 2002;Vartiainen et al., 2007).

In this chapter, the several aspects related to the nuclear actin presence and its importance in the regulation of gene expression will be reviewed.

2. Nuclear architecture and distribution of actin

The cell nucleus is a complex and multi-functional organelle, which displays a high degree of spatial organization and structural integrity. The most well characterized structural component of the cell nucleus is the nuclear lamina, mainly composed of A- and B-type laminas as well as lamina–associated proteins (Stewart et al., 2007). The laminas are evolutionarily conserved nuclear-specific intermediate filaments that are essential for many nuclear functions, including the maintenance of nuclear shape, DNA replication, transcription, chromatin organization, cell cycle regulation and apoptosis (Andres and Gonzalez, 2009;Vlcek and Foisner, 2007;Wiesel et al., 2008). Actin has been shown to interact with the c-terminus of A-type laminas (Sasseville and Langelier, 1998). A-type laminas are connected to the cytoskeleton by a linker of nucleoskeleton and cytoskeleton (LINC) complex found in the nuclear envelope. Connecting the A-type laminas to the cytoskeleton is necessary for nuclear migration and positioning within the cell as well as for transmitting mechanical signals from the cytoplasm to the nucleus (Starr, 2009;Tzur et al., 2006;Worman and Gundersen, 2006).

Two important components of the LINC complex are Sun domain proteins and Nesprins. Located on the outer nuclear membrane, Nesprin 1 and 2 can interact with F-actin as well as Sun 1 and Sun 2 located on the inner nuclear membrane. Sun proteins, in turn, bind to lamina A (Crisp et al., 2006;Ostlund et al., 2009). Emerin, a lamina-associated protein, is also important for nuclear structure and has been shown to bind to actin. The interactions between actin, lamina and emerin indicate that an actin-containing structural network exists at the nuclear envelope and is involved in maintaining the nuclear structure and nuclear functions (Fairley et al., 1999;Holaska and Wilson, 2007;Lattanzi et al., 2003). The importance of actin in nuclear assembly was demonstrated using *Xenopus* egg extracts in which nuclear assembly is initiated after fluorescence-labelled actin is added. Moreover, the nuclear assembly gets blocked by Latrunculin A, which binds to G-actin and inhibits F-actin formation, suggesting that F-actin is required for nuclear assembly. In addition, the interaction between actin and protein 4.1 is implicated in this process (Krauss et al., 2003;Krauss et al., 2002).

Actin has also been associated with the nuclear matrix (Capco et al., 1982;Okorokov et al., 2002;Valkov et al., 1989;Verheijen et al., 1986). The nuclear matrix is a network of proteins throughout the inside of nucleus, which provides a structural framework for maintaining spatial order within nucleus and for proper nuclear functions , such as DNA replication and repair, gene transcription, RNA splicing and transport (Berezney, 2002;Berezney et al., 1996;Hancock, 2000). It is tempting to speculate that nuclear actin acts as a component of intranuclear filament network (or nucleoskeleton) that is analogous to cytoskeleton. This was supported by a study showing a colocalization between actin and EAST (enhances adult sensory threshold), a structural protein of the nucleus. In a Drosophila model, EAST has been shown to be a ubiquitious nuclear protein forming a network throughout the nucleus (Wasser and Chia, 2000). A number of studies have also confirmed that an actin-containing filament network exists in the nucleus. Studies of the *Xenopus* oocyte nuclei using electron microscopy have found that filaments containing actin and protein 4.1 form a network that attach to Cajal bodies and other subnuclear organelles (Kiseleva *et al.*, 2004). In this manner, the meshwork of actin-containing filaments might contribute to the nuclear compartmentalization.

3. Regulation of nuclear actin

3.1 The form of actin in the nucleus

Actin has been shown to be involved in diverse nuclear processes; but how and in what form actin takes part in these events remains to be elucidated. It has been suggested that nuclear actin coexists as a monomer (G-actin), short oligomer and polymer structure (Gieni and Hendzel, 2009;McDonald *et al.*, 2006). These different forms of nuclear actin are believed to be required for a variety of processes in the nucleus. There has been a great body of evidence in support of the presence of G-actin in the nucleus (Pederson and Aebi, 2002;Pederson and Aebi, 2005). Firstly, a number of G-actin binding proteins have been identified in the nucleus, including cofilin, profilin, β-thymosin, gelsolin and gelsolin-like protein (Huff *et al.*, 2004;Pendleton *et al.*, 2003;Percipalle, 2009;Prendergast and Ziff, 1991;Skare *et al.*, 2003). Secondly, using DNase I affinity chromatography, actin can be copurified with RNA polymerase I and II machinery (Fomproix and Percipalle, 2004;Kukalev *et al.*, 2005;Obrdlik *et al.*, 2008). DNase I binds to G-actin with very high affinity and F-actin with low affinity (Zechel, 1980). This suggesting that actin co-precipitated with RNA polymerase I and II is likely to be present in its monomeric or short oligomeric form. Thirdly, monoclonal antibodies directed against epitopes which are unique to monomeric or dimeric actin, display distinctive immunostaining of the nucleus (Jockusch *et al.*, 2006). Fourthly, the nuclear lamina proteins, such as lamina A (Sasseville and Langelier, 1998), emerin (Lattanzi *et al.*, 2003), and nesprin (Zhang *et al.*, 2002a) form complexes with actin. Biochemical evidence reveals that G-actin is present in these complexes.

It has been very challenging to document polymerization status of actin in the nucleus. Phalloidin staining is the most common method used for detecting actin filaments in the cytoplasm. Under physiological conditions, nuclear actin present in most of the cells cannot be detected by phalloidin staining, which specifically recognizes actin filaments of at least seven subunits in length. However, under certain cellular stress conditions, distinctive actin rods (also called bundles or paracrystals) can be induced in the nucleus in a variety of cell types. These conditions include dimethyl sulfoxide (DMSO) treatment (Sanger *et al.*, 1980a;Sanger *et al.*, 1980b), heat shock (Iida *et al.*, 1986;Welch and Suhan, 1985), Latrunculin B treatment and ATP deletion (Pendleton *et al.*, 2003) as well as viral infection (Charlton and Volkman, 1991;Feierbach *et al.*, 2006). Cellular stress-induced formation of actin filaments seems to be caused by an increased nuclear actin level because nuclear translocation and accumulation of actin are also observed at the same time. This is supported by the observation that actin filaments exist in the *Xenopus* oocytes, which have a very high concentration of actin (~2mg/ml) due to the lack of nuclear export receptor, exportin 6 (Bohnsack *et al.*, 2006;Clark and Rosenbaum, 1979;Roeder and Gard, 1994;Stuven *et al.*, 2003). In addition, some nuclear-actin dependent functions, such as nuclear export of RNA and proteins (Hofmann *et al.*, 2001), nuclear envelope assembly (Krauss *et al.*, 2003), transcription (McDonald *et al.*, 2006) and intranuclear movement of Herpes simplex virus-1 capsid

(Forest et al., 2005) as well as movement of chromosome loci (Hu et al., 2008) can be inhibited by Latrunculin B, a drug that binds G-actin with high affinity and prevents polymerization and thus F-actin formation (Spector et al., 1989). These indirect evidence imply that some sort of polymerized actin exist in the nucleus to carry out corresponding nuclear functions. The presence of polymeric actin in the nucleus was also shown (McDonald et al., 2006) in living cells using fluorescence recovery after photobleaching (FRAP) experiments. In that study, FRAP, which allows to analyze the dynamic properties of GFP-actin in the nucleus, shows that both a fast recovery and a slow recovery GFP-actin exist in the nucleus. Moreover, the latter type of actin is sensitive to actin mutants and Latrunculin B. Therefore, the slow species represents a polymeric form of actin with distinctive dynamics which is quite different from the actin dynamics observed in the cytoplasm. Interestingly, recent studies provided evidence that the nuclear polymeric actin is important for RNA polymerase I-mediated transcription and transcriptional activation of HoxB genes by RNA polymerase II (Ferrai et al., 2009;Ye et al., 2008).

3.2 Regulation of nuclear translocation of actin

Extracellular stress can induce nuclear translocation of actin. Sanger and colleagues demonstrated that a disappearance of stress fibers from the cytoplasm and a reversible translocation of cytoplasmic actin into the nucleus occur after treatment of PtK2 and WI-38 cells with 10% DMSO (Sanger et al., 1980a;Sanger et al., 1980b). Courgeon and colleagues showed that heat shock causes actin to accumulate in the nucleus of *Drosophila* cells (Courgeon et al., 1993). In mast cells, entry of actin into the nucleus was induced by either treatment with Latrunculin B, or ATP depletion (Pendleton et al., 2003). Most recently, nuclear translocation of actin was found in HL-60 cells and human peripheral blood monocytes when differentiated to macrophages by phorbol 12 myristate 13-acetate (PMA) (Xu et al., 2010). These results suggest that actin is able to shuttle between the cytoplasm and the nucleus. To date, the molecular mechanism by which actin enters into the nucleus in response to cellular stress has not been established.

The nuclear envelope is a lipid bilayer that forms a barrier between the nuclear and cytoplasmic spaces. The traffic between nucleus and cytoplasm is mediated through nuclear pore complexes (NPCs) embedded in the nuclear envelope. NPCs allow passive diffusion of small molecules (such as ion and protein smaller than 40 kDa) but restrict the movement of larger molecules across the nuclear envelope. Macromolecules usually carry specific signals allowing them to access the nucleocytoplasmic transport machinery. Monomeric actin has a molecular weight of ~43 kDa, therefore it is unlikely to enter into nucleus by diffusion. Actin lacks a classical nuclear localization signal (NLS) and to date, no specific import receptor for actin has been identified. Therefore it most likely relies on an active carrier which guides it into the nucleus. Cofilin, an actin-binding protein, is suggested to be involved in the regulation of nuclear import of actin. Cofilin contains a NLS and it has been recognized as a component of intranuclear actin rods in response heat shock and DMSO treatment (Nishida et al., 1987). A study by Pendleton et al. showed that stress-induced nuclear accumulation of actin was blocked by an anti-cofilin antibody, demonstrating that cofilin is required for actin import into the nucleus (Pendleton et al., 2003).

For nuclear export, actin seems to use an active transport mechanism. The actin polypeptide has two well conserved nuclear export signals (NESs). In yeast, these two sequences were specifically recognized by chromosome region maintenance 1 (CRM1, also known as exportin 1), a general export receptor for cargos bearing leucine-rich export signals, and actin can then be rapidly removed from nucleus. Transfection of cells with mutant actin lacking NESs or inhibition of CRM1 by leptomycin B results in nuclear accumulation of actin (Wada et al., 1998). Exportin 6, a member of the importin β superfamily of transport receptor, is responsible for nuclear actin export in mammalian cells (Stuven et al., 2003). Knockdown of exportin 6 by RNA interference also leads to nuclear accumulation of actin and the formation of actin rods. Interestingly, exportin 6 recognizes the actin:profilin complex rather than actin or profilin individually, suggesting a difference in the form of actin being presented to CRM1 and to exportin 6.

So far, the exact roles of nuclear accumulation of actin in response to external signals remain to be understood. Nuclear actin controls transcription of its target genes through several different ways: (1) Actin specifically binds to a 27-nt repeat element in the intron 4 of the endothelial nitric oxide synthase gene to regulate its expression (Ou et al., 2005;Wang et al., 2002); (2) Actin participates in chromatin remodeling for gene activation as a component of the chromatin remodeling complex (Rando et al., 2002;Song et al., 2007;Zhao et al., 1998); (3) Actin plays a direct role in RNA transcription by being part of the pre-initiation complex with RNA polymerase II (Hofmann et al., 2004). (4) Actin participates in transcriptional elongation as a component of RNP particles. Therefore, it is tempting to speculate that under stress, actin translocates into nuclei to function as a transcriptional modulator, playing an important role in the regulation of gene transcription along with stress-activated transcription factor. This hypothesis is supported by recent studies showing that nuclear accumulation of actin is involved in transcriptional activation of SLC11A1 gene during macrophage-like differentiation of HL-60 cells induced by PMA (Xu et al., 2011;Xu et al., 2010).

3.3 Regulation of actin polymerization

It is believed that the concentration of nuclear actin is sufficient for spontaneous polymerization. Therefore, in order to have dynamic equilibrium of the different forms of actin, an active process preventing polymerization is required.

Many of the regulators known to control cytoplasmic actin dynamics have also been shown to be present in the nucleus (Table 1). These regulators include Arps such as Arp 2/3; and ABPs such as cofilin, profilin and CapG; and signalling molecules (see section 3.4). In humans, Arp2/3 represents a stable complex of two Arps (Arp2 and Arp3) and five other subunits including p16, p20, p21, p34, p41 (Deeks and Hussey, 2005;Welch et al., 1997). The Arp2/3 complex is capable of initiating *de novo* polymerization of actin and stimulating the formation of branched actin filaments when activated by members of Wiskott-Aldrich syndrome protein (WASP) family (Higgs and Pollard, 2001;Machesky and Insall, 1998;Pollard and Borisy, 2003;Volkmann et al., 2001). The WASP family members share a common C-terminal verprolin-cofilin-acidic (VCA) region. Polymerization of actin is initiated by the interaction of the VCA region with both Arp2/3 complex and an actin monomer, forming the first subunit of *de novo* actin polymer (Dayel and Mullins, 2004;Kim et al., 2000;Prehoda et al., 2000;Rohatgi et al., 1999). The potential role of Arp2/3 in the regulation of actin dynamics in the nucleus was suggested based on the viral infection studies, for example infection with baculovirus, results in accumulation of Arp2/3 complex in the nucleus, where it becomes activated by WASP-like virus protein p78/83. This event in turn results in Arp2/3-mediated actin polymerization that is essential for virus replication (Goley et al., 2006). Furthermore, it has been demonstrated that N-WASP and Arp2/3 complex associate with RNA polymerase II and regulate the efficiency of gene transcription.

Induction of actin polymerization through the N-WASP-Arp2/3 complex pathway has been shown to be required for efficient transcription by RNA polymerase II (Wu et al., 2006;Yoo et al., 2007). Importance of the Arp2/3 complex –mediated actin polymerization in other nuclear actin-dependent processes remains to be fully elucidated.

Protein	Roles in the nucleus	References
Arp 2/3	De novo actin polymerization	Higgs and Pollard, 2001
	Formation of Branched actin filaments	Pollard et al., 2003
	Associated with transcription by pol II	Wu et al., 2006; Yoo et al., 2007
N-WASP	Activating ARP2/3-mediated actin polymerization	Higgs and Pollard, 2001; Volkmann et al. 2001
	Regulating transcription by pol II	Wu et al., 2006; Yoo et al., 2007
Gelsolin	Serving actin polymers	Ocampo et al., 2005
	Androgen receptor co-activator	Nishimura et al., 2003
Flightless I	Chromosome remodelling	Archer et al., 2005
Supervillin	Nuclear receptor-induced transcription	Ting et al., 2002
Filamin	Androgen receptor action	Ozanne et al., 2000
CapG	Unknown	De Corte et al., 2004
Profilin	Nuclear export of actin mediated by exportin 6	Stuven et al., 2003
	Possible involvement in pre-mRNA splicing	Skare et al., 2003
Thymosin β4	Sequestering actin and blocking actin polymerization	Hannappel et al., 2007; Huff et al., 2004
Cofilin	Nuclear import of actin	Pendleton et al., 2003
	Repressor of the glucocorticoid receptor	Ruegg et al., 2004
	A component of nuclear actin-rods	Nishida et al., 1987
Emerin	Nuclear architecture	Holaska et al., 2004
Myo1c/NM1	Transcription	Hofmann et al., 2006 ; Ye et al., 2008
	Chromatin remodeling	Percipalle et al., 2006
Tropomodulin	Unknown	Kong and Kedes, 2004
Protein 4.1	Nuclear assembly	Krauss et al., 2003
Actinin	Nuclear receptor activator (actinin alpha 4)	Khurana et al., 2011
	Regulation of DNase Y activity (actinin alpha 4)	Liu et al., 2004
Spectrin II α	Involved in DNA repair	Sridharan et al., 2003
Paxillin	Stimulating DNA synthesis and	Dong et al., 2009
	Promoting cell proliferation	
CAP2	Unkown	Peche et al., 2007
CABP14	Possible role in cell division	Aroian et al., 1997

Table 1. Proteins known to modulate cytoplasmic actin dynamics exist in nucleus

Actin filaments capping proteins bind the barbed (or fast growing end) of an actin filament and therefore block filament assembly or promote disassembly at that end. In the cytoplasm, members of the gelsolin family are characterized by the ability to cap, sever and bundle actin filaments in a Ca^{2+}-dependent manner in the cytoplasm (Archer et al., 2005). Several members of gelsolin family has been detected in nucleus, including gelsolin (Nishimura et al., 2003;Salazar et al., 1999), CapG (De, V et al., 2004;Onoda et al., 1993), flightless (Lee et al., 2004) and supervillin (Wulfkuhle et al., 1999). In the nucleus, gelsolin has been found to be involved in chromosome decondensation by severing actin (Ocampo et al., 2005). Flightless I has been found to bind to actin and Arp BAF53, a subunit of mammalian chromatin remodelling complex, and negatively regulates actin polymerization (Archer et al., 2005). It is currently unclear whether other members regulate actin dynamics in the nucleus. Interestingly, many of them appear to function as transcriptional coactivators for nuclear hormone receptors (Gettemans et al., 2005).

Many G-actin binding proteins are also present in the nucleus. Thymosin β4 is the most abundant polypeptide of the β-thymosin family in the cytoplasm and regulates F-actin polymerization by sequestering polymers (Huff et al., 2004). In the nucleoplasm, thymosin β4 is present at a high level and suggested to sequester nuclear actin and block actin polymerization (Hannappel, 2007;Huff et al., 2004). In addition, it has been shown to interact with ATP-dependent DNA helicase II to regulate specific gene expression (Bednarek et al., 2008). Despite its small size (~4.9 kDa), Huff et al. showed that passive diffusion of thymosin β4 through the NPC can be ruled out (Huff et al., 2004), and its nuclear localization has been reported to be regulated by the DNA mismatch repair enzyme human mutL homolog 1 (hMLH1) (Brieger et al., 2007). Profilin is a small protein that binds specifically with G-actin. It enhances the nucleotide exchange on actin to convert ADP actin into ATP actin, which can readily be incorporated in to a growing filament. In the nucleus, formation of profilin-actin complex is required for nuclear export of actin through exportin 6 (Stuven et al., 2003), to avoid excess actin polymerization in the nucleus. This was supported by Bohnsack and colleagues' work (Bohnsack et al., 2006).

ADP/cofilins represent a family of small actin-regulatory proteins that bind to both actin monomers and filaments, and remove actin filaments by severing and depolymerising (Maciver and Hussey, 2002). Using fluorescence resonance energy transfer assay, they have been shown to bind to actin directly in the nucleus and at levels much higher than in the cytoplasm (Chhabra and dos Remedios, 2005). As mentioned in section 3.2, actin accumulates in the nucleus and forms intranuclear actin rods under a variety of cellular stress conditions. Cofilin has been recognized as a component of the actin rods (Gettemans et al., 2005). The high level of cofilin present in the nucleoplasm and in the actin rods might explain the reason why actin filaments appear to be restricted in the nucleus since the cofilin/actin structures cannot be stained with phalloidin (Nishida et al., 1987). The formation of nuclear actin rods is highly dynamic and is reversible when the cellular stress conditions are removed (Gieni and Hendzel, 2009), suggesting that cofilin might play a role in restricting the excess accumulation of polymeric actin, which otherwise could affect the polymeric actin-mediated nuclear process.

3.4 Signalling molecules regulating actin dynamics

The activities of ABPs are tightly controlled through various signalling pathways to ensure proper spatial and temporal regulation of actin dynamics in the cells. Several signalling molecules, including small GTPases, Ca^{2+} and phosphoinositides which display well-characterized effects on actin dynamics in the cytoplasm, are also found in the nucleus.

Small GTPases of the Rho family, such as Cdc42 and Rac1, have been found in the nucleus, (Williams, 2003). As discussed above, Arp2/3 is an important candidate for regulating nuclear actin polymerization and N-WASP, the most potent inducer of Arp2/3 -mediated actin nucleation remains to be the only member of the WASP family found in the nucleus (Suetsugu et al., 2001;Zalevsky et al., 2001). In the cytoplasm, N-WASP is activated by Cdc42, linking Rho family GTPase signalling with Arp2/3 -mediated actin polymerization (Rohatgi et al., 1999). N-WASP is also activated by Rac1, and both Cdc42- and Rac1-mediated stimulation of N-WASP activity is further enhanced by the presence of phosphatidylinositol 4,5-bisphosphate (PIP2). The functional significance of the presence of Rho GTPases in the nucleus is not fully known. Some downstream effectors of Rho family GTPase, such as LIM kinases (LIMK), has been shown to localize to the nucleus. LIMK can phosphorylate and inactivate cofilin, suggesting that Rho GTPase signalling pathway may play an important role in regulation of nuclear actin cytoskeleton. Rac1 was shown to shuttle in and out of the nucleus during the cell cycle and to accumulate in the nucleus in late G2 phase. In addition, GTP-bound Rac1 and a Rac1/Cdc42 GTPase activating- protein, MgcRacGAP, bind directly to phosphorylated transcription factors, STAT3 and STAT5, to mediate their translocation into the nucleus. Therefore, nuclear accumulation of Rac1 may also regulate actin polymerization influencing RNA polymerase II-mediated transcription.

Phosphoinositides (PIs) are major regulators of actin dynamics in the cytoplasm (Mao and Yin, 2007). PIs control actin polymerization by modulating the activity of regulatory proteins promoting actin assembly and inhibiting disassembly of actin filaments. For example, PIP2 activates nucleation of actin filaments induced by N-WASP-Arp2/3 complex and inhibits the actin-binding activity of cofilin (Hilpela *et al.*, 2004). PIP2 also binds and inhibits capping proteins, and seems to remove capping proteins from capped ends of actin filaments, which may help to stimulate actin assembly (Kim *et al.*, 2007). Based on the observations, one can speculate that PIs also modulate actin-binding activity of capping proteins in the nucleus. So far, the downstream targets of PI signalling remains poorly identified. Several studies have linked chromatin remodelling complexes with PIs. For example, PIP2 participates in the recruitment of mammalian chromatin remodelling complex, BRG1/BRM associated factor (BAF), to nuclear matrix-associated chromatin, upon activation of antigen receptor in T-lymphocytes (Zhao *et al.*, 1998). Further analysis has revealed that PIP2 can bind directly to BRG1, an ATPase subunit of the BAF complex, modulate the actin-binding activity of BRG1 (Rando *et al.*, 2002). Within the BAF complex, BRG1 is associated with β-actin and Arp BAF53 through two actin-binding domains. Interestingly, one of the acting-binding domains of BRG1 is required for PIP2 binding. Based on these findings, a model is designed in which interaction between PIP2 and BRG1 would essentially uncap β-actin or BAF53, thereby allowing them to interact with actin filaments in the nuclear matrix (Rando *et al.*, 2002).

In the cytoplasm, actin dynamics is also controlled by Ca^{2+} level. The activity of several ABPs, including members of gelsolin family, are regulated by Ca^{2+} influx (Archer *et al.*, 2005). For example, Ca^{2+} activates gelsolin to allow capping and severing of actin filaments. The importance of Ca^{2+}-regulated actin severing has been well-documented in platelet activation (Witke *et al.*, 1995). Gelsolin has six Ca^{2+} binding sites within domain S1-S6. When domains S5 and S6 are occupied by Ca^{2+} at submicromolar concentration gelsolin is activated to bind actin. However, for full activation of severing activity, higher Ca^{2+} concentrations are required most likely filling the sites on domains S1, S2 and S4 (Burtnick *et al.*, 2004;Choe *et al.*, 2002). It is clear that nuclear Ca^{2+} level is regulated which in turn regulates the activity of transcription factors, such as DREAM (Carrion *et al.*, 1999) and CREB (Chawla *et al.*, 1998). Likewise, it is possible that nuclear Ca^{2+} level could modulate the activity of actin-containing chromatin remodelling complex by controlling activity of certain nuclear ABPs.

4. Involvement of actin in chromatin remodelling

Eukaryotic DNA is tightly packaged into nucleosome repeats. Each nucleosome unit consist of a histone octamer core surrounded by a segment of 146 base pairs of double stranded

DNA. Histone octamer core is composed of two-molecule each of H2A, H2B, H3 and H4 proteins. This kind of packaging of genomic DNA in chromatin represents barriers that restrict access to a variety of DNA regulatory proteins involved in the processes of transcription, replication, DNA repair and recombination machinery. To overcome these barriers, eukaryotic cells possess a number of multiprotein complexes which can alter the chromatin structure and make DNA accessible. These complexes can be divided into two groups, histone-modifying enzymes and ATP-dependent chromatin remodelling complexes. The histone-modifying enzymes post-translationally modify the N-terminal tails of histone proteins through acetylation, phosphorylation, ubiquitination, ADP-ribosylation and methylation (Sterner and Berger, 2000;Wang et al., 2007). On the other hand, ATP-dependent chromatin remodelling complexes use the energy of ATP hydrolysis to disrupt the DNA-histone contact, move nucleosomes along DNA, and remove or exchange nucleosomes (Gangaraju and Bartholomew, 2007). Actin and Arps were first identified as integral components of the BAF complex, a mammalian SWI/SNF-like chromatin remodelling complex, that is involved in T-lymphocyte activation (Zhao et al., 1998). Since then, actin and Arps have been found to be present in a wide variety of chromatin remodelling and histone-modifying complexes (Figure 1C) in yeast, *Drosophila* and mammalian cells.

4.1 Actin–containing chromatin remodelling complex

The ATP-dependent chromatin remodelling complexes can be classified into at least four different families based on their central ATPases: SWI/SNF complex (or BAF complexes) with a SWI2/SNF2 ATPase; ISWI complex with an ISWI ATPase; Mi-2 (or CHD) complex containing a chromodomain-helicase-DNA binding protein ATPase; and INO80 complex with an INO80 ATPase (Farrants, 2008). Only the complexes of SWI/SNF and INO80 families have actin and Arps as subunits that are bound directly. Most members of SWI/SNF family contain actin and Arp4 homologues. In *Drosophilia*, the orthologous complexes BAP (Brahma associated proteins) and PBAP (polybromo-associated BAP) each contains actin and the Arp BAF55. In mammals, the orthologous complexes BAF, PBAF and p400 all contain actin and the Arp BAF53. The yeast SWI/SNF complex was the first to be discovered in *S. cerevisiae*. Many components of the SWI/SNF complex were initially identified in independent screens for genes that regulate mating-type switching (SWI genes) and sucrose non-fermenting (SNF genes) phenotype in yeasts (Abrams et al., 1986;Carlson et al., 1981;Nasmyth and Shore, 1987;Neigeborn and Carlson, 1984;Neigeborn and Carlson, 1987;Stern et al., 1984;Vignali et al., 2000). The yeast SWI/SNF complex contains 11 known subunits, of which the SWI2/SNF2 subunit possesses both chromatin remodelling and DNA-dependent ATPase activities. In yeast, the SWI/SNF complex and another orthologous complex RSC lack actin but were shown to contain two yeast specific Arps - Arp7 and Arp9 (Table 2). Interestingly, the yeast genome encodes both actin and Arp4 but they are replaced with novel Arps.

The INO80 family includes the yeast INO80 complex and its orthologues Pho-dINO80 (*Drosophila*) and INO80 (human); the yeast SWR1 complex and its orthologue SRCAP (human); and the yeast NuA4 complex and its orthologues TIP60 (*Drosophila*) and TRAAP/TIP60 (human) (Table 3). The yeast INO80 complex contains actin and Arp 4, Arp5 and Arp8, of which Arp5 and Arp8 and actin are conserved in *Drosophila* and mammals. Arp4 and actin are also components of yeast SWR1 and Nu4 complex. BAF53, a mammalian orthologue of Arp4, is present in the mammalian orthologous complex INO80, SRCAP, and TIP60 (Hargreaves and Crabtree, 2011).

Complex	SWI/SNF	RSC	BAP	PBAP	BAF	PBAF	nBAF	npBAF
Species	Yeast	Yeast	Drosophila	Drosophila	Human	Human	Mouse	Mouse
ATPase	Swi2/Snf2	Sth1	BRM	BRM	BRG1/hBRM	BRG1	BRG1	BRG1
Actin	No	No	β-actin	β-actin	β-actin	β-actin	β-actin	β-actin
ARP	ARP7, ARP9	ARP7, ARP9	BAP55 or BAP47	BAP55 or BAP47	BAF53a or BAF53b	BAF53a	BAF53b	BAF53a
Main subunits or or and/or			OSA	BAP170	BAF250a or BAF250b	BAF200	BAF250a or BAF250b or BAF200	BAF250a BAF250b BAF200
	Swi3	Rsc1, Rsc2, Rsc4 Rsc8	BAP155	Polybromo BAP155	BAF155 and/or	BAF180 BAF155 and/or	BAF155 and/or	BAF155
	Swp73	Rsc6	BAP60	BAP60	BAF170 BAF60a or BAF60b or BAF60c	BAF170 BAF60a	BAF170 BAF60a	BAF170 BAF60a
BAF47/SNF5	Snf5	Sfh1	BAP45/SNR1	BAP45/SNR1	BAF47/hSNF5	BAF47/hSNF5	BAF47/SNF5	
			BAP111	BAP111	BAF57	BAF57	BAF57	BAF57
Unique Subunits	Swi1, Swp82 Taf14, Snf6 Snf11	Rsc3, 5, 7, 9, 10, 30 Htl1, Lbd7, Rtt102						

Table 2. Complexes of the SWI/SNF family

Complex	INO80	Pho-dINO80	INO80	SWR1	SRCAP	NuA4	TIP60	TRAAP/TIP60
Species	Yeast	Drosophila	Human	Yeast	Human	Yeast	Drosophila	Human
ATPase	Ino80	dIno80	hINO80	Swr1	SRCAP		Domino	p400
Actin	Act1	dActin	β-actin	Act1	β-actin	Act1	Act87E	β-actin
ARP	ARP4, ARP5 ARP8	dARP5, dARP8	BAF53a, ARP5 ARP8	ARP4, ARP6	BAF53a, ARP6	ARP4	BAP53	BAF53a
Main subunits	Rvb1, Rvb2 Taf14 Ies 2 Ies 6	Reptin, Pontin	Tip49a, Tip49b Yaf9 hIes 2 hIes 6	Rvb1, Rvb2 Yaf9	Tip49a, Tip49b GAS41	Yaf9	Reptin, Pontin dGAS41	Tip49a, Tip49b GAS41
				Swc2 Swc4 Bdf1 Swc6	YL-1 DMAP1 Znf-HIT1	Swc4	dYL-1 dMAP1 dBrd8	YL-1 DMAP1 BRD8/TRCp120
						Tra1 Eaf3 Eaf6 Eaf7 Esa1 Epl1 Yng2	dTra1 dMRG15 dEaf6 dMRGBP dTip60 E(pc) dING3	TRAAP MRG15 FLJ11730 MRGBP Tip60 EPC1 ING3
Unique Subunits	Ies 1, Ies3, 4, 5 Nhp10	Pleiohomeotic	Amida, MCRS1, FLJ20309, UCH37 NFRKB, CCDC95	Swc3, 5, 7		Eaf1, 5		

Table 3. Complexes of the INO80 family

Actin was first identified in the mammalian BAF complex. Biochemical analysis indicated that actin is not only tightly bound to BRG1, the ATPase subunit of BAF, but also needed for the ATPase activity required for BAF association with chromatin (Zhao et al., 1998). To date, the molecular mechanisms that underlie the functions of actin and Arps in the chromatin remodelling remain largely unknown. Recently, a helicase-SANT-associated (HSA) domain was identified in the ATPase of several chromatin remodelling complexes. This domain is required for the binding of actin and Arps. Altering the HSA domain causes a loss of actin and Arps in these complexes and reduces ATPase activity, confirming the important role of actin and Arps in chromatin remodelling (Szerlong et al., 2008).

4.2 Actin recruits histone modifying enzymes

Actin has been identified as a component of pre-mRNA particles (pre-mRNPs) via binding to heterogeneous nuclear ribonucleoprotein particles (hnRNPs) in insects and mammals (Kukalev et al., 2005;Percipalle et al., 2003;Percipalle et al., 2002;Percipalle et al., 2001;Zhang et al., 2002b). In the dipteran insect *Chironomus tentans*, actin was found to bind directly to the nuclear protein HRP65-2 (HRP65 isoform 2). Disruption of this interaction by a competing peptide, which mimics the actin-binding motif of HRP65, inhibited RNA polymerase II-mediated transcription at the level of transcript elongation (Percipalle et al., 2003). The inhibitory effect of this peptide can be counteracted by a general inhibitor of histone deacetylases (trichostatin A), suggesting that actin-HRP65-2 interaction is involved in acetylation/deacetylation of histones. Indeed, HRP65-2 and actin were shown to form a complex with p2D10 *in vivo*. p2D10 is a histone H3 –specific acetyltransferase, a *C. tentans* ortholog of the largest subunit of the transcription factor TFIIIC (Sjolinder et al., 2005). Disruption of the interaction between HRP65-2 and actin releases p2D10 from RNA polymerase II-transcribing gene, coinciding with reduced H3 histone acetylation and inhibition of transcription, indicating that HRP65-actin interaction provides a molecular platform to recruit chromatin modifying factors to the transcribing genes allowing to maintain genes in an active state (Figure 1B) (Sjolinder et al., 2005).

Similarly, in human cells, the interaction between actin and hnRNP U, another component of pre-mRNPs, was also shown to be essential for RNA polymerase II-mediated transcription elongation. hnRNP U has been shown to bind to actin via a conserved actin-binding motif located at the C-terminus. Both actin and hnRNP U were shown to be associated with the phosphorylated C-terminal domain (CTD) of polymerase II and antibodies against either of these components are able to block transcription of class II genes (Kukalev et al., 2005). Furthermore, the actin - hnRNP U complex was shown to be required for the recruitment of histone acetyltransferase (HAT), PCAF, to actively transcribed genes, and they are all present at promoter and coding regions of constitutively expressed class II genes (Figure 1 B) (Obrdlik et al., 2008). It was previously shown that binding of hnRNP U to RNA polymerase II inhibited the phosphorylation of the CTD mediated by TFIIH, suggesting that actin-hnRNP U interaction might modify the inhibitory effect of hnRNP U on CTD phosphorylation, and that this modification is required for transcription elongation (Kim and Nikodem, 1999;Kukalev et al., 2005).

5. Involvement of actin in transcription machinery

Transcription is a process of synthesizing an RNA molecule from a sequence of DNA. The major steps of transcription include pre-initiation, initiation, elongation and termination.

Transcription is performed by an enzyme called RNA polymerase. Eukaryotic cells have three distinct classes of RNA polymerases characterized by the type of RNA they synthesize. RNA polymerase I is located in the nucleolus, a functionally highly specialized subnuclear compartment, and it is responsible for transcribing ribosomal RNA (rRNA). RNA polymerase II is located in nucleoplasm and responsible for synthesizing the precursors of messenger RNA (mRNA) and most small nuclear RNAs (snRNA) and microRNA (miRNA). RNA polymerase III is also located in nucleoplasm, and transcribes 5S rRNA, transfer RNA (tRNA), U6 snRNA and other small RNAs.

The first finding that demonstrated a role of nuclear actin in transcriptional process was documented by Smith et al. (Smith *et al.*, 1979). The authors found that actin co-purified with RNA polymerase II from the slime mold *Physarum polycephalum*. Following this original finding, the subsequent studies demonstrated that actin was present in transcriptionally active nuclear extracts from HeLa cells and calf thymus, and was able to initiate the transcription by RNA polymerase II *in vitro* (Egly *et al.*, 1984). Another study, published about the same time, showed that transcription of lampbrush chromosomes was inhibited when antibodies directed against actin or ABPs were microinjected into the nuclei of living oocytes of *Pleurodeles waltl*. This study provided first solid evidence for an association between actin and transcription (Scheer *et al.*, 1984). However, these two important findings were largely ignored and postulated as being artifacts of contamination. Although a number of key advances in the nuclear actin field occurred after 1984, skepticism of new data remained until two decades later when several studies finally provided convincing and non-contestable evidence for the involvement of actin in gene transcription (Grummt, 2006;Pederson and Aebi, 2002; Percipalle and Visa, 2006). However, many questions remain to be answered to fully understand the exact molecular mechanism of regulation of transcription by various forms of actin.

5.1 Role of actin and nuclear myosin 1 in gene transcription by RNA polymerase I

Actin has been shown to be present not only in mammalian nucleoplasm but also in nucleoli (Andersen *et al.*, 2005), suggesting a role for actin in transcription by RNA polymerase I. Indeed, co-immunoprecipitation and chromatin immunoprecipitation (ChIP) assays showed that actin is associated physically with RNA polymerase I and present on actively transcribing ribosomal genes at both the promoter and transcribed regions (Fomproix and Percipalle, 2004; Philimonenko *et al.*, 2004). Microinjection of anti-actin antibodies into the nuclei inhibited rRNA synthesis in living cells (Philimonenko *et al.*, 2004). Furthermore, *in vitro* transcription assays revealed that antibody against actin also inhibited rRNA synthesis in cell-free systems containing either naked rDNA or pre-assembled chromatin templates (Philimonenko *et al.*, 2004). Interestingly, anti-actin antibody did not affect the synthesis of initial trinucleotide (initiation phase) but inhibited the synthesis of run-off transcripts (elongation phase), indicating that actin is required for RNA polymerase I transcription in post-initiation steps.

Nuclear myosin 1 (NM1), a short-tailed myosin acting as an actin-dependent ATPase, has also been found in nucleoli (Fomproix and Percipalle, 2004), suggesting that actin and myosin might work together as actomyosin in transcription. NM1 and actin are present in a complex with RNA polymerase I (Fomproix and Percipalle, 2004; Philimonenko *et al.*, 2004). The same as actin, NM1 is present on the rDNA promoter and antibodies directly against

NM1 also inhibited RNA polymerase I transcription both in *in vivo* and *in vitro* transcription assays (Percipalle *et al.*, 2006;Philimonenko *et al.*, 2004). Previously, two contradictory findings were reported on this subject. One study showed that NM1 is not associated with the coding region (Philimonenko *et al.*, 2004); however another similar study using different anti-NM1 antibodies demonstrated that a fraction of nucleolar NM1 is associated with the coding region (Percipalle *et al.*, 2006). A possible reason for these discrepant findings could be that NM1 has different conformations during different steps of transcription, which could be recognized by different antibodies. Philimonenko and co-workers have shown that actin can directly interact with RNA polymerase I independent of whether or not it is engaged in transcription; however, NM1 binds to the transcription machinery via interaction with TIF-IA (Philimonenko *et al.*, 2004). TIF-IA is a RNA polymerase I-specific transcription initiation factor that mediates growth-dependent regulation of RNA polymerase I activity and rRNA transcription (Grummt, 2003). Based on these findings, one could speculate that actin and NM1 get close in proximity that they can interact with each other during the formation of transcription initiation complex and presumably activate RNA polymerase activity and rRNA synthesis. Ye and co-workers' studies provided further support to this hypothesis (Ye et al., 2008). In addition, NM1 was also found to be present on the coding region of 18S and 28S genes as a component of the chromatin remodelling complex B-WICH, which comprises the William syndrome transcription factor (WSTF) and SNF2h besides NM1, and is required for RNA polymerase I transcription activation and maintenance (Percipalle *et al.*, 2006). This suggests that MN1 is also implicated in the post-initiation phases of transcription. Recently, a study showed that knockdown of WSTF resulted in reduced recruitment of HATs at the rDNA which coincided with a lower level of histone acetylation.

5.2 Role of actin in transcription by RNA polymerase II

A number of studies have supported the role of actin in RNA polymerase II transcription machinery. Firstly, actin is co-purified with RNA polymerase II (Egly *et al.*, 1984;Hofmann *et al.*, 2004;Smith *et al.*, 1979), which maybe a general feature, as actin is also co-purified with RNA polymerase I (see 5.1) and III (see 5.3). Secondly, microinjection of anti-actin antibodies into the nuclei of *Xenopus* oocytes blocks chromosome condensation (Rungger *et al.*, 1979), and microinjecting antibodies directed against actin or Arps inhibit RNA polymerase II-mediated transcription (Hofmann *et al.*, 2004;Scheer *et al.*, 1984;Xu *et al.*, 2010). Thirdly, actin is a component of preinitiation complexes (Figure 1A), and the formations of preinitiation complexes are blocked by depletion of actin from nuclear extracts (Hofmann *et al.*, 2004). Fourthly, ChIP assays showed that actin can be recruited to the promoters of actively transcribed genes (Hofmann *et al.*, 2004;Xu *et al.*, 2010). For example, actin is absent from the promoters of the interferon-γ-inducible gene MHC2TA and interferon-α-inducible gene G1P3 before induction but it is associated with their promoters after gene induction. Recently, using ChIP-on chip assays, we have demonstrated that actin is recruited to a wide range of gene promoters during the PMA-induced macrophage-like differentiation of HL-60 cells. These data disprove the notion that actin might non-specifically interacts with the promoter regions. If this was the case, β-actin would have been found at the promoter of all genes even in the absence of induction. Even though significant progress has been made, the mechanisms of how actin is getting selectively recruited to the target genes still remains unknown.

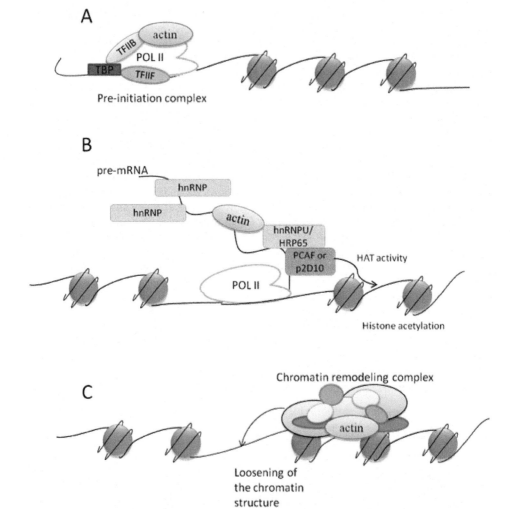

Fig. 1. Model for the function of actin in RNA polymerase II mediated transcription.

(A) Actin can interact with RNA polymerase II and is involved in the formation of preinitiation complex, and affect transcription directly. (B) Actin participates in the recruitment of histone modifying enzymes to protein-coding genes. Actin binds to hnRNP proteins and becomes incorporated into pre-mRNPs. Actin forms a complex with adaptor proteins, such as hnRNP U in mammals and HRP65 in *C. tentans*, and facilitates the recruitment of HATs, such as PCAF and p2D10 to the transcribing gene. HATs acetylate histones that maintain genes in active state. (C) Actin is implicated in chromatin remodelling as a component of ATP-dependent chromatin remodelling complexes. TBP, TATA binding protein; TFII B and TFII F, transcription factor II B and transcription factor IIF; Pol II, RNA polymerase II; hnRNP U, heterogeneous nuclear protein U; HAT, histone acetyltransferase.

5.3 Role of actin in RNA transcription by RNA polymerase III

Actin also plays an important role in RNA polymerase III transcription. Hu and colleagues have shown that actin can be co-purified with either tagged or endogenous RNA polymerase III, like with polymerase I and II (Hu et al., 2003;Hu et al., 2004). Moreover, β-actin was found to be associated with RNA polymerase III via direct protein-protein interactions with at least one of three RNA polymerase III subunits RPC3, RPABC2 and RPABC3. ChIP assays showed that β-actin is located at the promoter region of actively transcribed U6 gene, and can be dissociated from RNA polymerase III complex after inhibition of transcription by methane methylsulfonate (MMS), which resulted in the appearance of an inactivate RNA polymerase III. Notably, in an *in vitro* U6 transcription system, this inactive RNA polymerase III was not able to perform transcription, while adding exogenous β-actin reconstituted the transcriptional activity. These data indicate a crucial role for actin in RNA polymerase III-mediated transcription.

5.4 Regulation of subcellular localization and activity of transcription factor

It has been shown that nuclear actin is associated with the transcription activity of the serum response factor (SRF), a member of highly conserved MADX box family of transcription factors. SRF regulates the expression of immediate-early genes such as c-fos and actin, as well as muscle-specific genes, and these target genes are involved in cell growth, proliferation, differentiation and cytoskeletal organization (Miano, 2003). Myocardin and myocardin-related transcription factors (MRTFs) act as powerful cofactor of SRF in mammalian cells (Cen et al., 2004;Parmacek, 2007). MRTF-A (also known as MAL and MKL1) is a G-actin binding protein and its subcellular localization and activity are regulated by the concentration of monomeric actin (Miralles et al., 2003;Vartiainen et al., 2007). In serum-starved NIH 3T3 cells, MRTF-A predominantly resides in the cytoplasm, where it interact with G-actin via its N-terminal RPEL domain (Guettler et al., 2008;Miralles et al., 2003). Activation of RhoA signalling, by stimulation with serum for example, results in increased actin polymerization and decreased G-actin level, respectively. Sensing depletion of the G-actin pool, MRTF-A dissociates from G-actin and rapidly accumulates in nucleus. In the nucleus, MRTF-As physically associate with SRF, facilitating the binding of SRF to CArG box to activate the transcription of target genes (Du et al., 2004;Vartiainen et al., 2007;Zhou and Herring, 2005). MRTF-A contains an unusually long bipartite NLS located within the RPEL domain. Pawlowski and colleagues demonstrated that importin α/β heterodimer competitively binds to the RPEL domain with G-actin via interaction with NLS. Importantly, this binding was shown to mediate the nuclear import of MATF-A (Pawlowski et al., 2010). MRTF-A also binds G-actin in the nucleus and this association is required for export of MRTF-A from the nucleus (Guettler et al., 2008;Vartiainen et al., 2007). Furthermore, actin binding to MRTF-A in the nucleus inhibits the activity of the latter, and subsequently SRF-mediated transcription. Therefore, actin regulates SRF activity through modulating the sublocalization of MRTF-A and its activity within the nucleus (Vartiainen et al., 2007). (Figure 2).

Striated muscle activator of Rho signalling (STARS), a muscle-specific ABP, is capable of stimulating the transcription activity of SRF through a mechanism involving RhoA activation and actin polymerization. MRTFs, including MRTF-A and MRTF-B, were shown to serve as a linker between STARS stimulation and SRF activity (Arai et al., 2002). Studies have demonstrated that STARS can substitute for serum signalling and promote the nuclear traslocation of MRTF-A and MRTF-B, and subsequently activate the SRF-dependent transcription (Kuwahara et al., 2005). The STARS protein contains a conserved actin-binding domain within the 142 residues of C-terminus (Arai *et al.*, 2002). The C-terminal mutant of STARS, N233, which cannot bind actin, was unable to induce the nuclear translocation of MATF-A and MATF-B and to enhance the MRTF-mediated activation of SRF- dependent transcription. In contrast, the C-terminal 142 amino acids of STARS, which can bind actin , was shown to induce the nuclear accumulation of MRTFs and to synergistically enhance the MRTF-mediated transcription activation as efficiently as full-length STARS (Kuwahara *et al.*, 2005). In addition, stimulation of nuclear translocation of MRTFs by STARS can be inhibited by Latrunculin B. Similarly, inhibition of Rho A activity by treatment with C3 transferase or by the expression of dominant-negative Rho A prevents nuclear accumulation of MRTFs and subsequently MRTF-mediated SRF activation (Kuwahara *et al.*, 2005). Although Rho A activity is required in this process, it seems not to act as a downstream effector of STARS, since Rho A activity in STARS-transfected cells does not appear to be different from that observed in untransfected cells. However, it has been well documented that STARS requires Rho-actin signalling to evoke its stimulatory effects. As STARS binds to actin and induces actin polymerization (Arai *et al.*, 2002;Kuwahara *et al.*, 2005), it has been suggested that STARS stimulates the MRTF activity by inducing the dissociation of MRTF from actin and subsequently promoting its nuclear accumulation (Figure 2).

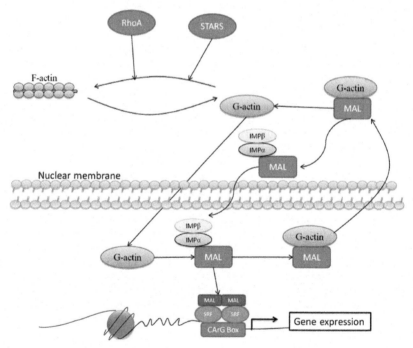

Fig. 2. Model for actin dynamics regulating SRF activity. Rho signalling (non-muscle cells) or STARS signalling (muscle cells) promote the assembly of F-actin from monomeric G-actin. Sensing depletion of G-actin, MAL dissociates from G-actin and is imported into nucleus through binding to heterodimer importin α/β. In the nucleus, MAL binds and activates SRF activity. G- actin also binds MAL in the nucleus and mediates the nuclear export of MAL. STARS, striated muscle activator of Rho signalling; MAL, also known as MRTF-A, myocardin-related transcription factor A; IMPα/β, importin α/β; SRF, serum response factor.

Interestingly, SRF activity is regulated by actin/MRTF interaction and in turn, SRF controls the transcription of many genes encoding actin isoforms and ABPs (Posern and Treisman, 2006;Sun et al., 2006). SRF-regulated genes can encode structural components of actin microfilament (for example, actin), effectors of actin turnover (for example, cofilin 1) as well as regulator of actin dynamics (for example, filamin A)(Olson and Nordheim, 2010).

6. Roles of actin in genome organization

It is well known that the eukaryotic nucleus is highly organized into morphologically and functionally distinct subnuclear compartments, which spatially separate different physiological processes. These compartments include chromatin territories, proteinaceous nuclear bodies (e.g. nucleolus, Cajal body, speckle and PML body), compartmentalized multiprotein complexes such as transcription factories, and nuclear pore complexes that regulate nucleocytoplasmic transport of certain molecules. Within the interphase nuclei, chromosomes are non-randomly organized, but occupy discrete regions, known as chromosome territories (CTs). This was first suggested by Rabel in 1885, but until almost one hundred years later, Cremer and his group carried out experiments to indirectly demonstrate the existence of CTs (Cremer et al., 1982). By using non-radioactive in situ hybridization, several groups unequivocally confirmed that the chromosomes are not distributed throughout the nucleus during the interphase, but confined to subnuclear domains (Borden and Manuelidis, 1988;Lichter et al., 1988;Schardin et al., 1985). Initial reports, based mainly on fluorescence in situ hybridization (FISH) demonstrated that these CTs are non-overlapping. However, Branco and co-workers (Branco and Pombo, 2006), using a high resolution cryo-FISH technique, revealed a significant intermingling of CTs in interphase, suggesting CTs interact more than previous thought. Interestingly, gene-rich CTs tend to occupy the interior of the nucleus, whereas many of the gene-poor chromosomes are associated with the nuclear periphery (Cremer et al., 2001;Croft et al., 1999). The gene-density-correlated radial organization of CTs was confirmed by analyses comprising all human chromosomes (Boyle et al., 2001). This non-random radial distribution of CTs was also observed in rodents (Neusser et al., 2007), cattle (Koehler et al., 2009) and birds (Habermann et al., 2001), suggesting that it has been evolutionary conserved. In addition, chromosome size-correlated radial arrangements have also been described (Sun et al., 2000) and such organization seems to occur in flat nuclei (Bolzer et al., 2005). However, the genome organization of eukaryotic cells is dynamic and chromosome arrangement can change in response to cellular signals.

An association of actin with various chromatin states has been shown by a number of studies (Milankov and De, 1993;Sauman and Berry, 1994). Bacterial actin was shown to be involved in plasmid and chromosome segregation (Becker et al., 2006;Moller-Jensen et al., 2007). In eukaryotic cell, chromosome segregation is generally driven by microtubules, however, in the oocyte of starfish, chromosome congression requires actin polymerization (Lenart et al., 2005). Mehta and colleagues have recently demonstrated that CTs relocate in quiescent human fibroblasts (Mehta et al., 2010). In the absence of serum stimulation, a number of chromosomes were observed to change position in the interphase nuclei; however, this kind of chromosome movement can be prevented by the inhibition of actin and myosin polymerization or knockdown of nuclear myosin 1β by RNA interference experiments. Inhibition of ATPase and/or GTPase also blocked the chromosome movement upon serum withdrawal, suggesting that this kind of chromosome movement is an energy-dependent active process. In an earlier study, using a novel tool to visualize chromatin movement in living cells, Chuang and co-workers (Chuang et al., 2006) reported long-range vectorial movements of chromatin exceeding 1μm. Furthermore, the authors found that repositioning of a chromatin locus was completely abolished in cells transfected with a nonpolymerizable actin mutant, whereas a mutant stabilizing filamentous actin accelerated locus redistribution. Transfection with a myosin mutant significantly delayed locus reposition. Dundr and colleagues (Dundr et al., 2007) analyzed the dynamic association between Cajal bodies and U2 snRNA gene. Upon transcriptional activation, the chromosome region containing U2 snRNA genes moved toward the Cajal bodies which are relatively stably positioned. Inactivated U2 snRNA genes do not associate with the Cajal bodies. Similarly, this process was also found to be actin-dependent.

Actin and NM1 are also required for estrogen-induced interchromosomal interactions. Two estrogen-regulated loci, TFF1 and GREB1, located in different chromosomes colocalize after stimulation with 17β-estradiol (Hu et al., 2008). This interaction was blocked by treatment of the cells with Latrunculin or jasplakinolide, which inhibit actin polymerization and depolymerisation, respectively. Depletion of actin or NM1 by siRNAs or nuclear microinjection of specific antibodies against NM1 also blocked the interaction. Furthermore, the inhibitory effect of anti-NM1 antibodies could be rescued by the expression of wild-type NM1, but not by the expression of NM1 mutant deficient in ATPase or actin-binding activity. The results demonstrate that the dynamics of nuclear actin affect the chromatin movement and gene positioning. How actin and NM1 cooperate to organize genome and facilitate the regulation of gene expression still needs to be studied in more detail.

7. Conclusion

The past decade has seen great advances in discovering the versatile functions of actin in the nucleus. Actin participates not only in the basal transcription mediated by all three RNA polymerases, as a component of RNA polymerase complex or pre-mRNP particles, but also in the transcriptional regulation as a component of chromosome remodelling complex. Actin also plays a role in movement, organization, and regulation of chromatin and activated genes in the nucleus. In addition, actin acts as a multifunctional brick in the nuclear architecture to help maintain nuclear shape, spatial order and nuclear functions. The challenge now is to understand the molecular mechanisms that underlie the many functions of actin in these nuclear processes.

In recent years, studies have shown that nuclear actin can exist as a polymeric form and that actin polymerization is implicated in transcription (Wu et al., 2006;Ye et al., 2008;Yoo et al., 2007). However, the form of actin in various complexes associated with transcription seem to be monomeric and its functions in theses complexes do not appear to require polymerization/depolymerisation dynamics. There are two possible explanations for why this is so. First, actin polymerization may maintain a proper G-actin pool for transcription. Second, the polymeric state affects the movement of gene loci (Hu et al., 2008), therefore, may affect the transcription. From this view point, it is important to address the dynamic behaviour of nuclear actin in order to understand how actin regulates transcription, for example, how actin polymerization is regulated inside the nucleus. What are the roles of

ABPs in the nucleus? It seems that many APBs have specific nuclear functions that are not related to the regulation of actin dynamics. Therefore, identifying novel ABPs is expected to be an important way to understand the regulation of actin polymerization. In addition, there is evidence to show the existence of a communication between cytoplasmic actin and nuclear actin pool (Vartiainen et al., 2007). But what are the signalling pathways that link cytoplasmic actin dynamics and nuclear actin behaviour? What are the mechanisms that controlling the nucleocytoplasmic shuttling of the actin? Although actin has functional NESs, it is always found to be exported in a complex with other cargo, for example with profilin and with MRTF-A.

Recently, studies have well documented that both nuclear actin and NM1 are implicated in the movement of CTs and chromosomal loci. The interactions of chromosomal loci with functional subnuclear domains, such as Cajal bodies, as well as interactions between distinct chromosomal loci are important for transcriptional regulation and genome-based nuclear processes. It is most likely that actin cooperates with NM1 as a motor to drive and direct the movement of chromosomal loci. Nevertheless, actin may also acts as a component of chromatin remodelling complex to relax the chromatin structure. New technologies need to be developed to investigate exact mechanisms of involvement of actin and NM1 in movement of chromosomal loci. In summary, there is still a long way to fully understanding the complexity of actin in the structural and functional networks of the nucleus.

8. References

Abrams, E., Neigeborn, L., and Carlson, M. (1986). Molecular analysis of SNF2 and SNF5, genes required for expression of glucose-repressible genes in Saccharomyces cerevisiae. Mol. Cell Biol. 6, 3643-3651.

Andersen, J.S., Lam, Y.W., Leung, A.K., Ong, S.E., Lyon, C.E., Lamond, A.I., and Mann, M. (2005). Nucleolar proteome dynamics. Nature. 433, 77-83.

Andres, V. and Gonzalez, J.M. (2009). Role of A-type lamins in signaling, transcription, and chromatin organization. J. Cell Biol. 187, 945-957.

Arai, A., Spencer, J.A., and Olson, E.N. (2002). STARS, a striated muscle activator of Rho signaling and serum response factor-dependent transcription. J. Biol. Chem. 277, 24453-24459.

Archer, S.K., Claudianos, C., and Campbell, H.D. (2005). Evolution of the gelsolin family of actin-binding proteins as novel transcriptional coactivators. Bioessays. 27, 388-396.

Aroian, R.V., Field, C., Pruliere, G., Kenyon, C., and Alberts, B.M. (1997). Isolation of actin-associated proteins from Caenorhabditis elegans oocytes and their localization in the early embryo. EMBO J. 16, 1541-1549.

Ascough, K.R. (2004). Endocytosis: Actin in the driving seat. Curr. Biol. 14, R124-R126.

Becker, E., Herrera, N.C., Gunderson, F.Q., Derman, A.I., Dance, A.L., Sims, J., Larsen, R.A., and Pogliano, J. (2006). DNA segregation by the bacterial actin AlfA during Bacillus subtilis growth and development. EMBO J. 25, 5919-5931.

Bednarek, R., Boncela, J., Smolarczyk, K., Cierniewska-Cieslak, A., Wyroba, E., and Cierniewski, C.S. (2008). Ku80 as a novel receptor for thymosin beta4 that mediates its intracellular activity different from G-actin sequestering. J. Biol. Chem. *283*, 1534-1544.

Berezney, R. (2002). Regulating the mammalian genome: the role of nuclear architecture. Adv. Enzyme Regul. *42:39-52.*, 39-52.

Berezney, R., Mortillaro, M., Ma, H., Meng, C., Samarabandu, J., Wei, X., Somanathan, S., Liou, W.S., Pan, S.J., and Cheng, P.C. (1996). Connecting nuclear architecture and genomic function. J. Cell Biochem. *62*, 223-226.

Blessing, C.A., Ugrinova, G.T., and Goodson, H.V. (2004). Actin and ARPs: action in the nucleus. Trends Cell Biol. *14*, 435-442.

Bohnsack, M.T., Stuven, T., Kuhn, C., Cordes, V.C., and Gorlich, D. (2006). A selective block of nuclear actin export stabilizes the giant nuclei of Xenopus oocytes. Nat. Cell Biol. *8*, 257-263.

Bolzer, A. *et al.* (2005). Three-dimensional maps of all chromosomes in human male fibroblast nuclei and prometaphase rosettes. PLoS. Biol. *3*, e157.

Borden, J. and Manuelidis, L. (1988). Movement of the X chromosome in epilepsy. Science. *242*, 1687-1691.

Boyle, S., Gilchrist, S., Bridger, J.M., Mahy, N.L., Ellis, J.A., and Bickmore, W.A. (2001). The spatial organization of human chromosomes within the nuclei of normal and emerin-mutant cells. Hum. Mol. Genet. *10*, 211-219.

Brakebusch, C. and Fassler, R. (2003). The integrin-actin connection, an eternal love affair. EMBO J. *22*, 2324-2333.

Branco, M.R. and Pombo, A. (2006). Intermingling of chromosome territories in interphase suggests role in translocations and transcription-dependent associations. PLoS. Biol. *4*, e138.

Brieger, A., Plotz, G., Zeuzem, S., and Trojan, J. (2007). Thymosin beta 4 expression and nuclear transport are regulated by hMLH1. Biochem. Biophys. Res. Commun. *364*, 731-736.

Burtnick, L.D., Urosev, D., Irobi, E., Narayan, K., and Robinson, R.C. (2004). Structure of the N-terminal half of gelsolin bound to actin: roles in severing, apoptosis and FAF. EMBO J. *23*, 2713-2722.

Capco, D.G., Wan, K.M., and Penman, S. (1982). The nuclear matrix: three-dimensional architecture and protein composition. Cell. *29*, 847-858.

Carlson, M., Osmond, B.C., and Botstein, D. (1981). Mutants of yeast defective in sucrose utilization. Genetics. *98*, 25-40.

Carrion, A.M., Link, W.A., Ledo, F., Mellstrom, B., and Naranjo, J.R. (1999). DREAM is a Ca^{2+}-regulated transcriptional repressor. Nature. *398*, 80-84.

Cen, B., Selvaraj, A., and Prywes, R. (2004). Myocardin/MKL family of SRF coactivators: key regulators of immediate early and muscle specific gene expression. J. Cell Biochem. *93*, 74-82.

Charlton, C.A. and Volkman, L.E. (1991). Sequential rearrangement and nuclear polymerization of actin in baculovirus-infected Spodoptera frugiperda cells. J. Virol. *65*, 1219-1227.

Chawla, S., Hardingham, G.E., Quinn, D.R., and Bading, H. (1998). CBP: a signal-regulated transcriptional coactivator controlled by nuclear calcium and CaM kinase IV. Science. *281*, 1505-1509.

Chen, M. and Shen, X. (2007). Nuclear actin and actin-related proteins in chromatin dynamics. Curr. Opin. Cell Biol. *19*, 326-330.

Chhabra, D. and dos Remedios, C.G. (2005). Cofilin, actin and their complex observed in vivo using fluorescence resonance energy transfer. Biophys. J. *89*, 1902-1908.

Choe, H., Burtnick, L.D., Mejillano, M., Yin, H.L., Robinson, R.C., and Choe, S. (2002). The calcium activation of gelsolin: insights from the 3A structure of the G4-G6/actin complex. J. Mol. Biol. *324*, 691-702.

Chuang, C.H., Carpenter, A.E., Fuchsova, B., Johnson, T., de, L.P., and Belmont, A.S. (2006). Long-range directional movement of an interphase chromosome site. Curr. Biol. *16*, 825-831.

Clark, T.G. and Rosenbaum, J.L. (1979). An actin filament matrix in hand-isolated nuclei of X. laevis oocytes. Cell. *18*, 1101-1108.

Courgeon, A.M., Maingourd, M., Maisonhaute, C., Montmory, C., Rollet, E., Tanguay, R.M., and Best-Belpomme, M. (1993). Effect of hydrogen peroxide on cytoskeletal proteins of Drosophila cells: comparison with heat shock and other stresses. Exp. Cell Res. *204*, 30-37.

Cremer, M., von, H.J., Volm, T., Brero, A., Kreth, G., Walter, J., Fischer, C., Solovei, I., Cremer, C., and Cremer, T. (2001). Non-random radial higher-order chromatin arrangements in nuclei of diploid human cells. Chromosome. Res. *9*, 541-567.

Cremer, T., Cremer, C., Baumann, H., Luedtke, E.K., Sperling, K., Teuber, V., and Zorn, C. (1982). Rabl's model of the interphase chromosome arrangement tested in Chinese hamster cells by premature chromosome condensation and laser-UV-microbeam experiments. Hum. Genet. *60*, 46-56.

Crisp, M., Liu, Q., Roux, K., Rattner, J.B., Shanahan, C., Burke, B., Stahl, P.D., and Hodzic, D. (2006). Coupling of the nucleus and cytoplasm: role of the LINC complex. J. Cell Biol. *172*, 41-53.

Croft, J.A., Bridger, J.M., Boyle, S., Perry, P., Teague, P., and Bickmore, W.A. (1999). Differences in the localization and morphology of chromosomes in the human nucleus. J. Cell Biol. *145*, 1119-1131.

Dayel, M.J. and Mullins, R.D. (2004). Activation of Arp2/3 complex: addition of the first subunit of the new filament by a WASP protein triggers rapid ATP hydrolysis on Arp2. PLoS. Biol. *2*, E91.

De, C., V, Van, I.K., Bruyneel, E., Boucherie, C., Mareel, M., Vandekerckhove, J., and Gettemans, J. (2004). Increased importin-beta-dependent nuclear import of the actin modulating protein CapG promotes cell invasion. J. Cell Sci. *117*, 5283-5292.

Deeks, M.J. and Hussey, P.J. (2005). Arp2/3 and SCAR: plants move to the fore. Nat. Rev. Mol. Cell Biol. *6*, 954-964.

dos Remedios, C.G., Chhabra, D., Kekic, M., Dedova, I.V., Tsubakihara, M., Berry, D.A., and Nosworthy, N.J. (2003). Actin binding proteins: regulation of cytoskeletal microfilaments. Physiol Rev. *83*, 433-473.

Dong, J.M., Lau, L.S., Ng, Y.W., Lim, L., and Manser, E. (2009). Paxillin nuclear-cytoplasmic localization is regulated by phosphorylation of the LD4 motif: evidence that nuclear paxillin promotes cell proliferation. Biochem. J. *418*, 173-184.

Du, K.L., Chen, M., Li, J., Lepore, J.J., Mericko, P., and Parmacek, M.S. (2004). Megakaryoblastic leukemia factor-1 transduces cytoskeletal signals and induces smooth muscle cell differentiation from undifferentiated embryonic stem cells. J. Biol. Chem. *279*, 17578-17586.

Dundr, M., Ospina, J.K., Sung, M.H., John, S., Upender, M., Ried, T., Hager, G.L., and Matera, A.G. (2007). Actin-dependent intranuclear repositioning of an active gene locus in vivo. J. Cell Biol. *179*, 1095-1103.

Egly, J.M., Miyamoto, N.G., Moncollin, V., and Chambon, P. (1984). Is actin a transcription initiation factor for RNA polymerase B? EMBO J. *3*, 2363-2371.

Fairley, E.A., Kendrick-Jones, J., and Ellis, J.A. (1999). The Emery-Dreifuss muscular dystrophy phenotype arises from aberrant targeting and binding of emerin at the inner nuclear membrane. J. Cell Sci. *112*, 2571-2582.

Farrants, A.K. (2008). Chromatin remodelling and actin organisation. FEBS Lett. *582*, 2041-2050.

Feierbach, B., Piccinotti, S., Bisher, M., Denk, W., and Enquist, L.W. (2006). Alpha-herpesvirus infection induces the formation of nuclear actin filaments. PLoS. Pathog. *2*, e85.

Ferrai, C., Naum-Ongania, G., Longobardi, E., Palazzolo, M., Disanza, A., Diaz, V.M., Crippa, M.P., Scita, G., and Blasi, F. (2009). Induction of HoxB transcription by retinoic acid requires actin polymerization. Mol. Biol. Cell. *20*, 3543-3551.

Fomproix, N. and Percipalle, P. (2004). An actin-myosin complex on actively transcribing genes. Exp. Cell Res. *294*, 140-148.

Forest, T., Barnard, S., and Baines, J.D. (2005). Active intranuclear movement of herpesvirus capsids. Nat. Cell Biol. *7*, 429-431.

Gangaraju, V.K. and Bartholomew, B. (2007). Mechanisms of ATP dependent chromatin remodeling. Mutat. Res. *618*, 3-17.

Gedge, L.J., Morrison, E.E., Blair, G.E., and Walker, J.H. (2005). Nuclear actin is partially associated with Cajal bodies in human cells in culture and relocates to the nuclear periphery after infection of cells by adenovirus 5. Exp. Cell Res. *303*, 229-239.

Gettemans, J., Van, I.K., Delanote, V., Hubert, T., Vandekerckhove, J., and De, C., V (2005). Nuclear actin-binding proteins as modulators of gene transcription. Traffic. *6*, 847-857.

Gieni, R.S. and Hendzel, M.J. (2009). Actin dynamics and functions in the interphase nucleus: moving toward an understanding of nuclear polymeric actin. Biochem. Cell Biol. *87*, 283-306.

Goley, E.D., Ohkawa, T., Mancuso, J., Woodruff, J.B., D'Alessio, J.A., Cande, W.Z., Volkman, L.E., and Welch, M.D. (2006). Dynamic nuclear actin assembly by Arp2/3 complex and a baculovirus WASP-like protein. Science. *%20;314*, 464-467.

Grummt, I. (2003). Life on a planet of its own: regulation of RNA polymerase I transcription in the nucleolus. Genes Dev. *17*, 1691-1702.

Grummt, I. (2006). Actin and myosin as transcription factors. Curr. Opin. Genet. Dev. *16*, 191-196.

Guettler, S., Vartiainen, M.K., Miralles, F., Larijani, B., and Treisman, R. (2008). RPEL motifs link the serum response factor cofactor MAL but not myocardin to Rho signaling via actin binding. Mol. Cell Biol. *28*, 732-742.

Habermann, F.A., Cremer, M., Walter, J., Kreth, G., von, H.J., Bauer, K., Wienberg, J., Cremer, C., Cremer, T., and Solovei, I. (2001). Arrangements of macro- and microchromosomes in chicken cells. Chromosome. Res. *9*, 569-584.

Hancock, R. (2000). A new look at the nuclear matrix. Chromosoma. *109*, 219-225.
Hannappel, E. (2007). beta-Thymosins. Ann. N. Y. Acad. Sci. *1112:21-37. Epub;%2007 Apr 27.*, 21-37.
Hargreaves, D.C. and Crabtree, G.R. (2011). ATP-dependent chromatin remodeling: genetics, genomics and mechanisms. Cell Res. *21*, 396-420.
Higgs, H.N. and Pollard, T.D. (2001). Regulation of actin filament network formation through ARP2/3 complex: activation by a diverse array of proteins. Annu. Rev. Biochem. *70:649-76.*, 649-676.
Hilpela, P., Vartiainen, M.K., and Lappalainen, P. (2004). Regulation of the actin cytoskeleton by PI(4, 5)P2 and PI(3, 4, 5)P3. Curr. Top. Microbiol. Immunol. *282:117-63.*, 117-163.
Hofmann, W. et al. (2001). Cofactor requirements for nuclear export of Rev response element (RRE)- and constitutive transport element (CTE)-containing retroviral RNAs. An unexpected role for actin. J. Cell Biol. *152*, 895-910.
Hofmann, W.A. et al. (2004). Actin is part of pre-initiation complexes and is necessary for transcription by RNA polymerase II. Nat. Cell Biol. *6*, 1094-1101.
Holaska, J.M. and Wilson, K.L. (2007). An emerin "proteome": purification of distinct emerin-containing complexes from HeLa cells suggests molecular basis for diverse roles including gene regulation, mRNA splicing, signaling, mechanosensing, and nuclear architecture. Biochemistry. *46*, 8897-8908.
Hu, P., Wu, S., and Hernandez, N. (2003). A minimal RNA polymerase III transcription system from human cells reveals positive and negative regulatory roles for CK2. Mol. Cell. *12*, 699-709.
Hu, P., Wu, S., and Hernandez, N. (2004). A role for beta-actin in RNA polymerase III transcription. Genes Dev. *18*, 3010-3015.
Hu, Q. et al. (2008). Enhancing nuclear receptor-induced transcription requires nuclear motor and LSD1-dependent gene networking in interchromatin granules. Proc. Natl. Acad. Sci. U. S. A. *105*, 19199-19204.
Huff, T., Rosorius, O., Otto, A.M., Muller, C.S., Ballweber, E., Hannappel, E., and Mannherz, H.G. (2004). Nuclear localisation of the G-actin sequestering peptide thymosin beta4. J. Cell Sci. *117*, 5333-5341.
Iida, K., Iida, H., and Yahara, I. (1986). Heat shock induction of intranuclear actin rods in cultured mammalian cells. Exp. Cell Res. *165*, 207-215.
Jockusch, B.M., Schoenenberger, C.A., Stetefeld, J., and Aebi, U. (2006). Tracking down the different forms of nuclear actin. Trends Cell Biol. *16*, 391-396.
Khurana, S., Chakraborty, S., Cheng, X., Su, Y.T., and Kao, H.Y. (2011). The actin-binding protein, actinin alpha 4 (ACTN4), is a nuclear receptor coactivator that promotes proliferation of MCF-7 breast cancer cells. J. Biol. Chem. *286*, 1850-1859.
Kim, A.S., Kakalis, L.T., Abdul-Manan, N., Liu, G.A., and Rosen, M.K. (2000). Autoinhibition and activation mechanisms of the Wiskott-Aldrich syndrome protein. Nature. *404*, 151-158.
Kim, K., McCully, M.E., Bhattacharya, N., Butler, B., Sept, D., and Cooper, J.A. (2007). Structure/function analysis of the interaction of phosphatidylinositol 4, 5-bisphosphate with actin-capping protein: implications for how capping protein binds the actin filament. J. Biol. Chem. *282*, 5871-5879.

Kim, M.K. and Nikodem, V.M. (1999). hnRNP U inhibits carboxy-terminal domain phosphorylation by TFIIH and represses RNA polymerase II elongation. Mol. Cell Biol. *19*, 6833-6844.

Kiseleva, E., Drummond, S.P., Goldberg, M.W., Rutherford, S.A., Allen, T.D., and Wilson, K.L. (2004). Actin- and protein-4.1-containing filaments link nuclear pore complexes to subnuclear organelles in Xenopus oocyte nuclei. J. Cell Sci. *117*, 2481-2490.

Koehler, D., Zakhartchenko, V., Froenicke, L., Stone, G., Stanyon, R., Wolf, E., Cremer, T., and Brero, A. (2009). Changes of higher order chromatin arrangements during major genome activation in bovine preimplantation embryos. Exp. Cell Res. *315*, 2053-2063.

Kong, K.Y. and Kedes, L. (2004). Cytoplasmic nuclear transfer of the actin-capping protein tropomodulin. J. Biol. Chem. *279*, 30856-30864.

Krauss, S.W., Chen, C., Penman, S., and Heald, R. (2003). Nuclear actin and protein 4.1: essential interactions during nuclear assembly in vitro. Proc. Natl. Acad. Sci. U. S. A. *100*, 10752-10757.

Krauss, S.W., Heald, R., Lee, G., Nunomura, W., Gimm, J.A., Mohandas, N., and Chasis, J.A. (2002). Two distinct domains of protein 4.1 critical for assembly of functional nuclei in vitro. J. Biol. Chem. *277*, 44339-44346.

Kukalev, A., Nord, Y., Palmberg, C., Bergman, T., and Percipalle, P. (2005). Actin and hnRNP U cooperate for productive transcription by RNA polymerase II. Nat. Struct. Mol. Biol. *12*, 238-244.

Kuwahara, K., Barrientos, T., Pipes, G.C., Li, S., and Olson, E.N. (2005). Muscle-specific signaling mechanism that links actin dynamics to serum response factor. Mol. Cell Biol. *25*, 3173-3181.

Lattanzi, G., Cenni, V., Marmiroli, S., Capanni, C., Mattioli, E., Merlini, L., Squarzoni, S., and Maraldi, N.M. (2003). Association of emerin with nuclear and cytoplasmic actin is regulated in differentiating myoblasts. Biochem. Biophys. Res. Commun. *303*, 764-770.

Lee, Y.H., Campbell, H.D., and Stallcup, M.R. (2004). Developmentally essential protein flightless I is a nuclear receptor coactivator with actin binding activity. Mol. Cell Biol. *24*, 2103-2117.

Lenart, P., Bacher, C.P., Daigle, N., Hand, A.R., Eils, R., Terasaki, M., and Ellenberg, J. (2005). A contractile nuclear actin network drives chromosome congression in oocytes. Nature. *436*, 812-818.

Lichter, P., Cremer, T., Borden, J., Manuelidis, L., and Ward, D.C. (1988). Delineation of individual human chromosomes in metaphase and interphase cells by in situ suppression hybridization using recombinant DNA libraries. Hum. Genet. *80*, 224-234.

Liu, Q.Y., Lei, J.X., LeBlanc, J., Sodja, C., Ly, D., Charlebois, C., Walker, P.R., Yamada, T., Hirohashi, S., and Sikorska, M. (2004). Regulation of DNaseY activity by actinin-alpha4 during apoptosis. Cell Death. Differ. *11*, 645-654.

Machesky, L.M. and Insall, R.H. (1998). Scar1 and the related Wiskott-Aldrich syndrome protein, WASP, regulate the actin cytoskeleton through the Arp2/3 complex. Curr. Biol. *8*, 1347-1356.

Maciver, S.K. and Hussey, P.J. (2002). The ADF/cofilin family: actin-remodeling proteins. Genome Biol. *3*, reviews3007.

Mao, Y.S. and Yin, H.L. (2007). Regulation of the actin cytoskeleton by phosphatidylinositol 4-phosphate 5 kinases. Pflugers Arch. *455*, 5-18.

McDonald, D., Carrero, G., Andrin, C., de, V.G., and Hendzel, M.J. (2006). Nucleoplasmic beta-actin exists in a dynamic equilibrium between low-mobility polymeric species and rapidly diffusing populations. J. Cell Biol. *172*, 541-552.

Mehta, I.S., Amira, M., Harvey, A.J., and Bridger, J.M. (2010). Rapid chromosome territory relocation by nuclear motor activity in response to serum removal in primary human fibroblasts. Genome Biol. *11*, R5.

Miano, J.M. (2003). Serum response factor: toggling between disparate programs of gene expression. J. Mol. Cell Cardiol. *35*, 577-593.

Milankov, K. and De, B.U. (1993). Cytochemical localization of actin and myosin aggregates in interphase nuclei in situ. Exp. Cell Res. *209*, 189-199.

Miralles, F., Posern, G., Zaromytidou, A.I., and Treisman, R. (2003). Actin dynamics control SRF activity by regulation of its coactivator MAL. Cell. *113*, 329-342.

Moller-Jensen, J., Ringgaard, S., Mercogliano, C.P., Gerdes, K., and Lowe, J. (2007). Structural analysis of the ParR/parC plasmid partition complex. EMBO J. *26*, 4413-4422.

Nasmyth, K. and Shore, D. (1987). Transcriptional regulation in the yeast life cycle. Science. *237*, 1162-1170.

Neigeborn, L. and Carlson, M. (1984). Genes affecting the regulation of SUC2 gene expression by glucose repression in Saccharomyces cerevisiae. Genetics. *108*, 845-858.

Neigeborn, L. and Carlson, M. (1987). Mutations causing constitutive invertase synthesis in yeast: genetic interactions with snf mutations. Genetics. *115*, 247-253.

Neusser, M., Schubel, V., Koch, A., Cremer, T., and Muller, S. (2007). Evolutionarily conserved, cell type and species-specific higher order chromatin arrangements in interphase nuclei of primates. Chromosoma. *116*, 307-320.

Nishida, E., Iida, K., Yonezawa, N., Koyasu, S., Yahara, I., and Sakai, H. (1987). Cofilin is a component of intranuclear and cytoplasmic actin rods induced in cultured cells. Proc. Natl. Acad. Sci. U. S. A. *84*, 5262-5266.

Nishimura, K. *et al.* (2003). Modulation of androgen receptor transactivation by gelsolin: a newly identified androgen receptor coregulator. Cancer Res. *63*, 4888-4894.

Obrdlik, A., Kukalev, A., Louvet, E., Farrants, A.K., Caputo, L., and Percipalle, P. (2008). The histone acetyltransferase PCAF associates with actin and hnRNP U for RNA polymerase II transcription. Mol. Cell Biol. *28*, 6342-6357.

Ocampo, J., Mondragon, R., Roa-Espitia, A.L., Chiquete-Felix, N., Salgado, Z.O., and Mujica, A. (2005). Actin, myosin, cytokeratins and spectrin are components of the guinea pig sperm nuclear matrix. Tissue Cell. *37*, 293-308.

Okorokov, A.L., Rubbi, C.P., Metcalfe, S., and Milner, J. (2002). The interaction of p53 with the nuclear matrix is mediated by F-actin and modulated by DNA damage. Oncogene. *21*, 356-367.

Olave, I.A., Reck-Peterson, S.L., and Crabtree, G.R. (2002). Nuclear actin and actin-related proteins in chromatin remodeling. Annu. Rev. Biochem. *71:755-81.*

Olson, E.N. and Nordheim, A. (2010). Linking actin dynamics and gene transcription to drive cellular motile functions. Nat. Rev. Mol. Cell Biol. *11*, 353-365.

Onoda, K., Yu, F.X., and Yin, H.L. (1993). gCap39 is a nuclear and cytoplasmic protein. Cell Motil. Cytoskeleton. *26*, 227-238.

Ostlund, C., Folker, E.S., Choi, J.C., Gomes, E.R., Gundersen, G.G., and Worman, H.J. (2009). Dynamics and molecular interactions of linker of nucleoskeleton and cytoskeleton (LINC) complex proteins. J. Cell Sci. *122*, 4099-4108.

Ou, H., Shen, Y.H., Utama, B., Wang, J., Wang, X., Coselli, J., and Wang, X.L. (2005). Effect of nuclear actin on endothelial nitric oxide synthase expression. Arterioscler. Thromb. Vasc. Biol. *25*, 2509-2514.

Ozanne, D.M., Brady, M.E., Cook, S., Gaughan, L., Neal, D.E., and Robson, C.N. (2000). Androgen receptor nuclear translocation is facilitated by the f-actin cross-linking protein filamin. Mol. Endocrinol. *14*, 1618-1626.

Parmacek, M.S. (2007). Myocardin-related transcription factors: critical coactivators regulating cardiovascular development and adaptation. Circ. Res. *100*, 633-644.

Pawlowski, R., Rajakyla, E.K., Vartiainen, M.K., and Treisman, R. (2010). An actin-regulated importin alpha/beta-dependent extended bipartite NLS directs nuclear import of MRTF-A. EMBO J. *29*, 3448-3458.

Peche, V. *et al.* (2007). CAP2, cyclase-associated protein 2, is a dual compartment protein. Cell Mol. Life Sci. *64*, 2702-2715.

Pederson, T. and Aebi, U. (2002). Actin in the nucleus: what form and what for? J. Struct. Biol. *140*, 3-9.

Pederson, T. and Aebi, U. (2005). Nuclear actin extends, with no contraction in sight. Mol. Biol. Cell. *16*, 5055-5060.

Pendleton, A., Pope, B., Weeds, A., and Koffer, A. (2003). Latrunculin B or ATP depletion induces cofilin-dependent translocation of actin into nuclei of mast cells. J. Biol. Chem. *278*, 14394-14400.

Percipalle, P. (2009). The long journey of actin and actin-associated proteins from genes to polysomes. Cell Mol. Life Sci. *66*, 2151-2165.

Percipalle, P., Fomproix, N., Cavellan, E., Voit, R., Reimer, G., Kruger, T., Thyberg, J., Scheer, U., Grummt, I., and Farrants, A.K. (2006). The chromatin remodelling complex WSTF-SNF2h interacts with nuclear myosin 1 and has a role in RNA polymerase I transcription. EMBO Rep. *7*, 525-530.

Percipalle, P., Fomproix, N., Kylberg, K., Miralles, F., Bjorkroth, B., Daneholt, B., and Visa, N. (2003). An actin-ribonucleoprotein interaction is involved in transcription by RNA polymerase II. Proc. Natl. Acad. Sci. U. S. A. *100*, 6475-6480.

Percipalle, P., Jonsson, A., Nashchekin, D., Karlsson, C., Bergman, T., Guialis, A., and Daneholt, B. (2002). Nuclear actin is associated with a specific subset of hnRNP A/B-type proteins. Nucleic Acids Res. *30*, 1725-1734.

Percipalle, P. and Visa, N. (2006). Molecular functions of nuclear actin in transcription. J. Cell Biol. *172*, 967-971.

Percipalle, P., Zhao, J., Pope, B., Weeds, A., Lindberg, U., and Daneholt, B. (2001). Actin bound to the heterogeneous nuclear ribonucleoprotein hrp36 is associated with Balbiani ring mRNA from the gene to polysomes. J. Cell Biol. *153*, 229-236.

Philimonenko, V.V. *et al.* (2004). Nuclear actin and myosin I are required for RNA polymerase I transcription. Nat. Cell Biol. *6*, 1165-1172.

Pollard, T.D. and Borisy, G.G. (2003). Cellular motility driven by assembly and disassembly of actin filaments. Cell. *112*, 453-465.

Posern, G. and Treisman, R. (2006). Actin' together: serum response factor, its cofactors and the link to signal transduction. Trends Cell Biol. *16*, 588-596.

Prehoda, K.E., Scott, J.A., Mullins, R.D., and Lim, W.A. (2000). Integration of multiple signals through cooperative regulation of the N-WASP-Arp2/3 complex. Science. *290*, 801-806.

Prendergast, G.C. and Ziff, E.B. (1991). Mbh 1: a novel gelsolin/severin-related protein which binds actin in vitro and exhibits nuclear localization in vivo. EMBO J. *10*, 757-766.

Rando, O.J., Zhao, K., Janmey, P., and Crabtree, G.R. (2002). Phosphatidylinositol-dependent actin filament binding by the SWI/SNF-like BAF chromatin remodeling complex. Proc. Natl. Acad. Sci. U. S. A. *99*, 2824-2829.

Ruegg, J., Holsboer, F., Turck, C., and Rein, T. (2004). Cofilin 1 is revealed as an inhibitor of glucocorticoid receptor by analysis of hormone-resistant cells. Mol. Cell Biol. *24*, 9371-9382.

Roeder, A.D. and Gard, D.L. (1994). Confocal microscopy of F-actin distribution in Xenopus oocytes. Zygote. *2*, 111-124.

Rohatgi, R., Ma, L., Miki, H., Lopez, M., Kirchhausen, T., Takenawa, T., and Kirschner, M.W. (1999). The interaction between N-WASP and the Arp2/3 complex links Cdc42-dependent signals to actin assembly. Cell. *97*, 221-231.

Rungger, D., Rungger-Brandle, E., Chaponnier, C., and Gabbiani, G. (1979). Intranuclear injection of anti-actin antibodies into Xenopus oocytes blocks chromosome condensation. Nature. *282*, 320-321.

Saitoh, N., Spahr, C.S., Patterson, S.D., Bubulya, P., Neuwald, A.F., and Spector, D.L. (2004). Proteomic analysis of interchromatin granule clusters. Mol. Biol. Cell. *15*, 3876-3890.

Salazar, R., Bell, S.E., and Davis, G.E. (1999). Coordinate induction of the actin cytoskeletal regulatory proteins gelsolin, vasodilator-stimulated phosphoprotein, and profilin during capillary morphogenesis in vitro. Exp. Cell Res. *249*, 22-32.

Sanger, J.W., Gwinn, J., and Sanger, J.M. (1980a). Dissolution of cytoplasmic actin bundles and the induction of nuclear actin bundles by dimethyl sulfoxide. J. Exp. Zool. *213*, 227-230.

Sanger, J.W., Sanger, J.M., Kreis, T.E., and Jockusch, B.M. (1980b). Reversible translocation of cytoplasmic actin into the nucleus caused by dimethyl sulfoxide. Proc. Natl. Acad. Sci. U. S. A. *77*, 5268-5272.

Sasseville, A.M. and Langelier, Y. (1998). In vitro interaction of the carboxy-terminal domain of lamin A with actin. FEBS Lett. *425*, 485-489.

Sauman, I. and Berry, S.J. (1994). An actin infrastructure is associated with eukaryotic chromosomes: structural and functional significance. Eur. J. Cell Biol. *64*, 348-356.

Schardin, M., Cremer, T., Hager, H.D., and Lang, M. (1985). Specific staining of human chromosomes in Chinese hamster x man hybrid cell lines demonstrates interphase chromosome territories. Hum. Genet. *71*, 281-287.

Scheer, U., Hinssen, H., Franke, W.W., and Jockusch, B.M. (1984). Microinjection of actin-binding proteins and actin antibodies demonstrates involvement of nuclear actin in transcription of lampbrush chromosomes. Cell. *39*, 111-122.

Shumaker, D.K., Kuczmarski, E.R., and Goldman, R.D. (2003). The nucleoskeleton: lamins and actin are major players in essential nuclear functions. Curr. Opin. Cell Biol. *15*, 358-366.

Sjolinder, M., Bjork, P., Soderberg, E., Sabri, N., Farrants, A.K., and Visa, N. (2005). The growing pre-mRNA recruits actin and chromatin-modifying factors to transcriptionally active genes. Genes Dev. *19*, 1871-1884.

Skare, P., Kreivi, J.P., Bergstrom, A., and Karlsson, R. (2003). Profilin I colocalizes with speckles and Cajal bodies: a possible role in pre-mRNA splicing. Exp. Cell Res. *286*, 12-21.

Smith, S.S., Kelly, K.H., and Jockusch, B.M. (1979). Actin co-purifies with RNA polymerase II. Biochem. Biophys. Res. Commun. *86*, 161-166.

Song, Z., Wang, M., Wang, X., Pan, X., Liu, W., Hao, S., and Zeng, X. (2007). Nuclear actin is involved in the regulation of CSF1 gene transcription in a chromatin required, BRG1 independent manner. J. Cell Biochem. *102*, 403-411.

Spector, I., Shochet, N.R., Blasberger, D., and Kashman, Y. (1989). Latrunculins--novel marine macrolides that disrupt microfilament organization and affect cell growth: I. Comparison with cytochalasin D. Cell Motil. Cytoskeleton. *13*, 127-144.

Sridharan, D., Brown, M., Lambert, W.C., McMahon, L.W., and Lambert, M.W. (2003). Nonerythroid alphaII spectrin is required for recruitment of FANCA and XPF to nuclear foci induced by DNA interstrand cross-links. J. Cell Sci. *116*, 823-835.

Starr, D.A. (2009). A nuclear-envelope bridge positions nuclei and moves chromosomes. J. Cell Sci. *122*, 577-586.

Stern, M., Jensen, R., and Herskowitz, I. (1984). Five SWI genes are required for expression of the HO gene in yeast. J. Mol. Biol. *178*, 853-868.

Sterner, D.E. and Berger, S.L. (2000). Acetylation of histones and transcription-related factors. Microbiol. Mol. Biol. Rev. *64*, 435-459.

Stewart, C.L., Roux, K.J., and Burke, B. (2007). Blurring the boundary: the nuclear envelope extends its reach. Science. *318*, 1408-1412.

Stuven, T., Hartmann, E., and Gorlich, D. (2003). Exportin 6: a novel nuclear export receptor that is specific for profilin.actin complexes. EMBO J. *22*, 5928-5940.

Suetsugu, S., Miki, H., Yamaguchi, H., Obinata, T., and Takenawa, T. (2001). Enhancement of branching efficiency by the actin filament-binding activity of N-WASP/WAVE2. J. Cell Sci. *114*, 4533-4542.

Suetsugu, S. and Takenawa, T. (2003). Regulation of cortical actin networks in cell migration. Int. Rev. Cytol. *229:245-86.*, 245-286.

Sun, H.B., Shen, J., and Yokota, H. (2000). Size-dependent positioning of human chromosomes in interphase nuclei. Biophys. J. *79*, 184-190.

Sun, Q., Chen, G., Streb, J.W., Long, X., Yang, Y., Stoeckert, C.J., Jr., and Miano, J.M. (2006). Defining the mammalian CArGome. Genome Res. *16*, 197-207.

Szerlong, H., Hinata, K., Viswanathan, R., Erdjument-Bromage, H., Tempst, P., and Cairns, B.R. (2008). The HSA domain binds nuclear actin-related proteins to regulate chromatin-remodeling ATPases. Nat. Struct. Mol. Biol. *15*, 469-476.

Ting, H.J., Yeh, S., Nishimura, K., and Chang, C. (2002). Supervillin associates with androgen receptor and modulates its transcriptional activity. Proc. Natl. Acad. Sci. U. S. A. *99*, 661-666.

Tzur, Y.B., Wilson, K.L., and Gruenbaum, Y. (2006). SUN-domain proteins: 'Velcro' that links the nucleoskeleton to the cytoskeleton. Nat. Rev. Mol. Cell Biol. *7*, 782-788.

Valkov, N.I., Ivanova, M.I., Uscheva, A.A., and Krachmarov, C.P. (1989). Association of actin with DNA and nuclear matrix from Guerin ascites tumour cells. Mol. Cell Biochem. *87*, 47-56.

Vartiainen, M.K., Guettler, S., Larijani, B., and Treisman, R. (2007). Nuclear actin regulates dynamic subcellular localization and activity of the SRF cofactor MAL. Science. *316*, 1749-1752.

Verheijen, R., Kuijpers, H., Vooijs, P., van, V.W., and Ramaekers, F. (1986). Protein composition of nuclear matrix preparations from HeLa cells: an immunochemical approach. J. Cell Sci. *80:103-22.*, 103-122.

Vignali, M., Hassan, A.H., Neely, K.E., and Workman, J.L. (2000). ATP-dependent chromatin-remodeling complexes. Mol. Cell Biol. *20*, 1899-1910.

Vlcek, S. and Foisner, R. (2007). Lamins and lamin-associated proteins in aging and disease. Curr. Opin. Cell Biol. *19*, 298-304.

Volkmann, N., Amann, K.J., Stoilova-McPhie, S., Egile, C., Winter, D.C., Hazelwood, L., Heuser, J.E., Li, R., Pollard, T.D., and Hanein, D. (2001). Structure of Arp2/3 complex in its activated state and in actin filament branch junctions. Science. *293*, 2456-2459.

Wada, A., Fukuda, M., Mishima, M., and Nishida, E. (1998). Nuclear export of actin: a novel mechanism regulating the subcellular localization of a major cytoskeletal protein. EMBO J. *17*, 1635-1641.

Wang, G.G., Allis, C.D., and Chi, P. (2007). Chromatin remodeling and cancer, Part I: Covalent histone modifications. Trends Mol. Med. *13*, 363-372.

Wang, J., Dudley, D., and Wang, X.L. (2002). Haplotype-specific effects on endothelial NO synthase promoter efficiency: modifiable by cigarette smoking. Arterioscler. Thromb. Vasc. Biol. *22*, e1-e4.

Wasser, M. and Chia, W. (2000). The EAST protein of drosophila controls an expandable nuclear endoskeleton. Nat. Cell Biol. *2*, 268-275.

Welch, M.D., DePace, A.H., Verma, S., Iwamatsu, A., and Mitchison, T.J. (1997). The human Arp2/3 complex is composed of evolutionarily conserved subunits and is localized to cellular regions of dynamic actin filament assembly. J. Cell Biol. *138*, 375-384.

Welch, W.J. and Suhan, J.P. (1985). Morphological study of the mammalian stress response: characterization of changes in cytoplasmic organelles, cytoskeleton, and nucleoli, and appearance of intranuclear actin filaments in rat fibroblasts after heat-shock treatment. J. Cell Biol. *101*, 1198-1211.

Wiesel, N., Mattout, A., Melcer, S., Melamed-Book, N., Herrmann, H., Medalia, O., Aebi, U., and Gruenbaum, Y. (2008). Laminopathic mutations interfere with the assembly, localization, and dynamics of nuclear lamins. Proc. Natl. Acad. Sci. U. S. A. *105*, 180-185.

Williams, C.L. (2003). The polybasic region of Ras and Rho family small GTPases: a regulator of protein interactions and membrane association and a site of nuclear localization signal sequences. Cell Signal. *15*, 1071-1080.

Witke, W., Sharpe, A.H., Hartwig, J.H., Azuma, T., Stossel, T.P., and Kwiatkowski, D.J. (1995). Hemostatic, inflammatory, and fibroblast responses are blunted in mice lacking gelsolin. Cell. *81*, 41-51.

Worman, H.J. and Gundersen, G.G. (2006). Here come the SUNs: a nucleocytoskeletal missing link. Trends Cell Biol. *16*, 67-69.

Wu, X., Yoo, Y., Okuhama, N.N., Tucker, P.W., Liu, G., and Guan, J.L. (2006). Regulation of RNA-polymerase-II-dependent transcription by N-WASP and its nuclear-binding partners. Nat. Cell Biol. *8*, 756-763.

Wulfkuhle, J.D., Donina, I.E., Stark, N.H., Pope, R.K., Pestonjamasp, K.N., Niswonger, M.L., and Luna, E.J. (1999). Domain analysis of supervillin, an F-actin bundling plasma membrane protein with functional nuclear localization signals. J. Cell Sci. *112*, 2125-2136.

Xu, Y.Z., Thuraisingam, T., Marino, R., and Radzioch, D. (2011). Recruitment of SWI/SNF complex is required for transcriptional activation of the SLC11A1 gene during macrophage differentiation of HL-60 cells. J. Biol. Chem. *286*, 12839-12849.

Xu, Y.Z., Thuraisingam, T., Morais, D.A., Rola-Pleszczynski, M., and Radzioch, D. (2010). Nuclear translocation of beta-actin is involved in transcriptional regulation during macrophage differentiation of HL-60 cells. Mol. Biol. Cell. *21*, 811-820.

Ye, J., Zhao, J., Hoffmann-Rohrer, U., and Grummt, I. (2008). Nuclear myosin I acts in concert with polymeric actin to drive RNA polymerase I transcription. Genes Dev. *22*, 322-330.

Yoo, Y., Wu, X., and Guan, J.L. (2007). A novel role of the actin-nucleating Arp2/3 complex in the regulation of RNA polymerase II-dependent transcription. J. Biol. Chem. *282*, 7616-7623.

Zalevsky, J., Lempert, L., Kranitz, H., and Mullins, R.D. (2001). Different WASP family proteins stimulate different Arp2/3 complex-dependent actin-nucleating activities. Curr. Biol. *11*, 1903-1913.

Zechel, K. (1980). Isolation of polymerization-competent cytoplasmic actin by affinity chromatography on immobilized DNAse I using formamide as eluant. Eur. J. Biochem. *110*, 343-348.

Zhang, Q., Ragnauth, C., Greener, M.J., Shanahan, C.M., and Roberts, R.G. (2002a). The nesprins are giant actin-binding proteins, orthologous to Drosophila melanogaster muscle protein MSP-300. Genomics. *80*, 473-481.

Zhang, S., Buder, K., Burkhardt, C., Schlott, B., Gorlach, M., and Grosse, F. (2002b). Nuclear DNA helicase II/RNA helicase A binds to filamentous actin. J. Biol. Chem. *277*, 843-853.

Zhao, K., Wang, W., Rando, O.J., Xue, Y., Swiderek, K., Kuo, A., and Crabtree, G.R. (1998). Rapid and phosphoinositol-dependent binding of the SWI/SNF-like BAF complex to chromatin after T lymphocyte receptor signaling. Cell. *95*, 625-636.

Zhou, J. and Herring, B.P. (2005). Mechanisms responsible for the promoter-specific effects of myocardin. J. Biol. Chem. *280*, 10861-10869.

2

Signaling of Receptor Tyrosine Kinases in the Nucleus

Sally-Anne Stephenson, Inga Mertens-Walker and Adrian Herington
Queensland University of Technology,
Australia

1. Introduction

Since the discovery of the first receptor tyrosine kinase (RTK) proteins in the late 1970s and early 1980s, many scientists have explored the functions of these important cell signaling molecules. The finding that these proteins are often deregulated or mutated in diseases such as cancers and diabetes, together with their potential as clinical therapeutic targets, has further highlighted the necessity for understanding the signaling functions of these important proteins. The mechanisms of RTK regulation and function have been recently reviewed by Lemmon & Schlessinger (2010) but in this review we instead focus on the results of several recent studies that show receptor tyrosine kinases can function from sub-cellular localisations, including in particular the nucleus, in addition to their classical plasma membrane location. Nuclear localisation of receptor tyrosine kinases has been demonstrated to be important for normal cell function but is also believed to contribute to the pathogenesis of several human diseases.

2. Classical signaling by receptor tyrosine kinases

The ability of a cell to receive signals from the outside, and deliver these inside so it can respond appropriately and in co-ordination with other cells, is required for the correct functioning of a multicellular organism as a whole. Cells communicate in two key ways – direct physical interaction or by way of communication molecules. These communication molecules, collectively called ligands, include those (eg steroid hormones, vitamins) that can pass directly through the lipid bilayer of the cell and interact with intracellular proteins and those such as protein hormones and peptide growth factors which cannot enter the cell directly. These latter ligands interact with plasma membrane-associated proteins called receptors to activate cascades of interactions between intracellular proteins that can result in a diverse range of responses and ultimately determine cell behaviour (Figure 1).

One large family of membrane receptors, the receptor tyrosine kinases (RTKs), is characterised by their intrinsic protein tyrosine kinase activity, an enzymatic function which catalyses the transfer of the γ phosphate of ATP to hydroxyl groups on tyrosine residues on target proteins (Hunter, 1998). Binding of the ligand stabilises dimers of the receptors to allow autophosphorylation *via* activation of the receptors' intrinsic tyrosine kinase activity that then initiates a network of sequentially acting components such as those of the

Ras/MAPK (mitogen-activated protein kinase) pathway, or single component systems, such as the STAT pathway. The combination of the activated signal transduction pathways constitute the mechanism by which this intracellular transfer of biochemical information is mediated and can determine the biological responses of cells to growth factors. Members of the RTK family play important roles in the control of most fundamental cellular processes including cell proliferation and differentiation, cell cycle, cell migration, cell metabolism and cell survival.

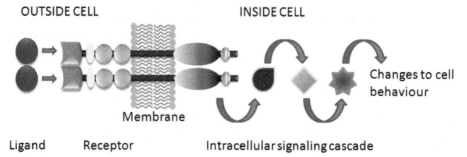

Fig. 1. Classical receptor tyrosine kinase signaling. Ligand binding stablilises dimers of the receptors within the plasma membrane. Autophosphorylation of one intracellular kinase domain by the other activates a signal transduction cascade into the cell so the cell can respond appropriately.

3. Protein structures of receptor tyrosine kinases

The general structure of RTK proteins is similar and all members of the RTK family have an intracellular kinase domain through which signaling is mediated by phosphorylation of tyrosine residues. In addition to the kinase domain, all RTKs have an extracellular domain, usually glycosylated, separated from the cytoplasmic part, containing the kinase domain, by a single hydrophobic transmembrane α helix. With the exception of the insulin (IR) and insulin-like growth factor (IGFR) receptor families, which are disulfide linked dimers of two polypeptide chains (α and β) that form a heterodimer (α2β2), RTKs are normally present as monomers in the cell membrane. Ligand binding induces receptor dimerisation resulting in autophosphorylation (the kinase domain of one RTK monomer cross-phosphorylates the other and *vice versa*). Receptor dimerisation is further stabilised by receptor:receptor interactions and the clustering of many receptors into lipid rich domains on the cell membrane (Pike, 2003). Further division of the 58 human RTKs into 20 different classes is based on similarities in primary structure, and the combinations of further functional domains in both extracellular and intracellular parts of the proteins (Figure 2).

4. Trafficking of receptor tyrosine kinases

Ligand activation of receptor tyrosine kinases present on the plasma membrane of cells promotes numerous downstream signal transduction pathways that result in cell responses including proliferation, migration and differentiation. Following ligand activation, virtually all receptor tyrosine kinases are rapidly endocytosed. This would allow the cell to discriminate new signals from old ones but it has been suggested that, because trafficking is a complex and highly regulated process, it is likely that endocytosis provides more than just a mechanism for removal of receptor-ligand complexes from the cell surface. Endocytosed receptors can be either recycled back to the membrane after disengagement of the ligand, or targeted for lysosomal degradation. Most receptor tyrosine kinases are internalised *via* clathrin-coated pits which then shed the clathrin and deliver the internalised receptor-ligand complexes to early endosomes. Bifurcation of receptor trafficking occurs in the early endosomes, allowing either recycling back to the plasma membrane or degradation through lysosomes. In some cases continued signaling from the endosomes has also been demonstrated (Ceresa & Schmid, 2000; Di Fiore & De Camilli, 2001; Wang et al., 2004a).

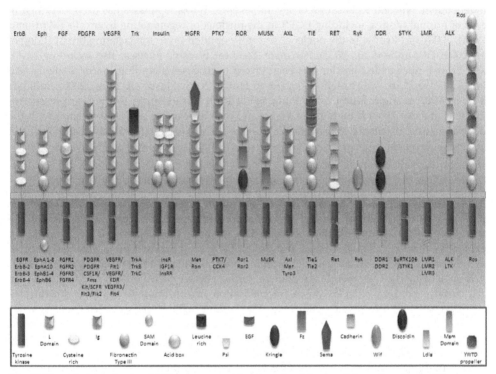

Fig. 2. Domain structures of 58 human receptor tyrosine kinases determines their sub-classification into 20 different families. The name of each family is shown above with the members listed below. A key indicates the various motifs common to individuals within that family.

Recent data also suggest that endocytosis controls sub-cellular localisation of activated receptors and their signaling complexes (Beguinot et al., 1984; Sorkin & Waters, 1993). For example, the prototypical receptor tyrosine kinase, the Epidermal Growth Factor Receptor (EGFR), has been found in caveoli, Golgi, endoplasmic reticulum, lysosome-like structures and nuclear envelopes (Carpentier et al., 1986; Lin et al., 2001). Given the continuity of the endomembrane system, linking endoplasmic reticulum, Golgi membranes, the plasma membrane, vesicles of both the endosomal and lysosomal systems and even the nuclear membrane, it is probably not surprising that receptors would be found within the membranes of these structures.

It also appears that endocytosis and trafficking of vesicles is involved in localisation of receptor tyrosine kinases to the nucleus. Nuclear localisation of receptor tyrosine kinases has emerged as a highly significant occurrence in the last decade, with reports indicating that the EGFR (ErbB-1 and -2), FGFR1 and IGF-IR can all translocate to the nucleus as full-length receptors or protein fragments devoid of the extracellular domain. In some cases this has been found to be ligand-dependent, within as early as 2 minutes of ligand stimulation, although there are also cases in which nuclear translocation appears to be ligand-independent. Nuclear localisation of several receptor tyrosine kinases has been identified in cells of normal tissues, including EGFR in the nucleus of regenerating liver cells (Marti and Wells, 2000) and ErbB-4 in the nuclei of secretory epithelium in the lactating breast (Long et al., 2003; Tidcombe et al., 2003). For many receptor tyrosine kinases, also including EGFR and ErbB-4, nuclear localisation has been linked to diseases including cancer, diabetes and inflammation (Citri & Yarden, 2006; Lo & Hung, 2006; Massie & Mills, 2006; Bublil & Yarden, 2007; Wang & Hung, 2009; Wang et al., 2010). For example, the nuclear presence of EGFR is associated with high grade breast and ovarian cancers and is associated with the development of resistance to some radio-, chemo- and monoclonal antibody-therapies (Lo et al., 2005a; Xia et al., 2009).

5. Mechanisms of receptor tyrosine kinase translocation to the nucleus

It has been hypothesised that in order for a receptor tyrosine kinase to translocate to the nucleus it must somehow 'escape' from the lipid bilayer of the cell surface and/or the trafficking of the endomembrane system. Exactly how this happens is only just being explored experimentally, but Wells & Marti (2002) have proposed three potential 'escape' mechanisms using EGFR as a model receptor tyrosine kinase. In the first, a mutant EGFR protein, lacking the transmembrane domain, forms a dimer with a wild-type receptor on the cell surface. Binding of EGF causes internalisation of the mutant-wild-type dimer *via* a clathrin-coated pit into an early endosome. The mutant EGFR is disassociated from the wild-type protein in the endosome and released into the cytosol, and from there it is transported into the nucleus. In the second scenario, full-length wild-type EGFR is trafficked from the plasma membrane to the endoplasmic reticulum, where it interacts with an accessory protein that removes it from the membrane for translocation into the nucleus. In the third, EGFR is targeted by proteases at the plasma membrane and an intracellular fragment translocates to the nucleus again by interaction with nuclear transport proteins. Recently, Liao & Carpenter (2007) provided support for the second scenario by showing that EGFR in the endosome associates with an accessory protein Sec61β, a component of the Sec61 translocon and is then retrotranslocated from the ER to the cytoplasm and from there translocated to the nucleus by nuclear transport proteins.

6. Nuclear localisation sequences and importins

Transport of proteins into the nucleus through the nuclear-pore-complex can be facilitated by the dedicated nuclear transport receptors of the β-karyopherin family which includes the importins (Gorlich and Kutay, 1999). Proteins translocated *via* importins contain nuclear localisation signals (NLS), a short stretch of amino acids that mediates the transport of proteins into the nucleus (Cokol et al., 2000). NLS motifs can be either monopartite, characterised by a cluster of basic residues preceded by a helix-breaking residue, or bipartite, where two clusters of basic residues are separated by 9-12 residues (Cokol et al., 2000). In the classical process of NLS-mediated nuclear translocation, an importin-α adaptor protein binds to a lysine-rich NLS in the cargo protein. An importin-β protein then binds to this importin-α/cargo complex through an NLS in the importin-α protein itself and guides the complex through the nuclear pore. Importin-β proteins are the key import mediators and can also bind non-classical NLS motifs, of which there are several types, to transport proteins without requiring importin-α interaction. In addition to basic NLSs, several other small epitopes have been identified that, when phosphorylated, can promote nuclear import (Nardozzi et al., 2010). These include the nuclear transport signal (NTS) of ERK1/2, which is a Ser-Pro-Ser (SPS) motif that, upon stimulation, is phosphorylated and functionally active as a binding site for the nuclear transport receptor importin-β7 (Chuderland et al., 2008).

7. Receptor tyrosine kinases reported to translocate to the nucleus

7.1 Epidermal Growth Factor Receptor (EGFR)/ErbB family

The Epidermal Growth Factor Receptor (EGFR) family of receptor tyrosine kinases, also known as ErbB (named after the viral oncogene v-erb-B2) or Human Epidermal growth factor Receptor (HER) receptors, contains four members: EGFR/ErbB-1/HER1, ErbB-2/HER2/Neu, ErbB-3/HER3 and ErbB-4/HER4. These receptors are expressed in various tissues of epithelial, mesenchymal and neuronal origin. Activation of ErbB receptors is controlled by the spatial and temporal expression of their 11 different ligands, all encoded by separate genes and all members of the EGF family of growth factors. These include EGF, epigen, transforming growth factor alpha (TGF-α), and amphiregulin, which bind EGFR; neuregulins (NRGs) 1,2,3,4, which bind ErbB-3 and/or ErbB-4, and betacellulin, heparin-binding EGF-like growth factor, and epiregulin, which bind EGFR and ErbB-4 (Riese & Stern, 1998). Ligand binding induces receptor dimerisation, and both homodimers and heterodimers with other ErbBs may be formed, and this then leads to the activation of a diverse range of downstream signaling pathways depending on the dimers and cross-activation of other ErbBs on the cell surface (Stern et al., 1986; Riese et al., 1995; Riese & Stern, 1998; Zaczek et al., 2005). Heterodimerisation is particularly important for signaling through ErbB-2, which lacks a conventional growth factor ligand, and ErbB-3, which has an inactive/impaired kinase domain.

Excessive EGFR, ErbB-2 and ErbB-3 signaling, as a result of receptor over-expression, mutations or autocrine stimulation, is a well known hallmark of a wide variety of solid tumours and leads to both increased cell proliferation and resistance to growth-inhibitory cytokines (Hynes & Lane, 2005). In contrast, ErbB-4 appears to be associated with growth suppression and improved patient prognosis in breast cancer (Jones, 2008; Muraoka-Cook et al., 2008). In addition, all four members of the ErbB family have a sub-membrane importin alpha-binding basic NLS that allows transport from the cytosol to the nucleus by the importin α/β complex. Consequently, ErbB proteins have been detected in the nucleus of both normal cells and cancer cells (Marti et al., 1991; Marti & Hug, 1995; Marti & Wells, 2000; Citri &Yarden, 2006; Lo & Hung, 2006; Massie & Mills, 2006; Bublil & Yarden, 2007; Wang &

Hung, 2009; Wang et al., 2010). In multiple cancer types, nuclear accumulation correlates with poor patient survival, tumor grade, and pathologic stage (Lo et al., 2005a; Psyrri et al., 2005; Junttila et al., 2005; Koumakpayi et al., 2006; Lo & Hung, 2006; Maatta et al., 2006; Hoshino et al., 2007; Xia et al., 2009; Hadzisejdic et al., 2010).

7.1.1 Epidermal Growth Factor Receptor (EGFR/ErbB-1/HER1)

Nuclear EGFR, and its ligands EGF and proTGF-α, were first observed in hepatocytes during liver regeneration (Raper et al., 1987; Marti et al., 1991; Marti & Hug, 1995; Marti & Wells, 2000; Grasl-Kraupp et al., 2002). Translocation of EGFR to the nucleus is also induced by DNA damage caused by irradiation (UV and ionizing) and cisplatin treatment but this appears to be ligand-independent (Dittmann et al., 2005; Xu et al., 2009). Full length EGFR is translocated into the nucleus through interactions with importin β-1, the nucleoporin protein Nup358 and proteins known to be involved in endocytotic internalisation of these proteins from the plasma membrane. Once in the nucleus, EGFR has three different roles depending on the initial signal, 1) as a direct regulator of gene transcription, 2) regulating cell proliferation and DNA replication *via* its kinase function, and 3) DNA repair and chemo- and radio-resistance through protein-protein interactions (Lin et al., 2001; Dittmann et al., 2005; Wang et al., 2006; Das et al., 2007; Kim et al., 2007; Wanner et al., 2008; Hsu & Hung, 2007). As a direct regulator of gene transcription, the C-terminal domain of EGFR directly interacts with the genome through binding and activating AT-rich sequences in the cyclin D1, nitric oxide synthetase (*iNOS*), *Aurora-A* and *B-myb* promoters (Liao and Carpenter, 2007; Lo, 2010). Nuclear EGFR interacts with STAT5 or STAT3 to transactivate the expression of the *Aurora-A* or *iNOS* genes respectively (Hung et al., 2008; Lo et al., 2005b). Nuclear EGFR can regulate cell proliferation and DNA replication by direct tyrosine phosphorylation of target proteins including chromatin bound proliferating cell nuclear antigen (PCNA) (Wang et al., 2006). EGFR kinase activity phosphorylates PCNA on tyrosine 211, stabilising the PCNA protein and stimulating DNA replication. In its third role, nuclear EGFR stimulates DNA repair by forming a direct protein-protein interaction with DNA-dependent protein kinase (DNA-PK) (Dittmann et al., 2005).

In addition to localisation to the plasma membrane and the nucleus, EGFR has also been found in the Golgi Apparatus, endoplasmic reticulum and the mitochondria (Carpentier et al., 1986; Lin et al., 2001; Boerner et al., 2004). EGFR was first reported in the mitochondria by Boerner et al., (2004) who found that in the presence of EGF, Src mediated the phosphorylation of EGFR residue Y845. EGFR phosphorylated at Y845 was found in the mitochondria and interacted with cytochrome c oxidase subunit II (CoxII) to possibly regulate cell survival. The method by which EGFR is translocated to the mitochondria is unknown, but was not related to endocytotsis of the EGFR protein and did not involve the function of Shc adaptor proteins (Yao et al., 2010). Furthermore, deletion studies showed that a putative mitochondrial-targeting signal between amino acids 646 and 660 was only partially responsible for migration (Boerner et al., 2004).

7.1.2 ErbB-2/HER2/Neu

Although ErbB-2 is catalytically active, it cannot bind the heregulin (HRG) ligand directly, but instead dimerises with either HRG-bound ErbB-3 or ErbB-4 to form a complex that is capable of signaling through either ErbB-2 or ErbB-4 (ErbB-3 is catalytically inactive/impaired) (Carraway et al., 1994). Upon HRG stimulation, cell-membrane embedded ErbB-2 migrates from the cell surface *via* early endosomes and is then either targeted to lysosomes for degradation, or recycled back to the surface. By an as yet undefined mechanism, ErbB-2 can also be removed from the lipid bilayer to form a complex with both importin β1 and EEA1 (Giri et al., 2005). This complex then moves through the nuclear pore complex into the nucleus. Once in the nucleus, ErbB-2 can form a complex with β-actin and RNA polymerase-1, enhancing binding of RNA pol 1 to rDNA, and progressing the early and elongation steps of transcription to expedite rRNA synthesis and protein translation (Li et al., 2011). The nuclear function of ErbB-2 would appear to be unrelated to its normal signaling role transduced through PI3-K and MEK/ERK because inhibitors to these kinases (LY294002 and U0126, respectively) did not affect the levels of 45S pre-rRNA in these cells. In addition to this role in expediting overall rRNA synthesis and protein translation, nuclear ErbB-2 has also been shown to bind to the promoter of the cyclooxygenase enzyme (*COX-2*) and up-regulate its expression. *COX-2* catalyzes the conversion of lipids to inflammatory prostaglandin and contributes to increased anti-apoptotic, pro-angiogenic, and metastatic potential in cancer cells (Vadlamudi et al., 1999; Howe et al., 2001; Gupta & DuBois, 2001; Half et al., 2002; Subbaramaiah et al., 2002; Turini & DuBois, 2002). The promoters of *PRPK, MMP16* and *DDX10* have also been identified as direct targets of nuclear ErbB-2 (Wang et al., 2004b).

7.1.3 ErbB-3/HER3

The kinase domain of ErbB-3 has been described as either catalytically inactive or impaired. Despite this ErbB-3 forms dimers with other ErbB receptors, and can recruit novel proteins to activate diverse signaling pathways (Guy et al., 1994; Zaczek et al., 2005). Intact ErbB-3 was detected in nuclei of prostate cancer cells in metastatic specimens (Koumakpayi et al., 2006; Cheng et al., 2007). Nuclear localisation was then studied in a model of prostate cancer using the MDA-PC 2b cells and this demonstrated that both the tumour microenvironment and androgen status influenced nuclear localisation of ErbB-3 in these cells (Cheng et al., 2007). Metastasis of prostate cancer cells to the bone and depletion of androgens from subcutaneous tumours both increased the nuclear translocation of ErbB-3. This also correlated with a decrease in cell proliferation. Once the tumours resumed aggressive growth, ErbB-3 then relocalised from the nucleus to the membrane and cytoplasm of the prostate cancer cells. This suggests that nuclear ErbB-3 may be involved in the progression of prostate cancer in bone after androgen-ablation therapy. ErbB-3 has also been identified in the nucleus, and possibly within the nucleolus, of both normal and malignant human mammary epithelial cells (Offterdinger et al., 2002). The role of nuclear ErbB-3 in these cells has not been determined but yeast two-hybrid approaches have been used to identify several transcription factors that associate with ErbB-3 including p23/p198 (Yoo & Hamburger, 1999), early growth response-1 (Thaminy et al., 2003) and the zinc finger protein ZNF207 (Thaminy et al., 2003) suggesting a gene regulation function. Finally, alternative transcription initiation of the ErbB-3 gene in Schwann cells leads to the production of a nuclear targeted variant of ErbB-3 that binds to chromatin and regulates the transcriptional activity of the ezrin and *HMGB1* genes (Adilakshmi et al., 2011).

7.1.4 ErbB-4/HER4

ErbB-4 has multiple functions during embryogenesis (Gassmann et al., 1995) and expression has recently been shown to be essential during breast development and lactation. In the lactating breast, ErbB-4 localizes to the nuclei of secretory epithelium (Long et al., 2003; Tidcombe et al., 2003). A unique proteolytic cleavage mechanism leads to the nuclear translocation of an intracellular fragment of ErbB-4. Cell membrane expressed ErbB-4 is successively cleaved by TACE/ADAM17, to release the ectodomain, and then γ-secretase to release an 80 kDa soluble intracellular fragment (s80) (Ni et al., 2001). This active kinase fragment binds to YAP (Yes-associated protein) which facilitates its translocation to the nucleus (Komuro et al., 2003). ErbB-4 also has three potential polycationic NLSs in its carboxy-terminal part which may provide an alternative route for nuclear translocation (Williams et al., 2004). The ErbB-4 s80 fragment functions as a nuclear chaperone for the STAT5A, co-translocating this transcription factor and regulating the expression of target genes including β-casein by binding with STAT5 to the β-casein promoter (Long et al., 2003; Williams et al., 2004). ErbB-4 also contains a nuclear export signal (NES) recognised by exportin proteins allowing transport of the protein out of the nucleus as well.

7.2 Fibroblast growth factor receptor family

The fibroblast growth factor (FGF) family consists of 18 secreted polypeptidic growth factors that bind to four high-affinity receptors (FGFR1-4) and assist in the regulation of cell proliferation, survival, migration and differentiation during development and in adult tissue homeostasis (Wesche et al., 2011). FGFs also bind to low-affinity heparan sulfate proteoglycans (HSPGs) present on most cells, which assist in the formation of the FGF-FGFR complex and protect the ligands from degradation. Overactivity of FGFR signaling is associated with several developmental disorders and cancer (Wesche et al., 2011).

7.2.1 FGFR1 (Fibroblast growth factor receptor 1)

Nuclear localisation of full length FGFR1 has been reported in astrocytes, glioma cells, neurons, fibroblasts and retinal cells and has been shown to be important for neuronal differentiation in the central nervous system (Stachowiak et al., 2003a; Stachowiak et al., 2003b). Nuclear accumulation is induced by many different stimuli including activation of acetylcholine receptors, stimulation of angiotensin II receptors, activation of adenylate cyclase or protein kinase C. Biotinylation of cell surface proteins showed that nuclear FGFR1 was unlikely to have been derived from the cell surface (Stachowiak et al., 1997; Peng et al., 2002). Because nuclear FGFR1 is glycosylated the suggestion is that the protein is at least partially processed through the ER-Golgi but that it is not stable in the endomembrane system and is released into the cytosol (Myers et al., 2003). It is also not clear how FGFR1 is then translocated to the nucleus as it lacks a typical NLS. However, several members of the fibroblast growth factor (FGF) family, including FGF-1 and FGF-2, lack signal peptide sequences and are therefore found in trace amounts, if at all, outside of cells. Some of these, for example FGF-2, have nuclear localisation sequences and are highly concentrated in the cell nucleus and it is believed that these FGF ligands act as chaperones for the translocation of receptors like FGFR1 into the nucleus (Myers et al., 2003). Although FGFR1 in the nucleus has been demonstrated to have FGF-regulated kinase activity and is phosphorylated, there appears to be limited co-localisation of FGF-2 and FGFR1 in the nucleus (Peng et al., 2002). Nuclear FGFR1 physically interacts with Ribosomal S6 Kinase isoform 1 (RSK1) and regulates its transcriptional activity (Hu et al., 2004). Target genes include *FGF-2, c-jun,* cyclin D1 and *MAP2,* genes that are involved in cell growth and differentiation (Reilly & Maher, 2001). FGFR1 has also been shown to be involved in the activation of the tyrosine hydroxylase promoter that is mediated through a cAMP responsive element (CRE) (Fang et al., 2005).

7.2.2 FGFR2

FGFR2 has been identified in the nuclei of quiescent Sertoli cells in the testes (Schmahl et al., 2004). In this study of FGF-9 knock-out mice, FGFR2 nuclear localisation was shown to correlate with male sex determination in the early gonads. The presence of FGFR2 in the nucleus coincides with the expression of the sex-determination gene *Sry* and the differentiation of progenitor cells in the gonads into Sertoli cells.

7.2.3 FGFR3

FGFR3 is a major negative regulator of linear bone growth and gain of function mutations cause the most common forms of dwarfism in humans as these are anti-proliferative (Colvin et al., 1996; Deng et al., 1996). Somatic mutations have been detected in several cancers where, by contrast, they are believed to drive proliferation and inhibit apoptosis (Trudel et al., 2004). Binding of FGF-1 to FGFR3 induces endocytosis *via* a dynamin/clathrin-mediated process to an endosomal compartment. Here the ectodomain is proteolytically cleaved possibly by an endosomal cathepsin although this has not yet been confirmed. The membrane anchored intracellular fragment is then cleaved in a second event by γ-secretase to generate a soluble intracellular domain that is released into the cytosol and can translocate to the nucleus. This requirement for endocytosis distinguishes FGFR3 proteolysis from that of most other RTKs.

7.3 VEGFR (Vascular endothelial growth factor receptor)

Cellular responses to the ligand vascular endothelial growth factor (VEGF) are activated through two structurally related receptors, VEGFR-1 (Flt-1) and VEGFR-2 (KDR) and are critically important in the regulation of endothelial cell growth and function (Cross et al., 2003). Stimulation of endothelial cells with VEGF induced the translocation of VEGFR-2, eNOS and caveolin-1 into the nucleus (Feng et al., 1999). The consequences of nuclear localisation of these three proteins have yet to be clarified. Non-endothelial expression of VEGFR-2 has also been reported (Stewart et al., 2003). A recent study by Susarla et al., (2011) identified VEGFR-2 expression on normal thyroid follicular cells. The VEGFR-2 expressed by these cells was phosphorylated and, although there was some staining in the cytoplasm, the highest concentration of VEGFR-2 was seen in most nuclei. VEGFR-1 and VEGFR-3 immunoreactivity was also seen predominantly in the nucleus with VEGFR-1 also localised at points of cell to cell contact. The role that VEGF receptors play in the nucleus has not been determined but the intranuclear staining was not co-incidental with chromatin and it is therefore unlikely that VEGFR proteins act as transcription factors.

7.4 Insulin receptor

Insulin is secreted by pancreatic β-cells in response to an increase in circulating glucose level to trigger tissues to increase glucose uptake and suppress hepatic glucose release. This biological action of insulin is initiated by binding to the insulin receptor InsR (Youngren, 2007). The presence of InsR in the nucleus was first reported in 1987 by Podlecki et al., but more recently this was further characterised by Rodrigues et al., (2008) who demonstrated that the insulin receptor appears in the nucleus of hepatocytes within 2.5 min of stimulation with insulin. This translocation event was associated with selective hydrolysis of nuclear PIP2 and formation of InsP3-dependent Ca^{2+} signaling within the nucleus that regulates glucose metabolism, gene expression and cell growth (Poenie et al., 1985; Hardingham et al., 1997; Nathanson et al., 1999; Pusl et al., 2002; Rodrigues et al., 2007). Nelson et al., (2011) have identified two potential gene targets for InsR in the nucleus, the early growth response 1 (*egr-1*) gene that is involved in the mitogenic response, and the glucokinase (*Gck*) gene which encodes a key metabolic enzyme.

7.5 IGF-1R (Insulin-like growth factor 1 receptor)

The insulin-like growth factor 1 receptor (IGF-1R) plays crucial roles in development and is often over-expressed in cancer. Stimulation with insulin-like growth factor 1 (IGF-I) or 2 (IGF-II) promotes cell proliferation, anti-apoptosis, angiogenesis, differentiation and development. Over-expression of IGF-1R is common in cancer but the mechanisms underlying the role of IGF-1R are not fully understood. Recently, Sehat et al., (2010) showed that IGF-I promotes the modification of IGF-1R by small ubiquitin-like modifier protein-1 (SUMO-1) and this then mediates translocation of IGF-1R to the nucleus. Nuclear import was also enhanced by stimulation with IGF-II but only modestly by insulin, in keeping with the affinity of IGF-1R for these ligands. Full length IGF-1Rα and IGF-1Rβ chains which make up the multi-subunit IGF-1R are found in the nucleus (Aleksic et al., 2010). Although it has been reported that IGF-1R binds to chromatin and acts directly as a transcriptional enhancer, direct transcriptional effects of nuclear IGF-1R are yet to be identified.

SUMOylation is initiated by a SUMO activating enzyme, such as SAE1 or SAE2, followed by a transfer of the active SUMO to Ubc9, the only known SUMO-conjugating enzyme, which then catalyses the transfer of SUMO to the target protein (Wilkinson and Henley, 2010). Seventy-five percent of known SUMO targets are modified within the consensus motif ψKxD/E where ψ is a hydrophic amino acid and x is any residue (Xu et al., 2008). Four SUMO isoforms have been identified in mammalian cells and SUMO-1 is the most widely studied member. Modification by SUMO-1 can result in a variety of functional consequences ranging from transcriptional repression (Garcia-Dominguez & Reyes, 2009) to DNA repair, mainly through targeting of p53 and BRCA1 (Bartek & Hodny, 2010), protein stability (Cai & Robertson, 2010) and cytoplasmic-nuclear shuttling (Salinas et al., 2004; Miranda et al., 2010; Sehat et al., 2010). Currently, IGF-1R is the only receptor tyrosine kinase for which nuclear translocation may be regulated by SUMOylation.

7.6 Eph receptors

Eph receptors are the largest group of transmembrane receptor tyrosine kinases with 14 human members divided into 2 subclasses, EphA (EphA1–EphA8, EphA10) and EphB (EphB1–EphB4, EphB6) (Pitulescu & Adams, 2010). Eph receptors are activated by their ligands the ephrins, proteins that are anchored to the plasma membrane of a neighbouring cell by either a glycosylphosphatidylinositol (GPI) anchor (type A) or a transmembrane amino acid sequence (type B). Eph-ephrin signaling plays important roles in neuronal and vascular development and many are over-expressed in various cancers (Flanagan & Vanderhaeghen, 1998; Adams & Klein, 2000; Stephenson et al., 2001; Lee et al., 2005; Pasquale, 2005; Chen et al., 2008).

To date only a single member of the Eph family, EphA4, has been reported in the nucleus (Kuroda et al., 2008). EphA4 is critically involved in development of neural tissue and more recently has been identified in hypertrophic chondrocytes and osteoblasts in the growth plate of developing mouse long bones (Kuroda et al., 2008). In the human osteoblastic cell line SaOS-2, EphA4 was found on the plasma membrane as expected, but also in the cytoplasm and in the nucleus. EphA4 accumulated in particular areas in the nucleus, but these were distinct from the nucleolus. It is not clear whether the EphA4 in the nucleus is full-length or a processed intracellular fragment and the role of EphA4 in the osteoblast nucleus has not been explored to date.

7.7 Ryk (Related to Receptor Tyrosine Kinase)

Ryk is a Wnt receptor that plays an important role in neurogenesis, neurite outgrowth, and axon guidance. Although a catalytically inactive receptor tyrosine kinase, Ryk is believed to signal *via* heterodimerisation with other receptor tyrosine kinases and has been shown to bind two members of the Eph receptor family, EphB2 and EphB3 (Halford et al., 2000). In neural progenitor cells, upon binding of Wnt3a, Ryk is cleaved at an intracellular site and the C-terminal cleavage product, Ryk ICD, translocates to the nucleus. Recently it was shown that Cdc37, a subunit of the molecular chaperone Hsp90 complex, binds to the Ryk ICD, promoting stabilization of the ICD fragment and providing the mechanism for nuclear translocation. Once in the nucleus, Ryk ICD regulates the expression of the key cell-fate determinants *Dlx2* (stimulated) and *Olig2* (inhibited) to promote GABAergic neuronal differentiation and inhibition of oligodendrocyte differentiation (Zhong et al., 2011).

7.8 Ror (RTK-like orphan receptor)

Ror1 and Ror2 receptor tyrosine kinases are involved in the development of mammalian central neurons (Paganioni & Ferreira, 2003; Paganioni & Ferreira, 2005). Although the ligand of Ror2 has been identified as Wnt-5A (Liu et al., 2008), Ror1 remains an orphan receptor protein tyrosine kinase without an identified interacting ligand. Tseng et al., (2010) used an in silico approach to predict receptor tyrosine kinases with likely nuclear localisation. Ror1 and Ror2 were identified in a panel that included receptors with known nuclear localisation including ErbB proteins, FGFR proteins and VEGFR proteins. The juxtamembrane domain of Ror1, responsible for nuclear localisation of this protein, was identified using deletion reporter constructs and the small GTPase Ran was identified as playing a key role in the nuclear transport. The function of Ror1 in the nucleus remains to be determined.

7.9 Trk (Tropomyocin Receptor Kinase)

Neurotrophins are a family of protein nerve growth factors that are critical for the development and functioning of the nervous system, regulating a wide range of biological processes. The receptors for neurotrophins are the Trk receptors - TrkA (or NTRK1), TrkB (or NTRK2), and TrkC (or NTRK3). Binding of neurotrophins to Trk receptors promotes both neuronal cell survival and death by activating signal transduction cascades including Ras/MAPK (mitogen-activated protein kinase) pathway and the PI3K (phosphatidylinositol 3-kinase) pathway. TrkA accumulates in the nucleus and on the mitotic apparatus of the human glioma cell line U251 after binding the neurotrophin ligand, nerve growth factor (NGF) (Gong et al., 2007). Translocation of phosphorylated TrkA is *via* carrier vesicles which sort and concentrate the receptors. These vesicles then interact with the nuclear envelope but how the TrkA protein is then removed from the membrane to move into the nucleoplasm is unclear. Once in the nucleus of the U251 glioma cells, TrkA co-localises with α-tubulin at the mitotic spindle. Interestingly, it has been shown that NGF co-localises with γ-tubulin at the centrosomes or spindle poles. Zhang et al., (2005) suggest that NGF concentrated to the centrosome can recruit its receptor TrkA from the nucleoplasm, activate the tyrosine kinase activity of the receptor to phosphorylate the tubulin and promote the mitotic spindle assembly that modulates the mitosis of human glioma cells.

7.10 HGFR (Hepatocyte growth factor receptor)

The HGFR family includes three members, MET, RON and SEA, produced mainly by cells of epithelial origin, which bind hepatocyte and hepatocyte-like growth factors secreted by mesenchymal cells, to regulate cell growth, cell motility, and morphogenesis (Comoglio & Boccaccio, 1996). Members of the HGFR family are described as oncoproteins because over-expression and/or abnormal activity correlates with the poor prognosis of many cancers (Accornero et al., 2010).

7.10.1 MET

Hepatocyte growth factor (HGF) secreted by stromal cells is a mitogenic factor and binds to MET on hepatocytes to activate pathways involved in cell proliferation, differentiation, and related activities that aid tissue regeneration in the liver. Other cell targets of HGF include epithelium, endothelium, myoblasts, spinal motor neurons, and hematopoietic cells. MET over-expression and hyper-activation are reported to correlate with metastatic ability of the tumor cells of several different tissue origins. Gomes et al., (2008) used the SkHep1 liver cell line to show that stimulation of cells with HGF caused the rapid translocation of phosphorylated MET from the plasma membrane to the nucleus, with peak levels detected after only 4 min of HGF exposure. Translocation of MET to the nucleus was mediated by binding of Gab1, an adaptor protein that contains a NLS for importin-driven translocation. In the nucleus, MET was shown to initiate nuclear Ca^{2+} signaling that stimulates cell proliferation (Rodrigues et al., 2007).

7.10.2 RON (Recepteur d'origine nantais)

RON is a receptor tyrosine kinase whose expression is highly restricted to cells of epithelial origin (Wang et al., 2010). Its ligand is the HGF-like macrophage stimulating protein (MSP) which stabilises two monomers of RON as a homodimer on the cell membrane. RON has been shown to be aberrantly expressed or mutated in many cancers including those from the bladder, breast, colon, lung, ovary, pancreas and prostate, particularly in aggressive tumours associated with poor patient survival (reviewed in Wang et al., 2010). Activated RON can promote c-Src activities that mediate cell-cycle progression, angiogenesis and survival of tumor cells (Danilkovitch-Miagkov et al., 2000; Feres et al., 2009). In bladder cancer cells, under conditions of serum starvation, RON has been shown to migrate from the cell membrane to the nucleus in a complex with EGFR with passage through the nuclear pore complex mediated by importins. In the nucleus, RON and EGFR co-operate in the transcriptional regulation of at least 134 different target genes known to participate in three stress-responsive networks: p53 (genes included *RBBP6, RB1, TP53BP2* and *JUN*), stress-activated protein kinase/c-jun N-terminal kinase (*JUN, MAPK8IP3, NFATC1* and *TRADD*) and phosphatidylinositol 3-kinase/Akt (*GHR, PPP2R3B* and *PRKCZ*) (Liu et al., 2010). Nuclear translocation of RON was therefore suggested to be a response to physiological stress. Furthermore, because MSP stimulation, homodimerisation and phosphorylation were not required for nuclear translocation, this is a ligand-independent response in these cells. A consensus sequence for binding nuclear RON was identified as GCA(G)GGGGCAGCG in genes that were both confirmed up-regulated (*FLJ46072, JUN, MLXIPL, NARG1* and *SSTR1*) and down-regulated (*RBBP6* and *POLRMT*) after serum starvation.

8. Conclusion

Although early reports of the presence of receptor tyrosine kinases in the nucleus of cells was met with scepticism, a significant collection of data now supports a role for many of these proteins in the nucleus of both normal and dysplastic cells. To date, 18 of the 58 human receptor tyrosine kinases have been found within nuclei and it is likely that more will be found. In general, the result of nuclear translocation of receptors is alterations to gene expression, but the full consequences of the presence of these proteins in the nucleus have yet to be determined. Only through further exploration can the complexity that nuclear localisation provides to receptor tyrosine kinase functions be determined.

9. References

Accornero, P, Pavone, LM, & Baratta, M. (2010) The scatter factor signaling pathways as therapeutic associated target in cancer treatment. *Current Medicinal Chemistry*, Vol. 17, No.25 (September 2010), pp. 2699-712, ISSN 0929-9673

Adams, RH, & Klein, R. (2000) Eph receptors and ephrin ligands, essential mediators of vascular development. *Trends in Cardiovascular Medicine*, Vol. 10, No. 5, (July 2000), pp. 183-188, ISSN 1050-1738

Adilakshmi, T, Ness-Myers, J, Madrid-Aliste, C, Fiser, Andras, & Tapinos, N. (2011) A nuclear variant of ErbB3 receptor tyrosine kinase regulates ezrin distribution and schwann cell myelination. *The Journal of Neuroscience*, Vol.31, No. 13 (March 2011), pp. 5106-5119, ISSN 0270-6474

Aleksic, T, Chitnis, MM, Perestenko, OV, Gao, S, Thomas, PH, Turner, GD, Protheroe, AS, Howarth, M, Macaulay, VM. (2010) Type 1 insulin-like growth factor receptor translocates to the nucleus of human tumor cells. *Cancer Research*, Vol. 70, No. 16, (August 2010), pp. 6412-6419, ISSN 0008-5472

Bartek, J, & Hodny, Z. (2010) SUMO boosts the DNA damage response barrier against cancer. *Cancer Cell*, Vol 17, No. 1, (January 2010), pp. 9-11, ISSN 1535-6108

Beguinot, L, Lyall, RM, Willingham, MC, & Pastan, I. (1984) Down-regulation of the epidermal growth factor receptor in KB cells is due to receptor internalization and subsequent degradation in lysosomes. *Proceedings of the National Academy of Sciences of the USA*, Vol. 81, No. 8, (April 1984), pp. 2384-2388, ISSN 1091-6490

Boerner, JL, Demory, ML, Silva, C, & Parsons, SJ. (2004) Phosphorylation of Y845 on the epidermal growth factor receptor mediates binding to the mitochondrial protein cytochrome c oxidase subunit II. *Molecular and Cellular Biology*, Vol. 24, No. 16, (August 2004), pp. 7059-7071, ISSN 0270-7306

Bublil, EM, & Yarden, Y. (2007) The EGF receptor family: spearheading a merger of signaling and therapeutics. *Current Opinion in Cell Biology*, Vol. 19, No. 2, (April 2007), pp. 124-134, ISSN 0955-0674

Cai, Q, & Robertson, ES. (2010) Ubiquitin/SUMO modification regulates VHL protein stability and nucleocytoplasmic localization. *PLoS One*, Vol. 5, No. 9, (September 2010), pii. e12636, ISSN 1932-6203

Carpentier, JL, Rees, AR, Gregoriou, M, Kris, R, Schlessinger, J & Orci, L. (1986) Subcellular distribution of the external and internal domains of the EGF receptor in A-431 cells. *Experimental Cell Research*, Vol. 166, No. 2, (October 1986), pp. 312-326, ISSN 0014-4827

Carraway, KL 3rd, Sliwkowski, MX, Akita, R, Platko, JV, Guy, PM, Nuijens, A, Diamonti, AJ, Vandlen, RL, Cantley, LC, & Cerione, RA. (1994) The erbB3 gene product is a receptor for heregulin. *Journal of Biological Chemistry*, Vol. 269, No. 19, (May 1994), pp. 14303-14306, ISSN 0021-9258

Ceresa, BP, & Schmid, SL. (2000) Regulation of signal transduction by endocytosis. *Current Opinion in Cell Biology*, Vol. 12, No. 2, (April 2000), pp. 204 -210, ISSN 0955-0674

Chen, J, Zhuang, G, Frieden, L, & Debinski, W. (2008) Eph receptors and Ephrins in cancer: common themes and controversies. *Cancer Research*, Vol. 68, No. 24, (December 2008), pp.10031-10033, ISSN 1538-7445

Cheng, CJ, Ye, XC, Vakar-Lopez, F, Kim, J, Tu, SM, Chen, DT, Navone, NM, Yu-Lee, LY, Lin, SH, & Hu, MC. (2007) Bone microenvironment and androgen status modulate subcellular localization of ErbB3 in prostate cancer cells. *Molecular Cancer Research*, Vol. 5, No. 7, (July 2007), pp. 675-684, ISSN 1541-7786

Chuderland, D, Konson, A, & Seger, R. (2008) Identification and characterization of a general nuclear translocation signal in signaling proteins. *Molecular Cell*, Vol. 31, No. 6 (September 2008), pp. 850-861, ISSN 10197-2765

Citri, A & Yarden, Y. (2006) EGF-ERBB signalling: towards the systems level. *Nature Reviews Molecular Cell Biology*, Vol. 7, No. 7, (July 2006), pp. 505-516, ISSN 1471-0072

Cokol, M, Nair, R, & Rost, B. (2000) Finding nuclear localization signals. *EMBO Reports*, Vol. 1, No. 5 (November 2000), pp. 411-415, ISSN 1469-3178

Colvin, JS, Bohne, BA, Harding, GW, McEwen, DG, & Ornitz, DM. (1996) Skeletal overgrowth and deafness in mice lacking fibroblast growth factor receptor 3. *Nature Genetics*, Vol. 12, No. 4, (April 1996), pp. 390-397, ISSN 1061-4036

Comoglio, PM, & Boccaccio, C. (1996) The HGF receptor family: unconventional signal transducers for invasive cell growth. *Genes to Cells*, Vol. 1, No. 4, (April 1996), pp. 347-354, ISSN 1365-2443

Cross, MJ, Dixelius, J, Matsumoto, T, & Claesson-Welsh, L. (2003) VEGF receptor signal transduction. *Trends in Biochemical Sciences*, Vol. 23, No. 9, (September 2003), pp. 488 – 494. ISSN 0968-0004

Danilkovitch-Miagkova,A, Angeloni, D, Skeel, A, Donley, S, Lerman, M, & Leonard, EJ. (2000) Integrin-mediated RON growth factor receptor phosphorylation requires tyrosine kinase activity of both the receptor and c-Src. *Journal of Biological Chemistry*, Vol. 275, No. 20, (May 2000), pp. 14783–15147, ISSN 0021-9258

Das, AK, Chen, BP, Story, MD, Sato, M, Minna, JD, Chen, DJ, Nirodi, CS. (2007) Somatic mutations in the tyrosine kinase domain of epidermal growth factor receptor (EGFR) abrogate EGFR-mediated radioprotection in non-small cell lung carcinoma. *Cancer Research*, Vol. 67, No. 11, (June 2007), pp. 5267-5274, ISSN 1538-7445

Deng, C, Wynshaw-Boris, A, Zhou, F, Kuo, A, & Leder, P. (1996) Fibroblast growth factor receptor 3 is a negative regulator of bone growth. *Cell*, Vol. 84, No. 6, (March 1996), pp. 911-921, ISSN 0092-8674

Dittmann, K, Mayer, C, Fehrenbacher, B, Schaller, M, Raju, U, Milas, L, Chen,DJ, Kehlbach, R, & Rodemann, HP. (2005) Radiation-induced epidermal growth factor receptor nuclear import is linked to activation of DNA-dependent protein kinase. *Journal of Biological Chemistry*, Vol. 280, No. 35, (September 2005), pp. 31182-31189, ISSN 0021-9258

Di Fiore, PP, & De Camilli, P. (2001) Endocytosis and signalling: An inseparable partnership. *Cell*, Vol. 106, No. 1, (July 2001), pp. 1– 4, ISSN 0092-8674

Fang, X, Stachowiak, EK, Dunham-Ems, SM, Klejbor, I, Stachowiak, MK (2005) Control of CREB-binding protein signaling by nuclear fibroblast growth factor receptor-1: a novel mechanism of gene regulation. *Journal of Biological Chemistry*, Vol. 280, No. 31, (August 2005), pp. 28451-28462, ISSN 0092-8674

Feng, Y, Venema, VJ, Venema, RC, Tsai, N, & Caldwell, RB. (1999) VEGF induces nuclear translocation of Flk-1/KDR, endothelial nitric oxide synthase, and caveolin-1 in vascular endothelial cells. *Biochemical and Biophysical Research Communications,* Vol. 256, No. 1, (March 1999), pp. 192-197, ISSN 0006-291X

Feres, KJ, Ischenko, I, & Hayman, MJ. (2009) The RON receptor tyrosine kinase promotes MSP-independent cell spreading and survival in breast epithelial cells. *Oncogene*, Vol. 28, No. 2, (January 2009), pp. 279–288, ISSN 0950-9232

Flanagan, JG, & Vanderhaeghen, P. (1998) The ephrins and Eph receptors in neural development. *Annual Review of Neuroscience*, Vol. 21, (1998), pp. 309–345, ISSN 1545-4126

Garcia-Dominguez, M, & Reyes, JC. (2009) SUMO association with repressor complexes, emerging routes for transcriptional control. *Biochimica et Biophysica Acta*, Vol. 1789, No. 6-8, (June-August 2009), pp. 451-459, ISSN 0006-3002

Gassmann, M, Casagranda, F, Orioli, D, Simon, H, Lai, C, Klein, R, Lemke, G. (1995) Aberrant neural and cardiac development in mice lacking the ErbB4 neuregulin receptor. *Nature*, Vol. 378. No. 6555, (November 1995), pp. 390-394, ISSN 0028-0836

Giri, DK, Ali-Seyed, M, Li, LY, Lee, DF, Ling, P, Bartholomeusz, G, Wang, SC, & Hung, MC. (2005) Endosomal transport of ErbB-2:mechanism for nuclear entry of the cell surface receptor. *Molecular and Cellular Biology*, Vol. 25, No. 24, (December 2005), pp. 11005–11018, ISSN 1471-0072

Gomes, DA, Rodrigues, MA, Leite, MF, Gomez, MV, Varnai, P, Balla, T, Bennett, AM, & Nathanson, MH. (2008) c-Met must translocate to the nucleus to initiate calcium signals. *Journal of Biological Chemistry*, Vol. 283, No. 7, (February2008), pp. 4344-4351.

Gong, A, Zhang, Z, Xiao, D, Yang, Y, Wang, Y, & Chen Y. (2007) Localization of phosphorylated TrkA in carrier vesicles involved in its nuclear translocation in U251 cell line. *Science in China. Series C, Life Science*. Vol. 50, No. 2, (April 2007), pp. 141-146. ISSN 1674-7305

Gorlich, D, & Kutay, U. (1999) Transport between the cell nucleus and the cytoplasm. *Annual Review Cell and Developmental Biology*, Vol.15, (1999), pp. 607–660, ISSN 1081-0706

Grasl-Kraupp, B, Schausberger, E, Hufnagl, K, Gerner, C, Low-Baselli, A, Rossmanith, W, Parzefall, W, & Schulte-Hermann, R. (2002) A novel mechanism for mitogenic signaling via pro-transforming growth factor alpha within hepatocyte nuclei. *Hepatology* Vol. 3, No. 6, (June 2002), pp. 1372-1380, ISSN 1527-3350

Gupta, RA, & DuBois, RN. (2001) Colorectal cancer prevention and treatment by inhibition of cyclooxygenase-2. *Nature Reviews Cancer*, Vol. 1, No. 1, (October 2001), pp. 11–21, ISSN 1474-175X

Guy, PM, Platko, JV, Cantley, LC, Cerione, RA, & Carraway, KL III (1994). Insect cell-expressed p180erbB3 possesses an impaired tyrosine kinase activity. *Proceedings of the National Academy of Sciences of the USA*, Vol. 91, No. 17, (August 1994), pp. 8132-8136. ISSN 1091-6490

Hadzisejdic, I, Mustac, E, Jonjic, N, Petkovic, M, & Grahovac, B. (2010) Nuclear EGFR in ductal invasive breast cancer: correlation with cyclin-D1 and prognosis. *Modern Pathology*, Vol. 23, No. 2, (March 2010), pp. 392–403, ISSN 0893-3952

Half, E, Tang, X, Gwyn, K, Sahin, A, Wathen, K, & Sinicrope, FA. (2002) Cyclooxygenase-2 expression in human breast cancers and adjacent ductal carcinoma in situ. *Cancer Research*, Vol. 62, No. 6, (March 2002), pp. 1676–1681, ISSN 1538-7445

Halford, MM, Armes, J, Buchert, M, Meskenaite, V, Grail, D, Hibbs, ML, Wilks, AF, Farlie, PG, Newgreen, DF, Hovens, CM, & Stacker, SA (2000) Ryk-deficient mice exhibit craniofacial defects associated with perturbed Eph receptor crosstalk. *Nature Genetics*, Vol. 25, No. 4, (August 2000), pp. 414–418, ISSN 1061-4036

Hardingham, GE, Chawla, S, Johnson, CM, & Bading, H. (1997) Distinct functions of nuclear and cytoplasmic calcium in the control of gene expression. *Nature*, Vol. 385, No. 6613, (January 1997), pp. 260–265, ISSN 0028-0836

Hoshino, M, Fukui, H, Ono, Y, Sekikawa, A, Ichikawa, K, Tomita, S, Imai, Y, Imura, J, Hiraishi, H, Fujimori, T. (2007) Nuclear expression of phosphorylated EGFR is associated with poor prognosis of patients with esophageal squamous cell carcinoma. *Pathobiology*, Vol. 74, No. 1, (January 2007), pp. 15–21, ISSN 1015-2008

Howe, LR, Subbaramaiah, K, Brown, AM, & Dannenberg, AJ. (2001) Cyclooxygenase-2: a target for the prevention and treatment of breast cancer. *Endocrine Related Cancer*, Vol. 8, No. 2, (June 2001), pp. 97–114, ISSN 1351-0088

Hsu, SC, & Hung, MC (2007) Characterization of a novel tripartite nuclear localization sequence in the EGFR family. *Journal of Biological Chemistry*, Vol. 282, No. 14, (May 2007), pp. 10432-10440, ISSN 0021-9258

Hu, Y, Fang, X, Dunham, SM, Prada, C, Stachowiak, EK, & Stachowiak, MK (2004) 90-kDa ribosomal S6 kinase is a direct target for the nuclear fibroblast growth factor receptor 1 (FGFR1): role in FGFR1 signaling. *Journal of Biological Chemistry*, Vol. 279, No. 28, (July 2004), pp. 29325-29335, ISSN 0021-9258

Hung, LY, Tseng, JT, Lee, YC, Xia, W, Wang, YN, Wu, ML, Chuang, YH, Lai, CH, Chang, WC. (2008) Nuclear epidermal growth factor receptor (EGFR) interacts with signal transducer and activator of transcription 5 (STAT5) in activating Aurora-A gene expression. *Nucleic Acids Research*, Vol. 36, Np. 13, (August 2008), pp. 4337-4351, ISSN 1362-4962

Hunter T. (1998) The Croonian lecture, 1997. The phosphorylation of proteins on tyrosine: its role in cell growth and disease. *Philosophical Transactions of the Royal Society London. Series B, Biological Sciences*, Vol. 353, No. 1368, (April 1998), pp. 583–605, ISSN 1471-2970

Hynes, NE & Lane, HA. (2005) ERBB receptors and cancer: the complexity of targeted inhibitors. *Nature Reviews Cancer*, Vol. 5, No. 5 *(May 2005), pp.* 341-354, ISSN 1474-175X

Jones, FE. (2008) HER4 intracellular domain (4ICD) activity in the developing mammary gland and breast cancer. *Journal of Mammary Gland Biology and Neoplasia*, Vol. 13, No. 2, (June 2008), pp. 247-258, ISSN 1573-7039

Junttila, TT, Sundvall, M, Lundin, M, Lundin, J, Tanner, M, Harkonen, P, Joensuu H, Isola, J, Elenius, K. (2005) Cleavable ErbB4 isoform in estrogen receptor-regulated growth of breast cancer cells. *Cancer Research*, Vol. 65, No. 4, (February 2005), pp. 1384–1393, ISSN 1538-7445

Kim, J, Jahng, WJ, Di Vizio, D, Lee, JS, Jhaveri, R, Rubin, MA, Shisheva, A, & Freeman, MR. (2007) The phosphoinositide kinase PIKfyve mediates epidermal growth factor receptor trafficking to the nucleus. *Cancer Research*, Vol. 67, No. 19, (October 2007), pp. 9229-9237, ISSN 1538-7445

Komuro, A, Nagai, M, Navin, NE, & Sudol, M. (2003) WW domain-containing protein YAP associates with ErbB-4 and acts as a co-transcriptional activator for the carboxyl-terminal fragment of ErbB-4 that translocates to the nucleus. *Journal of Biological Chemistry*, Vol. 278, No. 35, (August 2003), pp. 33334-33341, ISSN 0021-9258

Koumakpayi, IH, Diallo, JS, Le Page, C, Lessard, L, Gleave, M, Bégin, LR, Mes-Masson, AM, Saad, F. (2006) Expression and nuclear localization of ErbB3 in prostate cancer. *Clinical Cancer Research*, Vol. 12, No. 9, (May 2006), pp. 2730-2737, ISSN 1557-3265

Kuroda, C, Kubota, S, Kawata, K, Aoyama, E, Sumiyoshi, K, Oka, M, Inoue, M, Minagi, S, & Takigawa, M. (2008) Distribution, gene expression, and functional role of EPHA4 during ossification. *Biochemical and Biophysical Research Communications*, Vol. 374, No. 1, (September 2008), pp. 22-27, ISSN 0006-291X

Lee, YC, Perren JR, Douglas, EL, Raynor, MP, Bartley, MA, Bardy, PG, & Stephenson, SA. (2005) Investigation of the expression of the EphB4 receptor tyrosine kinase in prostate carcinoma. *BMC Cancer*, Vol 5, (September 2005), pp. 119, ISSN 1471-2407

Lemmon, MA & Schlessinger, J. (2010) Cell signaling by receptor tyrosine kinases. *Cell*, Vol. 141, No. 7 (June 2010), pp. 1117 – 1134, ISSN 0092-8674

Li, LY, Chen, H, Hsieh, YH, Wang, YN, Chu, HJ, Chen, YH, Chen, HY, Chien, PJ, Ma, HT, Tsai, HC, Lai, CC, Sher, YP, Lien, HC, Tsai, CH, & Hung, MC. (2011) Nuclear ErbB2 enhances translation and cell growth by activating transcription of ribosomal RNA genes. *Cancer Research*, Vol. 71, No. 12, (June 2011), pp. 4269-4279, ISSN 0008-5472

Liao, HJ, Carpenter, G. (2007) Role of the Sec61 translocon in EGF receptor trafficking to the nucleus and gene expression. *Molecular Biology of the Cell*, Vol. 18, No. 3, (March 2007), pp. 1064-1072, ISSN 1059-1524

Lin, SY, Makino, K, Xia, W, Matin, A, Wen, Y, Kwong, KY, Bourguignon, L, & Hung, MC. (2001) Nuclear localization of EGF receptor and its potential new role as a transcription factor. *Nature Cell Biology*, Vol. 3, No. 9, (September 2001), pp. 802-808, ISSN 1465-7392

Liu, HS, Hsu, PY, Lai, MD, Chang, HY, Ho, CL, Cheng, HL, Chen, HT, Lin, YJ, Wu, TJ, Tzai, TS, & Chow, NH. (2010) An unusual function of RON receptor tyrosine kinase as a transcriptional regulator in cooperation with EGFR in human cancer cells. *Carcinogenesis*, Vol. 31, No. 8, (August 2010), pp. 1456-1464, ISSN 0143-3334

Liu, Y, Rubin, B, Bodine, PV, & Billiard, J. (2008) Wnt5a induces homodimerization and activation of Ror2 receptor tyrosine kinase. *Journal of Cellular Biochemistry*, Vol. 105, No. 2, (October 2008), pp. 497-502, ISSN 0730-2312

Lo, HW. (2010) Nuclear mode of the EGFR signaling network: biology, prognostic value, and therapeutic implications. *Discovery Medicine*, Vol. 10, No. 50, (July 2010), pp. 44-51, ISSN 1539-6509

Lo, HW, Xia, W, Wei, Y, Ali-Seyed, M, Huang, SF, & Hung, MC. (2005a) Novel prognostic value of nuclear epidermal growth factor receptor in breast cancer. *Cancer Research*, Vol. 65, No. 1, (January 2005), pp. 338-348, ISSN 1538-7445

Lo, HW, Hsu, SC, Ali-Seyed, M, Gunduz, M, Xia, W, Wei, Y, Bartholomeusz, G, Shih, JY, & Hung, MC. (2005b) Nuclear interaction of EGFR and STAT3 in the activation of the iNOS/NO pathway. *Cancer Cell*, Vol. 7, No. 6, (June 2005), pp. 575-589, ISSN 1535-6108

Lo, HW, & Hung, MC. (2006) Nuclear EGFR signalling network in cancers: linking EGFR pathway to cell cycle progression, nitric oxide pathway and patient survival. *British Journal of Cancer*, Vol. 94, No. 2, (January 2006), pp. 184–188, ISSN 0007-0920

Long, W, Wagner, K-U, Lloyd, KCK, Binart, N, Shillingford, JM, Hennighausen, L, & Jones, FE. (2003) Impaired differentiation and lactational failure in ErbB4-deficient mammary glands identify ERBB4 as an obligate mediator of Stat5. *Development*, Vol. 130, No. 21, (November 2003), pp. 5257-5268, ISSN 1011-6370

Maatta, JA, Sundvall, M, Junttila,TT, Peri, L, Laine, VJ, Isola, J, Egeblad, M, & Elenius, K. (2006) Proteolytic cleavage and phosphorylation of a tumor-associated ErbB4 isoform promote ligand-independent survival and cancer cell growth. *Molecular Biology of the Cell*, Vol. 17, No. 1, (January 2006), pp. 67–79, ISSN 1059-1524

Marti, U, Burwen, SJ, Well,s A, Barker ME, Huling S, Feren AM, Jones, AL (1991) Localization of epidermal growth factor receptor in hepatocyte nuclei. *Hepatology*, Vol.13, No. 1, (January 1991), pp. 15-20, ISSN 1527-3350

Marti, U, & Hug, M. (1995) Acinar and cellular distribution and mRNA expression of the epidermal growth factor receptor are changed during liver regeneration. *Journal of Hepatology*, Vol. 23, No. 3, (September 1995), pp. 318-327 ISSN 0168-8278

Marti, U, & Wells, A. (2000) The nuclear accumulation of a variant epidermal growth factor receptor (EGFR) lacking the transmembrane domain requires coexpression of a full-length EGFR. *Molecular Cell Biology Research Communications*, Vol. 3, No. 1 (January 2000), pp. 8-14, ISSN 1522-4724

Massie, C & Mills, IG. (2006) The developing role of receptors and adaptors. *Nature Reviews Cancer*, Vol. 6, No. 5, (May 2006), pp. 403-409, ISSN 1474-175X

Miranda, KJ, Loeser, RF, & Yammani, RR. (2010) Sumoylation and nuclear translocation of S100A4 regulate IL-1beta-mediated production of matrix metalloproteinase-13. *Journal of Biological Chemistry*, Vol. 285, No. 41, (October 2010), pp. 31517-31524.

Muraoka-Cook, RS, Feng, SM, Strunk, KE, Earp 3rd HS. (2008) ErbB4/HER4: role in mammary gland development, differentiation and growth inhibition. *Journal of Mammary Gland Biology and Neoplasia*, Vol. 13, No. 2, (June 2008), pp. 235-246, ISSN 1083-3021

Myers, JM, Martins, GG, Ostrowski, J, & Stachowiak, MK. (2003) Nuclear trafficking of FGFR1: a role for the transmembrane domain. *Journal of Cellular Biochemistry*, Vol. 88, No. 6, (April 2003), pp. 1273-1291, ISSN 0730-2312

Nardozzi, JD, Lott, K, & Cingolani, G. (2010) Phosphorylation meets nuclear import: a review. *Cell Communication and Signaling*. Vol. 23, No. 8, (December 2010), pp. 32, ISSN 1478-811X

Nathanson, MH, Rios-Velez, L, Burgstahler, AD, & Mennone, A. (1999) Communication via gap junctions modulates bile secretion in the isolated perfused rat liver. *Gastroenterology*, Vol. 116, No. 5, (May 1999), pp. 1176–1183, ISSN 0016-5085

Nelson, JD, LeBoeuf, RC & Bomsztyk, K. (2011) Direct recruitment of insulin receptor and ERK signaling cascade to insulin-inducible gene loci. *Diabetes*, Vol. 60, No. 1, (January 2011), pp. 127–137, ISSN 0021-1797

Ni, CY, Murphy, MP, Golde, TE & Carpenter, G. (2001) γ-Secretase cleavage and nuclear localization of ErbB-4 receptor tyrosine kinase. *Science*, Vol. 294, No. 5549, (October 2001), pp. 2179–2181, ISSN 0036-8075

Offterdinger, M, Schöfer, C, Weipoltshammer, K, & Grunt, TW. (2002) c-erbB-3: a nuclear protein in mammary epithelial cells. *Journal of Cell Biology*, Vol. 157, No. 6, (June 2002), pp. 929-939, ISSN 0021-9525

Paganoni, S, & Ferreira, A. (2003) Expression and subcellular localization of Ror tyrosine kinase receptors are developmentally regulated in cultured hippocampal neurons. *Journal of Neuroscience Research*, Vol. 73. No. 4, (August 2003), pp. 429-440, ISSN 0360-4012

Paganoni, S, & Ferreira, A. (2005) Neurite extension in central neurons: a novel role for the receptor tyrosine kinases Ror1 and Ror2. *Journal of Cell Science*, Vol. 118, No. 2, (January 2005), pp. 433-446, ISSN 1477-9137

Pasquale, EB. Eph receptor signalling casts a wide net on cell behaviour. *Nature Reviews Molecular Cell Biology*, Vol. 6, No. 6, (June 2005), pp. 462-475, ISSN 1471-0072

Peng, Hu, Myers, J, Fang, X, Stachowiak, EK, Maher, PA, Martins, GG, Popescu, G, Berezney, R, & Stachowiak, MK. (2002) Integrative nuclear FGFR1 signaling (INFS) pathway mediates activation of the tyrosine hydroxylase gene by angiotensin II, depolarization, and protein kinase C. *Journal of Neurochemistry*, Vol. 81, No. 3, (May 2002), pp. 506-524, ISSN 1471-4159

Pike, LJ. (2003) Lipid rafts: bringing order to chaos. *Journal of Lipid Research*, Vol. 44, No. 4, (April 2003), pp. 655-667, ISSN 0022-2275

Pitulescu, ME, & Adams, RH. (2010) Eph/ephrin molecules--a hub for signaling and endocytosis. *Genes and Development*. Vol 24, No. 22, (November 2010), pp. 2480-2492, ISSN 0890-9369

Podlecki, DA, Smith, RM, Kao, M, Tsai, P, Huecksteadt, T, Brandenburg, D, Lasher, RS, Jarett, L, & Olefsky, JM. (1987) Nuclear translocation of the insulin receptor. A possible mediator of insulin's long term effects. *Journal of Biological Chemistry*, Vol. 262, No. 7. (March 1987), pp. 3362-3368, ISSN 0021-9258

Poenie, M, Alderton, J, Tsien, RY, & Steinhardt, RA. (1985) Changes of free calcium levels with stages of the cell division cycle. *Nature*, Vol. 315, No. 6015, (May 1985), pp. 147-149, ISSN 0028-0836

Psyrri, A, Yu, Z, Weinberger, PM, Sasaki, C, Haffty, B, Camp, R, Rimm, D & Burtness, BA (2005) Quantitative determination of nuclear and cytoplasmic epidermal growth factor receptor expression in oropharyngeal squamous cell cancer by using automated quantitative analysis. *Clinical Cancer Research*, Vol. 11, No. 16, (August 2005), pp. 5856-5862, ISSN 1557-3265

Pusl, T, Wu, JJ, Zimmerman, TL, Zhang, L, Ehrlich, BE, Berchtold, MW, Hoek, JB, Karpen, SJ, Nathanson, MH, & Bennett, AM. (2002) Epidermal growth factor-mediated activation of the ETS domain transcription factor Elk-1 requires nuclear calcium. *Journal of Biological Chemistry*, Vol. 277, No. 30, (July 2002), pp. 27517-27527, ISSN 0021-9258

Raper, SE, Burwen, SJ, Barker, ME, & Jones, AL. (1987) Translocation of epidermal growth factor to the hepatocyte nucleus during rat liver regeneration. *Gastroenterology*, Vol. 92, No. 5, (May 1987), pp. 1243-1250, ISSN 0016-5085

Reilly, JF & Maher PA. (2001) Importin β-mediated nuclear import of fibroblast growth factor receptor: role in cell proliferation. *Journal of Cell Biology*. Vol. 152, No. 6, (March 2001), pp. 1307-1312, ISSN 0021-9525

Riese, II DJ, & Stern, DF. (1998) Specificity within the EGF/ErbB receptor family signaling network. *BioEssays*, Vol. 20, No. 1, (January 1998), pp. 41-48 ISSN 1521-1878

Riese, II DJ, van Raaij, TM, Plowman, GD, Andrews, GC, & Stern, DF. (1995) Cellular response to neuregulins is governed by complex interactions of the ErbB receptor family. *Molecular and Cellular Biology*, Vol. 15, No. 10, (October 1995), pp. 5770-5776, ISSN 1098-5549

Rodrigues, MA, Gomes, DA, Leite, MF, Grant, W, Zhang, L, Lam, W, Cheng, YC, Bennett, AM, & Nathanson, MH. (2007) Nucleoplasmic calcium is required for cell proliferation. *Journal of Biological Chemistry*, Vol. 282, No. 23 (June 2007), pp. 17061-17068, ISSN 0021-9258

Rodrigues, MA, Gomes, DA, Andrade, VA, Leite, MF, & Nathanson MH. (2008) Insulin induces calcium signals in the nucleus of rat hepatocytes. *Hepatology*. Vol. 48, No. 5, (November 2008), pp. 1621-1631, ISSN 1527-3350

Salinas, S, Briançon-Marjollet, A, Bossis, G, Lopez, MA, Piechaczyk, M, Jariel-Encontre, I, Debant, A, & Hipskind, RA. (2004) SUMOylation regulates nucleo-cytoplasmic shuttling of Elk-1. *Journal of Cell Biology*, Vol. 165, No. 6. (June 2004), pp. 767-773, ISSN 0021-9525

Schmahl, J, Kim, Y, Colvin, JS, Ornitz, DM, & Capel B. (2004) Fgf9 induces proliferation and nuclear localization of FGFR2 in Sertoli precursors during male sex determination. Development 2004, 131(15):3627-3636. *Cancer Research*, Vol. 71, No. 12, (August 2004), pp. 4269-4279, ISSN 1538-7445

Sehat, B, Tofigh, A, Lin, Y, Trocmé, E, Liljedahl, U, Lagergren, J, & Larsson, O. (2010) SUMOylation mediates the nuclear translocation and signaling of the IGF-1 receptor. *Science Signaling*, Vol. 3, No. 108, (February 2010), pp. ra10, ISSN 1945-0877

Sorkin, A, & Waters, CM (1993) Endocytosis of growth factor receptors. *Bioessays*, Vol. 15, No. 6, (June 1993), pp. 375-382, ISSN 1521-1878

Stachowiak, MK, Moffett, J, Maher, PA, Tucholski, J, & Stachowiak, EK. (1997) Growth factor regulation of cell growth and proliferation in the nervous system. A new intracrine nuclear mechanism. *Molecular Neurobiology*, Vol. 15, No. 3, (December 1997), pp. 1-27 ISSN 0893-7648

Stachowiak, MK, Fang, X, Myers, JM, Dunham, SM, Berezney, R, Maher, PA, & Stachowiak, EK. (2003a) Integrative nuclear FGFR1 signaling (INFS) as a part of a universal "feed-forward-and-gate" signaling module that controls cell growth and differentiation. *Journal of Cellular Biochemistry*, Vol. 90, No. 4, (November 2003), pp. 662-691, ISSN 0730-2312

Stachowiak, EK, Fang, X, Myers, J, Dunham, S, Stachowiak, MK. (2003b) cAMP induced differentiation of human neuronal progenitor cells is mediated by nuclear fibroblast growth factor receptor-1 (FGFR1). *Journal of Neurochemistry*, Vol 84, No. 6, (March 2003), pp. 1296-1312, ISSN 1471-4159

Stephenson, SA, Slomka, S, Douglas, EL, Hewett, PJ, & Hardingham, JE. (2001) Receptor protein tyrosine kinase EphB4 is up-regulated in colon cancer. *BMC Molecular Biology*, Vol. 2, (December 2001), pp. 15, ISSN 1471-2199

Stern, DF, Heffernan, PA, & Weinberg, RA. (1986) p185, a product of the neu protooncogene, is a receptor-like protein associated with tyrosine kinase activity. *Molecular and Cellular Biology*, Vol. 6, No. 5, (May 1986), pp. 1729-1740, ISSN 1098-5549

Stewart, M, Turley, H, Cook, N, Pezzella, F, Pillai, G, Ogilvie, D, Cartlidge, S, Paterson, D, Copley, C, Kendrew, J, Barnes, C, Harris, AL, & Gatter, KC. (2003) The angiogenic receptor KDR is widely distributed in human tissues and tumours and relocates intracellularly on phosphorylation. An immunohistochemical study. *Histopathology.* Vol. 43, No. 1, (July 2003), pp. 33-39, ISSN 1365-2559

Subbaramaiah, K, Norton, L, Gerald, W, and Dannenberg, AJ. (2002) Cyclooxygenase-2 is overexpressed in Her-2/neu-positive breast cancer: evidence for involvement of AP-1 and PEA3. *Journal of Biological Chemistry*, Vol. 277, No. 21, (May 2002), pp. 18649-18657, ISSN 1083-351X

Susarla, R, Gonzalez, AM, Watkinson, JC, & Eggo, MC. (2011) Expression of receptors for VEGFS on normal human thyroid follicular cells and their role in follicle formation. *Journal of Cellular Physiology*, (July 2011), DOI 10.1002/jcp.22930, ISSN 1097-4652

Thaminy, S, Auerbach, D, Arnoldo, A, & Stagljar, I. (2003) Identification of novel ErbB3-interacting factors using the split-ubiquitin membrane yeast two-hybrid system. *Genome Research*, Vol. 13, No. 7, (July 2003), pp. 1744-1753, ISSN 1549-5469

Tidcombe, H, Jackson-Fisher, A, Mathers, K, Stern, DF, Gassmann, M, & Golding, JP. (2003) Neural and mammary gland defects in ErbB4 knockout mice genetically rescued from embryonic lethality. *Proceedings of the National Academy of Sciences of the USA*, Vol. 100, No. 14, (July 2003), pp. 8281-8286, ISSN 1091-6490

Trudel, S, Ely, S, Farooqi, Y, Affer, M, Robbiani, DF, Chesi, M, Bergsagel, PL. (2004) Inhibition of fibroblast growth factor receptor 3 induces differentiation and apoptosis in t(4;14) myeloma. *Blood*, Vol. 103, No. 9, (May 2004) pp. 3521-8, ISSN 0006-4971

Tseng, HC, Lyu, PC, Lin, WC. (2010) Nuclear localization of orphan receptor protein kinase (Ror1) is mediated through the juxtamembrane domain. *BMC Cell Biology*, Vol. 11, (June 2010), pp. 48, ISSN 1471-2121

Turini, ME, & DuBois, RN. (2002) Cyclooxygenase-2: a therapeutic target. *Annual Review of Medicine*, Vol. 53, (2002), pp. 35-57, ISSN 0066-4219

Vadlamudi, R, Mandal, M, Adam, L, Steinbach, G, Mendelsohn, J, & Kumar, R. (1999) Regulation of cyclooxygenase-2 pathway by HER2 receptor. *Oncogene*, Vol. 18, No. 2, (January 1999), pp. 305-314, ISSN 0950-9232

Wang, MH, Padhye, SS, Guin, S, Ma, Q, & Zhou, YQ. (2010) Potential therapeutics specific to c-MET/RON receptor tyrosine kinases for molecular targeting in cancer therapy. *Acta Pharmacologica Sinica*, Vol. 31, No. 9, (September 2010), pp. 1181-1188, ISSN 1745-7254

Wang, SC, Nakajima, Y, Yu, YL, Xia, W, Chen, CT, Yang, CC, McIntush, EW, Li, LY, Hawke, DH, Kobayashi, R, & Hung, MC. (2006) Tyrosine phosphorylation controls PCNA function through protein stability. *Nature Cell Biology*, Vol. 8, No. 12, (December 2006), pp. 1359-1368, ISSN 1465-7392

Wang, Y, Pennock, SD, Chen, X, Kazlauskas, A, & Wang, Z. (2004a) Platelet-derived growth factor receptor-mediated signal transduction from endosomes. *Journal of Biological Chemistry*, Vol. 279, No.9, (February 2004), pp. 8038 – 8046, ISSN 0021-9258

Wang, SC, Lien, HC, Xia, W, Chen, IF, Lo, HW, Wang, Z, Ali-Seyed, M, Lee, DF, Bartholomeusz, G, Ou-Yang, F, Giri, DK, & Hung MC (2004b) Binding at and transactivation of the COX-2 promoter by nuclear tyrosine kinase receptor ErbB-2. *Cancer Cell.* Vol. 6, No. 3, (September 2004), pp. 251-261, ISSN 1535-6108

Wang, SC, & Hung, MC. (2009) Nuclear translocation of the EGFR family membrane tyrosine kinase receptors. *Clinical Cancer Research*, Vol. 15, No. 21, (November 2009), pp. 6484–6489, ISSN 1078-0432

Wanner, G, Mayer, C, Kehlbach, R, Rodemann, HP, & Dittmann, K. (2008) Activation of protein kinase C epsilon stimulates DNA-repair via epidermal growth factor receptor nuclear accumulation. *Radiotherapy and Oncology*, Vol 86, No. 3, (March 2008), pp. 383-390, ISSN 0167-8140

Wells, A, & Marti, U. (2002) Signalling shortcuts: cell-surface receptors in the nucleus? *Nature Reviews Molecular Cell Biology*, Vol. 3, No. 9, (September 2002), pp. 697-702, ISSN 1471-0072

Wesche, J, Haglund, K, & Haugsten, EM. (2011) Fibroblast growth factors and their receptors in cancer. *Biochemical Journal*, Vol. 437, No. 2, (July 2011), pp. 199–213, ISSN 0264-6021

Wilkinson, KA & Henley, JM. (2010) Mechanisms, regulation and consequences of protein SUMOylation. *Biochemistry Journal*, Vol. 428, No. 2, (May 2010), pp. 133-145, ISSN 0264-6021

Williams, CC, Allison, JG, Vidal, GA, Burow, ME, Beckman, BS, Marrero, L, Jones, FE. (2004) The ERBB4/HER4 receptor tyrosine kinase regulates gene expression by functioning as a STAT5A nuclear chaperone. *Journal Cell Biology*, Vol. 167, No. 3, (November 2004), pp. 469-478, ISSN 0021-9258

Xia, W, Wei, Y, Du, Y, Liu, J, Chang, B, Yu, YL, et al.,. Nuclear expression of epidermal growth factor receptor is a novel prognostic value in patients with ovarian cancer. *Molecular Carcinogenesis*, Vol. 48, No. 7, (July 2009), pp. 610-617, ISSN 1098-2744

Xu, J, He, Y, Qiang, B, Yuan, J, Peng, X, & Pan, XM. (2008) A novel method for high accuracy sumoylation site prediction from protein sequences. *BMC Bioinformatics*, Vol. 9, (January 2008), pp. 8, ISSN 1471-2105

Xu, Y, Shao, Y, Zhou, J, Voorhees, JJ, & Fisher, GJ. (2009) Ultraviolet irradiation-induces epidermal growth factor receptor (EGFR) nuclear translocation in human keratinocytes. *Journal of Cell Biochemistry*, Vol. 107, No. 5, (August 2009), pp. 873-880, ISSN 0021-9258

Yao, Y, Wang, G, Li, Z, Yan, B, Guo, Y, Jiang, X, & Xi, J. (2010) Mitochondrially localized EGFR is independent of its endocytosis and associates with cell viability. *Acta Biochimica et Biophysica Sinica (Singapore)*, Vol. 42, No. 11, (November 2010), pp. 763-770, ISSN 1672-9145

Yoo, JY, & Hamburger, AW. (1999) Interaction of the p23/p198 protein with ErbB-3. *Gene*, Vol. 229, No. 1-2, (March 1999), pp. 215-221, ISSN 0378-1119

Youngren JF. (2007) Regulation of insulin receptor function. *Cellular and Molecular Life Sciences*, Vol. 64, No. 7-8, (April 2007), pp. 873 – 891, ISSN 1420-682X

Zaczek, A, Brandt, B, & Bielawski, KP. (2005) The diverse signaling network of EGFR, HER2, HER3 and HER4 tyrosine kinase receptors and the consequences for therapeutic approaches. *Histology and Histopathology*, Vol. 20, No. 3, (July 2005), pp. 1005-1015, ISSN 0213-3911

Zhang, Z, Yang, Y, Gong, A, Wang, C, Liang, Y, & Chen Y. (2005) Localization of NGF and TrkA at mitotic apparatus in human glioma cell line U251. *Biochemical and*

Biophysical Research Communications, Vol. 337, No. 1, (November 2005), pp. 68-74, ISSN 0006-291X

Zhong, J, Kim, H, Lyu, J, Yoshikawa, K, Nakafuku, M & Lu1, W (2011) The Wnt receptor Ryk controls specification of GABAergic neurons versus oligodendrocytes during telencephalon development. *Development*, Vol. 138, No. 3, (February 2011), pp. 409-419, ISSN 1011-6370

3

The Kinetochore and Mitosis: Focus on the Regulation and Correction Mechanisms of Chromosome-to-Microtubule Attachments

Rita M. Reis[1] and Hassan Bousbaa[1,2]
[1]Centro de Investigação em Ciências da Saúde (CICS),
Instituto Superior de Ciências da Saúde Norte/CESPU,
[2]Centro de Química Medicinal da Universidade do Porto (CEQUIMED-UP),
Portugal

1. Introduction

During mitosis, accurate segregation of the chromosomes duplicated at the S phase of the cell cycle relies on the successful of their proper attachment to microtubules of the mitotic spindle and their alignment at the metaphase plate before the onset of anaphase. Attachment errors or failure to align can lead to chromosome gains or losses, a condition known as chromosome instability (CIN) which is a common feature amongst many cancers.

The kinetochore, a multiprotein structure assembled on the centromeres, plays a key role in chromosome attachment and high fidelity of mitotic chromosome segregation between daughter cells.

The kinetochore structure provides the attachment site of microtubule polymers to chromosome. Each mitotic chromosome has a pair of kinetochores, positioned on opposite sides of the centromere, that allow for its bipolar attachment to the spindle. Besides this role in attachment, the kinetochore controls the metaphase-anaphase transition by inhibiting chromatid separation until all chromosomes are properly attached and aligned at the metaphase plate, thereby ensuring equitable sharing of chromosomes upon cell division.

During the last decade, considerable progress has been made in understanding the molecular composition of the kinetochore, shedding light on the higher order organization of the kinetochore and on it activity in mediating chromosome attachment to spindle microtubules and in regulating the fidelity of the metaphase to anaphase transition. We address these aspects, concentrating in particular on the role that the kinetochores have in sensing and resolving aberrant kinetochore-microtubule attachments and in preventing chromosome missegregation.

2. The events of Mitosis

In 1879, Walther Flemming was the first to describe in great detail cell division. The Fleming's great discovery was that during the cell division one longitudinal half of each chromosome goes in each direction, so that each daughter nucleus is formed from a complete set of longitudinal halves. In 1882 he termed this process Mitosis (from the Greek mitos "thread-metamorphosis") (Baker, 1949; Paweletz, 2001).

2.1 Mitotic phases

Nowadays, it is known that Mitosis is a complex and highly regulated process, used by eukaryotes to generate two identical daughter cells from one original mother cell. This particular phase of the cell cycle, ensures that the cell faithfully segregates the sister chromatids, of the duplicated chromosomes, into the daughter cells, producing two cells that are identical to one another and to the original parent cell. Conventionally, mitosis is divided into five stages: Prophase, Prometaphase, Metaphase, Anaphase and Telophase (Fig. 1).

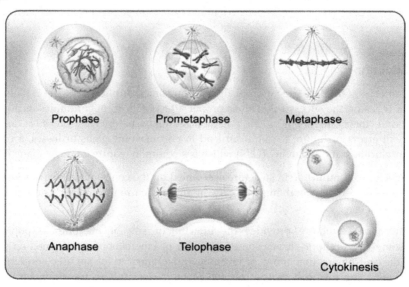

Fig. 1. Mitotic stages and Cytokinesis of animal cells. Prophase: chromatin condenses into chromosomes and centrosomes start to separate; Prometaphase: begins with nuclear envelope breakdown, chromosomes are captured by microtubules that are growing from opposite poles of the spindle; Metaphase: all chromosomes align at the equator of the cell; Anaphase: sister chromatids move to opposite poles; Telophase: sister chromatids reach the poles, decondense and nuclear membrane reforms around each identical daughter nuclei. Simultaneously, the division of the cytoplasm generates two independent daughter cells by a process called Cytokinesis.

Prophase, is marked by the appearance of condensed chromosomes that emerge as two identical filaments (sister chromatids, produced in the S phase of the cell cycle) and the centrosomes begin to separate to opposite sites initiating the mitotic spindle assembly. The next stage, Prometaphase, starts with the nuclear envelope breakdown allowing that microtubules, from the opposite centrosomes, occupy all the cytoplasmic space and fully organize the mitotic spindle. These microtubules bind the chromosomes and help them to reach and align at the equatorial cell region; at this stage the cell has achieved the Metaphase. In Metaphase, chromosomes are aligned at the middle of the spindle forming the metaphase plate, awaiting the signal to the next stage, Anaphase. This transition, from Metaphase to Anaphase, is the most critical event of the cell cycle. This event occurs when the cohesion between sister chromatids is dissolved and the separated sisters migrate to opposite poles of the spindle. The last stage of mitosis is Telophase, during which the spindle disassembles, the chromosomes decondense and the nuclear envelop reappears. Simultaneously, the division of the cytoplasm generates two independent interphase daughter cells by a process called Cytokinesis. If these events do not occur properly and in the right sequence, the newly formed cells either die or carry on genetic damages that can lead ultimately to cancer.

2.2 Regulation of Mitosis

The timing and coordination of Mitosis progression relies mostly on mechanisms of protein Phosphorylation and Proteolysis (Nigg, 2001). While Phosphorylation is a reversible protein modification, and thus ideal for the control of reversible mitotic processes such as spindle assembly, Proteolysis, in contrast, by its chemical irreversibility, is more appropriate for controlling events that must not be reversed such as sister-chromatid separation (King et al., 1996; Morgan, 2007). The Mitosis-Promoting Factor (MPF), first called Maturation-Promoting Factor, was described as the "entity" whose activity controls entry into Mitosis (Masui & Markert, 1971). Now, it is well-established that the MPF is a heterodimer composed by one molecule of cyclin B and one of Cdc2 (cell division cycle). Later, Cdc2 became known as Cdk1 since it was established, at the Cold Spring Harbor Symposium on the Cell Cycle in 1991, that kinases that are associated with cyclins should be called "cyclin dependent kinases" or Cdks (Doree & Hunt, 2002). Cyclin B-Cdk1 (MPF) accumulates before entry into mitosis and its activation leads to phosphorylation of several substrates responsible for the morphological changes that occur in early stages of mitosis such as nuclear envelop breakdown, centrosome separation, spindle assembly, chromosome condensation, and endoplasmic reticulum and Golgi fragmentation. However, proteolysis-mediated disassembly of Cyclin-Cdk1 complexes is required for mitotic exit and cytokinesis (Nigg, 2001). Besides the direct role in regulating Cdk activity by controlling cyclin levels, proteolysis also drives cell cycle progression by directly triggering some key cell cycle events such as sister chromatids separation at metaphase-anaphase transition, thus providing directionality of the cell cycle (King et al., 1996). Therefore, these two mechanisms, phosphorylation and proteolysis, are interdependent since proteolysis events are controlled by phosphorylation and the mitotic kinases (M-Cdks) are inactivated by proteolytic destruction of cyclins (King et al., 1996; Morgan, 2007; Nigg, 2001).

The major events of mitosis are sister-chromatid separation and segregation. If these processes do not occur accurately the result would be the production of cells with extra or missing chromosomes, a state known as aneuploidy, which is a common characteristic of cancer cells (Holland & Cleveland, 2009; Schvartzman et al., 2010). To avoid the occurrence of aneuploidy, cells have developed a control system called mitotic checkpoint or Spindle Assembly Checkpoint (SAC), which prevents the cell entry in anaphase until all chromosomes are correctly aligned, forming the metaphase plate, with proper attachment to the mitotic spindle (Rieder et al., 1995) and under a certain tension (Nicklas et al., 1995). When these conditions are satisfied, SAC is turned off. This checkpoint arrests cells in mitosis by blocking protein degradation. With its inactivation, the Anaphase Promoting Complex or Cyclosome (APC/C), an ubiquitin protein ligase whose activity depends on the activator protein Cdc20, targets Securin and Cyclin B for ubiquitylation and posterior proteolysis through the 26S proteasome. Destruction of Securin turns on Separase which cleaves the cohesion complex that holds sister chromatids together; destruction of Cyclin B leads to anaphase onset (Zachariae & Nasmyth, 1999).

2.3 The mitotic spindle

In order to congress and align at the center of the cell, and then segregate its sister chromatids to opposite poles, chromosomes use the mitotic spindle. The mitotic spindle is organized in a symmetric and fusiform structure composed of microtubules, polymers made of α- and β-tubulin heterodimers that being all oriented in the same way create a polar nature with β-tubulin exposed at one end (plus-end) and α-tubulin at the other end (minus-end) (Desai & Mitchison, 1997). Depending on the position of the microtubule plus-ends, spindle microtubules can be divided into three classes: astral-microtubules, interpolar-microtubules and kinetochore-microtubules; all contribute to the bipolarity of the mitotic spindle. Astral microtubules, emanate from the spindle poles and radiate out throughout the cytoplasm with the plus-ends interacting with the cell cortex. Interpolar microtubules, extend from the spindle poles to the spindle midzone where their plus-ends form an interdigitating system that connects the two spindle poles. Kinetochore microtubules, connect the spindle poles to chromosomes with the minus ends near the poles and the plus ends binding laterally or end-on, specifically to the kinetochores (an intricate protein complex raised on the centromeric DNA) (Hayden et al., 1990; Merdes & De Mey, 1990; Rieder & Alexander, 1990). These kinetochore-microtubules form a morphologically distinct bundle denominated K-fiber (kinetochore-fiber), made of up to 30 kinetochore-attached microtubules in higher eukaryotes, which is directly involved in chromosome congression and sister-chromatid segregation (Rieder & Salmon, 1998).

The ability of spindle microtubules to quickly assemble and disassemble (dynamic instability) provides them the necessary behavior to capture chromosomes. This statement becomes the basis of the first and favorite model for spindle assembly in systems with centrosomes, and is known as "search and capture" (Kirschner & Mitchison, 1986; Mitchison & Kirschner, 1984). This model postulates that, when nuclear envelop breaks down, chromosomes become accessible to microtubules radiated from centrosomes , which, through their dynamic nature, randomly explore the cytoplasm until capture a kinetochore, laterally or with the plus-end, forming an attachment that stabilizes the microtubule (Mitchison et al., 1986; Nicklas & Kubai, 1985). Although the plus-ends of microtubules can bind directly the kinetochore, the first contact usually occurs laterally. After binding one of the unattached sister kinetochores, the chromosome is rapidly transported along the side of the microtubule towards the spindle pole, in a mechanism that is independent of microtubule depolymerization and involves the minus-end directed motor protein Dynein (Rieder & Alexander, 1990; Yang et al., 2007). Since the spindle pole has a high microtubule density, additional microtubules from the same pole will attach to the kinetochore, and the plus-ends of laterally associated microtubules shorten until reach the end-on binding, resulting in a stable K-fiber (Rieder, 2005). Since this model is based on "search and capture" of astral microtubules that radiate from centrosomes, it is not valid to cells lacking centrosomes, like higher plants and many animal oocytes. An alternative model, called "spindle self organization", proposes that microtubules are nucleated in the vicinity of chromatin, elongate and then, helped by motor proteins, are sorted into a bipolar array and focused at the poles (Walczak et al., 1998). Indeed, it was shown that cells with centrosomes, besides the "search and capture" mechanism, also use this chromosome-driven K-fiber formation for spindle assembly (Khodjakov et al., 2000; Maiato et al., 2004). In these "combined" system, the K-fibers nucleated from chromosomes are integrated at the spindle assisted by astral microtubules that search, capture and transport them toward the pole in a dynein-dependent manner (Khodjakov et al., 2003; Maiato et al., 2004). Several studies, in *Xenopus* extracts and in mammalian cells, show that the chromosome-driven microtubule formation relies in a Ran-GTP concentration gradient around mitotic chromosomes, which in turn induces a gradient of proteins that regulate microtubule nucleation/dynamics, by dissociating them from importin-β (Fuller, 2010; Kalab et al., 2002; Tulu et al., 2006). Furthermore, studies in mammalian somatic cells support a model in which the kinetochores are the key players in this chromosome-mediated spindle assembly (O'Connell et al., 2009). Another mechanism that contributes to spindle formation is the "search and transport", in which peripheral microtubules, similarly to K-fibers nucleated from chromosomes, are transported through aster microtubules to the poles, where they are incorporated into the spindle (O'Connell & Khodjakov, 2007; Tulu et al., 2003).

For a successful cell division, chromosomes must interact with spindle microtubules. Through their dynamic behavior, microtubules allow that chromosomes congress to the equatorial region of the spindle forming the metaphase plate and, are responsible for the segregation of sister chromatids to opposite poles. There are three major forces acting on chromosomes that drivethese movements: (1) The "Polar ejection force" or "Polar wind" that is generated by non-kinetochore microtubules and pushes chromosomes away from the spindle poles (anti-poleward movement) (Kapoor & Compton, 2002; Rieder et al., 1986); (2) "Microtubule flux", which consists in a flow of tubulin subunits from the spindle equator to the spindle poles as consequence of polymerization at the plus-ends, depolymerization at minus-ends, and translocation of the entire microtubule toward the spindle pole (Cassimeris, 2004; Mitchison, 1989); and (3) "Pacman" mechanism, in which the kinetochores catalyze the depolymerization of kinetochore-microtubule plus-ends resulting in chromosomal movement to the spindle poles by "chewing up" the microtubule track (Gorbsky et al., 1987; Mitchison et al., 1986).

3. The kinetochore structure

In 1894, Metzner was the first investigator to describe the "kinetic region", a specific chromatin area, located at the primary constriction on each side of the chromosome, that leads the way of sister chromatids during the poleward motion (Metzner, 1894; Rieder, 2005; Schrader, 1944). Later, in 1934, Sharp coined these structures as "kinetochores", from the Greek 'kineto-' meaning 'move' and '-chore' meaning 'means for distribution' (Rieder, 2005; Schrader, 1936). In 1960, Bill Brinkley was the first to describe the mammalian kinetochore structure as a trilaminar proteinaceous disc structure that flanked the centromere: an electron-dense inner plate located on the surface of the centromeric heterochromatin, separated from an electron dense outer plate by a lighter middle layer (Brinkley & Stubblefield, 1966). However, using high-pressure frozen specimens, this electron-translucent middle layer is not visible, suggesting that it is an artifact produced during the classical EM fixation and/or dehydration procedures (McEwen et al., 1998). In 1967, Jokelainen shows the existence of a corona of electron opaque substance that covers the outer kinetochore layer, which after microtubule binding, becomes hard to detect by EM (Cassimeris et al., 1990; Jokelainen, 1967)(Fig. 2).

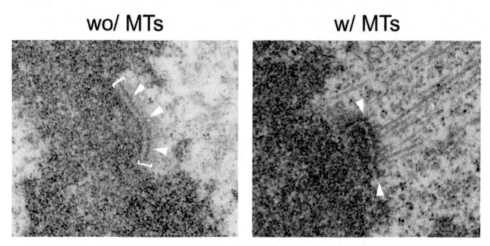

Fig. 2. Electron micrographs of kinetochores, from PTK1 cells, in absence and presence of microtubules. The trilaminar structure of kinetochore (brackets) is well defined without microtubules (wo/ MTs) and becomes distorted with microtubules (w/ MTs) embedded within the outer kinetochore plate (arrowheads). Courtesy of Dr. Helder Maiato (Maiato et al., 2006).

These mature kinetochores, with the trilaminar structure, occurs only in prometaphase after nuclear envelop breakdown. The kinetochore is built on the centromeric region of each sister chromatid by the assembly of multiprotein complexes (Fig. 3). In early G1 the typical layer conformation disappears giving rise to a condensed structure that unfolds in late G1 forming a linear, bead-like conformation that persists until S-phase, where it transforms into a loose fibrous bundle that duplicates at late S-phase. In late G2, pre-kinetochores refold into two separated and condensed structures. During prophase, these duplicated pre-kinetochores differentiate at the primary constriction of the sister chromatids originating the kinetochore layers at the time of nuclear envelop breakdown, completing the cycle (He & Brinkley, 1996).

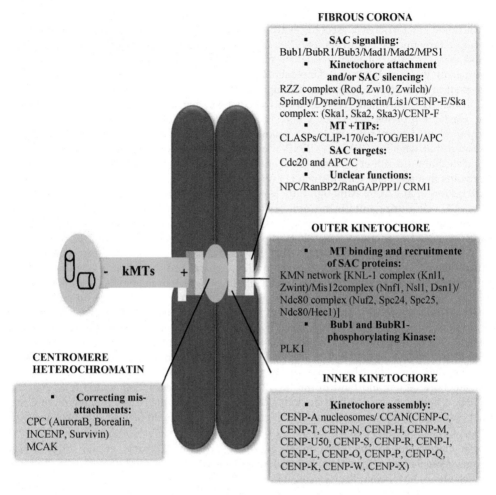

Fig. 3. Overview of protein complexes that build the kinetochore in animal cells. The kinetochore is built on the centromere as a trilaminar protein-rich structure: the inner kinetochore, the outer kinetochore and the fibrous corona. The inner and outer kinetochore layers are separated by the translucent interzone (not depicted). Proteins at centromere heterochromatin and at each kinetochore layer are indicated; the list is not exhaustive as it is continuously growing. Adapted from (Silva et al., 2011).

Kinetochores localize at the constriction region on each sister chromatid, assembling on the centromere, from the Greek 'centro-' (meaning 'central') and '-mere' (meaning 'part'), a specific chromatin region with distinct epigenetic marks (Cheeseman & Desai, 2008). In this particular region, the nucleosomal histone H3 is replaced by the variant CENP-A (CENtromere Protein A)(Vafa & Sullivan, 1997). Although it seems consensual that CENP-A is essential for specifying the site for kinetochore formation, the role of CENP-A in kinetochore assembly is still unclear (Bowers & Mellone, 2011). Some studied, using overexpression/mistargeting of CENP-A show that CENP-A is not sufficient to originate functional kinetochores (Gascoigne et al., 2011; Van Hooser et al., 2001). Other studies, using ectopic targeting of HJURP (a CENP-A chromatin assembly factor) (Barnhart et al., 2011) and using *in vitro* kinetochores (Guse et al., 2011), demonstrate that CENP-A is indeed sufficient to form functional kinetochores. Besides CENP-A, several other proteins are present at vertebrate centromeres throughout the cell cycle. They were named CCAN for Constitutive Centromere Associated Network (Cheeseman & Desai, 2008). CCAN comprises at least 16 proteins, CENP-C, CENP-H, CENP-I, CENP-K-U, CENP-W, and CENP-X, grouped into several subcomplexes (Amano et al., 2009; Santaguida & Musacchio, 2009). It is known that whereas CENP-C, CENP-N and CENP-K associate specifically with CENP-A nucleosomes (Carroll et al., 2009; Guse et al., 2011), CENP-T/W complex are DNA-binding proteins that associate with histone H3 nucleosomes in the centromeric region (Guse et al., 2011; Hori et al., 2008). Although only few proteins are present at centromeres throughout the cell cycle, during mitosis this number increases substantially, indicating the existence of a complex assembly regulatory process. Gascoigne et al. have contributed with a piece of this puzzle, by demonstrating that phosphorylation of CENP-T by CDK can control kinetochore assembly in vertebrates. They saw that CENP-T becomes phosphorylated in G2, has a maximum at metaphase and drops until anaphase, and show that preventing CENP-T phosphorylation abolishes the recruitment of Ndc80 (Hec1 in mammals) (Gascoigne et al., 2011). Ndc80 is part of the Ndc80 complex that in turns is part of the so called KMN network, the microtubule-binding core of kinetochore, composed by the two-subunit Knl1 complex (containing Knl1/Blinkin and Zwint), the four-subunit complex Mis12 (containing Mis12, Dsn1, Nnf1 and Nsl1) and the four-subunit Ndc80 complex (containing Ndc80, Nuf2, Spc24 and Spc25). In the Ndc80 complex, Nuf2 and Ndc80 localize to the outer kinetochore region and bind directly to microtubules, and Spc24 and Spc25 localize at the inner kinetochore region and bind to the Mis12 and Knl1 complexes (Santaguida & Musacchio, 2009). A recent study, demonstrate that a conserved motif in the N-terminal region of Cenp-C binds directly to the Mis12 complex (Screpanti et al., 2011). In fact, several proteins from the CCAN, such as CENP-C, CENP-H, CENP-K, CENP-I, CENP-T/W, and CENP-X have been implicated in the recruitment of KMN proteins (Petrovic et al., 2010). These connections between the CCAN network and the KMN network link the inner to the outer kinetochore. KMN complex was considered as the essential core of the kinetochore. Besides the direct binding of centromeres to microtubules, it interacts, directly or indirectly, with proteins that are involved in crucial mitotic functions such as, regulation of the activity of spindle assembly checkpoint and kinetochore-microtubule interactions (Cheeseman & Desai, 2008; Przewloka & Glover, 2009). It allows, directly or indirectly, the recruitment/interaction of regulatory proteins such as Microtubule associated proteins (like CLASP, CLIP170, EB1, APC), motor proteins (like CENP-E, dynein/dynactin complex), protein involved in spindle assembly checkpoint (like Mad1, Mad2, Bub1, Bub3, BubR1, Bub1, RZZ complex, Mps1). It also interacts with the SKA (Spindle and Kinetochore Associated) complex, thought to be involved in stable end-on kinetochore-microtubule attachment; and with the CPC (Chromosome Passenger Complex- Aurora B, INCENP, Borealin and Survivin) complex, among others (Fig. 3). Indeed, depletion of any protein from the KMN network, in all eukaryotes, result in abnormal kinetochore structure or in a lack of kinetochore-microtubule attachment (Lampert & Westermann, 2011; Przewloka & Glover, 2009; Santaguida & Musacchio, 2009).

4. Kinetochore functions

The Kinetochores are dynamic structures with more than 100 proteins organized into networks of several complexes of probably intertwined functions during cell division (Cheeseman & Desai, 2008). They attach sister-chromatids to microtubules of the bipolar mitotic spindle and position the chromosome at the spindle equator; they inhibit anaphase onset until all chromosomes are properly attached and aligned at the metaphase plate; and provide correction mechanism of erroneous attachments. Next, we address the overall mechanisms and the molecular link between chromosome attachment to spindle microtubules, correction of attachment errors, and the spindle assembly checkpoint.

4.1 Kinetochores mediate bipolar attachment to the spindle

High-fidelity chromosome segregation at the onset of anaphase relies on the success of sister kinetochore bi-orientation on the mitotic spindle (Tanaka, 2010). Chromosome bi-orientation is a step-wise process that sequentially involves initial interaction of a kinetochore with the lateral surface of a microtubule; transport of the kinetochore to a spindle pole; conversion from the lateral to the end-on attachment; and attachment of the sister kinetochore to microtubules from the opposite spindle pole (Fig. 4).

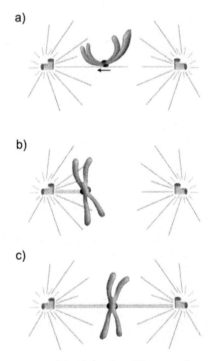

Fig. 4. Steps towards chromosome bi-orientation. Upon nuclear envelope breakdown, chromosome bi-orientation sequentially involves a) initial interaction of a kinetochore with the lateral surface of a microtubule and transport of the kinetochore to a spindle pole; b) conversion from the lateral to the end-on attachment; and c) attachment of the sister kinetochore to microtubules from the opposite spindle pole to achieve bipolar attachment.

As stated above, initial encounter of spindle pole microtubules with the kinetochore is mediated by the so-called "search and capture" process, guided by a Ran-GTP concentration gradient around the chromosome (Clarke & Zhang, 2008; Wollman et al., 2005). Additionally, and to avoid further delay in this initial encounter, kinetochore can nucleate microtubules which, by interacting with spindle-pole nucleated microtubules along their length, facilitate kinetochore loading onto centrosome-nucleated microtubules (Kitamura et al., 2010). The first step of the "search and capture" process is the interaction of the kinetochore with the lateral surface of a microtubule, called the lattice, followed by chromosome sliding along the microtubule lattice towards a spindle pole. The poleward kinetochore transport, powered by the minus end-directed dynein motor protein complex, brings chromosomes scattered throughout the cytoplasm to the mitotic spindle area (Sharp et al., 2000; Yang et al., 2007). Dynein binding to kinetochores requires the protein Spindly which in turn requires the RZZ complex [made of the proteins Rough-deal (ROD), Zeste-white (ZW10), and Zwilch] to localize to kinetochore (Chan et al., 2009; Griffis et al., 2007; Karess, 2005). The connection to the essential microtubule-binding core of the kinetochore is mediated by the protein Zwint-1 that links the RZZ complex to the KMN network (Wang et al., 2004).

During the association of the kinetochore with the microtubule lattice, shrinking of the microtubule plus-end leads to kinetochore tethering at the microtubule plus-end (end-on attachment) and to its further transport towards the spindle pole. End-on attachments are more robust than lateral attachments and are critical for bi-orientation and for the generation of load-bearing attachments (Joglekar et al., 2010). The mechanism of the conversion from a lateral into an end-on attachment remains unclear. Proteins thought to be required for this conversion include the *C. elegans* RZZ complex and Spindly/SPDL-1 (Gassmann et al., 2008), the vertebrate Ska1-3 (Gaitanos et al., 2009; Guimaraes & Deluca, 2009), and the *Saccharomyces cerevisiae* Ndc80 loop region (Tanaka, 2010).

The chromosome reaches the spindle pole with one kinetochore end-on attached to k-fiber microtubules from that pole and its sister kinetochore unattached, a state known as monotelic attachment. Aided by the back-to-back kinetochore geometry, the unattached kinetochore becomes attached when captured by microtubule searching from the opposite pole, thereby leading to chromosome bi-orientation (amphitelic attachment) (Fig. 4). Subsequently, k-fiber microtubule shrinking and elongation promote congression of the chromosomes towards the spindle equator in order to form the metaphase plate (Silva et al., 2011). As an additional mechanism that leads to bi-orientation, chromosomes can be transported towards the spindle equator by gliding alongside microtubules attached to other already bi-oriented chromosomes, driven by kinetochore-bound CENP-E, a plus end-directed microtubule motor of the kinesin-7 family (Kapoor et al., 2006).

4.2 Correcting aberrant kinetochore-microtubule attachments

Although the back-to-back orientation of sister kinetochores imposes a geometric constraint that favors chromosome bi-orientation, errors in kinetochore-microtubule attachments are frequent due to the stochastic nature of the search and capture mechanism. Such errors include monotelic (one kinetochore unattached while its sister attached to one spindle pole), syntelic (two sister kinetochores bound to microtubules from the same pole), and merotelic (one sister kinetochore bound to microtubules from both poles) attachments (Fig. 5) (Silva et al., 2011). Most of these errors occur at the beginning of prometaphase and, if left uncorrected, would lead to unequal chromosome segregation and aneuploidy (Kops et al., 2005).

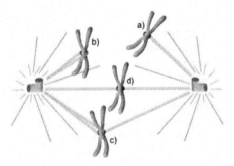

Fig. 5. Possible attachment errors during chromosome bi-orientation event. Errors include a) monotelic, with one kinetochore unattached while its sister attached to one spindle pole; b) syntelic, with two sister kinetochores bound to microtubules from the same pole; and c) merotelic, with one sister kinetochore bound to microtubules from both poles attachments. These errors are generally converted to d) amphitelic attachments, the only correct attachment configuration, by the error correction machinery.

Error correction is the result of biochemical changes induced by mechanical forces acting on the kinetochore. Bi-oriented kinetochores (amphitelic attachment) are under tension that results from the pulling forces of spindle microtubules in opposite directions. Cohesion between sister centromeres is necessary to generate this tension (Tanaka et al., 1999). Tension across the centromere stabilizes kinetochore-microtubule attachments, as evidenced by Bruce Nicklas in his classical micromanipulation experiments with insect spermatocytes (Nicklas & Koch, 1969; Nicklas & Ward, 1994). Tension artificially applied with a glass microneedle on grasshopper spermatocyte chromosomes stabilizes unipolar attachments by increasing the occupancy of microtubule attachment sites.

The first evidence of the translation of mechanical forces (tension) acting on the kinetochore into biochemical changes was provided by the identification of the Ipl-1 kinase in the budding yeast *Saccharomyces cerevisiae* (Biggins et al., 1999). In Ipl-1 defective yeast cells, kinetochores can interact with microtubules but sister-kinetochores often fail to bi-orient, suggesting that Ipl-1 promotes bi-orientation. Interestingly, in the same mutant, kinetochore-microtubule attachments are stabilized in the absence of tension, suggesting that Ipl-1 promotes bi-orientation by destabilizing tensionless attachments (Dewar et al., 2004; Tanaka et al., 2002). Defective Aurora B kinase (the mammalian functional homolog of Ipl-1) induces an increase in syntelic attachments, suggesting a similar work in mammalian cells (Hauf et al., 2003; Kallio et al., 2002). Aurora B localizes to the inner centromere, as the catalytic component of the chromosome passenger complex (CPC), together with the binding partners INCENP, Survivin, and Borealin (Ruchaud et al., 2007). While the molecular mechanism of error detection and correction is not fully understood, a current model proposes that Aurora B kinase promotes the turnover of erroneous kinetochore-microtubule attachments through phosphorylation of its substrates at the outer kinetochore, in a tension-dependent manner (Lampson & Cheeseman, 2011) . In this so-called spatial model, the distance between Aurora B and its outer kinetochore substrates is a critical determinant for their phosphorylation. When tension is low or absent, outer kinetochore substrates are phosphorylated due to their proximity to an Aurora B kinase activity gradient around the inner centromere (Wang et al., 2011), which destabilizes kinetochore-microtubule attachments. For instance, in higher eukaryotes, Aurora B phosphorylation of the Ndc80 complex was reported to weaken its affinity to microtubules, while phosphorylation of the microtubule depolymerase kinesin MCAK activates its depolymerase activity (Cheeseman et al., 2006; Lan et al., 2004). Both changes promote destabilization of tensionless kinetochore-microtubule attachments, providing a further opportunity for chromosome to bi-orient. Bi-orientation locates outer kinetochore substrates away from the Aurora B activity gradient and close to the opposing and constitutively active phosphatase. In this way, dephosphorylation of Ndc80 and MCAK stabilizes amphitelic kinetochore-microtubule attachments. Therefore, tension provides the determinant for the error correction machinery to distinguish syntelic and merotelic from amphitelic attachments. Once all chromosomes become bi-oriented and aligned at the metaphase plate, Aurora B leaves the inner centromere and become concentrated on the spindle midzone. This presumably prevents turnover of kinetochore-microtubules attachments during anaphase, in which tension is reduced.

4.3 The spindle assembly checkpoint

Cells must be maintained in mitosis, by preventing securin and cyclin B degradation, until all chromosomes are properly bi-oriented and aligned at the metaphase plate. This is achieved by the evolutionary conserved mechanism called The "Spindle Assembly Checkpoint" (SAC) that prevents the E3 ubiquitin ligase APC/C (Anaphase Promoting Complex/Cyclosome) from targeting securin and cyclin B for degradation by the 26S proteasome, as long as unattached or improperly attached chromosomes are present (Silva et al., 2011).

Components of the SAC are conserved from yeast to human and include the core SAC proteins Mad1, Mad2, Bub1, Bub3, Mad3/BubR1, and the protein kinases Mps1 and Aurora B (Musacchio & Salmon, 2007). It is widely accepted that the SAC inhibits the APC/C through the Mad2 protein that sequesters Cdc20 an essential activator of the APC/C individually or in an inhibitory complex with BubR1 and Bub3, forming the "Mitotic Checkpoint Complex" (MCC) (Kallio et al., 2002; Sudakin et al., 2001). Unattached kinetochore is believed to provide the platform for the generation of this diffusible "wait anaphase" signal that inhibits mitosis. Indeed, checkpoint proteins dynamically associate with unattached kinetochores, reflecting the catalytic assembly and then release and diffusion of MCC. However this model is not universal as MCC formation does not require a kinetochore in yeast (Fraschini et al., 2001).

SAC becomes satisfied once all chromosomes are correctly bi-oriented at the metaphase plate. In order for cell to proceed to anaphase, the production of MCC must be halted and existing MCC must be disassembled. This is referred to as "SAC silencing", thought to be mediated by several mechanisms at least in mammalian cells (Fuller & Stukenberg, 2009; Vanoosthuyse & Hardwick, 2009). Kinetochore-mediated production of MCC is suggested to be stopped by Dynein-driven transport of SAC proteins from attached kinetochores towards spindle poles, along microtubules (Howell et al., 2001); and through dissociation of Mad2/Cdc20 complex due to competition binding of the protein $p31^{comet}$ to the dimerization interface of Mad2. Disassembly of existing MCC is promoted by $p31^{comet}$- and UbcH10-dependent Cdc20 autoubiquitination (Reddy et al., 2007; Stegmeier et al., 2007), and proteasomal degradation (Ma & Poon, 2011). The coordination between these mechanisms during SAC inactivation and mitotic exit remains unclear. Once SAC is turned off, APC/C becomes activated and targets securin and cyclin B for degradation by the proteasome, thereby promoting sister-chromatid separation and exit from mitosis, respectively.

4.4 Relationship between microtubule attachment, attachment error correction, and SAC

The role of SAC in delaying anaphase onset, in the presence of unattached or improperly attached kinetochores, foresees the existence of a dynamic relationship between microtubule attachment, error correction machinery, and checkpoint signaling. This relationship could efficiently assure that attachment errors are detected; checkpoint signals are produced; and attachment errors are corrected only at the kinetochore platform, being all these activities in the right place at the right time.

A molecular link between microtubule attachment machinery and SAC activity has been suggested by the phenotype of yeasts with defective Ndc80 (Ndc80, Nuf2, Spc24, and Spc25) complex, a component of the KMN supercomplex (the core microtubule-binding interface of kinetochores). These mutants fail to attach chromosomes to the spindle and to activate SAC (Burke & Stukenberg, 2008; McCleland et al., 2003). In metazoans, the relationship between kinetochore-microtubule binding and SAC signaling is further suggested by the observation that KMN, through the Ndc80 complex, is required for kinetochore assembly of SAC proteins as well as for the generation of a SAC signal (Burke & Stukenberg, 2008). In addition, Blinkin (the human homologue of yeast kinetochore protein Spc105) directs Bub1 and BubR1 to kinetochores through interaction with their TPR domains (Bolanos-Garcia et al., 2011; Kiyomitsu et al., 2007). Preventing interaction between the Bubs and Blinkin, either by siRNA or by point mutation of the TPR domain of Bubs, abolishes the generation of SAC signals, suggesting that Blinkin has a role both in microtubule attachment and SAC signaling (Kiyomitsu et al., 2007). The SAC proteins Bub1, BubR1, Bub3, and Mps1 were themselves involved in the regulation of microtubule attachment, beside their role in SAC signaling (Logarinho & Bousbaa, 2008).

The highly conserved serine/threonine Aurora B kinase provides the main link between the error correction machinery and SAC. By destabilizing tensionless kinetochore-microtubule attachments, Aurora B creates unattached kinetochores that can be filled with checkpoint proteins to generate SAC signals (Burke & Stukenberg, 2008). This in turn allows time for error correction and bi-orientation. The direct involvement of Aurora B activity in SAC control is suggested by its requirement for SAC protein recruitment to unattached kinetochores (tensionless) artificially created in cells treated with the microtubule-depolymerizing drug nocodazole (Kallio et al., 2002). Moreover, recent studies strongly suggest that Aurora B directly contributes to SAC signaling independently of its error correction activity (Santaguida et al., 2011).

5. Conclusion

Accurate chromosome segregation during mitosis relies on the activities of the kinetochore. Here, we highlighted the event of mitosis and focused on the structure and functions of kinetochore in chromosome attachment, chromosome movement, error correction, and the generation of inhibitory signals that prevent anaphase in the presence of attachment errors.

The basic mechanisms of these kinetochore functions, their interplays, and regulatory pathways remain under investigation. Elucidating these mechanisms is crucial for future progress and is relevant to cancer aetiology and therapy. Indeed, failure in any of these kinetochore functions can lead to chromosome missegregation, with chromosome losses and gains, which may contributes to the aneuploidy phenotype that characterizes many cancers (Kops et al., 2005; Thompson et al., 2010).

6. Acknowledgment

H.B. is supported by grant 02-GCQF-CICS-2011N, from Cooperativa de Ensino Superior Politécnico e Universitário (CESPU); through national funds from FCT – Fundação para a Ciência e a Tecnologia under the project CEQUIMED - PEst-OE/SAU/UI4040/2011 and also by FEDER funds through the COMPETE program under the project FCOMP-01-0124-FEDER-011057. We apologize to all the authors whose work could not be referred due to space limitation.

7. References

Amano, M.; Suzuki, A.; Hori, T.; Backer, C.; Okawa, K.; Cheeseman, I.M. & Fukagawa, T. (2009). The CENP-S complex is essential for the stable assembly of outer kinetochore structure. *J Cell Biol,* Vol. 186, N° 2, pp. (173-182)

Baker, J.R. (1949). The cell-theory; a restatement, history, and critique. *Q J Microsc Sci,* Vol. 90, N° 1, pp. (87-108)

Barnhart, M.C.; Kuich, P.H.; Stellfox, M.E.; Ward, J.A.; Bassett, E.A.; Black, B.E. & Foltz, D.R. (2011). HJURP is a CENP-A chromatin assembly factor sufficient to form a functional de novo kinetochore. *J Cell Biol,* Vol. 194, N° 2, pp. (229-243)

Biggins, S.; Severin, F.F.; Bhalla, N.; Sassoon, I.; Hyman, A.A. & Murray, A.W. (1999). The conserved protein kinase Ipl1 regulates microtubule binding to kinetochores in budding yeast. *Genes Dev,* Vol. 13, N° 5, pp. (532-544)

Bolanos-Garcia, V.M.; Lischetti, T.; Matak-Vinkovic, D.; Cota, E.; Simpson, P.J.; Chirgadze, D.Y.; Spring, D.R.; Robinson, C.V.; Nilsson, J. & Blundell, T.L. (2011). Structure of a Blinkin-BUBR1 Complex Reveals an Interaction Crucial for Kinetochore-Mitotic Checkpoint Regulation via an Unanticipated Binding Site. *Structure,* Vol., N°

Bowers, S.R. & Mellone, B.G. (2011). Starting from scratch: de novo kinetochore assembly in vertebrates. *Embo J,* Vol. 30, N° 19, pp. (3882-3884)

Brinkley, B.R. & Stubblefield, E. (1966). The fine structure of the kinetochore of a mammalian cell in vitro. *Chromosoma,* Vol. 19, N° 1, pp. (28-43)

Burke, D.J. & Stukenberg, P.T. (2008). Linking kinetochore-microtubule binding to the spindle checkpoint. *Dev Cell,* Vol. 14, N° 4, pp. (474-479)

Carroll, C.W.; Silva, M.C.; Godek, K.M.; Jansen, L.E. & Straight, A.F. (2009). Centromere assembly requires the direct recognition of CENP-A nucleosomes by CENP-N. *Nat Cell Biol,* Vol. 11, N° 7, pp. (896-902)

Cassimeris, L. (2004). Cell division: eg'ing on microtubule flux. *Curr Biol,* Vol. 14, N° 23, pp. (R1000-1002)

Cassimeris, L.; Rieder, C.L.; Rupp, G. & Salmon, E.D. (1990). Stability of microtubule attachment to metaphase kinetochores in PtK1 cells. *J Cell Sci,* Vol. 96 (Pt 1), N° pp. (9-15)

Chan, Y.W.; Fava, L.L.; Uldschmid, A.; Schmitz, M.H.; Gerlich, D.W.; Nigg, E.A. & Santamaria, A. (2009). Mitotic control of kinetochore-associated dynein and spindle orientation by human Spindly. *J Cell Biol*, Vol. 185, N° 5, pp. (859-874)

Cheeseman, I.M.; Chappie, J.S.; Wilson-Kubalek, E.M. & Desai, A. (2006). The conserved KMN network constitutes the core microtubule-binding site of the kinetochore. *Cell*, Vol. 127, N° 5, pp. (983-997)

Cheeseman, I.M. & Desai, A. (2008). Molecular architecture of the kinetochore-microtubule interface. *Nat Rev Mol Cell Biol*, Vol. 9, N° 1, pp. (33-46)

Clarke, P.R. & Zhang, C. (2008). Spatial and temporal coordination of mitosis by Ran GTPase. *Nat Rev Mol Cell Biol*, Vol. 9, N° 6, pp. (464-477)

Desai, A. & Mitchison, T.J. (1997). Microtubule polymerization dynamics. *Annu Rev Cell Dev Biol*, Vol. 13, N° pp. (83-117)

Dewar, H.; Tanaka, K.; Nasmyth, K. & Tanaka, T.U. (2004). Tension between two kinetochores suffices for their bi-orientation on the mitotic spindle. *Nature*, Vol. 428, N° 6978, pp. (93-97)

Doree, M. & Hunt, T. (2002). From Cdc2 to Cdk1: when did the cell cycle kinase join its cyclin partner? *J Cell Sci*, Vol. 115, N° Pt 12, pp. (2461-2464)

Fraschini, R.; Beretta, A.; Sironi, L.; Musacchio, A.; Lucchini, G. & Piatti, S. (2001). Bub3 interaction with Mad2, Mad3 and Cdc20 is mediated by WD40 repeats and does not require intact kinetochores. *Embo J*, Vol. 20, N° 23, pp. (6648-6659)

Fuller, B.G. (2010). Self-organization of intracellular gradients during mitosis. *Cell Div*, Vol. 5, N° 1, pp. (5)

Fuller, B.G. & Stukenberg, P.T. (2009). Cell division: righting the check. *Curr Biol*, Vol. 19, N° 14, pp. (R550-553)

Gaitanos, T.N.; Santamaria, A.; Jeyaprakash, A.A.; Wang, B.; Conti, E. & Nigg, E.A. (2009). Stable kinetochore-microtubule interactions depend on the Ska complex and its new component Ska3/C13Orf3. *Embo J*, Vol. 28, N° 10, pp. (1442-1452)

Gascoigne, K.E.; Takeuchi, K.; Suzuki, A.; Hori, T.; Fukagawa, T. & Cheeseman, I.M. Induced ectopic kinetochore assembly bypasses the requirement for CENP-A nucleosomes. *Cell*, Vol. 145, N° 3, pp. (410-422)

Gascoigne, K.E.; Takeuchi, K.; Suzuki, A.; Hori, T.; Fukagawa, T. & Cheeseman, I.M. (2011). Induced ectopic kinetochore assembly bypasses the requirement for CENP-A nucleosomes. *Cell*, Vol. 145, N° 3, pp. (410-422)

Gassmann, R.; Essex, A.; Hu, J.S.; Maddox, P.S.; Motegi, F.; Sugimoto, A.; O'Rourke, S.M.; Bowerman, B.; McLeod, I.; Yates, J.R., 3rd; Oegema, K.; Cheeseman, I.M. & Desai, A. (2008). A new mechanism controlling kinetochore-microtubule interactions revealed by comparison of two dynein-targeting components: SPDL-1 and the Rod/Zwilch/Zw10 complex. *Genes Dev*, Vol. 22, N° 17, pp. (2385-2399)

Gorbsky, G.J.; Sammak, P.J. & Borisy, G.G. (1987). Chromosomes move poleward in anaphase along stationary microtubules that coordinately disassemble from their kinetochore ends. *J Cell Biol*, Vol. 104, N° 1, pp. (9-18)

Griffis, E.R.; Stuurman, N. & Vale, R.D. (2007). Spindly, a novel protein essential for silencing the spindle assembly checkpoint, recruits dynein to the kinetochore. *J Cell Biol*, Vol. 177, N° 6, pp. (1005-1015)

Guimaraes, G.J. & Deluca, J.G. (2009). Connecting with Ska, a key complex at the kinetochore-microtubule interface. *Embo J*, Vol. 28, N° 10, pp. (1375-1377)

Guse, A.; Carroll, C.W.; Moree, B.; Fuller, C.J. & Straight, A.F. (2011). In vitro centromere and kinetochore assembly on defined chromatin templates. *Nature*, Vol. 477, N° 7364, pp. (354-358)

Hauf, S.; Cole, R.W.; LaTerra, S.; Zimmer, C.; Schnapp, G.; Walter, R.; Heckel, A.; van Meel, J.; Rieder, C.L. & Peters, J.M. (2003). The small molecule Hesperadin reveals a role for Aurora B in correcting kinetochore-microtubule attachment and in maintaining the spindle assembly checkpoint. *J Cell Biol*, Vol. 161, N° 2, pp. (281-294)

Hayden, J.H.; Bowser, S.S. & Rieder, C.L. (1990). Kinetochores capture astral microtubules during chromosome attachment to the mitotic spindle: direct visualization in live newt lung cells. *J Cell Biol*, Vol. 111, N° 3, pp. (1039-1045)

He, D. & Brinkley, B.R. (1996). Structure and dynamic organization of centromeres/prekinetochores in the nucleus of mammalian cells. *J Cell Sci*, Vol. 109 (Pt 11), N° pp. (2693-2704)

Holland, A.J. & Cleveland, D.W. (2009). Boveri revisited: chromosomal instability, aneuploidy and tumorigenesis. *Nat Rev Mol Cell Biol*, Vol. 10, N° 7, pp. (478-487)

Hori, T.; Amano, M.; Suzuki, A.; Backer, C.B.; Welburn, J.P.; Dong, Y.; McEwen, B.F.; Shang, W.H.; Suzuki, E.; Okawa, K.; Cheeseman, I.M. & Fukagawa, T. (2008). CCAN makes multiple contacts with centromeric DNA to provide distinct pathways to the outer kinetochore. *Cell*, Vol. 135, N° 6, pp. (1039-1052)

Howell, B.J.; McEwen, B.F.; Canman, J.C.; Hoffman, D.B.; Farrar, E.M.; Rieder, C.L. & Salmon, E.D. (2001). Cytoplasmic dynein/dynactin drives kinetochore protein transport to the spindle poles and has a role in mitotic spindle checkpoint inactivation. *J Cell Biol*, Vol. 155, N° 7, pp. (1159-1172)

Joglekar, A.P.; Bloom, K.S. & Salmon, E.D. (2010). Mechanisms of force generation by end-on kinetochore-microtubule attachments. *Curr Opin Cell Biol*, Vol. 22, N° 1, pp. (57-67)

Jokelainen, P.T. (1967). The ultrastructure and spatial organization of the metaphase kinetochore in mitotic rat cells. *J Ultrastruct Res*, Vol. 19, N° 1, pp. (19-44)

Kalab, P.; Weis, K. & Heald, R. (2002). Visualization of a Ran-GTP gradient in interphase and mitotic Xenopus egg extracts. *Science*, Vol. 295, N° 5564, pp. (2452-2456)

Kallio, M.J.; McCleland, M.L.; Stukenberg, P.T. & Gorbsky, G.J. (2002). Inhibition of aurora B kinase blocks chromosome segregation, overrides the spindle checkpoint, and perturbs microtubule dynamics in mitosis. *Curr Biol*, Vol. 12, N° 11, pp. (900-905)

Kapoor, T.M. & Compton, D.A. (2002). Searching for the middle ground: mechanisms of chromosome alignment during mitosis. *J Cell Biol*, Vol. 157, N° 4, pp. (551-556)

Kapoor, T.M.; Lampson, M.A.; Hergert, P.; Cameron, L.; Cimini, D.; Salmon, E.D.; McEwen, B.F. & Khodjakov, A. (2006). Chromosomes can congress to the metaphase plate before biorientation. *Science*, Vol. 311, N° 5759, pp. (388-391)

Karess, R. (2005). Rod-Zw10-Zwilch: a key player in the spindle checkpoint. *Trends Cell Biol*, Vol. 15, N° 7, pp. (386-392)

Khodjakov, A.; Cole, R.W.; Oakley, B.R. & Rieder, C.L. (2000). Centrosome-independent mitotic spindle formation in vertebrates. *Curr Biol*, Vol. 10, N° 2, pp. (59-67)

Khodjakov, A.; Copenagle, L.; Gordon, M.B.; Compton, D.A. & Kapoor, T.M. (2003). Minus-end capture of preformed kinetochore fibers contributes to spindle morphogenesis. *J Cell Biol*, Vol. 160, N° 5, pp. (671-683)

King, R.W.; Deshaies, R.J.; Peters, J.M. & Kirschner, M.W. (1996). How proteolysis drives the cell cycle. *Science*, Vol. 274, N° 5293, pp. (1652-1659)

Kirschner, M. & Mitchison, T. (1986). Beyond self-assembly: from microtubules to morphogenesis. *Cell*, Vol. 45, N° 3, pp. (329-342)

Kitamura, E.; Tanaka, K.; Komoto, S.; Kitamura, Y.; Antony, C. & Tanaka, T.U. (2010). Kinetochores generate microtubules with distal plus ends: their roles and limited lifetime in mitosis. *Dev Cell*, Vol. 18, N° 2, pp. (248-259)

Kiyomitsu, T.; Obuse, C. & Yanagida, M. (2007). Human Blinkin/AF15q14 is required for chromosome alignment and the mitotic checkpoint through direct interaction with Bub1 and BubR1. *Dev Cell*, Vol. 13, N° 5, pp. (663-676)

Kops, G.J.; Kim, Y.; Weaver, B.A.; Mao, Y.; McLeod, I.; Yates, J.R., 3rd; Tagaya, M. & Cleveland, D.W. (2005). ZW10 links mitotic checkpoint signaling to the structural kinetochore. *J Cell Biol*, Vol. 169, N° 1, pp. (49-60)

Lampert, F. & Westermann, S. (2011). A blueprint for kinetochores - new insights into the molecular mechanics of cell division. *Nat Rev Mol Cell Biol*, Vol. 12, N° 7, pp. (407-412)

Lampson, M.A. & Cheeseman, I.M. (2011). Sensing centromere tension: Aurora B and the regulation of kinetochore function. *Trends Cell Biol*, Vol. 21, N° 3, pp. (133-140)

Lan, W.; Zhang, X.; Kline-Smith, S.L.; Rosasco, S.E.; Barrett-Wilt, G.A.; Shabanowitz, J.; Hunt, D.F.; Walczak, C.E. & Stukenberg, P.T. (2004). Aurora B phosphorylates centromeric MCAK and regulates its localization and microtubule depolymerization activity. *Curr Biol*, Vol. 14, N° 4, pp. (273-286)

Logarinho, E. & Bousbaa, H. (2008). Kinetochore-microtubule interactions "in check" by Bub1, Bub3 and BubR1: The dual task of attaching and signalling. *Cell Cycle*, Vol. 7, N° 12, pp. (1763-1768)

Ma, H.T. & Poon, R.Y. (2011). Orderly inactivation of the key checkpoint protein mitotic arrest deficient 2 (MAD2) during mitotic progression. *J Biol Chem*, Vol. 286, N° 15, pp. (13052-13059)

Maiato, H.; Hergert, P.J.; Moutinho-Pereira, S.; Dong, Y.; Vandenbeldt, K.J.; Rieder, C.L. & McEwen, B.F. (2006). The ultrastructure of the kinetochore and kinetochore fiber in Drosophila somatic cells. *Chromosoma*, Vol. 115, N° 6, pp. (469-480)

Maiato, H.; Rieder, C.L. & Khodjakov, A. (2004). Kinetochore-driven formation of kinetochore fibers contributes to spindle assembly during animal mitosis. *J Cell Biol*, Vol. 167, N° 5, pp. (831-840)

Masui, Y. & Markert, C.L. (1971). Cytoplasmic control of nuclear behavior during meiotic maturation of frog oocytes. *J Exp Zool*, Vol. 177, N° 2, pp. (129-145)

McCleland, M.L.; Gardner, R.D.; Kallio, M.J.; Daum, J.R.; Gorbsky, G.J.; Burke, D.J. & Stukenberg, P.T. (2003). The highly conserved Ndc80 complex is required for kinetochore assembly, chromosome congression, and spindle checkpoint activity. *Genes Dev*, Vol. 17, N° 1, pp. (101-114)

McEwen, B.F.; Hsieh, C.E.; Mattheyses, A.L. & Rieder, C.L. (1998). A new look at kinetochore structure in vertebrate somatic cells using high-pressure freezing and freeze substitution. *Chromosoma*, Vol. 107, N° 6-7, pp. (366-375)

Merdes, A. & De Mey, J. (1990). The mechanism of kinetochore-spindle attachment and polewards movement analyzed in PtK2 cells at the prophase-prometaphase transition. *Eur J Cell Biol*, Vol. 53, N° 2, pp. (313-325)

Metzner, R. (1894). Beitrage zur granulalehre. I. Kern und kerntheilung. *Arch Anat Physiol*, Vol., N° pp. (309–348)

Mitchison, T.; Evans, L.; Schulze, E. & Kirschner, M. (1986). Sites of microtubule assembly and disassembly in the mitotic spindle. *Cell*, Vol. 45, N° 4, pp. (515-527)

Mitchison, T. & Kirschner, M. (1984). Dynamic instability of microtubule growth. *Nature*, Vol. 312, N° 5991, pp. (237-242)

Mitchison, T.J. (1989). Polewards microtubule flux in the mitotic spindle: evidence from photoactivation of fluorescence. *J Cell Biol*, Vol. 109, N° 2, pp. (637-652)

Morgan, D.O. (2007). The cell cycle: principles of control (London, New Science Press Ltd).

Musacchio, A. & Salmon, E.D. (2007). The spindle-assembly checkpoint in space and time. *Nat Rev Mol Cell Biol*, Vol. 8, N° 5, pp. (379-393)

Nicklas, R.B. & Koch, C.A. (1969). Chromosome micromanipulation. 3. Spindle fiber tension and the reorientation of mal-oriented chromosomes. *J Cell Biol*, Vol. 43, N° 1, pp. (40-50)

Nicklas, R.B. & Kubai, D.F. (1985). Microtubules, chromosome movement, and reorientation after chromosomes are detached from the spindle by micromanipulation. *Chromosoma*, Vol. 92, N° 4, pp. (313-324)

Nicklas, R.B. & Ward, S.C. (1994). Elements of error correction in mitosis: microtubule capture, release, and tension. *J Cell Biol*, Vol. 126, N° 5, pp. (1241-1253)

Nicklas, R.B.; Ward, S.C. & Gorbsky, G.J. (1995). Kinetochore chemistry is sensitive to tension and may link mitotic forces to a cell cycle checkpoint. *J Cell Biol*, Vol. 130, N° 4, pp. (929-939)

Nigg, E.A. (2001). Mitotic kinases as regulators of cell division and its checkpoints. *Nat Rev Mol Cell Biol*, Vol. 2, N° 1, pp. (21-32)

O'Connell, C.B. & Khodjakov, A.L. (2007). Cooperative mechanisms of mitotic spindle formation. *J Cell Sci*, Vol. 120, N° Pt 10, pp. (1717-1722)

O'Connell, C.B.; Loncarek, J.; Kalab, P. & Khodjakov, A. (2009). Relative contributions of chromatin and kinetochores to mitotic spindle assembly. *J Cell Biol*, Vol. 187, N° 1, pp. (43-51)

Paweletz, N. (2001). Walther Flemming: pioneer of mitosis research. *Nat Rev Mol Cell Biol*, Vol. 2, N° 1, pp. (72-75)

Petrovic, A.; Pasqualato, S.; Dube, P.; Krenn, V.; Santaguida, S.; Cittaro, D.; Monzani, S.; Massimiliano, L.; Keller, J.; Tarricone, A.; Maiolica, A.; Stark, H. & Musacchio, A. (2010). The MIS12 complex is a protein interaction hub for outer kinetochore assembly. *J Cell Biol*, Vol. 190, N° 5, pp. (835-852)

Przewloka, M.R. & Glover, D.M. (2009). The kinetochore and the centromere: a working long distance relationship. *Annu Rev Genet*, Vol. 43, N° pp. (439-465)

Reddy, S.K.; Rape, M.; Margansky, W.A. & Kirschner, M.W. (2007). Ubiquitination by the anaphase-promoting complex drives spindle checkpoint inactivation. *Nature*, Vol. 446, N° 7138, pp. (921-925)

Rieder, C.L. (2005). Kinetochore fiber formation in animal somatic cells: dueling mechanisms come to a draw. *Chromosoma*, Vol. 114, N° 5, pp. (310-318)

Rieder, C.L. & Alexander, S.P. (1990). Kinetochores are transported poleward along a single astral microtubule during chromosome attachment to the spindle in newt lung cells. *J Cell Biol*, Vol. 110, N° 1, pp. (81-95)

Rieder, C.L.; Cole, R.W.; Khodjakov, A. & Sluder, G. (1995). The checkpoint delaying anaphase in response to chromosome monoorientation is mediated by an inhibitory signal produced by unattached kinetochores. *J Cell Biol*, Vol. 130, N° 4, pp. (941-948)

Rieder, C.L.; Davison, E.A.; Jensen, L.C.; Cassimeris, L. & Salmon, E.D. (1986). Oscillatory movements of monooriented chromosomes and their position relative to the spindle pole result from the ejection properties of the aster and half-spindle. *J Cell Biol*, Vol. 103, N° 2, pp. (581-591)

Rieder, C.L. & Salmon, E.D. (1998). The vertebrate cell kinetochore and its roles during mitosis. *Trends Cell Biol*, Vol. 8, N° 8, pp. (310-318)

Ruchaud, S.; Carmena, M. & Earnshaw, W.C. (2007). Chromosomal passengers: conducting cell division. *Nat Rev Mol Cell Biol*, Vol. 8, N° 10, pp. (798-812)

Santaguida, S. & Musacchio, A. (2009). The life and miracles of kinetochores. *Embo J*, Vol. 28, N° 17, pp. (2511-2531)

Santaguida, S.; Vernieri, C.; Villa, F.; Ciliberto, A. & Musacchio, A. (2011). Evidence that Aurora B is implicated in spindle checkpoint signalling independently of error correction. *Embo J*, Vol. 30, N° 8, pp. (1508-1519)

Schrader, F. (1936). Kinetochore or Spindle Fibre Locus in Amphiuma tridactylum. *Biological Bulletin* Vol. 70, N° pp. (484-498)

Schrader, F. (1944). Mitosis. The Movement of Chromosomes in Cell Division (New York, Columbia University Press).

Schvartzman, J.M.; Sotillo, R. & Benezra, R. (2010). Mitotic chromosomal instability and cancer: mouse modelling of the human disease. *Nat Rev Cancer*, Vol. 10, N° 2, pp. (102-115)

Screpanti, E.; De Antoni, A.; Alushin, G.M.; Petrovic, A.; Melis, T.; Nogales, E. & Musacchio, A. (2011). Direct binding of Cenp-C to the Mis12 complex joins the inner and outer kinetochore. *Curr Biol*, Vol. 21, N° 5, pp. (391-398)

Sharp, D.J.; Rogers, G.C. & Scholey, J.M. (2000). Cytoplasmic dynein is required for poleward chromosome movement during mitosis in Drosophila embryos. *Nat Cell Biol*, Vol. 2, N° 12, pp. (922-930)

Silva, P.; Barbosa, J.; Nascimento, A.V.; Faria, J.; Reis, R. & Bousbaa, H. (2011). Monitoring the fidelity of mitotic chromosome segregation by the spindle assembly checkpoint. *Cell Prolif*, Vol. 44, N° 5, pp. (391-400)

Stegmeier, F.; Rape, M.; Draviam, V.M.; Nalepa, G.; Sowa, M.E.; Ang, X.L.; McDonald, E.R., 3rd; Li, M.Z.; Hannon, G.J.; Sorger, P.K.; Kirschner, M.W.; Harper, J.W. & Elledge, S.J. (2007). Anaphase initiation is regulated by antagonistic ubiquitination and deubiquitination activities. *Nature*, Vol. 446, N° 7138, pp. (876-881)

Sudakin, V.; Chan, G.K. & Yen, T.J. (2001). Checkpoint inhibition of the APC/C in HeLa cells is mediated by a complex of BUBR1, BUB3, CDC20, and MAD2. *J Cell Biol*, Vol. 154, N° 5, pp. (925-936)

Tanaka, T.; Cosma, M.P.; Wirth, K. & Nasmyth, K. (1999). Identification of cohesin association sites at centromeres and along chromosome arms. *Cell*, Vol. 98, N° 6, pp. (847-858)

Tanaka, T.U. (2010). Kinetochore-microtubule interactions: steps towards bi-orientation. *Embo J*, Vol. 29, N° 24, pp. (4070-4082)

Tanaka, T.U.; Rachidi, N.; Janke, C.; Pereira, G.; Galova, M.; Schiebel, E.; Stark, M.J. & Nasmyth, K. (2002). Evidence that the Ipl1-Sli15 (Aurora kinase-INCENP) complex promotes chromosome bi-orientation by altering kinetochore-spindle pole connections. *Cell,* Vol. 108, N° 3, pp. (317-329)

Thompson, S.L.; Bakhoum, S.F. & Compton, D.A. (2010). Mechanisms of chromosomal instability. *Curr Biol,* Vol. 20, N° 6, pp. (R285-295)

Tulu, U.S.; Fagerstrom, C.; Ferenz, N.P. & Wadsworth, P. (2006). Molecular requirements for kinetochore-associated microtubule formation in mammalian cells. *Curr Biol,* Vol. 16, N° 5, pp. (536-541)

Tulu, U.S.; Rusan, N.M. & Wadsworth, P. (2003). Peripheral, non-centrosome-associated microtubules contribute to spindle formation in centrosome-containing cells. *Curr Biol,* Vol. 13, N° 21, pp. (1894-1899)

Vafa, O. & Sullivan, K.F. (1997). Chromatin containing CENP-A and alpha-satellite DNA is a major component of the inner kinetochore plate. *Curr Biol,* Vol. 7, N° 11, pp. (897-900)

Van Hooser, A.A.; Ouspenski, II; Gregson, H.C.; Starr, D.A.; Yen, T.J.; Goldberg, M.L.; Yokomori, K.; Earnshaw, W.C.; Sullivan, K.F. & Brinkley, B.R. (2001). Specification of kinetochore-forming chromatin by the histone H3 variant CENP-A. *J Cell Sci,* Vol. 114, N° Pt 19, pp. (3529-3542)

Vanoosthuyse, V. & Hardwick, K.G. (2009). Overcoming inhibition in the spindle checkpoint. *Genes Dev,* Vol. 23, N° 24, pp. (2799-2805)

Walczak, C.E.; Vernos, I.; Mitchison, T.J.; Karsenti, E. & Heald, R. (1998). A model for the proposed roles of different microtubule-based motor proteins in establishing spindle bipolarity. *Curr Biol,* Vol. 8, N° 16, pp. (903-913)

Wang, E.; Ballister, E.R. & Lampson, M.A. (2011). Aurora B dynamics at centromeres create a diffusion-based phosphorylation gradient. *J Cell Biol,* Vol. 194, N° 4, pp. (539-549)

Wang, H.; Hu, X.; Ding, X.; Dou, Z.; Yang, Z.; Shaw, A.W.; Teng, M.; Cleveland, D.W.; Goldberg, M.L.; Niu, L. & Yao, X. (2004). Human Zwint-1 specifies localization of Zeste White 10 to kinetochores and is essential for mitotic checkpoint signaling. *J Biol Chem,* Vol. 279, N° 52, pp. (54590-54598)

Wollman, R.; Cytrynbaum, E.N.; Jones, J.T.; Meyer, T.; Scholey, J.M. & Mogilner, A. (2005). Efficient chromosome capture requires a bias in the 'search-and-capture' process during mitotic-spindle assembly. *Curr Biol,* Vol. 15, N° 9, pp. (828-832)

Yang, Z.; Tulu, U.S.; Wadsworth, P. & Rieder, C.L. (2007). Kinetochore dynein is required for chromosome motion and congression independent of the spindle checkpoint. *Curr Biol,* Vol. 17, N° 11, pp. (973-980)

Zachariae, W. & Nasmyth, K. (1999). Whose end is destruction: cell division and the anaphase-promoting complex. *Genes Dev,* Vol. 13, N° 16, pp. (2039-2058)

4

Drosophila: A Model System That Allows *in vivo* Manipulation and Study of Epithelial Cell Polarity

Andrea Leibfried[1] and Yohanns Bellaïche[2]
[1]EMBL Heidelberg,
[2]Institut Curie Paris,
[1]Germany
[2]France

1. Introduction

Epithelia are specialized tissues that emerged early in evolution to subdivide the body into distinct parts and to form barriers by lining both the outside (skin) and the inside cavities and lumen of bodies. Functions of epithelial cells include secretion, absorption, protection, transcellular transport, sensation detection and selective permeability. To achieve these different tasks, several types of epithelial tissues emerged during evolution, all of them having one feature in common that is indispensable for their function: the cells are attached to each other in order to form a layer that works as a barrier. Epithelial cells thus show a ordered morphology, they are polarized. Disruption of polarity is an important feature of epithelial cancers that accounts for more than 90% of fatal malignancies in adults, rendering the understanding of the fundamental processes needed for polarity an ongoing subject of high interest.

Epithelial cells are highly polarized and the cells are oriented so that their external, so-called apical, surfaces face the outside or a central lumen and the internal, or basolateral, surfaces are in contact with other cells and the basement membrane. These characteristics were discovered using histological and thus descriptive analyses. With the advent of molecular genetics, biochemistry and molecular cell biology our knowledge about polarization processes increased dramatically, allowing for a better understanding of the mechanisms used by epithelial cells to fulfill their specialized tasks and to act as barriers. Thus, the polarization is not only reflected in the morphology of the cells, but also in the positioning of their organelles and in the apico-basal (AB) localization of polarity protein complexes at the plasma membrane. The plasma membrane can be divided into an apical, a junctional and a basolateral domain, each domain comprising its own set of polarity proteins that are widely conserved in eukaryotes. This AB polarity is required for formation of functional epithelial tissues. The asymmetrical deployment of proteins is mediated through subcellular trafficking and the polarized localization of transcripts. Genetic studies have revealed that the polarity protein complexes function in a sequential but interdependent manner to regulate the establishment and maintenance of cellular polarity.

The proteins involved in epithelial cell polarization are largely conserved between species, as mentioned above, even between vertebrates and invertebrates. Much of our hitherto knowledge stems from studies performed in model organisms and until now it is impossible to culture whole epithelia for a long time. Existing *in vitro* cell culture systems give important insights into epithelial cell function, but mechanisms following biological input from the living and developing organism are obviously missed in these systems. Therefore, the fruit fly, *Drosophila*, provides an excellent model system for studying columnar epithelial cells (Fig. 1), allowing their static and especially dynamic analysis in the context of a whole living organism. Furthermore, state of the art microscopy imaging techniques can be easily coupled to the extensive genetic tools available in the *Drosophila* system, allowing for *in vivo* analysis and the concomitant dissection of pathways needed for epithelial cell polarity.

In this chapter we would like to introduce *Drosophila* as a model system to study epithelial cell function, establishment and maintenance. First, we will show the similarities and differences between invertebrate and vertebrate columnar epithelial cells. Second, we will depict the current knowledge of polarity protein complex regulation, highlighting the achievements derived from work in *Drosophila*. Finally, we will provide examples that nicely show how easy *in vivo* studies on epithelial cells can be performed using *Drosophila*.

This chapter will not only give a background on epithelial cell polarity regulation, but it will highlight the importance of whole organismal studies for the understanding of epithelial tissues. Furthermore it will reveal the value of a model system like the fruit fly when deciphering mechanisms underlying biological processes.

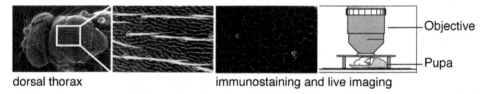

dorsal thorax   immunostaining and live imaging

Fig. 1. Dorsal thorax of *Drosophila melanogaster*.

The dorsal thorax serves as a model system for epithelial maintenance and can easily be analyzed by immunostaining or non-invasive live imaging, using for example the dorsal thorax of the immobile, developing pupa.

2. Establishment of epithelial cell polarization in metazoans

Epithelial cells have an adhesive belt that encircles the cells apically, the zonula adherens (ZA). It assembles from the aggregation of spot adherens junctions. Basally, integrin-based focal contacts connect the epithelial cells to the basement membrane. Vertebrate epithelial cells develop a tight junction (TJ) apical to the ZA, which impedes intercellular diffusion and forms a region of close membrane contacts. In *Drosophila*, the functional equivalent to TJs is the septate junction (SJ) that lies basal to the ZA. A domain with similar protein composition as found in TJs is located apically of the ZA in the subapical region (SAR). Only a single junction, the *C. elegans* apical junction (CeAJ), has been identified in *C. elegans*, which resembles the ZA of *Drosophila* and vertebrates (Knust & Bossinger, 2002; Tepass et al., 2001) (Fig. 2). Thus, the overall structure of epithelial cells is highly conserved between species. The polarity protein complex located close to the ZA, the junctional Par complex, has been shown to play an essential role in the establishment of epithelial polarity.

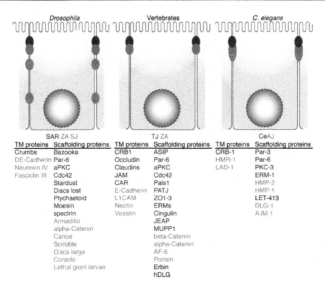

Fig. 2. (adapted from Knust & Bossinger, 2002). Epithelial cell characteristics and typical polarity protein composition for *Drosophila*, vertebrate and *C. elegans* epithelial cells. The transmembrane (TM) and scaffolding proteins are color-coded depending on their localization. SAR: subapical region; ZA: zonula adherens; TJ: tight junction; CeAJ: *C. elegans* apical junction.

2.1 The junctional Par complex

About 20 years ago, Kemphues and colleagues identified six Par (partitioning defective) proteins, Par-1 to Par-6, and an atypical protein kinase C (aPKC, known as PKC-3 in *C. elegans* and homolog to human PKCζ) in a screen for partition defective cell division in the one cell stage embryo in *C. elegans*, resulting all in loss of the anterior-posterior axis of the embryo when mutated (Cheng et al., 1995; Kemphues et al., 1988; Kirby et al., 1990). Five of the Par proteins - all but Par-2 - as well as aPKC are highly conserved throughout the animal kingdom and are needed for cell polarization. Par-1, Par-4 (also known as LKB1) and aPKC are kinases, Par-3 and Par-6 are PDZ (PSD95, DlgA, ZO-1)-domain containing proteins and Par-5 is a 14-3-3 protein. Par-3, Par-6 and aPKC form a complex localized at the anterior cell cortex of the one cell stage *C. elegans* embryo, while Par-1 and Par-2 remain at the posterior cortex and Par-4 and Par-5 localize uniformly at the cortex. The polarized localization of these proteins is triggered upon fertilization by sperm entry, which enriches the RacGAP CYK-4 (Cytokinesis defect-4) at the posterior pole to give a spatial cue for polarity. CYK-4 functions as a GTPase activating protein (GAP) for small GTPases like Rho, Cdc42 or Rac. Thus its localized activity leads to a gradient of acto-myosin via the inactivation of small GTPases, which distributes the Par proteins in a polarized manner (Jenkins et al., 2006) (Fig. 3). One of the essential functions of the Par complex comprised of Par-3, Par-6 and aPKC, is to set-up epithelial polarity in metazoans in response to the formation of initial cell-cell contact or discrete membrane domains, whereas the basolateral Par complex (Par-1) functions to promote the expansion of the lateral membrane. In general, Par proteins have been found to regulate cell polarization in many different contexts in diverse animals: in epithelia, in directed cell migration, in polarized cells like neurons or in self-renewing cells; suggesting that they form part of an ancient and fundamental mechanism of cell polarization.

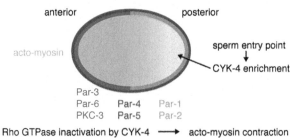

Fig. 3. Par protein distribution in the one cell stage *C. elegans* embryo.

Sperm entry marks the posterior pole and leads to an enrichment in CYK-4 protein, which functions as a GAP for Cdc42, Rho and Rac, thus leading to a gradient of acto-myosin and to the spatial restriction of the different Par complexes.

2.2 The apical Crumbs complex

The apical region of polarized epithelial cells harbors Crumbs (Crb), a transmembrane protein with 30 EGF-like repeats in the extracellular domain that binds with its intracellular domain to the MAGUK (membrane-associated guanylate kinase) protein Pals1 (protein associated with Lin7; also known as Stardust in *Drosophila*). Crb also recruits PATJ (Pals1-associated tight junction protein) into the most-apical complex (Tepass & Knust, 1993). Pals1 and PATJ are cytoplasmic scaffolding proteins with several protein-protein interaction domains including L27 domains, SH3 (Src homology 3) domains, guanylate kinase (GUK) domains and PDZ domain. Overexpression or siRNA-mediated down-regulation of any of the three components in mammalian epithelial cells leads to defects in AB polarity formation suggesting the requirement of the complex as a whole for the proper development of cell polarity (Roh et al., 2003; Shin et al., 2005; Straight et al., 2004).

2.3 Basolateral domain: Lethal giant larvae, Discs large and Scribble

The basolaterally localized proteins Lethal giant larvae (Lgl), Discs large (Dlg) and Scribble (Scrib) cooperatively regulate cell polarity, junction formation and cell growth in epithelial cells. All three genes were identified as tumor suppressors in *Drosophila*. Lgl has at least four WD40 repeats, forms homo-oligomers and associates with the cytoskeleton by binding to non-muscle myosin II. Dlg contains three PDZ domains, a SH3 domain and a GUK domain. Scrib is a LAP (leucine-rich repeats and PDZ domain) protein, containing 16 leucine-rich-repeats (LRRs) and 4 PDZ domains (Yamanaka & Ohno, 2008).

2.4 Adherens junctions proteins

The adherens junctions harbor the E-Cadherin/β-Catenin/α-Catenin (E-Cad/β-Cat/α-Cat) complex that is indispensable for establishing and maintaining cell-cell adhesion. Cadherins are a large family of transmembrane glycoproteins that form homophilic, calcium-dependent interactions with neighboring cells (Gumbiner et al., 1988). E-Cadherin is the predominant epithelial isoform of cadherin. Its extracellular domain is composed of five ectodomain modules (EC1-EC5), with the most membrane-distal module (EC1) mediating binding with the E-Cad on the adjacent cell. Calcium ions bind between the EC domains to promote a rod-like conformation required for trans-interactions (Fig. 4). The cytoplasmic tail binds to β-Cat (known as Armadillo, Arm in *Drosophila*) in order to link E-Cad to α-Cat in metazoans (Pacquelet & Rørth, 2005). Arm is furthermore linked to the Par complex in *Drosophila* by binding to Bazooka (Baz), the *Drosophila* Par-3 homolog (Capaldo & Macara, 2007; Oda et al., 1994).

Stable epithelial adhesions require the F-actin network (Cavey et al., 2008) and the E-Cad/β-Cat/α-Cat complex is linked to actin via α-Cat, vinculin and α-actinin. It is a matter of debate whether the α-Cat-actin interaction indeed stabilizes the AJ complex since it has been shown in mammalian cells that α-Cat binds only in its homodimeric state to actin, whereas its binding to β-Cat is restricted to the monomeric form. Thus, the interaction of the E-Cad/β-Cat/α-Cat complex with the actin cytoskeleton is dynamic by way of association and dissociation of α-Cat with the complex and with actin filaments. Homodimeric α-Cat directly regulates actin-filament organization by suppressing Arp2/3 mediated actin polymerization, most likely by competing with the Arp2/3 complex for binding to actin filaments (Drees et al., 2005; Yamada et al., 2005). The observations made by Drees, Yamada and colleagues thus suggests a 'dynamic stabilization' of the AJ complex through actin-α-Cat interactions. In the light of a developing epithelial tissue, where a modulation of the cell-cell contacts needs to take place in order to allow for cell rearrangements and growth without compromising epithelial integrity and barrier properties (Classen et al., 2005), such a dynamic interaction between the stabilizing actin cytoskeleton and the AJ complex seems to be indispensable. Thus, this suggests the existence of adjustable mechanisms stabilizing the adherens junctions proteins at the junctions allowing for plasticity of the epithelial tissue; mechanisms that needed to be further elucidated *in vivo* to completely understand epithelial development and maintenance and associated E-Cad transport and turnover.

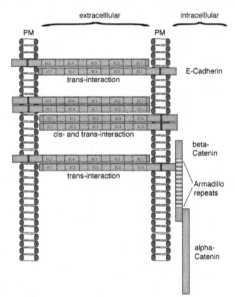

Fig. 4. E-Cadherin/β-Catenin/α-Catenin interaction.

β-Catenin binds to the cytoplasmic tail of E-Cadherin. α-Catenin binds to the N-terminal region of β-Catenin. E-Cadherin molecules can interact with each other in the extracellular space in cis and in trans, both interactions need Ca^{2+}. PM: plasma membrane.

3. Establishment of epithelial cell polarization in *Drosophila*

In *Drosophila*, the syncytial blastoderm gives rise to the first epithelium during cellularization. Membrane invaginations form, the so-called furrow canals, which already display discrete domains apical and lateral to the nuclei (Foe & Alberts, 1983; Mavrakis et al., 2009; Schejter & Wieschaus, 1993). Baz localizes below aPKC and Par-6 as the epithelium forms and its positioning is independent of aPKC and Par-6 but dependent on cytoskeletal cues given by the apical scaffold and dynein-mediated basal-to-apical transport as well as by cues that still need to be elucidated. Baz recruits Par-6 and aPKC, and subsequently Baz and Par-6 recruit Crb and PATJ respectively. aPKC in turn stabilizes apical Crb. Par-6 positioning is dependent on aPKC and on activated Cdc42 (Harris & Peifer, 2005). A transient basal adherens junction made up of ZA proteins E-Cad, Arm, α-Cat and PATJ forms and its assembly is coupled to correct Baz positioning (Hunter & Wieschaus, 2000). Once cellularization is complete, the lateral dispersed spot-like adherens junctions coalesce apically to form the belt-like ZA. In the SAR, Crb binds to Stardust (Sdt, Pals1 homolog of *Drosophila*) via its C-terminus and recruits PATJ into the complex. Crb also provides a link to the apical membrane cytoskeleton, which might reinforce the ZA, by binding to *Drosophila* Moesin (membrane-organizing extension spike protein) via its FERM (4.1, Ezrin, Radixin, Moesin) binding site and to spectrin (Médina et al., 2002). Spectrin is a tetrameric actin crosslinking protein. The spectrin cytoskeleton composition differs in the apical and basolateral domain of epithelial cells, thus giving spatial cues. The basolateral localization of the proteins Lgl, Dlg and Scrib depends on the presence of each of the three proteins and a failure of localization leads to the expansion of the apical Crb complex into more lateral domains. Consequently, the ZA does not form correctly, leading to multilayering of cells. Dlg and Scrib localize to the basolateral domain and the septate junctions just below the ZA, whereas Lgl co-localizes only partially with them below the septate junctions. Lgl is also found in the cytoplasm (Bilder & Perrimon, 2000; Bilder et al., 2000; Woods & Bryant, 1991).

Thus, epithelial polarity establishment has been extensively studied in *Drosophila*, which is due to the fact that the syncytial blastoderm-to-epithelium transition can be easily studied in wild-type and mutant conditions. Freshly laid eggs are collected from the flies and subsequently staged to observe cellularization (defects) at different timepoints.

4. Establishment of epithelial cell polarization in vertebrates

The TJs of vertebrate epithelial cells separate apical and basolateral membrane domains and harbor the transmembrane proteins occludin, claudin family members and junctional adhesion molecules (JAMs) (Fig. 2). All TJ proteins interact directly with cytoplasmic, PDZ-domain containing proteins like ZO-1, ZO-2, ZO-3 (Zonula occludens 1-3), Par-3 and Pals1. These cytoplasmic proteins recruit other cytoskeletal (F-actin) or signaling molecules (Mertens et al., 2005).

Many results explaining epithelial polarity establishment in vertebrates rely on a cell-based system: In MDCK (Madin-Darby canine kidney) cells, ZO-1 binds to the C-terminus of claudins and JAM via two of its PDZ domains and Par-3 directly associates with the C-terminus of JAM. Subsequently, and similar to what has been shown in *Drosophila*, Par-6, aPKC and Cdc42 are recruited to Par-3 and form a complex that is needed for the development of normal tight junctions (Afonso & Henrique, 2006; Chen & Macara, 2005; Joberty et al., 2000). Claudins play a central role in polarity establishment, since they also interact with the Crb complex: PATJ, whose expression depends on Pals1, interacts with claudin and ZO-3 via two of its PDZ domains and with Pals1 via its N-terminus. Pals1 interacts with its PDZ domain with the cytoplasmic tail of Crb, which reinforces the association of Crb with the Par complex (Lemmers et al., 2003; Straight et al., 2004).

5. Regulation of cell polarity in the epithelium

5.1 Regulation of the junctional Par complex

As described above, Par-3/Baz recruits Par-6 and aPKC in vertebrates and *Drosophila* and this has been confirmed by ectopic expression of Par-3, which leads to ectopic Par-6 and aPKC recruitment in mammalian cells (Joberty et al., 2000), indicating a general mechanism for this recruitment. Par-6 also plays an important role in Par complex localization and activation since it interacts with Cdc42 via its PDZ and semi-CRIB domain, with Par-3/Baz via its PDZ domain and with aPKC through its PB1 (Phox and Bem1) domain. Concomitant binding to aPKC and Cdc42 causes a conformational change in Cdc42-GTP, leading to aPKC activation (Hutterer et al., 2004; Joberty et al., 2000). Activated aPKC in turn is needed for correct Par-3 localization in mammalian cells and has been shown to spatially regulate the basolateral protein complex in *Drosophila*. aPKC phosphorylates Par-3 which in turn dissociates from the Par complex, possibly allowing the regulative interaction of aPKC with other proteins and thus leading to correct cell-cell contact formation and epithelial polarization (Joberty et al., 2000; Nagai-Tamai et al., 2002). At the basolateral side, Par-6 localization is excluded by Lgl in *Drosophila*, however the exclusion of the Par complex in mammalian cells has been shown to be regulated by Par-1 kinase activity: Par-1 phosphorylates Par-3, which leads to the binding of Par-5 (14-3-3) to Par-3 and thus inhibits the formation of the Par complex by blocking Par-3 oligomerization and binding to aPKC (Benton & St Johnston, 2003). Par-1 activation is regulated by Par-4 through phosphorylation of the activation loop of the Par-1 kinase domain (Suzuki & Ohno, 2006) and it is inactivated by phosphorylation by aPKC, which causes binding to Par-5 (14-3-3) and inhibition of plasma membrane binding in the apical domain (Hurov et al., 2004) (Fig. 5).

Proteins that have so far not been linked to epithelial polarity can of course also affect the functionality of the junctional Par complex, and in the upcoming years new regulators will for sure be identified. One example for such a regulation is protein phosphatase-2A (PP2A), a heterotrimeric serine/threonine phosphatase with broad substrate specificity and diverse cellular functions like cell growth and regulation of the cytoskeleton. PP2A inhibits aPKC function and dephosphorylates TJ components thereby triggering junction disassembly in mammalian cells (Nunbhakdi-Craig et al., 2002) (Fig. 5).

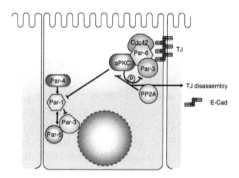

Fig. 5. Regulation of the junctional Par complex in mammalian cells.

aPKC inhibits the function of Par-1, which inactivates Par-3 in the basolateral domain. Par-4 activates Par-1. Cdc42 binds to Par-6 and thereby activates the Par complex. aPKC phosphorylates Par-3, which then dissociates from the Par complex. PP2A blocks aPKC function and leads thus to TJ disassembly. TJ: tight junction; PP2A: protein phosphatase-2A.

5.2 Regulation of the junctional E-Cad/β-Cat/α-Cat complex

The junctional E-Cadherin/β-Catenin/α-Catenin complex is needed for cell-cell adhesion in epithelial cells and its localization and maintenance must thus be strictly regulated. In mammalian cells, disassembly of the apical junctional complex is driven by reorganization of apical F-actin involving cofilin-1-dependent depolymerization and Arp2/3-assisted repolymerization as well as myosin II mediated contraction (Ivanov et al., 2004). Actin organization, myosin II phosphorylation and therefore localization and regulation of gene transcription and E-Cad localization is affected by Rho, which counteracts Cdc42 and Rac activity and thus inhibits AJ formation (Sturge et al., 2006). Cdc42, Rac and Rho are indispensable for epithelial polarity regulation at the junctional domain. Cdc42 promotes Par complex formation in vertebrates and *Drosophila* as depicted above and it promotes TJ development by activation of Rac in mammalian cells. Rac is also activated at the junctional domain through the GEF Tiam1 (T-cell lymphoma invasion and metastasis-1), which also binds to Par-3, providing a link between Rac and the junctional polarity protein complex. This activation is needed to counteract Rho activity, since Rho favors TJ disassembly through Rho kinase (ROCK) mediated myosin II phosphorylation (van Leeuwen et al., 1999). In *Drosophila*, Cdc42/Par-6/aPKC furthermore regulate E-Cad endocytosis, by recruiting and interacting with the actin and dynamin machinery needed for vesicle scission (Georgiou et al., 2008, Harris & Tepass, 2008, Leibfried et al., 2008) (Fig. 6).

Myosin II is a motor that converts chemical energy of ATP into mechanical forces, mediating the contractility of the actin cytoskeleton. It is activated by phosphorylation of its light chain through ROCK or MLCK (myosin light chain kinase). Rho activity is therefore down-regulated at the AJs in polarized epithelial cells via the interaction of the GAP p190RhoGAP and the catenin p120-Catenin. p190RhoGAP translocation to the AJs is mediated by Rac activity, a process taking also place in *Drosophila* epithelial cells, thus suggesting a general and not organism-specific down-regulation of Rho at the adherens junctions (Magie et al., 2002; Wildenberg et al., 2006). In mammalian cells, Rac has been shown to recruit actin to sites of cell-cell contacts, where it leads to the internalization of the E-Cad/Cat complex

(Akhtar & Hotchin, 2001). Therefore, Rac seems to have a dual role in E-Cad/Cat complex maintenance by recruiting p190RhoGAP and at the same time leading to E-Cad/Cat internalization at the junctions.

Echinoid (Ed) is a component of *Drosophila* AJs that stabilizes the adhesion complex through cooperation with E-Cad and by linking the AJs to actin filaments. The C-terminal PDZ domain of Ed furthermore binds to Baz, which leads to a strong linkage between the Par complex and the AJs, since E-Cad is also bound to Baz via its interaction with Arm (Laplante & Nilson, 2006; Wei et al., 2005). Moreover, actin filaments are organized by a pathway that is regulated by Bitesize, a synaptogamin-like protein that binds to Moesin and PIP2 at the apical domain, leading to stabilization of E-Cad at the AJs (Pilot et al., 2006). These results show that the regulation of the adhesion complex depends strongly on small GTPases and the underlying acto-myosin network as well as on protein-protein interactions between adhesion and junctional polarity proteins. Thus they form part of an ancient and fundamental mechanism of cell polarization.

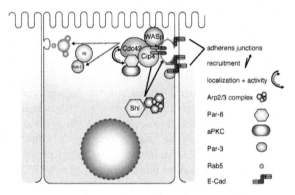

Fig. 6. Regulation of E-Cad endocytosis at the AJs in *Drosophila* epithelial cells.

Cdc42 most likely recruits together with Par-6 and aPKC the actin regulators Cip4 and WASp as well as dynamin, which is needed for endocytic vesicle scission, therefore allowing for correct endocytosis of junctional material.

5.3 Regulation of the apical Crumbs complex

The correct localization of the Crb complex depends on the Crb/Pals1/PATJ proteins themselves, on motor proteins, the Par complex and on regulation of the cytoskeleton. Most of our current knowledge is based on studies conducted on *Drosophila* epithelial cells: The apical localization of the Crb complex is highly dependent on its individual components. Crb localization depends on PATJ and on the transport of Crb protein and transcript by a cytoplasmic dynein complex. PATJ localization is in turn partially dependent on apical Crb localization, possibly resulting in a positive feedback loop between Crb and PATJ targeting (Horne-Badovinac & Bilder, 2008; Li et al., 2008; Michel et al., 2005; Tanentzapf et al., 2000). This positive regulation might be antagonized during later stages of epithelial development by the FERM protein Yurt in order to prevent an expansion of the apical domain. Yurt is localized at the basolateral domain, also in mammalian epithelial cells, but Crb recruits it to the apical membrane late during epithelial development where it counteracts Crb activity (Laprise et al.,

2006). Crb also recruits Pals1 to the apical domain and the same dynein complex used for Crb apical targeting can transport Pals1 transcript. The junctional Par complex regulates the Crb complex by phosphorylation. aPKC can interact with Crb in presence of PATJ and Pals1. As a consequence, Crb gets phosphorylated which is indispensable for correct membrane targeting of Crb and PATJ and for Crb activity (Sotillos et al., 2004). PATJ localization and ZA formation depend furthermore on Par-6 activity (Hutterer et al., 2004).

The apical and basolateral complex regulate each other and loss of the basolateral complex leads to the expansion of the Crb complex and loss of polarity, which can be rescued by the concomitant loss of the Crb complex (Bilder et al., 2003; Tanentzapf & Tepass, 2003). The discrete membrane domains are interdependent not only during epithelial establishment but also during its maintenance. For example loss of adherens junctions proteins E-Cad and Arm leads to loss of Crb and apical polarity via the disruption of the lateral spectrin and actin cytoskeleton (Tanentzapf et al., 2000), therefore resulting in a complete loss of polarity.

5.4 Regulation of the basolateral Lgl complex

Scrib stabilizes the AJ complex in the junctional domain, but Lgl localization is accurately restricted to the basolateral domain by aPKC. Lgl, aPKC and Par-6 can interact, leading to the phosphorylation of Lgl by aPKC at three conserved serine residues in mammalian cells and *Drosophila* (Betschinger et al., 2003). Phosphorylation inactivates Lgl on the apical side and inhibits its binding to the plasma membrane in both mammalian and *Drosophila* epithelial cells by changing the conformation of the protein (Betschinger et al., 2005; Plant et al., 2003). Lgl and aPKC seem to mutually regulate each other, since loss of Lgl (leading to an overproliferation phenotype) can be suppressed by concomitant loss of aPKC in *Drosophila* (Rolls et al., 2003) (Fig. 7). In mammalian cells, Lgl phosphorylation by aPKC is furthermore facilitated by the concomitant interaction of Lgl with P32 (Bialucha et al., 2007). Thus, the interplay between junctional and basolateral proteins maintains distinct membrane domains in polarized epithelial cells.

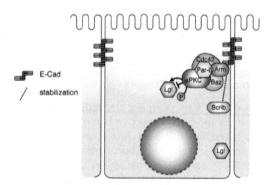

Fig. 7. Regulation of basolateral Lgl and stabilization of E-Cad by Scrib in *Drosophila* epithelial cells.

aPKC phosphorylates Lgl, which impedes the binding of Lgl to the cortex in the junctional and apical domain. Scrib leads to the stabilization of E-Cad. Arm: Armadillo; Baz: Bazooka; Lgl: Lethal giant larvae; Scrib: Scribble.

5.5 Regulation of epithelial polarity via TGF-β and EGF signaling

TGF-β and EGF signaling pathways are two out of several important signaling pathways needed for the correct development of an organism. They have an influence on epithelial polarity and can lead for example to epithelial to mesenchymal transition (EMT). Therefore, the inhibition of both pathways is important for epithelial polarity maintenance.

Transforming growth factor beta (TGF-β; Decapentaplegic, Dpp in *Drosophila*) controls proliferation and cellular differentiation in most cells and TGF-β signaling can lead to loss of cell polarity: in mammalian cells, the TGF-β downstream effector Rho controls EMT by changing the actin cytoskeleton (Ozdamar et al., 2005), whereas the downstream transcription factors of the Snail family concomitantly lead to down-regulation of claudins, occludin, E-Cad and Crb (Xu et al., 2009). Therefore, to maintain epithelial polarity, Par-6 interacts with the TGF-β receptor, which phosphorylates Par-6 so that it can recruit the E3 ubiquitin ligase Smurf1. Smurf1 leads to degradation of Rho and thus to a block of EMT. Snail protein activation is blocked by Smad6 and Smad7, thereby preventing down-regulation of TJ and apical proteins.

Epidermal growth factor (EGF) signaling also plays an important role in the regulation of cell growth, proliferation and differentiation and it can modulate mammalian TJ formation, which allows for concerted dissociation and re-establishment of cell-cell adhesion essential for morphogenesis. Two different pathways achieve this modulation. EGF signaling activates Src (sarcoma) family kinases that phosphorylate TJ proteins, including ZO-1, ZO-2, occludin, E-Cad and Par-3, leading to positive and negative regulation of TJ formation in mammalian cells (Chen et al., 2002; Shen et al., 2008). The down-regulation of E-Cad by Src promotes EMT because it alters E-Cad trafficking by redirecting E-Cad from a recycling pathway to a lysosomal pathway (Shen et al., 2008). In both mammalian and *Drosophila* cells, EGF signaling also induces the MAPK (mitogen-activated protein kinase) pathway which leads to E-Cad and claudin expression and subsequent translocation of junctional proteins from the cytoplasm to cell-cell contacts (O'Keefe et al., 2007; Wang et al., 2006). E-Cad and EGF receptor interact at cell-cell contacts to negatively regulate the MAPK pathway in mammalian cells, suggesting a general negative feedback loop to regulate adhesion and junctional integrity (Qian et al., 2004). These results suggest that EGF signaling needs to be tightly controlled to promote either EMT or the stabilization of junctions and they emphasize the need of a strict regulation of signaling pathways in the cell in order to maintain epithelial integrity.

5.6 Regulation of epithelial polarity via phosphatidylinositol signaling

Recent studies integrate phosphatidylinositol-phosphate signaling to polarization in both mammalian and *Drosophila* epithelial cells. Phosphatidylinositol-phosphate signaling was mainly known to regulate cell size by interacting with the insulin receptor, but Martin-Belmonte, von Stein and colleagues propose that it also enhances polarity establishment.

PDZ domains can bind to phosphatidylinositol lipid membranes and *Drosophila* Baz binds to both, phosphatidylinositol lipid membranes and to the phosphatase PTEN (phosphatase and tensin homolog), thereby possibly recruiting this protein to the apical domain (von Stein et al., 2005). PTEN converts PIP3 (Phosphatidylinositol-(3,4,5)-triphosphate) to PIP2 (Phosphatidylinositol-(4,5)-bisphosphate) and this leads to Cdc42 recruitment via Annexin-2 and subsequently to aPKC recruitment to the apical domain in mammalian cells (Martin-Belmonte et al., 2007). PIP3 is produced by activation of PI3K (phosphatidylinositol-3 kinase) at the adherens junctions, as shown in *Drosophila* epithelial cells, which locally activates Cdc42 and recruits more E-Cad to the junctions (von Stein et al., 2005). This suggests a general model where the Par complex recruits PTEN by binding to phosphatidylinositol lipid membranes, and where PTEN converts PI3K-produced PIP3 to PIP2. This might mediate the establishment of epithelial polarity by the recruitment of Cdc42, aPKC and E-Cad (Fig. 8).

PIP2 and PIP3 regulate furthermore the actin cytoskeleton: PIP2 binds to actin-associated proteins that link the actin cytoskeleton to the plasma membrane or to proteins that are involved in the initiation of *de novo* actin polymerization, and PIP3 activates WASp family proteins (like WASp, WAVE and WASH proteins) and the Arp2/3 complex via interaction with Rho GTPases (Fig. 8). These results point toward a spatio-temporal fine-regulation of Cdc42 recruitment during polarization and a coeval regulation of cell polarity establishment and cytoskeleton controlled by the balance between PIP2 and PIP3, leading to an enhanced effect for both Cdc42-mediated pathways.

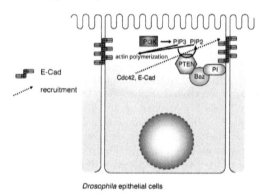

Fig. 8. Regulation of epithelial polarity via phosphatidylinositol signaling.

Baz can bind to PI and to PTEN. PTEN mediates PIP3 to PIP2 conversion, which leads to the recruitment of Cdc42 and E-Cad. PIP3, which is generated by PI3K activity in the apical domain, and PIP2 lead to actin polymerization at the junctions. PI: phosphatidylinositol; PIP3: Phosphatidylinositol-(3,4,5)-triphosphate; PIP2: Phosphatidylinositol-(4,5)-bisphosphate; PI3K: phosphatidylinositol-3 kinase; Baz: Bazooka.

5.7 Regulation of epithelial polarity via the acto-myosin cytoskeleton

Cell polarity also requires fine regulation of the cytoskeleton. Actin is needed for the furrow-canal formation during cellularization in *Drosophila* and AJ proteins are linked to the actin cytoskeleton. Furthermore, spectrin has a crucial role in anchoring Crb at the apical domain (Tanentzapf et al., 2000). In general, the regulation of the actin cytoskeleton is dependent on the activity of Rac and Cdc42, which both lead to actin nucleation upon their activation. Downstream of Rac and Cdc42 are WASp family proteins, which are activated through binding to the Cdc42 and Rac-binding domain (CRIB). WASp protein activity leads to actin nucleation by activation of the Arp2/3 complex. In mammalian cells, the binding of Cdc42-GTP and PIP2 to WASp synergistically enhances the activation of the protein (Parsons et al., 2005) and WASp upregulates the GEF activity of the Cdc42-

specific GEF intersectin (Malacombe et al., 2006), suggesting a positive feedback loop between Cdc42 and WASp activity. Cdc42 is also directly linked to epithelial polarity via the junctional Par complex (see 5.2). Downstream of Cdc42 is furthermore PAK1, which promotes microtubule formation, thus localizing actin nucleation and microtubule formation to the same confined space (Parsons et al., 2005), possibly needed for correct plasma membrane identity in polarized cells. Cell polarity is furthermore maintained by the AMP-activated protein kinase (AMPK), which alters the acto-myosin cytoskeleton in response to energetic stress situations. High AMP levels lead to a conformational change of the protein and Par-4 phosphorylates AMPK to activate it as indicated by biochemical data (Hawley et al., 2003). In *Drosophila*, Dystroglycan, which is localized at the basal domain, where it interacts with the extra-cellular matrix protein Perlecan, transduces a signal from the cellular energy sensor AMPK to myosin II, thus activating myosin II and regulating AB polarity (Mirouse et al., 2009). To conclude, regulation of both the actin cytoskeleton and myosin II by GTPases and polarity proteins seems to be indispensable for maintaining AB polarity.

6. *In vivo* studies on epithelial cells using *Drosophila*

As depicted above, most of our hitherto knowledge of epithelial polarity stems from genetic and biochemical studies performed in cell culture or on model systems. Though extensive, these two approaches cannot reveal all mechanisms underlying the process of polarization. Cell polarity is not only needed for the establishment and maintenance of the single cell, but also for the correct development of the whole multicellular organism. Some events, like spatio-temporal protein or organelle localization can only be captured when analyzing a whole, developing, epithelium. Whole tissue analysis can be easily performed using *Drosophila*: epithelia can be imaged *in vivo* by confocal microscopy (see Fig. 1) while the animal develops. This imaging techniques and the vast genetic tools available (e.g. gene knock-down or over-expression, expression of fluorescent tagged proteins in the living fly) have allowed for the dissection of new mechanisms regulating epithelial polarity.

6.1 Junction-cytoskeleton interaction dissected *in vivo*

The interaction of α-Catenin with E-Cad and the cortical actin control both stability and remodeling of adhesion. How this occurs is, as mentioned in section 2.4., not elucidated. Live imaging of *Drosophila* embryos expressing fluorescent E-Cad and actin (F-actin) revealed that E-Cad is not evenly distributed around the adhesion belt in epithelial cells as previously expected. Stable microdomains intersperse with mobile domains. Laser nano-ablation of actin and FRAP (Fluorescence Recovery After Photobleach) and photo-conversion experiments for E-Cad show that the stability and mobility of these microdomains depend on two actin populations, a stable network and one that rapidly turns over (Cavey et al., 2008).

Further *in vivo* studies using FRAP and nano-ablation show that the myosin-II forces needed for actin remodeling at the junctions is not as previously assumed based on polarized activity of junctional myosin-II, but by the polarized flow of medial actomyosin pulses towards a specific junctional domain (Rauzi et al., 2010).

6.2 Planar cell polarity mechanisms dissected *in vivo*

The morphogenesis and function of an epithelial tissue relies on the precise arrangement of its constituent cells. Tissue patterning and organization during development depends on the establishment of concentration gradients of signaling molecules along the tissue that furthermore can lead to the formation of polarized structures in the plane of the epithelium (like hairs). This type of polarization of a field of cells is referred to as planar cell polarity (PCP), where the spatial information that organizes planar polarity is transmitted locally from one cell to the next (for review see Seifert & Mlodzik, 2007). Thus, epithelial tissues not only show an apico-basal polarity, but also a positional oriented appearance in the plane in order to generate polarized structures and to orient themselves in a directed fashion. An evolutionary conserved set of genes control establishment of planar polarity in flies and vertebrates, the core Frizzled/PCP (Fz/PCP) factors Flamingo (Fmi), Strabismus (Stbm), Dishevelled (Dsh), Diego (Dgo), Fat, Dachsous (Ds) and Prickle (Pk) (Jenny et al., 2005). The Fz/PCP factors polarize a field of cells along a specific axis. Local differences in Fz activity between neighboring cells provide directional information required for planar polarity, in other words a gradient of the morphogen is needed for correct PCP. As a result, wing hairs point away from the site of highest Fz activity in *Drosophila* (Adler et al., 1997).

Drosophila wing epithelial cells are irregularly arranged throughout most of development, but they become hexagonally packed shortly before hair formation. PCP proteins promote hexagonal packing in the *Drosophila* wing by polarizing membrane trafficking (Classen et al., 2005). Planar polarity arises during growth due to a cell flow that is triggered by tension arising from the wing hinge contraction. *In vivo* imaging of *Drosophila* wing epithelial cells expressing fluorescent polarity proteins Stbm and E-Cad led to these observations, that would be very difficult to dissect *in situ* or *ex vivo* (Aigouy et al., 2010).

6.3 Asymmetric cell division of epithelial cells dissected *in vivo*

Planar cell polarity proteins, amongst others, are also needed to align the mitotic spindle correctly in order to allow for oriented mitosis to occur. Live imaging of the spindle in the *Drosophila* dorsal thorax using fluorescent microtubule components have revealed the mechanism that keeps the spindle in the correct plane and correct apico-basal tilt. This positioning controls the correct asymmetric cell division needed in the mechanosensory organ precursor cell, which resides in the dorsal thorax. Mutations in either Fz/Dsh or the NuMa homolog Mud as well as mutations in Ric-8, a guanine nucleotide-exchange factor for heterotrimeric G proteins, result in the mis-orientation of the spindle during division and the subsequent mal-formation of the sensory organ (David et al., 2005, Ségalen et al., 2010). The alignment of the spindle during mitosis can only be captured using *in vivo* studies and the underlying mechanism would have stayed unknown without the use of *Drosophila* for *in vivo* imaging and the vast genetic techniques available for this model system.

7. Concluding remarks

Epithelial cell function relies in both vertebrates and invertebrates on the tight regulation of the underlying polarity protein machinery, which is highly conserved (Leibfried, 2009). This regulation has been analyzed in epithelial cells in cell culture, but an analysis in a whole living organism, integrating all cellular and environmental cues that an epithelium is exposed to, remains an interesting task, comprising the truth about epithelial function and the regulatory networks needed for it. The use of *Drosophila* as a model system allows us to better study epithelial establishment, maintenance and plasticity in the context of a whole organism. Furthermore, green-fluorescent-protein (GFP) and its derivates give us the opportunity to analyze epithelial cell function in a spatio-temporal manner by live imaging. Thus, *Drosophila* will also in the future help to better understand epithelial establishment, maintenance and plasticity thanks to today's microscopy imaging and manipulation techniques, the extensive genetic tools available and the feasibility to study the epithelium in the context of the whole organism.

8. References

Adler, P.N., Krasnow, R.E., & Liu, J. (1997). Tissue polarity points from cells that have higher Frizzled levels towards cells that have lower Frizzled levels. *Curr Biol*, Vol. 7, No.12, pp. 940-949, ISSN 0960-9822

Afonso, C., & Henrique, D. (2006). PAR3 acts as a molecular organizer to define the apical domain of chick neuroepithelial cells. *J Cell Sci*, Vol. 119, No.Pt 20, pp. 4293-4304, ISSN 0021-9533

Aigouy, B., Farhadifar, R., Staple, D.B., Sagner, A., Röper, J.C., Jülicher, F., & Eaton, S. (2010). Cell flow reorients the axis of planar polarity in the wing epithelium of Drosophila. *Cell*, Vol. 142, No.5, pp. 773-786, ISSN 1097-4172

Akhtar, N., & Hotchin, N.A. (2001). RAC1 regulates adherens junctions through endocytosis of E-cadherin. *Mol Biol Cell*, Vol. 12, No.4, pp. 847-862, ISSN 1059-1524

Benton, R., & St Johnston, D. (2003). Drosophila PAR-1 and 14-3-3 inhibit Bazooka/PAR-3 to establish complementary cortical domains in polarized cells. *Cell*, Vol. 115, No.6, pp. 691-704, ISSN 0092-8674

Betschinger, J., Mechtler, K., & Knoblich, J.A. (2003). The Par complex directs asymmetric cell division by phosphorylating the cytoskeletal protein Lgl. *Nature*, Vol. 422, No.6929, pp. 326-330, ISSN 0028-0836

Betschinger, J., Eisenhaber, F., & Knoblich, J.A. (2005). Phosphorylation-induced autoinhibition regulates the cytoskeletal protein Lethal (2) giant larvae. *Curr Biol*, Vol. 15, No.3, pp. 276-282, ISSN 0960-9822

Bialucha, C.U., Ferber, E.C., Pichaud, F., Peak-Chew, S.Y., & Fujita, Y. (2007). p32 is a novel mammalian Lgl binding protein that enhances the activity of protein kinase Czeta and regulates cell polarity. *J Cell Biol*, Vol. 178, No.4, pp. 575-581, ISSN 0021-9525

Bilder, D., & Perrimon, N. (2000). Localization of apical epithelial determinants by the basolateral PDZ protein Scribble. *Nature*, Vol. 403, No.6770, pp. 676-680, ISSN 0028-0836

Bilder, D., Li, M., & Perrimon, N. (2000). Cooperative regulation of cell polarity and growth by Drosophila tumor suppressors. *Science*, Vol. 289, No.5476, pp. 113-116, ISSN 0036-8075

Bilder, D., Schober, M., & Perrimon, N. (2003). Integrated activity of PDZ protein complexes regulates epithelial polarity. *Nat Cell Biol*, Vol. 5, No.1, pp. 53-58, ISSN 1465-7392

Capaldo, C.T., & Macara, I.G. (2007). Depletion of E-cadherin disrupts establishment but not maintenance of cell junctions in Madin-Darby canine kidney epithelial cells. *Mol Biol Cell*, Vol. 18, No.1, pp. 189-200, ISSN 1059-1524

Cavey, M., Rauzi, M., Lenne, P.F., & Lecuit, T. (2008). A two-tiered mechanism for stabilization and immobilization of E-cadherin. *Nature*, Vol. 453, No.7196, pp. 751-756, ISSN 1476-4687

Chen, X., & Macara, I.G. (2005). Par-3 controls tight junction assembly through the Rac exchange factor Tiam1. *Nat Cell Biol*, Vol. 7, No.3, pp. 262-269, ISSN 1465-7392

Chen, Y.H., Lu, Q., Goodenough, D.A., & Jeansonne, B. (2002). Nonreceptor tyrosine kinase c-Yes interacts with occludin during tight junction formation in canine kidney epithelial cells. *Mol Biol Cell*, Vol. 13, No.4, pp. 1227-1237, ISSN 1059-1524

Cheng, N.N., Kirby, C.M., & Kemphues, K.J. (1995). Control of cleavage spindle orientation in Caenorhabditis elegans: the role of the genes par-2 and par-3. *Genetics*, Vol. 139, No.2, pp. 549-559, ISSN 0016-6731

Classen, A.K., Anderson, K.I., Marois, E., & Eaton, S. (2005). Hexagonal packing of Drosophila wing epithelial cells by the planar cell polarity pathway. *Dev Cell*, Vol. 9, No.6, pp. 805-817, ISSN 1534-5807

David, N.B., Martin, C.A., Segalen, M., Rosenfeld, F., Schweisguth, F., & Bellaiche, Y. (2005). Drosophila Ric-8 regulates Galphai cortical localization to promote Galphai-dependent planar orientation of the mitotic spindle during asymmetric cell division. *Nature cell biology*, Vol. 7, No.11, pp. 1083-1090,

Drees, F., Pokutta, S., Yamada, S., Nelson, W.J., & Weis, W.I. (2005). Alpha-catenin is a molecular switch that binds E-cadherin-beta-catenin and regulates actin-filament assembly. *Cell*, Vol. 123, No.5, pp. 903-915, ISSN 0092-8674

Foe, V.E., & Alberts, B.M. (1983). Studies of nuclear and cytoplasmic behaviour during the five mitotic cycles that precede gastrulation in Drosophila embryogenesis. *J Cell Sci*, Vol. 61, pp. 31-70, ISSN 0021-9533

Georgiou, M., Marinari, E., Burden, J., & Baum, B. (2008). Cdc42, Par6, and aPKC regulate Arp2/3-mediated endocytosis to control local adherens junction stability. *Curr Biol*, Vol. 18, No.21, pp. 1631-1638, ISSN 0960-9822

Gumbiner, B., Stevenson, B., & Grimaldi, A. (1988). The role of the cell adhesion molecule uvomorulin in the formation and maintenance of the epithelial junctional complex. *J Cell Biol*, Vol. 107, No.4, pp. 1575-1587, ISSN 0021-9525

Harris, K.P., & Tepass, U. (2008). Cdc42 and Par proteins stabilize dynamic adherens junctions in the Drosophila neuroectoderm through regulation of apical endocytosis. *The Journal of cell biology*, Vol. 183, No.6, pp. 1129-1143, ISSN 1540-8140

Harris, T.J., & Peifer, M. (2005). The positioning and segregation of apical cues during epithelial polarity establishment in Drosophila. *J Cell Biol*, Vol. 170, No.5, pp. 813-823, ISSN 0021-9525

Hawley, S.A., Boudeau, J., Reid, J.L., Mustard, K.J., Udd, L., Makela, T.P., Alessi, D.R., & Hardie, D.G. (2003). Complexes between the LKB1 tumor suppressor, STRAD/and MO25/are upstream kinases in the AMP-activated protein kinase cascade. *J Biol*, Vol. 2, No.4, pp. 28-21,

Horne-Badovinac, S., & Bilder, D. (2008). Dynein regulates epithelial polarity and the apical localization of stardust A mRNA. *PLoS Genet*, Vol. 4, No.1e8) ISSN 1553-7404

Hunter, C., & Wieschaus, E. (2000). Regulated expression of nullo is required for the formation of distinct apical and basal adherens junctions in the Drosophila blastoderm. *J Cell Biol*, Vol. 150, No.2, pp. 391-401, ISSN 0021-9525

Hurov, J.B., Watkins, J.L., & Piwnica-Worms, H. (2004). Atypical PKC phosphorylates PAR-1 kinases to regulate localization and activity. *Curr Biol*, Vol. 14, No.8, pp. 736-741, ISSN 0960-9822

Hutterer, A., Betschinger, J., Petronczki, M., & Knoblich, J.A. (2004). Sequential roles of Cdc42, Par-6, aPKC, and Lgl in the establishment of epithelial polarity during Drosophila embryogenesis. *Dev Cell*, Vol. 6, No.6, pp. 845-854, ISSN 1534-5807

Ivanov, A.I., McCall, I.C., Parkos, C.A., & Nusrat, A. (2004). Role for actin filament turnover and a myosin II motor in cytoskeleton-driven disassembly of the epithelial apical junctional complex. *Mol Biol Cell*, Vol. 15, No.6, pp. 2639-2651, ISSN 1059-1524

Jenkins, N., Saam, J.R., & Mango, S.E. (2006). CYK-4/GAP provides a localized cue to initiate anteroposterior polarity upon fertilization. *Science (New York, N.Y.)*, Vol. 313, No.5791, pp. 1298-1301, ISSN 1095-9203

Jenny, A., Reynolds-Kenneally, J., Das, G., Burnett, M., & Mlodzik, M. (2005). Diego and Prickle regulate Frizzled planar cell polarity signalling by competing for Dishevelled binding. *Nat Cell Biol*, Vol. 7, No.7, pp. 691-697, ISSN 1465-7392

Joberty, G., Petersen, C., Gao, L., & Macara, I.G. (2000). The cell-polarity protein Par6 links Par3 and atypical protein kinase C to Cdc42. *Nat Cell Biol*, Vol. 2, No.8, pp. 531-539, ISSN 1465-7392

Kemphues, K.J., Priess, J.R., Morton, D.G., & Cheng, N.S. (1988). Identification of genes required for cytoplasmic localization in early C. elegans embryos. *Cell*, Vol. 52, No.3311) ISSN 1097-4172

Kirby, C., Kusch, M., & Kemphues, K. (1990). Mutations in the par genes of Caenorhabditis elegans affect cytoplasmic reorganization during the first cell cycle. *Dev Biol*, Vol. 142, No.1, pp. 203-215, ISSN 0012-1606

Knust, E., & Bossinger, O. (2002). Composition and formation of intercellular junctions in epithelial cells. *Science (New York, N.Y.)*, Vol. 298, No.5600, pp. 1955-1959,

Laplante, C., & Nilson, L.A. (2006). Differential expression of the adhesion molecule Echinoid drives epithelial morphogenesis in Drosophila. *Development*, Vol. 133, No.16, pp. 3255-3264, ISSN 0950-1991

Laprise, P., Beronja, S., Silva-Gagliardi, N.F., Pellikka, M., Jensen, A.M., McGlade, C.J., & Tepass, U. (2006). The FERM protein Yurt is a negative regulatory component of the Crumbs complex that controls epithelial polarity and apical membrane size. *Dev Cell*, Vol. 11, No.3, pp. 363-374, ISSN 1534-5807

Leibfried, A., Fricke, R., Morgan, M.J., Bogdan, S., & Bellaiche, Y. (2008). Drosophila Cip4 and WASp define a branch of the Cdc42-Par6-aPKC pathway regulating E-cadherin endocytosis. *Curr Biol*, Vol. 18, No.21, pp. 1639-1648, ISSN 0960-9822

Leibfried, A. (2009). *Polarized transport of DE-Cadherin in Drosophila epithelial cells* (doctoral thesis, Université Pierre et Marie Curie), Paris, France.

Lemmers, C., Michel, D., Lane-Guermonprez, L., Delgrossi, M.H., Médina, E., Arsanto, J.P., & Le Bivic, A. (2004). CRB3 binds directly to Par6 and regulates the morphogenesis of the tight junctions in mammalian epithelial cells. *Mol Biol Cell*, Vol. 15, No.3, pp. 1324-1333, ISSN 1059-1524

Li, F., Schiemann, A.H., & Scott, M.J. (2008). Incorporation of the noncoding roX RNAs alters the chromatin-binding specificity of the Drosophila MSL1/MSL2 complex. *Molecular and cellular biology*, Vol. 28, No.4, pp. 1252-1264, ISSN 1098-5549

Magie, C.R., Pinto-Santini, D., & Parkhurst, S.M. (2002). Rho1 interacts with p120ctn and alpha-catenin, and regulates cadherin-based adherens junction components in Drosophila. *Development*, Vol. 129, No.16, pp. 3771-3782, ISSN 0950-1991

Malacombe, M., Ceridono, M., Calco, V., Chasserot-Golaz, S., McPherson, P.S., Bader, M.F., & Gasman, S. (2006). Intersectin-1L nucleotide exchange factor regulates secretory granule exocytosis by activating Cdc42. *EMBO J*, Vol. 25, No.15, pp. 3494-3503, ISSN 0261-4189

Martin-Belmonte, F., Gassama, A., Datta, A., Yu, W., Rescher, U., Gerke, V., & Mostov, K. (2007). PTEN-mediated apical segregation of phosphoinositides controls epithelial morphogenesis through Cdc42. *Cell*, Vol. 128, No.2, pp. 383-397, ISSN 0092-8674

Mavrakis, M., Rikhy, R., & Lippincott-Schwartz, J. (2009). Plasma membrane polarity and compartmentalization are established before cellularization in the fly embryo. *Dev Cell*, Vol. 16, No.1, pp. 93-104, ISSN 1534-5807

Médina, E., Williams, J., Klipfell, E., Zarnescu, D., Thomas, G., & Le Bivic, A. (2002). Crumbs interacts with moesin and beta(Heavy)-spectrin in the apical membrane skeleton of Drosophila. *J Cell Biol*, Vol. 158, No.5, pp. 941-951, ISSN 0021-9525

Mertens, A.E., Rygiel, T.P., Olivo, C., van der Kammen, R., & Collard, J.G. (2005). The Rac activator Tiam1 controls tight junction biogenesis in keratinocytes through binding to and activation of the Par polarity complex. *J Cell Biol*, Vol. 170, No.7, pp. 1029-1037, ISSN 0021-9525

Michel, D., Arsanto, J.P., Massey-Harroche, D., Béclin, C., Wijnholds, J., & Le Bivic, A. (2005). PATJ connects and stabilizes apical and lateral components of tight junctions in human intestinal cells. *J Cell Sci*, Vol. 118, No.Pt 17, pp. 4049-4057, ISSN 0021-9533

Mirouse, V., Christoforou, C.P., Fritsch, C., St Johnston, D., & Ray, R.P. (2009). Dystroglycan and perlecan provide a basal cue required for epithelial polarity during energetic stress. *Developmental cell*, Vol. 16, No.1, pp. 83-92, ISSN 1878-1551

Nagai-Tamai, Y., Mizuno, K., Hirose, T., Suzuki, A., & Ohno, S. (2002). Regulated protein-protein interaction between aPKC and PAR-3 plays an essential role in the polarization of epithelial cells. *Genes Cells*, Vol. 7, No.11, pp. 1161-1171, ISSN 1356-9597

Nunbhakdi-Craig, V., Machleidt, T., Ogris, E., Bellotto, D., White, C.L., & Sontag, E. (2002). Protein phosphatase 2A associates with and regulates atypical PKC and the epithelial tight junction complex. *J Cell Biol*, Vol. 158, No.5, pp. 967-978, ISSN 0021-9525

O'Keefe, D.D., Prober, D.A., Moyle, P.S., Rickoll, W.L., & Edgar, B.A. (2007). Egfr/Ras signaling regulates DE-cadherin/Shotgun localization to control vein morphogenesis in the Drosophila wing. *Developmental biology*, Vol. 311, No.1, pp. 25-39, ISSN 1095-564X

Oda, H., Uemura, T., Harada, Y., Iwai, Y., & Takeichi, M. (1994). A Drosophila homolog of cadherin associated with armadillo and essential for embryonic cell-cell adhesion. *Developmental biology*, Vol. 165, No.2, pp. 716-726,

Ozdamar, B., Bose, R., Barrios-Rodiles, M., Wang, H.R., Zhang, Y., & Wrana, J.L. (2005). Regulation of the polarity protein Par6 by TGFbeta receptors controls epithelial cell plasticity. *Science (New York, N.Y.)*, Vol. 307, No.5715, pp. 1603-1609, ISSN 1095-9203

Pacquelet, A., Lin, L., & Rorth, P. (2003). Binding site for p120/delta-catenin is not required for Drosophila E-cadherin function in vivo. *J Cell Biol*, Vol. 160, No.3, pp. 313-319, ISSN 0021-9525

Parsons, M., Monypenny, J., Ameer-Beg, S.M., Millard, T.H., Machesky, L.M., Peter, M., Keppler, M.D., Schiavo, G., Watson, R., Chernoff, J., Zicha, D., Vojnovic, B., & Ng, T. (2005). Spatially distinct binding of Cdc42 to PAK1 and N-WASP in breast carcinoma cells. *Mol Cell Biol*, Vol. 25, No.5, pp. 1680-1695, ISSN 0270-7306

Pilot, F., Philippe, J.M., Lemmers, C., & Lecuit, T. (2006). Spatial control of actin organization at adherens junctions by a synaptotagmin-like protein Btsz. *Nature*, Vol. 442, No.7102, pp. 580-584, ISSN 1476-4687

Plant, P.J., Fawcett, J.P., Lin, D.C., Holdorf, A.D., Binns, K., Kulkarni, S., & Pawson, T. (2003). A polarity complex of mPar-6 and atypical PKC binds, phosphorylates and regulates mammalian Lgl. *Nat Cell Biol*, Vol. 5, No.4, pp. 301-308, ISSN 1465-7392

Qian, X., Karpova, T., Sheppard, A.M., McNally, J., & Lowy, D.R. (2004). E-cadherin-mediated adhesion inhibits ligand-dependent activation of diverse receptor tyrosine kinases. *EMBO J*, Vol. 23, No.8, pp. 1739-1748, ISSN 0261-4189

Rauzi, M., Lenne, P.F., & Lecuit, T. (2010). Planar polarized actomyosin contractile flows control epithelial junction remodelling. *Nature*, Vol. 468, No.7327, pp. 1110-1114, ISSN 1476-4687

Roh, M.H., Fan, S., Liu, C.J., & Margolis, B. (2003). The Crumbs3-Pals1 complex participates in the establishment of polarity in mammalian epithelial cells. *J Cell Sci*, Vol. 116, No.Pt 14, pp. 2895-2906, ISSN 0021-9533

Rolls, M.M., Albertson, R., Shih, H.P., Lee, C.Y., & Doe, C.Q. (2003). Drosophila aPKC regulates cell polarity and cell proliferation in neuroblasts and epithelia. *J Cell Biol*, Vol. 163, No.5, pp. 1089-1098, ISSN 0021-9525

Schejter, E.D., & Wieschaus, E. (1993). Functional elements of the cytoskeleton in the early Drosophila embryo. *Annual review of cell biology*, Vol. 9, pp. 67-99, ISSN 0743-4634

Ségalen, M., Johnston, C.A., Martin, C.A., Dumortier, J.G., Prehoda, K.E., David, N.B., Doe, C.Q., & Bellaïche, Y. (2010). The Fz-Dsh planar cell polarity pathway induces oriented cell division via Mud/NuMA in Drosophila and zebrafish. *Developmental cell*, Vol. 19, No.5, pp. 740-752, ISSN 1878-1551

Seifert, J.R., & Mlodzik, M. (2007). Frizzled/PCP signalling: a conserved mechanism regulating cell polarity and directed motility. *Nat Rev Genet*, Vol. 8, No.2, pp. 126-138, ISSN 1471-0056

Shen, Y., Hirsch, D.S., Sasiela, C.A., & Wu, W.J. (2008). Cdc42 Regulates E-cadherin Ubiquitination and Degradation through an Epidermal Growth Factor Receptor to Src-mediated Pathway. *J Biol Chem*, Vol. 283, No.8, pp. 5127-5137, ISSN 0021-9258

Shin, K., Straight, S., & Margolis, B. (2005). PATJ regulates tight junction formation and polarity in mammalian epithelial cells. *J Cell Biol*, Vol. 168, No.5, pp. 705-711, ISSN 0021-9525

Sotillos, S., Díaz-Meco, M.T., Caminero, E., Moscat, J., & Campuzano, S. (2004). DaPKC-dependent phosphorylation of Crumbs is required for epithelial cell polarity in Drosophila. *J Cell Biol*, Vol. 166, No.4, pp. 549-557, ISSN 0021-9525

Straight, S.W., Shin, K., Fogg, V.C., Fan, S., Liu, C.J., Roh, M., & Margolis, B. (2004). Loss of PALS1 expression leads to tight junction and polarity defects. *Mol Biol Cell*, Vol. 15, No.4, pp. 1981-1990, ISSN 1059-1524

Sturge, J., Wienke, D., & Isacke, C.M. (2006). Endosomes generate localized Rho-ROCK-MLC2-based contractile signals via Endo180 to promote adhesion disassembly. *J Cell Biol*, Vol. 175, No.2, pp. 337-347, ISSN 0021-9525

Suzuki, A., & Ohno, S. (2006). The PAR-aPKC system: lessons in polarity. *J Cell Sci*, Vol. 119, No.Pt 6, pp. 979-987, ISSN 0021-9533

Tanentzapf, G., Smith, C., McGlade, J., & Tepass, U. (2000). Apical, lateral, and basal polarization cues contribute to the development of the follicular epithelium during Drosophila oogenesis. *J Cell Biol*, Vol. 151, No.4, pp. 891-904, ISSN 0021-9525

Tanentzapf, G., & Tepass, U. (2003). Interactions between the crumbs, lethal giant larvae and bazooka pathways in epithelial polarization. *Nat Cell Biol*, Vol. 5, No.1, pp. 46-52, ISSN 1465-7392

Tepass, U., & Knust, E. (1993). Crumbs and stardust act in a genetic pathway that controls the organization of epithelia in Drosophila melanogaster. *Dev Biol*, Vol. 159, No.1, pp. 311-326, ISSN 0012-1606

Tepass, U., Tanentzapf, G., Ward, R., & Fehon, R. (2001). Epithelial cell polarity and cell junctions in Drosophila. *Annual review of genetics*, Vol. 35, pp. 747-784, ISSN 0066-4197

van Leeuwen, F.N., van Delft, S., Kain, H.E., van der Kammen, R.A., & Collard, J.G. (1999). Rac regulates phosphorylation of the myosin-II heavy chain, actinomyosin disassembly and cell spreading. *Nat Cell Biol*, Vol. 1, No.4, pp. 242-248, ISSN 1465-7392

von Stein, W., Ramrath, A., Grimm, A., Müller-Borg, M., & Wodarz, A. (2005). Direct association of Bazooka/PAR-3 with the lipid phosphatase PTEN reveals a link between the PAR/aPKC complex and phosphoinositide signaling. *Development*, Vol. 132, No.7, pp. 1675-1686, ISSN 0950-1991

Wang, C.W., Hamamoto, S., Orci, L., & Schekman, R. (2006). Exomer: A coat complex for transport of select membrane proteins from the trans-Golgi network to the plasma membrane in yeast. *J Cell Biol*, Vol. 174, No.7, pp. 973-983, ISSN 0021-9525

Wei, S.Y., Escudero, L.M., Yu, F., Chang, L.H., Chen, L.Y., Ho, Y.H., Lin, C.M., Chou, C.S., Chia, W., Modolell, J., & Hsu, J.C. (2005). Echinoid is a component of adherens junctions that cooperates with DE-Cadherin to mediate cell adhesion. *Dev Cell*, Vol. 8, No.4, pp. 493-504, ISSN 1534-5807

Wildenberg, G.A., Dohn, M.R., Carnahan, R.H., Davis, M.A., Lobdell, N.A., Settleman, J., & Reynolds, A.B. (2006). p120-catenin and p190RhoGAP regulate cell-cell adhesion by coordinating antagonism between Rac and Rho. *Cell*, Vol. 127, No.5, pp. 1027-1039, ISSN 0092-8674

Woods, D.F., & Bryant, P.J. (1991). The discs-large tumor suppressor gene of Drosophila encodes a guanylate kinase homolog localized at septate junctions. *Cell*, Vol. 66, No.3, pp. 451-464, ISSN 0092-8674

Xu, J., Lamouille, S., & Derynck, R. (2009). TGF-beta-induced epithelial to mesenchymal transition. *Cell research*, Vol. 19, No.2, pp. 156-172, ISSN 1748-7838

Yamada, S., Pokutta, S., Drees, F., Weis, W.I., & Nelson, W.J. (2005). Deconstructing the cadherin-catenin-actin complex. *Cell*, Vol. 123, No.5, pp. 889-901, ISSN 0092-8674

Yamanaka, T., & Ohno, S. (2008). Role of Lgl/Dlg/Scribble in the regulation of epithelial junction, polarity and growth. *Frontiers in bioscience: a journal and virtual library*, Vol. 13, pp. 6693-6707, ISSN 1093-4715

5

Development and Cell Polarity of the *C. elegans* Intestine

Olaf Bossinger[1] and Michael Hoffmann[2]
[1]*Institute of Molecular and Cellular Anatomy (MOCA),
RWTH Aachen University, Aachen,*
[2]*Department of General Pediatrics, University Children's Hospital,
Heinrich-Heine-University Düsseldorf, Düsseldorf,
Germany*

1. Introduction

1.1 The nematode *C. elegans* as a model organism

Much of our knowledge on development of multicellular organisms and the underlying cellular and molecular processes is derived from the studies of model organisms, like *C. elegans*, *Drosophila*, *Xenopus*, zebrafish and mouse. These model organisms were selected based on their amenability to experimental studies.

In 1963, Sydney Brenner realized that "Part of the success of molecular genetics was due to the use of extremely simple organisms which could be handled in large numbers: bacteria and bacterial viruses." He further argued "…..that the future of molecular biology lies in the extension of research to other fields of biology, notably development and the nervous system". Thus, he proposed to the Medical Research Council: "we want a multicellular organism which has a short life cycle, can be easily cultivated, and is small enough to be handled in large numbers, like a micro-organism. It should have relatively few cells, so that exhaustive studies of lineage and patterns can be made, and should be amenable to genetic analysis.

We think we have a good candidate in the form of a small nematode worm, Caenorhabditis……" (cited after: Wood, 1988).

C. elegans genetics started in October 1967 with Sydney Brenner's first mutant hunt, which produced two mutants showing a "dumpy" and a "variable abnormal" phenotype (Brenner, 2009). In 1974, the article entitled "The genetics of *Caenorhabditis elegans*" (Brenner, 1974) reported a study of 300 EMS-induced mutants and a map of about 100 genes on six linkage groups, which provided an excellent starting point for future *C. elegans* research.

Since that time many key steps towards the total description of *C. elegans* have been undertaken:

- complete description of cellular development (cell lineage, Fig.2) from egg to adult (Sulston and Horvitz, 1977; Sulston et al., 1983)
- complete description of the nervous system: all branches and connections determined (White et al., 1986)

- first use of green fluorescent protein as a marker for gene expression in a multicellular organism (Chalfie et al., 1994; Hunt-Newbury et al., 2007)
- first draft genome sequence of a multicellular organism completed (The_C_elegans_Sequencing_Consortium, 1998)
- basic mechanism of double-stranded(ds) RNA-mediated interference worked out (Fire et al., 1998)
- nearly all predicted genes tested for function by RNAi (Fraser et al., 2000; Gönczy et al., 2000; Kamath et al., 2003)
- comprehensive databases on *WormBase* (http://www.wormbase.org) (Harris et al., 2010), *Wormatlas* (http://www.wormatlas.org) and *WormBook* (http://www.wormbook.org).

In the 1990s, the popularity of *C. elegans* climbed sharply, as indicated by the increase in the number of research publications per year. Thirteen and 744 research articles were published in 1974 and 2009, respectively (Han, 2010). Over the past decade, research on the nematode *C. elegans* was granted three Nobel prizes for groundbreaking discoveries such as programmed cell death (apoptosis), dsRNA-mediated interference and the use of the green fluorescent protein. The Nobel prize for Physiology or Medicine went to H. Robert Horvitz, John Sulston and Sydney Brenner in 2002 (Brenner, 2003; Horvitz, 2003; Sulston, 2003) and to Andrew Fire and Craig Mello in 2006 (Fire, 2007; Mello, 2007). The Nobel prize for Chemistry went to Martin Chalfie (with Osamu Shimamura and Roger Tsien) in 2008 (Chalfie, 2009; Tsien, 2009).

Caenorhabditis elegans is a small, free-living nematode (Blaxter, 2011) that survives by feeding primarily on bacteria. In the laboratory *C. elegans* normally grows at temperatures between 12 °C and 26 °C on agar plates, which are seeded with *E. coli* bacteria as a food source (Fig.1A). The animals can also be grown in liquid culture for biochemical analyses. Starved worm cultures retain their viability for months and strains can be frozen and stored at -80 °C or lower (http://www.cbs.umn.edu/CGC/). Such frozen stocks are stable for > 40 years. *C. elegans* is an important model system for biological research in many fields including genomics, cell biology, neuroscience and aging (http://www.wormbook.org). Among its many advantages for study are its short life cycle (Fig.1B), compact genome (100 x 10^6 base pairs, Fig.1C), invariance in cell number and anatomy, ease of propagation and small size. The simplicity and invariance permit complete and exhaustive descriptions. There are two *C. elegans* sexes: a self-fertilizing hermaphrodite (Fig.1A) and a male. The adult body plan is anatomically simple with about 1031 and 959 somatic cells in hermaphrodites and males, respectively. The *C. elegans* hermaphrodite produces a large number of progeny per adult (> 200) and is amenable to genetic crosses. *C. elegans* can be examined at the cellular level *in vivo* by Nomarski differential interference contrast microscopy, because it is transparent throughout its life cycle. The life cycle is temperature dependent and by a temperature shift from 16 °C to 25 °C the time needed for development can be accelarated about 100 % (Fig.1B).

Since 1974, when Sydney Brenner published his pioneering genetic screen (Brenner, 1974), researchers have developed increasingly powerful methods for identifying genes and genetic pathways in *C. elegans* (Jorgensen and Mango, 2002). The long history of *C. elegans* as a genetic model organism means that there are a large number of mutants available. The *C. elegans* Genetics Center (CGC) houses the community collection of *C. elegans* mutant strains and related nematode strains (http://www.cbs.umn.edu/CGC/). Due to the efforts of the *C. elegans* Gene Knockout Consortium (http://www.celeganskoconsortium.omrf.org/) in the United States and Canada and the National BioResource Project in Japan, deletion alleles have been obtained for about 5,500 out of 20,000 predicted genes (Mitani, 2009; Moerman and Barstead, 2008).

Working with existing mutants can be advantageous for several reasons: **First**, temperature-sensitive conditional alleles allow the analysis of otherwise lethal mutations. They may also provide a way of analyzing gene function during a specific developmental process. **Second**, genetic mutants avoid inconsistencies sometimes observed in RNAi phenotypes that may arise from variability in the bacterial expression of dsRNA or from the amount of bacteria ingested by the worm strain used. **Third**, genetic alleles may encode partially functional proteins or gain-of-function gene products, thus providing additional information about the structure-function features of the gene product.

To further analyze the function of a gene product, it is often helpful to have a complete loss-of-function allele. If such a mutant is not available, there are three knock-out consortiums (see above) that are generating large collections of deletion alleles for the *C. elegans* community. If a knock-out of your gene-of-interest does not exist, one can request a new screen through the websites. With new approaches to generate targeted deletion mutants and to control gene expression the arsenal of methods to investigate gene functions in *C. elegans* is growing (Boulin and Bessereau, 2007; Calixto et al., 2010b; Frokjaer-Jensen et al., 2010; Robert and Bessereau, 2007).

Obtaining strains containing heritable null mutations in every gene (see above) is complementary to RNAi, a so-called reverse genetics approach (Baylis and Vazquez-Manrique, 2011). RNAi in *C. elegans* (Fig.3) was first described in the 1990s (Guo and Kemphues, 1996) and quickly became an important laboratory tool for investigating gene function. RNAi is easily achieved in the worm and the availability of the genome sequence (The_C_elegans_Sequencing_Consortium, 1998) helped to make RNAi the reverse genetics tool of choice, particularly for genome-wide studies of developmental processes (Fraser et al., 2000; Gönczy et al., 2000; Kamath et al., 2003; Sönnichsen et al., 2005). The effectiveness of RNAi in *C. elegans* is even maintained during spaceflight (Etheridge et al., 2011). RNAi seems to be an evolutionary conserved cellular response to dsRNA, and the mechanism is thought to originate from an ancient endogenous defense mechanism against viral and other heterologous dsRNAs (Lu et al., 2005; Schott et al., 2005; Wilkins et al., 2005). In mammalian cells, introduction of dsRNAs longer than 30 bp activates antiviral pathways, leading to nonspecific inhibition of translation and cytotoxic responses.

To inactivate gene expression in early *C. elegans* embryos and to analyze the resulting phenotype, worms can e.g. be fed bacteria expressing dsRNA corresponding to the gene of interest (Fig.3). Because the adult hermaphrodite continuously produces oocytes, pre-existing mRNA is eliminated with each egg that is laid. Embryos born early after the initiation of RNAi are only mildly depleted of the gene product whereas embryos born later are usually highly depleted. The time required for efficient depletion varies among target genes, but generally 24 - 30 hours after the initiation of feeding, mRNA levels are reduced significantly, protein levels are almost undetectable and phenotypes are visible.

A problem often arises when looking for phenotypes by RNAi in later embryogenesis. If the gene product of interest is involved in a developmental process prior to the one to be observed or in multiple cell types, making specificity of the phenotype unclear. Worm strains that are sensitive to RNAi only in a particular tissue have now been generated (Calixto et al., 2010a; McGhee et al., 2009; Pilipiuk et al., 2009; Qadota et al., 2007). One strategy relies on a genetic background that is resistant to RNAi due to a mutation in an essential RNA processing protein, e.g. RDE-1 (Fig.3) and complementation in the tissue of interest by tissue-specific promoter induction of wild-type protein. Tissue-specific RNAi largely circumvents the problems mentioned but does rely on having promoters that turn on early enough in the tissue to have sufficient depletion by the developmental stage of interest. Nevertheless, RNAi has a few intrinsic limitations. First, RNAi efficiency is sensitive to the experimental conditions, and the result can be variable. Second, residual gene expression persists to an extent that is difficult to predict for a given gene. Third, some tissues are partially resistant to RNAi (Zhuang and Hunter, 2011).

In summary, the discovery of RNAi has led to a much greater reliance on the reverse genetics approach but with the advent of next-generation DNA sequencing technologies and the ensuing ease of whole-genome sequencing are reviving the use of classical genetics to investigate *C. elegans* development (Bowerman, 2011; Hobert, 2010).

1.2 Introduction to epithelial tissues

Epithelia are polarized tissues (Fig.4A) that outline the cavities (e.g. the digestive tract) and surfaces (e.g. the epidermis, Fig.4B-C) of the body (de Santa Barbara et al., 2003; Fuchs, 2007; Noah et al., 2011). They are specialized for secretion, absorption, protection or sensory functions. Polarization of epithelial cells is manifested by distinct apical and basolateral membrane domains, which are separated by cell junctions that form belt-like structures around the apex of the cells (Fig.4A; Knust and Bossinger, 2002; Nelson, 2003; Nelson, 2009; Weisz and Rodriguez-Boulan, 2009). Epithelial cell junctions serve the adhesion, communication, vectorial transport, and morphogenetic properties of epithelia. Two of the most important features for the functions of epithelia are to create a diffusion barrier between two biological compartments and to build a cell adhesion system between their cells. Cell-cell adhesion is regulated by cell-specific mechanical and biochemical constraints. For instance, fibroblasts and neuronal cells are involved in more labile and plastic interactions, whereas endothelial and epithelial cells require a strong adhesion.

During the process of epithelial polarization the organization and maintenance of the boundary between apical and basolateral membranes must be regulated. In vertebrate epithelia, this fence function is established by a specific intercellular junction, the tight junction (TJ; Anderson and Van Itallie, 2009; Ebnet, 2008; Eckert and Fleming, 2008; Tsukita et al., 2001). TJs are the most apical cell junction in vertebrate epithelia and lie adjacent to the more basally localized zonula adherens (ZA; Harris and Tepass, 2010; Wang and Margolis, 2007). TJs provide a fence to lateral diffusion of membrane proteins and a barrier to the diffusion of molecules in between the individual epithelial cells. In invertebrates, TJs have not been found thus far. However, a region just apical to the ZA in *Drosophila* epithelia harbors a probably larger protein complex, called the subapical region (SAR; Bulgakova and Knust, 2009). It has been suggested that one of the functions of this protein complex is the fence function of vertebrate TJs (Müller, 2000; Wodarz et al., 2000). In many invertebrate epithelia the paracellular transport through the epithelium is controlled by a unique invertebrate structure, the septate junction (SJ; Müller and Bossinger, 2003). In the nematode *C. elegans* SJ (Lints and Hall, 2009) have thus far only been found in the spermatheca epithelium (Pilipiuk et al., 2009), raising the interesting question as to how embryonic epithelia in these animals maintain a diffusion barrier. Claudins with four transmembrane domains are major cell adhesion molecules working at TJs in vertebrates. In *C. elegans* four claudin-related proteins (CLC-1 to -4) exist and two of them, CLC-1 and CLC-2, seem to be involved in the pharynx and epidermis barrier, respectively (Asano et al., 2003).

2. Development and differentiation of the *C. elegans* embryonic intestine

The *C. elegans* digestive tract is one of the most complex portions of the nematode anatomy and is composed of a large variety of tissues and cell types (Altun and Hall, 2009c; Bird and Bird, 1991; Kormish et al., 2010; White, 1988). It forms a separate epithelial tube running inside the cylindrical body wall, separated from it by the pseudocoelomic body cavity, and placed parallel to the gonad. The *C. elegans* digestive tract is divided into the foregut (stomodeum; buccal cavity and the pharynx; Altun and Hall, 2009d; Mango, 2007), the midgut (intestine; Altun and Hall, 2009b; McGhee, 2007), and the hindgut (proctodeum; rectum and anus in hermaphrodites and cloaca in males; Altun and Hall, 2009a) and contains a total of 127 cells (Schnabel et al., 1997; Sulston et al., 1983). In comparison to human digestive tracts, it lacks both an intestine-sheathing innervated muscle layer and a renewable/regenerating stem cell population. In *C. elegans*, ingested *E. coli* bacteria flow through the digestive tract by the muscular pumping and peristalsis of the pharynx at the anterior end, and the waste material is discarded through the opening of the anus at the posterior end by the action of the enteric muscles. Developmentally, the intestine (midgut) is endodermal in origin, deriving clonally from the E-lineage whereas the foregut and hindgut have a mixed lineage from ectodermal and mesodermal origins (Fig.2).

The *C. elegans* intestine is a large organ (~ 1/3 of the somatic tissue) that carries out multiple functions executed by distinct organs in higher eukaryotes (McGhee, 2007): digestion of food, absorption of processed nutrients, synthesis and storage of macromolecules, nurturing of oocytes by producing yolk, and initiation of an innate immune response to pathogens (Kimble and Sharrock, 1983; Schulenburg et al., 2004). Remarkably, despite a large increase in tissue volume during larval and adult development (Fig.1B), the intestine continues to grow without further cell or nuclei divisions. Intestinal cells become binucleate and polyploid during post-embryonic development. By the adult stage, the intestine is composed of only 20 (Fig.5A-E) cells with a total of 30-34 nuclei, which have increased their ploidy to 32C (Hedgecock and White, 1985; Sulston and Horvitz, 1977). Age-related changes in the intestine include the loss of critical nuclei, the degradation of intestinal microvilli, and changes in the size, shape, and cytoplasmic contents of the intestine (McGee et al., 2011).

The intestinal epithelium consists of 20 cells that are mostly positioned as bilaterally symmetric pairs to form a long tube around a lumen. Each of these cell pairs forms an intestinal ring (II-IX int rings). The anteriormost intestinal ring (int ring I) is an exception and is comprised of four cells (Fig.5E Leung et al., 1999; Sulston et al., 1983). The intestine is composed of large cells, with distinct apical, lateral and basal membrane domains. Each intestinal cell forms part of the intestinal lumen at its apical pole (Fig.5E'-E'') and contains a basal lamina at its basal pole (Kramer, 2005), whose constituents are either made by the intestine itself (laminin α and β nidogen/entactin) or by the muscle and somatic gonad (type IV collagen). Many microvilli extend into the lumen from the apical surface, forming a brush border. The microvilli are anchored into a strong cytoskeletal network of intermediate filaments at their base, called the terminal web. The core of each microvillus has a bundle of actin filaments that connects to this web (Bossinger et al., 2004; Carberry et al., 2009; Hüsken et al., 2008; MacQueen et al., 2005). Each intestinal cell is sealed laterally to its neighbors by large apical adherens junctions and connects to the neighboring intestinal cells via gap junctions on the lateral sides (Altun et al., 2009; Bossinger and Schierenberg, 1992; Cox and Hardin, 2004; Hardin and Lockwood, 2004; Labouesse, 2006; Michaux et al., 2001).

The molecular and cellular events that lead to the formation of the intestinal epithelial tube have been described and reviewed in great detail elsewhere. In brief, these events include the correct specification and asymmetric division of the intestinal founder cell EMS (Bossinger and Schierenberg, 1996; Goldstein, 1992; Han, 1997; Kormish et al., 2010; Schierenberg, 1987; see Fig.2 for further details), the ingression of the intestinal precursor cell Ea and Ep during gastrulation (Fig.5B-C; Chisholm, 2006; Putzke and Rothman, 2003; Rohrschneider and Nance, 2009; Sawyer et al., 2009; Schierenberg, 2005; Schierenberg, 2006), the cytoplasmic polarization of intestinal primordial cells (Fig.5D; Achilleos et al., 2010; Bossinger et al., 2001; Leung et al., 1999; Totong et al., 2007), the formation of apical adherens junction and the generation of the future lumen within the primordium (Fig.5E'-E"; Leung et al., 1999), the intercalation of specific sets of cells (Hoffmann et al., 2010; Leung et al., 1999), the invariant 'twist' in the anterior of the intestinal primordium (Hermann et al., 2000), and finally the differentiation of the late embryonic, larval and adult intestine that has been proposed to be under the control of the GATA-factor ELT-2 (McGhee et al., 2009; McGhee et al., 2007; Pauli et al., 2006).

3. Apicobasal polarity complexes in the *C. elegans* intestine

From genetic studies on *Drosophila* ectoderm and mammalian culture cells, it appears that at least four spatially restricted membrane associated protein-scaffolds are required for regulating the maturation of the ZA in epithelial cells: the PAR-3–PAR-6–aPKC (PPC) complex, the Crumbs–Stardust–Patj complex, the Scribble–Dlg–Lgl complex, and the Yurt–Coracle group (Betschinger et al., 2003; Bilder et al., 2003; Harris and Peifer, 2005; Harris and Peifer, 2007; Krahn et al., 2010a; Krahn et al., 2010b; Laprise et al., 2009; Plant et al., 2003; Tanentzapf and Tepass, 2003; Yamanaka et al., 2003).

In the *C. elegans* embryo, a single electron-dense structure, the "*C. elegans* apical junction" (CeAJ, McMahon et al., 2001), is a prerequisite for correct epithelial cell functions (Cox and Hardin, 2004; Labouesse, 2006; Lynch and Hardin, 2009; Michaux et al., 2001; Müller and Bossinger, 2003). The CeAJ is a belt-like junctional structure that encircles the apex of polarized epithelial cells and resembles the ZA in other systems. By immunohistochemistry, the apicolateral membrane domain can be subdivided into four subdomains (Fig.6): the PPC together with the *Drosophila* Crumbs homolog CRB-1 and the multi PDZ-domain containing protein MAGI-1 (Achilleos et al., 2010; Aono et al., 2004; Bossinger et al., 2001; Stetak and Hajnal, 2011; Totong et al., 2007), the catenin–cadherin complex (CCC; Costa et al., 1998; Grana et al., 2010; Kwiatkowski et al., 2010), the DLG-1–AJM-1 complex (DAC; Bossinger et al., 2001; Firestein and Rongo, 2001; Köppen et al., 2001; Lockwood et al., 2008; McMahon et al., 2001) and the LET-413 protein (Bossinger et al., 2004; Legouis et al., 2000; Legouis et al., 2003; Lockwood et al., 2008; Pilipiuk et al., 2009; Segbert et al., 2004).

Epithelial polarization of the *C. elegans* intestine can be subdivided into three processes, **first** the appearance of junctional complexes, i.e. the CCC and DAC (Köppen et al., 2001;

Kwiatkowski et al., 2010; Lockwood et al., 2008) at the future apical pole (Achilleos et al., 2010), **second** the assembly of a junctional belt around the apex of epithelial cells (Totong et al., 2007), **third** and **fourth** the maintenance of epithelial cell polarity (Bossinger et al., 2004; Legouis et al., 2000) and cell-cell adhesion (Segbert et al., 2004; van Fürden et al., 2004).

4. Targeting of junctional complexes

At the end of the *C. elegans* proliferation phase, when the intestinal primordium consists of 16, so-called E-cells (E[16], Fig.5D), foci of the CCC and DAC accumulate at the apical surface (Fig.7A-C) under the control of *par-3* and *let-413* gene functions, respectively (Fig.1C-E; Achilleos et al., 2010; Legouis et al., 2000). In very elegant experiments, a targeted protein degradation strategy was used to remove both maternal and zygotic PAR-3 (*par-3M/Z*) from *C. elegans* embryos before epithelial polarization starts (Achilleos et al., 2010; Totong et al., 2007).

While localization of the CCC is mainly PAR-3 regulated, the DAC is under control of PAR-3 and LET-413. Interestingly, apical but not basolateral localization of LET-413 in intestinal primordial cells seems to be PAR-3 dependent too (Achilleos et al., 2010), suggesting that PAR-3 presumably acts via LET-413 to promote apical targeting of the DAC (Fig.7I). Consistent with this idea in *let-413(RNAi)* embryos the DAC reaches its apical position less efficiently (compare Figs.1E and 1F; Köppen et al., 2001; Legouis et al., 2000; McMahon et al., 2001; Segbert et al., 2004), a phenotype reminescent of embryos depleted for maternal and zygotic PAR-3 (Achilleos et al., 2010).

Using RNAi to deplete PAR-3 and LET-413 in developing larvae of *C. elegans*, Aono et al. (2004) and Pilipiuk et al. (2009) only discovered a requirement for these proteins in spermathecal development but not in other epithelia. Spermathecal precursor cells are born during larval development and differentiate into an epithelial tube for the storage of sperm. In PAR-3 and LET-413–depleted worms, the distribution of the DAC and apical microfilaments are severely affected in spermathecal cells, suggesting that the primary defect is in the organization of the apical domain.

How PAR-3 and LET-413 become localized apically in intestinal primordial and spermathecal cells is not known. In *Drosophila* membrane targeting of Bazooka/PAR-3 is mediated by direct binding to phosphoinositide lipids (Krahn et al., 2010b). Recent deletion and point mutation analyses of three LAP proteins, using *C. elegans* LET-413, human Erbin and human Scribble demonstrate that their LRR domain is crucial for membrane targeting (Legouis et al., 2003). Importantly, functional studies of LET-413 in *C. elegans* show that the LRR domain but not the PDZ domain is necessary for LET-413 to function during embryogenesis (Legouis et al., 2003).

5. Assembly of the junctional belt

During the early morphogenesis phase of *C. elegans*, the assembly of junctional complexes into an adhesive belt encircling the apex of epithelial cells (Figs.5E'',7F) depends on LET-413, DLG-1 and PAR-6 gene functions (black arrows in Fig.7I). In mid-morphogenesis of *let-413(RNAi)* embryos, long stretches of normal DAC localization form at the subapical cortex of epithelial cells, which are separated intermittently by gaps completely lacking DAC (Legouis et al., 2000). In contrast, the AJM-1 pattern in DLG-1 depleted embryos is characterized by small aggregates separated by large regions in which AJM-1 is almost completely missing (Bossinger et al., 2001; McMahon et al., 2001). In *let-414;dlg-1(RNAi)* embryos AJM-1 localization is nearly completely abolished (Köppen et al., 2001).

The N-terminal leucine-rich repeats of LET-413, which mediate basolateral localization, show good similarity with the Ras-interacting protein SUR-8 (Legouis et al., 2003). Among the small GTPase families, the Rab proteins are well known for their role in vesicle trafficking (Jordens et al., 2005) and it has been postulated that many charcteristics of LET-413 qualify this protein for acting as a docking platform in a trafficking pathway, which is controlled by small GTPases and ensures assembly of the CeAJ (Legouis et al., 2000).

For several reasons, and consistent with data from cell culture (see above), we do not favor the F-actin network as a major player in early steps of CeAJ biogenesis. First, *C. elegans* mutants defective in components of the CCC show severe defects in actin filament bundling without interfering with the formation of an adhesive junctional belt (Costa et al., 1998). Second, depletion of ERM-1, the only Ezrin-Radixin-Moesin homolog in *C. elegans*, almost completely abolishes establishment of the F-actin network in the apical cortex. Nevertheless, the CeAJ continuously forms around the apex of intestinal cells (van Fürden et al., 2004). Third, both described phenotypes are quite different from *let-413/dlg-1* induced defects, in which clustering of CeAJ proteins becomes the predominant phenotype (Bossinger et al., 2001).

There are nine α-tubulins (TBA-1-9) and six β-tubulins (TBB-1-6) in the *C. elegans* genome. Microtubules (MTs) are oriented circumferentially in dorsal and ventral epidermal cells, but are less well-organized in lateral seam cells (Costa et al., 1998). During organogenesis of the *C. elegans* intestine, MTs are concentrated near the apical cortex, where they appear to emerge in a fountain-like array and extend along the lateral surfaces of the cells. Numerous MTs are in the vicinity of the centrosomes, suggesting that there might be a MT organizing center at the apical cortex (Leung et al., 1999). By contrast, in many other epithelial cells most MTs are noncentrosomal and align along the apicobasal polarity axis. They create asymmetry by orienting their minus- and plus-ends towards the apical and basal membrane domains, respectively (Bacallao et al., 1989; Bre et al., 1990).

The polarized MT cytoskeleton in the *C. elegans* embryonic intestine is ideally suited to transport vesicles from the basally located Golgi toward the apical surface (Leung et al., 1999). During *Drosophila* cellularization, strong MT nucleation from apical centrosomes is likely necessary for the assembly of lateral MTs that promote the apical transport of lipids/proteins to form cell membranes and the initial apical positioning of AJs (Harris and Peifer, 2005; Lecuit and Wieschaus, 2000; Papoulas et al., 2005). In the *C. elegans* intestine, centrosomal MTs might also help direct the symmetric positioning of the CeAJ around the subapical domain. MT motors have been previously implicated in AJ assembly. For example, dynein interacts with β-catenin and may tether MTs to AJs assembling between cultured epithelial cells (Ligon et al., 2001). Kinesin transports AJs proteins to nascent AJs in cell culture (Chen et al., 2003; Mary et al., 2002) and MKLP-1/ZEN-4 is required for apical targeting of AJM-1 in the *C. elegans* pharynx epithelium (Portereiko et al., 2004). During early epithelial development in *Drosophila* positioning of Bazooka/PAR-3 relies on cytoskeletal cues, including an apical scaffold and dynein-mediated basal-to-apical transport (Harris and Peifer, 2005).

The similarity of *let-413* and *dlg-1* phenotypes and the fact that many CeAJ proteins show comparable phenotypes after depletion of LET-413 and DLG-1 is remarkable. These observations suggest that both proteins might somehow control the release of vesicles from MTs, either by providing a docking platform as discussed for LET-413 (see above) or by directly interacting with motor proteins. In *Drosophila* neuroblasts, Discs large, kinesin Khc-73, and astral MTs induce cortical polarization of Pins/Gαi. Khc-73 localizes to astral MT plus ends, and Dlg/Khc-73 and Dlg/Pins coimmunoprecipitate, suggesting that MTs induce Pins/Gαi cortical polarity through Dlg/Khc-73 interactions (Siegrist and Doe, 2005). In *C. elegans*, the clustering of CeAJ proteins after interfering with *let-413* and *dlg-1* gene functions would then indicate a jam in vesicular trafficking.

6. Maintenance of epithelial cell polarity

During late morphogenesis of *let-413* mutant or RNAi embryos, apical membrane markers in the epidermis as well as in the intestine progressively spread into the lateral membrane, suggesting that LET-413 acts to maintain polarity (Bossinger et al., 2004; Köppen et al., 2001; McMahon et al., 2001).

Surprisingly, worms treated with *let-413(RNAi)* during larval and adult life are sterile and exhibit spermathecal defects but otherwise develop normally, suggesting that depletion of LET-413 level does not restrict the function of major epithelia, like the pharynx, the intestine, or the hypodermis (Pilipiuk et al., 2009). How this function is maintained during post-embryonic development in *C. elegans* remains puzzling and might depend upon so far unidentified proteins that either completely replace LET-413 function or act redundantly.

7. Maintenance of cell-cell adhesion

During *C. elegans* morphogenesis, only double-knockdowns, e.g. HMR-1/E-cadherin + SAX-7/L1CAM (Hoffmann et al., in preparation), HMP-1/α-catenin + DLG-1 (Segbert et al., 2004), or HMR-1/cadherin + ERM-1 (van Fürden et al., 2004) give rise to intestinal cell-cell adhesion defects. HMR-1/E-cadherin and SAX-7/L1CAM also function redundantly in blastomere compaction and non-muscle myosin accumulation during *C. elegans* gastrulation (Grana et al., 2010). Interestingly, early embryonic and epithelial cells lacking PAR-6 can separate from one another inappropriately (Nance, 2003; Totong et al., 2007). Hence, PAR-6 seems to function reiteratively to control cell-cell adhesion in the *C. elegans* embryo. While *par-6* gene function clearly interferes with the correct localization of the CCC and DAC in intestinal primordial cells (Totong et al., 2007) this relationship still has to be demonstrated for early embryogenesis. The enhancement of hypodermal defects through functional loss of the DAC in mutations of *vab-9* (encoding a claudin homolog orthologous to human brain cell membrane protein 1; Simske et al., 2003) is another example of functional redundancy concerning cell-cell adhesion in the *C. elegans* embryo.

In summary, these genetic data suggest that cell-cell adhesion in the intestine is regulated by at least two redundant systems, which both act at the level of cell adhesion molecules, linker proteins and cytoskeletal organizers.

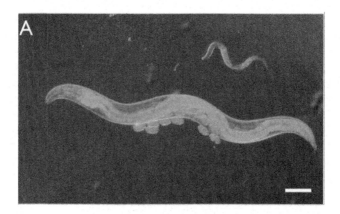

B. Developmental time (h) and length (μm) of *C. elegans* larva (L1-L4) and adult at different temperatures (°C)

STAGE	25°C	20°C	16°C
egg laid	0 h / 55 μm	0 h / 55 μm	0 h / 55 μm
egg hatches (L1)	8-9 h / 250 μm	10-12 h / 250 μm	16-18 h / 250 μm
first molt (L2)	18 h / 380 μm	26 h / 370 μm	36.5 h / 360 μm
second molt (L3)	25.5 h / 510 μm	34.5 h / 480 μm	48 h / 490 μm
third molt (L4)	31 h / 620 μm	43.5 h / 640 μm	60 h / 650 μm
fourth molt (young adult)	39 h / 940 μm	56 h / 850 μm	75 h / 900 μm
egg-laying begins (adult)	~ 47 h / 1110 μm	~ 65 h / 1060 μm	~90 h / 1150 μm
egg-laying ends (old adult)	~ 88 h	~ 96 h	~ 180 h

based on Byerly et al., 1976

C. *C. elegans* genome

Base pairs:	100,267,233
Coding Sequences: (11,068,632 base pairs)	25244 (100%) (20470 from protein-coding genes)
Confirmed:	12052 (47.7%)
Partially confirmed:	11172 (44.3%)
Predicted:	2020 (8.0%)

(based on http://wiki.wormbase.org/index.php/WS227)

Fig. 1. *Caenorhabditis elegans* development and genome

(A) Shows a DIC micrograph of a *C. elegans* larva (top) an adult hermaphrodite (middle) and embryos (bottom) maintained on agar plates with *E. coli* as food source (scale bar: 100 μm). (B) The table summarizes the developmental time (in hours) of *C. elegans* at different temperatures (°C), starting with the eggs released from the mother's uterus (0 h), completing embryogenesis (8-18 h), passing through four larval stages (L1-L4) and finally reaching adulthood (47-90 h). The length of the egg, larva and adult at each stage is given in micrometers (μm). (C) provides a short summary of the *C. elegans* genome (The_C_elegans_Sequencing_Consortium, 1998) that contains 100,267,633 base pairs and is estimated to have 25244 coding sequences (CDS) from which 47.7% have been confirmed (every base of every exon has transcription evidence). 44.3% CDS are partially confirmed (some, but not all exon bases are covered) and 8.0% CDS show no transcriptional evidence at all. Recent meta-analysis of results from four orthology prediction programs has yielded a set of 7633 *C. elegans* genes ("OrthoList") having human orthologs (Shaye and Greenwald, 2011).

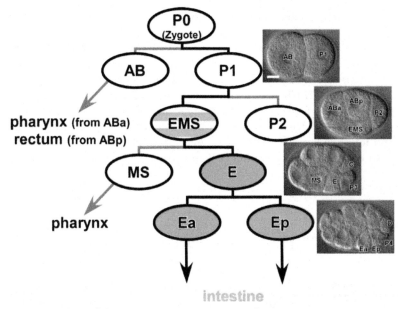

Fig. 2. Early cell lineage of *C. elegans*

The *C. elegans* one-cell embryo, also called zgote or P0, is a widely studied model of cell polarity (summarized in, Cowan and Hyman, 2004b; Gönczy, 2008; Nance and Zallen, 2011). The unfertilized oocyte has no developmentally significant polarity. Polarity is established shortly after fertilization in response to a signal contributed by the sperm (Cowan and Hyman, 2004a). This signal leads to the establishment of two distinct cortical domains defining the anterior-posterior axis of the embryo. The one-cell embryo divides asymmetrically according to the axis such that one cell inherits the anterior cortical domain and the other cell inherits the posterior domain. The division is also physically asymmetric: the volume of the posterior P1-cell is approximately half that of the anterior AB-cell (see DIC micrograph). The resulting cells are already functionally distinct. The anterior AB-cell proceeds along a differentiation pathway producing ectoderm (hypodermis, pharynx, and neurons). The posterior P1-cell re-establishes anterior-posterior polarity and again divides asymmetrically (into P2 and EMS; see DIC micrograph) in a stem cell-like mode of division. These stem cell-like divisions establish the founder cells for the somatic lineages of the worm (AB, MS, E, C and D; see DIC micrographs) and maintaining a single stem cell (P4; see DIC micrographs) for the germline, which finally produces sperms and oocytes in the adult hermaphrodite.

The complete *C. elegans* digestive tract consists of three "organs" derived from four distinct embryonic cell lineages (Sulston et al., 1983): pharynx (57 cells from ABa; 38 cells from MS), intestine (20 cells from E; green), and rectum (11 cells from ABp; Sewell et al., 2003). Only the intestine is a pure clone of 20 E-cells; the three other lineages produce cells both inside and outside of the digestive tract. The intestine is one of the few cell lineages in the *C. elegans* embryo where a plausible sequence of direct molecular interactions can be proposed throughout the life cycle (Kormish et al., 2010; McGhee, 2007), beginning with maternally-derived factors in the cytoplasm of the early embryo (e.g. SKN-1 and SYS-1/POP-1), progressing through a small number of zygotic transcription factors (e.g. END-1/3 and ELT-2), and ending with the transcription of e.g. vitellogenin genes in the adult intestine. ELT-2 has been proposed to participate directly in the regulation of most intestinal genes expressed from the E^2 cell stage (Ea and Ep, see DIC micrograph) and later (McGhee et al., 2009; McGhee et al., 2007). The molecular mechanisms that lead to the asymmetric division of the EMS blastomere (green striated) into a larger MS- and a smaller E blastomere (see DIC micrograph) and the correct specification of their cell fates, central to the formation of the pharynx and intestine has been describe in great detail elsewhere (Maduro, 2010; Mango, 2007; Sugioka et al., 2011). **Orientation** (DIC micrographs): anterior, left, dorsal top; scale bar: 10 μm.

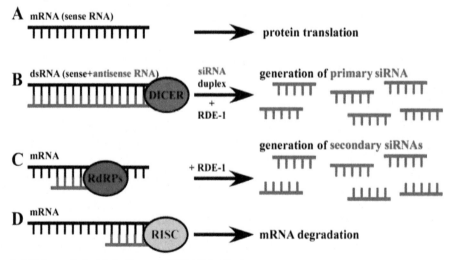

Fig. 3. RNA-mediated interference (RNAi) in *C. elegans*

Over the last decades, RNAi has been found not only be effective in *C. elegans* but also in other organisms and cell culture. The cartoon depicts a very simplified scheme of the exogenous RNAi-mechanism in *C. elegans* (for detailed reviews see: Ahringer, 2006; Fischer, 2010; Maine, 2008) that leads to targeted destabilization of endogenous, homologous mRNA molecules by double stranded RNA (dsRNA; Fire et al., 1998). **(A)** In a cell, RNA is used as a "messenger" (mRNA) to carry genetic information from the nucleus into the cytoplasm, where it is translated into proteins. **(B)** In *C. elegans*, exogenous dsRNA can be either applied by injection, "feeding" or "soaking" (Maeda et al., 2001; Mello et al., 1991; Timmons and Fire, 1998). dsRNA is then cut into ~22 nt primary siRNAs by a protein complex containing the RNAse III enzyme Dicer (DCR-1) and the dsRNA binding protein RDE-4 (Ketting et al., 2001; Tabara et al., 2002). The Argonaute protein RDE-1 (Tabara et al., 1999) binds siRNAs and seems only required for their stability (Parrish and Fire, 2001). Finally, RDE-1 slicer activity removes the passenger strand from the guide strand in the siRNA duplex (Steiner et al., 2009), which is necessary to allow guide-strand accessibility to the mRNA target. **(C)** RNAi in *C. elegans* includes an amplification step (Alder et al., 2003; Fire et al., 1998). The mRNA that is targeted by siRNAs serves as a template for the generation of secondary siRNAs mediated by RNA-dependent RNA polymerases (RdRPs). Secondary siRNAs are always antisense and have 5' triphosphates instead of the 5' monophosphate characteristic of Dicer cleavage. Secondary siRNAs are made by unprimed RNA synthesis by RdRPs, which are recruited to the target mRNA bound to the primary siRNA in complex with RDE-1 (Pak and Fire, 2007; Sijen et al., 2007). *In vitro* studies suggest that secondary siRNA generation is Dicer-independent (Aoki et al., 2007). **(D)** siRNAs present in the cell are associated with an effector complex called the RISC (RNA-induced silencing complex). In *C. elegans* multiple such complexes exist (Caudy et al., 2003; Chan et al., 2008; Gu et al., 2007), which finally drive mRNA destabilization.

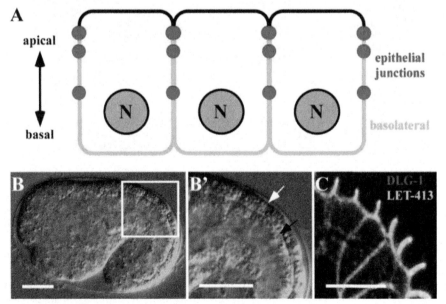

Fig. 4. Epithelial cell polarity and junctions

(A) Epithelial cells in general show a pronounced apicobasal polarity that becomes manifested by the establishment of apical (black) and basolateral (green) membrane domains that differ in the compositions of proteins and lipids. A hallmark of epithelial differentiation is the assembly of junctional complexes (red) along the lateral membrane domain, which fulfill different functions during epithelial development. **(B-B')** Shows a DIC micrograph of a *C. elegans* embryo during the elongation phase (B), focusing on two epithelia (B'), the epidermis (white arrow) and the intestine (black arrow). **(C)** Depicts an immunofluorescence micrograph of an embryo in B' stained against junctional protein DLG-1 (red) and basolateral protein LET-413 (green). See text for further details. **Orientation** (B-C'): anterior, left, dorsal top (A-E''); scale bar: 10 μm.

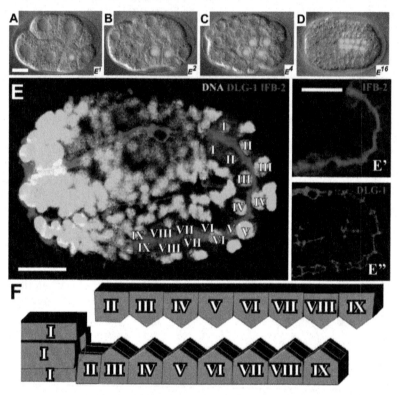

Fig. 5. Development and differentiation of the *C. elegans* embryonic intestine

The *C. elegans* intestine, the whole endoderm of the animal, consists of only 20 cells, which derive from a single somatic founder cell, the so-called E-cell (Deppe et al., 1978; Leung et al., 1999; Sulston et al., 1983). **(A-D)** Shows a series of DIC micrographs with E-cell nuclei colored in green. The E-cell is born at the 8-cell stage (A) and with the beginning of gastrulation (24-cell stage), 2 E-cells (E^2) ingress into the embryo (B) where they further undergo cell divisions (C, 4 E-cells, E^4). The ingression of Ea and Ep cells depends on correct cell fate specification and polarization of the machinery that orchestrates cell shape changes and cell migration (Lee and Goldstein, 2003; Sawyer et al., 2011). Among these, PAR-3 and

PAR-6 proteins regulate apical accumulation of myosin heavy chain, and a Wnt-Frizzled signaling pathway modulates contraction of the actomyosin network that drives apical constriction and finally leads to correct ingression of endodermal precursor cells (Cabello et al., 2010; Grana et al., 2010; Lee et al., 2006). Gastrulation in *C. elegans* later continues with the internalization of other cells including mesoderm and germline progenitors (Chisholm and Hardin, 2005; Nance et al., 2005). During early morphogenesis, the intestinal precursor cells (E^{16}) start to polarize (D, 16 E-cells, E^{16}, only 10 E-cells in focal plane) and finally an intestinal tube of 20 E-cells forms during ongoing morphogenesis of *C. elegans*. **(E-E'')** Shows micrographs of a mid-morphogenesis stage (similar to D) stained against DNA (E, green, YoYo), the intestinal-specific intermediate filament protein IFB-2 localized in the apical cortex (E', blue, mabMH33), and the junctional protein DLG-1 (E'', red, anti-DLG-1 antibodies). **(F)** The cartoon depicts the organization of the intestinal epithelial tube in nine units (I-IX), which are connected by the CeAJ (red). **Orientation** (A-E''): anterior, left, dorsal top (A-E''); scale bar: 10 μm.

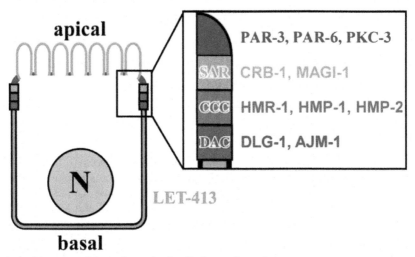

Fig. 6. Apical junctional complexes in the *C. elegans* intestine

Epithelia of the *C. elegans* embryo contain a single electron-dense apical junction (about 250 nm; Carberry et al., 2009; Müller and Bossinger, 2003), also referred to as "*C. elegans* apical junction" (CeAJ; McMahon et al., 2001) that has been subdivided into distinct parts by immunohistochemistry. In the basal part of the CeAJ, the DLG-1–AJM-1 complex (DAC; Köppen et al., 2001; Lockwood et al., 2008) is organized, while more apically the catenin-cadherin complex (CCC; Costa et al., 1998; Kwiatkowski et al., 2010), consisting of the proteins HMR-1 (E-cadherin), HMP-1 (α-catenin) and HMP-2 (β-catenin) can be found. The subapical region harbours the proteins MAGI-1 and probably CRB-1 (Bossinger et al., 2001; Stetak and Hajnal, 2011). By immunofluorescence analysis all these proteins show a typical, "junctional" staining pattern (e.g. DLG-1, Fig.5E'') that reflects the correct formation of the CeAJ within the embryonic intestine. Most apically, the PAR-3–PAR-6–PKC-3 complex (PPC; Achilleos et al., 2010; Leung et al., 1999; Totong et al., 2007) is localized, showing a more "cortical" staining pattern, comparable to that of intermediate filament proteins (e.g. IFB-2, Fig.5E').

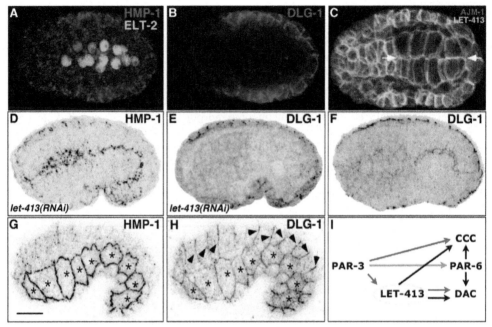

Fig. 7. Establishment of cell polarity and assembly of junctional complexes during development of the C. elegans intestine

(A-C) Early morphogenesis stages showing immunofluorescences (IF) of the catenin–cadherin complex (CCC, blue in A, anti-HMP-1/α-catenin IF), the intestine-specific GATA-factor ELT-2 (green in A, anti-GFP IF; McGhee et al., 2009; McGhee et al., 2007), the DLG-1–AJM-1 complex (DAC, red in B and C, anti-DLG-1/Discs large IF and anti-AJM-1 IF), and the LET-413/SCRIB protein (green in C, anti-CFP). (D-E) Mid morphogenesis stages after RNAi (Fire et al., 1998) against *let-413* gene function displaying anti-HMP-1 and anti-DLG-1 IFs. (F) During early morphogenesis stage, the C. elegans apical junction (CeAJ) forms around the apex of intestinal primordial cells (anti-DLG-1 IF). (G-H) IF analysis shows that the CCC (G) but not the DAC (H) moves away from the CeAJ (arrows in H) prior to the onset of cell fusion in the dorsal hypodermis (Oren-Suissa and Podbilewicz, 2007; Oren-Suissa and Podbilewicz, 2010). In contrast, both complexes clearly localize at the CeAJ in lateral seam cells (asterisks in G,H). (I) Schematic drawing of key players involved in epithelial polarization (colored arrows), formation of the junctional belt around the apex (black arrows) and maintenance of cell polarity (yellow circle). PAR-3 is a PDZ domain-containing protein orthologous to mammalian atypical PKC isotype-specific interacting protein (ASIP) and *Drosophila* Bazooka. PAR-6 contains PB1, CRIB and PDZ domains and is also conserved in *Drosophila* and mammals. LET-413 belongs to the LAP (LRR (for leucine-rich repeats) and PDZ (for PSD-95/Discs-large/ZO-1)) protein family and contains one PDZ domain and 16 LRR (Bilder et al., 2000; Legouis et al., 2000; Legouis et al., 2003). The DLG-1–AJM-1 complex (DAC; Köppen et al., 2001; Lockwood et al., 2008) is composed of DLG-1/Discs large (a MAGUK with three PDZ, one SH3, and one GUK domain) and AJM-1 (apical junction molecule) a coiled-coil protein. See text for further explanations. **Orientation**

(A-H): anterior (left), dorsal (top); scale bar: 10 μm. A-B, C and G-H: photo courtesy of Tobias Wiesenfahrt, Jennifer Pilipiuk and Eva Horzowski, respectively.

8. References

Achilleos, A., Wehman, A. M. and Nance, J. (2010). PAR-3 mediates the initial clustering and apical localization of junction and polarity proteins during C. elegans intestinal epithelial cell polarization. *Development* 137, 1833-42.

Ahringer, J. (2006). Reverse genetics. In *WormBook*,
(ed. The_Celegans_Research_Community): WormBook
http://dx.doi.org/doi:10.1895/wormbook.1.47.1.

Alder, M. N., Dames, S., Gaudet, J. and Mango, S. E. (2003). Gene silencing in Caenorhabditis elegans by transitive RNA interference. *RNA* 9, 25-32.

Altun, Z. F., Chen, B., Wang, Z. W. and Hall, D. H. (2009). High resolution map of Caenorhabditis elegans gap junction proteins. *Dev Dyn* 238, 1936-50.

Altun, Z. F. and Hall, D. H. (2009a). Alilmentary system, rectum and anus. In *WormAtlas*, (ed. The_Celegans_Research_Community). WormAtlas
http://dx.doi.org/doi:10.3908/wormatlas.1.5.

Altun, Z. F. and Hall, D. H. (2009b). Alimentary system, intestine. In *WormAtlas*, (ed. The_Celegans_Research_Community): WormAtlas
http://dx.doi.org/doi:10.3908/wormatlas.1.4.

Altun, Z. F. and Hall, D. H. (2009c). Alimentary system, overview. In *WormAtlas*, (ed. The_Celegans_Research_Community): WormAtlas
http://dx.doi.org/doi:10.3908/wormatlas.1.2.

Altun, Z. F. and Hall, D. H. (2009d). Alimentary system, pharynx. In *WormAtlas*, (ed. The_Celegans_Research_Community): WormAtlas
http://dx.doi.org/doi:10.3908/wormatlas.1.3.

Anderson, J. M. and Van Itallie, C. M. (2009). Physiology and function of the tight junction. *Cold Spring Harbor perspectives in biology* 1, a002584.

Aoki, K., Moriguchi, H., Yoshioka, T., Okawa, K. and Tabara, H. (2007). In vitro analyses of the production and activity of secondary small interfering RNAs in C. elegans. *EMBO J* 26, 5007-19.

Aono, S., Legouis, R., Hoose, W. A. and Kemphues, K. J. (2004). PAR-3 is required for epithelial cell polarity in the distal spermatheca of C. elegans. *Development* 131, 2865-74.

Asano, A., Asano, K., Sasaki, H., Furuse, M. and Tsukita, S. (2003). Claudins in Caenorhabditis elegans: their distribution and barrier function in the epithelium. *Curr Biol* 13, 1042-6.

Bacallao, R. L., Antony, C., Dotti, C., Karsenti, E., Stelzer, E. and Simons, K. (1989). The subcellular organization of Madin-Darby canine kidney cells during the formation of a polarized epithelium. *J Cell Biol* 109, 2817-32.

Baylis, H. A. and Vazquez-Manrique, R. P. (2011). Reverse genetic strategies in Caenorhabditis elegans: towards controlled manipulation of the genome. *ScientificWorldJournal* 11, 1394-410.

Betschinger, J., Mechtler, K. and Knoblich, J. A. (2003). The Par complex directs asymmetric cell division by phosphorylating the cytoskeletal protein Lgl. *Nature* 422, 326-30.

Bilder, D., Birnbaum, D., Borg, J. P., Bryant, P., Huigbretse, J., Jansen, E., Kennedy, M. B., Labouesse, M., Legouis, R., Mechler, B. et al. (2000). Collective nomenclature for LAP proteins. *Nature cell biology* 2, E114.

Bilder, D., Schober, M. and Perrimon, N. (2003). Integrated activity of PDZ protein complexes regulates epithelial polarity. *Nat Cell Biol* 5, 53-58.

Bird, A. F. and Bird, J. (1991). The structure of nematodes. California: Academic Press.

Blaxter, M. (2011). Nematodes: the worm and its relatives. *PLoS Biol* 9, e1001050.

Bossinger, O., Fukushige, T., Claeys, M., Borgonie, G. and McGhee, J. D. (2004). The apical disposition of the Caenorhabditis elegans intestinal terminal web is maintained by LET-413. *Dev Biol* 268, 448-456.

Bossinger, O., Klebes, A., Segbert, C., Theres, C. and Knust, E. (2001). Zonula adherens formation in Caenorhabditis elegans requires dlg-1, the homologue of the Drosophila gene discs large. *Dev Biol* 230, 29-42.

Bossinger, O. and Schierenberg, E. (1992). Cell-cell communication in the embryo of Caenorhabditis elegans. *Dev Biol* 151, 401-9.

Bossinger, O. and Schierenberg, E. (1996). Early embryonic induction in C. elegans can be inhibited with polysulfated hydrocarbon dyes. *Dev Biol* 176, 17-21.

Boulin, T. and Bessereau, J. (2007). Mos1-mediated insertional mutagenesis in Caenorhabditis elegans. *Nat Protoc* 2, 1276-87.

Bowerman, B. (2011). The near demise and subsequent revival of classical genetics for investigating Caenorhabditis elegans embryogenesis: RNAi meets next-generation DNA sequencing. *Mol Biol Cell* 22, 3556-8.

Bre, M. H., Pepperkok, R., Hill, A. M., Levilliers, N., Ansorge, W., Stelzer, E. H. and Karsenti, E. (1990). Regulation of microtubule dynamics and nucleation during polarization in MDCK II cells. *J Cell Biol* 111, 3013-21.

Brenner, S. (1974). The genetics of Caenorhabditis elegans. *Genetics* 77, 71-94.

Brenner, S. (2003). Nature's gift to science (Nobel lecture). *Chembiochem* 4, 683-7.

Brenner, S. (2009). In the beginning was the worm. *Genetics* 182, 413-5.

Bulgakova, N. A. and Knust, E. (2009). The Crumbs complex: from epithelial-cell polarity to retinal degeneration. *J Cell Sci* 122, 2587-96.

Cabello, J., Neukomm, L. J., Gunesdogan, U., Burkart, K., Charette, S. J., Lochnit, G., Hengartner, M. O. and Schnabel, R. (2010). The Wnt pathway controls cell death engulfment, spindle orientation, and migration through CED-10/Rac. *PLoS Biol* 8, e1000297.

Calixto, A., Chelur, D., Topalidou, I., Chen, X. and Chalfie, M. (2010a). Enhanced neuronal RNAi in C. elegans using SID-1. *Nature methods* 7, 554-9.

Calixto, A., Ma, C. and Chalfie, M. (2010b). Conditional gene expression and RNAi using MEC-8-dependent splicing in C. elegans. *Nat Methods* 7, 407-11.

Carberry, K., Wiesenfahrt, T., Windoffer, R., Bossinger, O. and Leube, R. E. (2009). Intermediate filaments in Caenorhabditis elegans. *Cell Motil Cytoskeleton* 66, 852-64.

Caudy, A. A., Ketting, R. F., Hammond, S. M., Denli, A. M., Bathoorn, A. M., Tops, B. B., Silva, J. M., Myers, M. M., Hannon, G. J. and Plasterk, R. H. (2003). A micrococcal nuclease homologue in RNAi effector complexes. *Nature* 425, 411-4.

Chalfie, M. (2009). GFP: lighting up life (Nobel Lecture). *Angewandte Chemie* 48, 5603-11.

Chalfie, M., Tu, Y., Euskirchen, G., Ward, W. and Prasher, D. (1994). Green fluorescent protein as a marker for gene expression. *Science* 263, 802-5.

Chan, S. P., Ramaswamy, G., Choi, E. Y. and Slack, F. J. (2008). Identification of specific let-7 microRNA binding complexes in Caenorhabditis elegans. *RNA* 14, 2104-14.

Chen, X., Kojima, S., Borisy, G. G. and Green, K. J. (2003). p120 catenin associates with kinesin and facilitates the transport of cadherin-catenin complexes to intercellular junctions. *J Cell Biol* 163, 547-57.

Chisholm, A. D. (2006). Gastrulation: Wnts signal constriction. *Curr Biol* 16, R874-6.

Chisholm, A. D. and Hardin, J. (2005). Epidermal morphogenesis. In *WormBook*, (ed. The_Celegans_Research_Community), pp. 1-22: WormBook http://dx.doi.org/doi:10.1895/wormbook.1.35.1.

Costa, M., Raich, W., Agbunag, C., Leung, B., Hardin, J. and Priess, J. (1998). A putative catenin-cadherin system mediates morphogenesis of the Caenorhabditis elegans embryo. *J Cell Biol* 141, 297-308.

Cowan, C. R. and Hyman, A. (2004a). Centrosomes direct cell polarity independently of microtubule assembly in C. elegans embryos. *Nature* 431, 92-6.

Cowan, C. R. and Hyman, A. A. (2004b). Asymmetric cell division in C. elegans: cortical polarity and spindle positioning. *Annual review of cell and developmental biology* 20, 427-53.

Cox, E. A. and Hardin, J. (2004). Sticky worms: adhesion complexes in C. elegans. *J Cell Sci* 117, 1885-97.

de Santa Barbara, P., van den Brink, G. R. and Roberts, D. J. (2003). Development and differentiation of the intestinal epithelium. *Cell Mol Life Sci* 60, 1322-32.

Deppe, U., Schierenberg, E., Cole, T., Krieg, C., Schmitt, D., Yoder, B. K. and von Ehrenstein, G. (1978). Cell lineages of the embryo of the nematode Caenorhabditis elegans. *Proc Natl Acad Sci U S A* 75, 376-80.

Ebnet, K. (2008). Organization of multiprotein complexes at cell-cell junctions. *Histochem Cell Biol* 130, 1-20.

Eckert, J. J. and Fleming, T. P. (2008). Tight junction biogenesis during early development. *Biochim Biophys Acta* 1778, 717-728.

Etheridge, T., Nemoto, K., Hashizume, T., Mori, C., Sugimoto, T., Suzuki, H., Fukui, K., Yamazaki, T., Higashibata, A., Szewczyk, N. J. et al. (2011). The effectiveness of RNAi in Caenorhabditis elegans is maintained during spaceflight. *PLoS ONE* 6, e20459.

Fire, A., Xu, S., Montgomery, M., Kostas, S., Driver, S. and Mello, C. (1998). Potent and specific genetic interference by double-stranded RNA in Caenorhabditis elegans. *Nature* 391, 806-11.

Fire, A. Z. (2007). Gene silencing by double-stranded RNA (Nobel Lecture). *Angewandte Chemie* 46, 6966-84.

Firestein, B. L. and Rongo, C. (2001). DLG-1 Is a MAGUK Similar to SAP97 and Is Required for Adherens Junction Formation. *Mol Biol Cell* 12, 3465-75.

Fischer, S. E. (2010). Small RNA-mediated gene silencing pathways in C. elegans. *The international journal of biochemistry & cell biology* 42, 1306-15.

Fraser, A. G., Kamath, R. S., Zipperlen, P., Martinez-Campos, M., Sohrmann, M. and Ahringer, J. (2000). Functional genomic analysis of C. elegans chromosome I by systematic RNA interference. *Nature* 408, 325-30.

Frokjaer-Jensen, C., Davis, M. W., Hollopeter, G., Taylor, J., Harris, T. W., Nix, P., Lofgren, R., Prestgard-Duke, M., Bastiani, M., Moerman, D. G. et al. (2010). Targeted gene deletions in C. elegans using transposon excision. *Nat Methods* 7, 451-3.
Fuchs, E. (2007). Scratching the surface of skin development. *Nature* 445, 834-42.
Goldstein, B. (1992). Induction of gut in Caenorhabditis elegans embryos. *Nature* 357, 255-7.
Gönczy, P. (2008). Mechanisms of asymmetric cell division: flies and worms pave the way. *Nat Rev Mol Cell Biol* 9, 355-66.
Gönczy, P., Echeverri, G., Oegema, K., Coulson, A., Jones, S. J., Copley, R. R., Duperon, J., Oegema, J., Brehm, M., Cassin, E. et al. (2000). Functional genomic analysis of cell division in C. elegans using RNAi of genes on chromosome III. *Nature* 408, 331-6.
Grana, T. M., Cox, E. A., Lynch, A. M. and Hardin, J. (2010). SAX-7/L1CAM and HMR-1/cadherin function redundantly in blastomere compaction and non-muscle myosin accumulation during Caenorhabditis elegans gastrulation. *Developmental biology* 344, 731-44.
Gu, S. G., Pak, J., Barberan-Soler, S., Ali, M., Fire, A. and Zahler, A. M. (2007). Distinct ribonucleoprotein reservoirs for microRNA and siRNA populations in C. elegans. *RNA* 13, 1492-504.
Guo, S. and Kemphues, K. J. (1996). A non-muscle myosin required for embryonic polarity in Caenorhabditis elegans. *Nature* 382, 455-8.
Han, M. (1997). Gut reaction to Wnt signaling in worms. *Cell* 90, 581-4.
Han, M. (2010). Advancing biology with a growing worm field. *Developmental dynamics : an official publication of the American Association of Anatomists* 239, 1263-4.
Hardin, J. and Lockwood, C. (2004). Skin tight: cell adhesion in the epidermis of Caenorhabditis elegans. *Curr Opin Cell Biol* 16, 486-92.
Harris, T. J. and Peifer, M. (2005). The positioning and segregation of apical cues during epithelial polarity establishment in Drosophila. *J Cell Biol* 170, 813-23.
Harris, T. J. and Peifer, M. (2007). aPKC controls microtubule organization to balance adherens junction symmetry and planar polarity during development. *Dev Cell* 12, 727-38.
Harris, T. J. and Tepass, U. (2010). Adherens junctions: from molecules to morphogenesis. *Nat Rev Mol Cell Biol* 11, 502-14.
Harris, T. W., Antoshechkin, I., Bieri, T., Blasiar, D., Chan, J., Chen, W. J., De La Cruz, N., Davis, P., Duesbury, M., Fang, R. et al. (2010). WormBase: a comprehensive resource for nematode research. *Nucleic Acids Res* 38, D463-7.
Hedgecock, E. M. and White, J. G. (1985). Polyploid tissues in the nematode Caenorhabditis elegans. *Dev Biol* 107, 128-33.
Hermann, G. J., Leung, B. and Priess, J. R. (2000). Left-right asymmetry in C. elegans intestine organogenesis involves a LIN-12/Notch signaling pathway. *Development* 127, 3429-40.
Hobert, O. (2010). The impact of whole genome sequencing on model system genetics: get ready for the ride. *Genetics* 184, 317-9.
Hoffmann, M., Segbert, C., Helbig, G. and Bossinger, O. (2010). Intestinal tube formation in Caenorhabditis elegans requires vang-1 and egl-15 signaling. *Dev Biol* 339, 268-279.
Horvitz, H. R. (2003). Worms, life, and death (Nobel lecture). *Chembiochem* 4, 697-711.

Hunt-Newbury, R., Viveiros, R., Johnsen, R., Mah, A., Anastas, D., Fang, L., Halfnight, E., Lee, D., Lin, J., Lorch, A. et al. (2007). High-throughput in vivo analysis of gene expression in Caenorhabditis elegans. *PLoS Biol* 5, e237.

Hüsken, K., Wiesenfahrt, T., Abraham, C., Windoffer, R., Bossinger, O. and Leube, R. (2008). Maintenance of the intestinal tube in Caenorhabditis elegans: the role of the intermediate filament protein IFC-2. *Differentiation* 76, 881-896.

Jordens, I., Marsman, M., Kuijl, C. and Neefjes, J. (2005). Rab proteins, connecting transport and vesicle fusion. *Traffic* 6, 1070-7.

Jorgensen, E. M. and Mango, S. E. (2002). The art and design of genetic screens: Caenorhabditis elegans. *Nat Rev Genet* 3, 356-69.

Kamath, R. S., Fraser, A. G., Dong, Y., Poulin, G., Durbin, R., Gotta, M., Kanapin, A., Le Bot, N., Moreno, S., Sohrmann, M. et al. (2003). Systematic functional analysis of the Caenorhabditis elegans genome using RNAi. *Nature* 421, 231-7.

Ketting, R. F., Fischer, S. E., Bernstein, E., Sijen, T., Hannon, G. J. and Plasterk, R. H. (2001). Dicer functions in RNA interference and in synthesis of small RNA involved in developmental timing in C. elegans. *Genes Dev* 15, 2654-9.

Kimble, J. and Sharrock, W. J. (1983). Tissue-specific synthesis of yolk proteins in Caenorhabditis elegans. *Dev Biol* 96, 189-96.

Knust, E. and Bossinger, O. (2002). Composition and formation of intercellular junctions in epithelial cells. *Science* 298, 1955-9.

Köppen, M., Simske, J. S., Sims, P. A., Firestein, B. L., Hall, D. H., Radice, A. D., Rongo, C. and Hardin, J. D. (2001). Cooperative regulation of AJM-1 controls junctional integrity in Caenorhabditis elegans epithelia. *Nat Cell Biol* 3, 983-91.

Kormish, J. D., Gaudet, J. and McGhee, J. D. (2010). Development of the C. elegans digestive tract. *Current opinion in genetics & development* 20, 346-54.

Krahn, M. P., Buckers, J., Kastrup, L. and Wodarz, A. (2010a). Formation of a Bazooka-Stardust complex is essential for plasma membrane polarity in epithelia. *The Journal of cell biology* 190, 751-60.

Krahn, M. P., Klopfenstein, D. R., Fischer, N. and Wodarz, A. (2010b). Membrane targeting of Bazooka/PAR-3 is mediated by direct binding to phosphoinositide lipids. *Current Biology* 20, 636-42.

Kramer, J. M. (2005). Basement membranes. In *WormBook*, (ed. The_Celegans_Research_Community), pp. 1-15: WormBook http://dx.doi.org/doi:10.1895/wormbook.1.16.1.

Kwiatkowski, A. V., Maiden, S. L., Pokutta, S., Choi, H. J., Benjamin, J. M., Lynch, A. M., Nelson, W. J., Weis, W. I. and Hardin, J. (2010). In vitro and in vivo reconstitution of the cadherin-catenin-actin complex from Caenorhabditis elegans. *Proc Natl Acad Sci U S A* 107, 14591-6.

Labouesse, M. (2006). Epithelial junctions and attachments. In WormBook, (ed. The_Celegans_Research_Community): WormBook http://dx.doi.org/doi:10.1895/wormbook.1.56.1.

Laprise, P., Lau, K. M., Harris, K. P., Silva-Gagliardi, N. F., Paul, S. M., Beronja, S., Beitel, G. J., McGlade, C. J. and Tepass, U. (2009). Yurt, Coracle, Neurexin IV and the Na(+),K(+)-ATPase form a novel group of epithelial polarity proteins. *Nature* 459, 1141-5.

Lecuit, T. and Wieschaus, E. (2000). Polarized insertion of new membrane from a cytoplasmic reservoir during cleavage of the Drosophila embryo. *J Cell Biol* 150, 849-60.

Lee, J. and Goldstein, B. (2003). Mechanisms of cell positioning during C. elegans gastrulation. *Development* 130, 307-20.

Lee, J., Marston, D. J., Walston, T., Hardin, J., Halberstadt, A. and Goldstein, B. (2006). Wnt/Frizzled signaling controls C. elegans gastrulation by activating actomyosin contractility. *Curr Biol* 16, 1986-97.

Legouis, R., Gansmuller, A., Sookhareea, S., Bosher, J. M., Baillie, D. L. and Labouesse, M. (2000). LET-413 is a basolateral protein required for the assembly of adherens junctions in Caenorhabditis elegans. *Nat Cell Biol* 2, 415-422.

Legouis, R., Jaulin-Bastard, F., Schott, S., Navarro, C., Borg, J. P. and Labouesse, M. (2003). Basolateral targeting by leucine-rich repeat domains in epithelial cells. *EMBO Rep* 4, 1096-1100.

Leung, B., Hermann, G. J. and Priess, J. R. (1999). Organogenesis of the Caenorhabditis elegans intestine. *Dev Biol* 216, 114-34.

Ligon, L. A., Karki, S., Tokito, M. and Holzbaur, E. L. (2001). Dynein binds to beta-catenin and may tether microtubules at adherens junctions. *Nat Cell Biol* 3, 913-7.

Lints, R. and Hall, D. H. (2009). Reproductive system, somatic gonad. In *WormAtlas*, (ed. The_Celegans_Research_Community). WormAtlas http://dx.doi.org/doi:10.3908/wormatlas.1.22.

Lockwood, C. A., Lynch, A. M. and Hardin, J. (2008). Dynamic analysis identifies novel roles for DLG-1 subdomains in AJM-1 recruitment and LET-413-dependent apical focusing. *J Cell Sci* 121, 1477-87.

Lu, R., Maduro, M. F., Li, F., Li, H. L., Broitman-Maduro, G., Li, W. and Ding, S. W. (2005). Animal virus replication and RNAi-mediated antiviral silencing in Caenorhabditis elegans. *Nature* 436, 1040-3.

Lynch, A. M. and Hardin, J. (2009). The assembly and maintenance of epithelial junctions in C. elegans. *Front Biosci* 14, 1414-32.

MacQueen, A. J., Baggett, J. J., Perumov, N., Bauer, R. A., Januszewski, T., Schriefer, L. and Waddle, J. A. (2005). ACT-5 is an essential Caenorhabditis elegans actin required for intestinal microvilli formation. *Molecular biology of the cell* 16, 3247-59.

Maduro, M. F. (2010). Cell fate specification in the C. elegans embryo. *Dev Dyn* 239, 1315-29.

Maeda, I., Kohara, Y., Yamamoto, M. and Sugimoto, A. (2001). Large-scale analysis of gene function in Caenorhabditis elegans by high-throughput RNAi. *Curr Biol* 11, 171-6.

Maine, E. M. (2008). Studying gene function in Caenorhabditis elegans using RNA-mediated interference. *Brief Funct Genomic Proteomic* 7, 184-94.

Mango, S. E. (2007). The C. elegans pharynx: a model for organogenesis. In *WormBook*, (ed. The_Celegans_Research_Community): WormBook http://dx.doi.org/doi:10.1895/wormbook.1.129.1.

Mary, S., Charrasse, S., Meriane, M., Comunale, F., Travo, P., Blangy, A. and Gauthier-Rouviere, C. (2002). Biogenesis of N-cadherin-dependent cell-cell contacts in living fibroblasts is a microtubule-dependent kinesin-driven mechanism. *Mol Biol Cell* 13, 285-301.

McGee, M. D., Weber, D., Day, N., Vitelli, C., Crippen, D., Herndon, L. A., Hall, D. H. and Melov, S. (2011). Loss of intestinal nuclei and intestinal integrity in aging C. elegans. *Aging Cell* 10, 699-710.

McGhee, J. D. (2007). The C. elegans intestine. In *WormBook*, (ed. The_Celegans_Research_Community): WormBook http://dx.doi.org/doi:10.1895/wormbook.1.133.1.

McGhee, J. D., Fukushige, T., Krause, M. W., Minnema, S. E., Goszczynski, B., Gaudet, J., Kohara, Y., Bossinger, O., Zhao, Y., Khattra, J. et al. (2009). ELT-2 is the predominant transcription factor controlling differentiation and function of the C. elegans intestine, from embryo to adult. *Dev Biol* 327, 551-65.

McGhee, J. D., Sleumer, M. C., Bilenky, M., Wong, K., McKay, S. J., Goszczynski, B., Tian, H., Krich, N. D., Khattra, J., Holt, R. A. et al. (2007). The ELT-2 GATA-factor and the global regulation of transcription in the C. elegans intestine. *Dev Biol* 302, 627-45.

McMahon, L., Legouis, R., Vonesch, J. L. and Labouesse, M. (2001). Assembly of C. elegans apical junctions involves positioning and compaction by LET-413 and protein aggregation by the MAGUK protein DLG-1. *J Cell Sci* 114, 2265-77.

Mello, C. C. (2007). Return to the RNAi world: rethinking gene expression and evolution (Nobel Lecture). *Angewandte Chemie* 46, 6985-94.

Mello, C. C., Kramer, J. M., Stinchcomb, D. T. and Ambros, V. (1991). Efficient gene transfer in C.elegans: extrachromosomal maintenance and integration of transforming sequences. *EMBO J* 10, 3959-70.

Michaux, G., Legouis, R. and Labouesse, M. (2001). Epithelial biology: lessons from Caenorhabditis elegans. *Gene* 277, 83-100.

Mitani, S. (2009). Nematode, an experimental animal in the national BioResource project. *Exp Anim* 58, 351-6.

Moerman, D. G. and Barstead, R. J. (2008). Towards a mutation in every gene in Caenorhabditis elegans. *Brief Funct Genomic Proteomic* 7, 195-204.

Müller, H. A. (2000). Genetic control of epithelial cell polarity: lessons from Drosophila. *Dev Dyn* 218, 52-67.

Müller, H. A. and Bossinger, O. (2003). Molecular networks controlling epithelial cell polarity in development. *Mech Dev* 120, 1231-56.

Nance, J. (2003). C. elegans PAR-3 and PAR-6 are required for apicobasal asymmetries associated with cell adhesion and gastrulation. *Development* 130, 5339-5350.

Nance, J., Lee, J. Y. and Goldstein, B. (2005). Gastrulation in C. elegans. In *WormBook*, (ed. The_Celegans_Research_Community), pp. 1-13: WormBook http://dx.doi.org/doi:10.1895/wormbook.1.23.1

Nance, J. and Zallen, J. A. (2011). Elaborating polarity: PAR proteins and the cytoskeleton. *Development* 138, 799-809.

Nelson, W. J. (2003). Adaptation of core mechanisms to generate cell polarity. *Nature* 422, 766-74.

Nelson, W. J. (2009). Remodeling epithelial cell organization: transitions between front-rear and apical-basal polarity. *Cold Spring Harbor perspectives in biology* 1, a000513.

Noah, T. K., Donahue, B. and Shroyer, N. F. (2011). Intestinal development and differentiation. *Exp Cell Res* 317, 2702-10.

Oren-Suissa, M. and Podbilewicz, B. (2007). Cell fusion during development. *Trends Cell Biol* 17, 537-46.

Oren-Suissa, M. and Podbilewicz, B. (2010). Evolution of programmed cell fusion: common mechanisms and distinct functions. *Dev Dyn* 239, 1515-28.

Pak, J. and Fire, A. (2007). Distinct populations of primary and secondary effectors during RNAi in C. elegans. *Science* 315, 241-4.

Papoulas, O., Hays, T. S. and Sisson, J. C. (2005). The golgin Lava lamp mediates dynein-based Golgi movements during Drosophila cellularization. *Nat Cell Biol* 7, 612-8.

Parrish, S. and Fire, A. (2001). Distinct roles for RDE-1 and RDE-4 during RNA interference in Caenorhabditis elegans. *RNA* 7, 1397-402.

Pauli, F., Liu, Y., Kim, Y. A., Chen, P. J. and Kim, S. H. (2006). Chromosomal clustering and GATA transcriptional regulation of intestine-expressed genes in C. elegans. *Development* 133, 287-95.

Pilipiuk, J., Lefebvre, C., Wiesenfahrt, T., Legouis, R. and Bossinger, O. (2009). Increased IP3/Ca2+ signaling compensates depletion of LET-413/DLG-1 in C. elegans epithelial junction assembly. *Dev Biol* 327, 34-47.

Plant, P. J., Fawcett, J. P., Lin, D. C., Holdorf, A. D., Binns, K., Kulkarni, S. and Pawson, T. (2003). A polarity complex of mPar-6 and atypical PKC binds, phosphorylates and regulates mammalian Lgl. *Nat Cell Biol* 5, 301-8.

Portereiko, M. F., Saam, J. and Mango, S. E. (2004). ZEN-4/MKLP1 is required to polarize the foregut epithelium. *Curr Biol* 14, 932-41.

Putzke, A. P. and Rothman, J. H. (2003). Gastrulation: PARtaking of the bottle. *Curr Biol* 13, R223-5.

Qadota, H., Inoue, M., Hikita, T., Koppen, M., Hardin, J. D., Amano, M., Moerman, D. G. and Kaibuchi, K. (2007). Establishment of a tissue-specific RNAi system in C. elegans. *Gene* 400, 166-73.

Robert, V. J. and Bessereau, J. L. (2007). Targeted engineering of the Caenorhabditis elegans genome following Mos1-triggered chromosomal breaks. *EMBO J* 26, 170-83.

Rohrschneider, M. R. and Nance, J. (2009). Polarity and cell fate specification in the control of Caenorhabditis elegans gastrulation. *Dev Dyn* 238, 789-96.

Sawyer, J. M., Glass, S., Li, T., Shemer, G., White, N. D., Starostina, N. G., Kipreos, E. T., Jones, C. D. and Goldstein, B. (2011). Overcoming Redundancy: an RNAi Enhancer Screen for Morphogenesis Genes in Caenorhabditis elegans. *Genetics* 188, 549-64.

Sawyer, J. M., Harrell, J. R., Shemer, G., Sullivan-Brown, J., Roh-Johnson, M. and Goldstein, B. (2009). Apical constriction: A cell shape change that can drive morphogenesis. *Dev Biol* 34, 5-19.

Schierenberg, E. (1987). Reversal of cellular polarity and early cell-cell interaction in the embryos of Caenorhabditis elegans. *Dev Biol* 122, 452-63.

Schierenberg, E. (2005). Unusual cleavage and gastrulation in a freshwater nematode: developmental and phylogenetic implications. *Dev Genes Evol* 215, 103-8.

Schierenberg, E. (2006). Embryological variation during nematode development. In *WormBook*, (ed. The_Celegans_Research_Community), pp. 1-13: WormBook http://dx.doi.org/doi:10.1895/wormbook.1.55.1.

Schnabel, R., Hutter, H., Moerman, D. G. and Schnabel, H. (1997). Assessing normal embryogenesis in Caenorhabditis elegans using a 4D microscope: variability of development and regional specification. *Dev Biol* 184, 234-65.

Schott, D. H., Cureton, D. K., Whelan, S. P. and Hunter, C. P. (2005). An antiviral role for the RNA interference machinery in Caenorhabditis elegans. *Proc Natl Acad Sci U S A* 102, 18420-4.
Schulenburg, H., Kurz, C. L. and Ewbank, J. J. (2004). Evolution of the innate immune system: the worm perspective. *Immunol Rev* 198, 36-58.
Segbert, C., Johnson, K., Theres, C., van Fürden, D. and Bossinger, O. (2004). Molecular and functional analysis of apical junction formation in the gut epithelium of Caenorhabditis elegans. *Dev Biol* 266, 17-26.
Sewell, S. T., Zhang, G., Uttam, A. and Chamberlin, H. M. (2003). Developmental patterning in the Caenorhabditis elegans hindgut. *Dev Biol* 262, 88-93.
Shaye, D. D. and Greenwald, I. (2011). OrthoList: A Compendium of C. elegans Genes with Human Orthologs. *PLoS ONE* 6, e20085.
Siegrist, S. E. and Doe, C. Q. (2005). Microtubule-induced pins/galphai cortical polarity in Drosophila neuroblasts. *Cell* 123, 1323-35.
Sijen, T., Steiner, F. A., Thijssen, K. L. and Plasterk, R. H. (2007). Secondary siRNAs result from unprimed RNA synthesis and form a distinct class. *Science* 315, 244-7.
Simske, J. S., Köppen, M., Sims, P. A., Hodgkin, J., Yonkof, A. and Hardin, J. (2003). The cell junction protein VAB-9 regulates adhesion and epidermal morphology in C. elegans. *Nat Cell Biol* 5, 619-25.
Sönnichsen, B., Koski, L. B., Walsh, A., Marschall, P., Neumann, B., Brehm, M., Alleaume, A. M., Artelt, J., Bettencourt, P., Cassin, E. et al. (2005). Full-genome RNAi profiling of early embryogenesis in Caenorhabditis elegans. *Nature* 434, 462-9.
Steiner, F. A., Okihara, K. L., Hoogstrate, S. W., Sijen, T. and Ketting, R. F. (2009). RDE-1 slicer activity is required only for passenger-strand cleavage during RNAi in Caenorhabditis elegans. *Nat Struct Mol Biol* 16, 207-11.
Stetak, A. and Hajnal, A. (2011). The C. elegans MAGI-1 protein is a novel component of cell junctions that is required for junctional compartmentalization. *Dev Biol* 350, 24-31.
Sugioka, K., Mizumoto, K. and Sawa, H. (2011). Wnt Regulates Spindle Asymmetry to Generate Asymmetric Nuclear beta-Catenin in C. elegans. *Cell* 146, 942-54.
Sulston, J. and Horvitz, H. (1977). Post-embryonic cell lineages of the nematode, Caenorhabditis elegans. *Dev Biol* 56, 110-56.
Sulston, J., Schierenberg, E., White, J. and Thomson, J. (1983). The embryonic cell lineage of the nematode Caenorhabditis elegans. *Dev Biol* 100, 64-119.
Sulston, J. E. (2003). Caenorhabditis elegans: the cell lineage and beyond (Nobel lecture). *Chembiochem* 4, 688-96.
Tabara, H., Sarkissian, M., Kelly, W. G., Fleenor, J., Grishok, A., Timmons, L., Fire, A. and Mello, C. C. (1999). The rde-1 gene, RNA interference, and transposon silencing in C. elegans. *Cell* 99, 123-132.
Tabara, H., Yigit, E., Siomi, H. and Mello, C. C. (2002). The dsRNA binding protein RDE-4 interacts with RDE-1, DCR-1, and a DExH-box helicase to direct RNAi in C. elegans. *Cell* 109, 861-71.
Tanentzapf, G. and Tepass, U. (2003). Interactions between the crumbs, lethal giant larvae and bazooka pathways in epithelial polarization. *Nat Cell Biol* 5, 46-52.
The_C_elegans_Sequencing_Consortium. (1998). Genome sequence of the nematode C. elegans: a platform for investigating biology. *Science* 282, 2012-8.
Timmons, L. and Fire, A. (1998). Specific interference by ingested dsRNA. *Nature* 395, 854.

Totong, R., Achilleos, A. and Nance, J. (2007). PAR-6 is required for junction formation but not apicobasal polarization in C. elegans embryonic epithelial cells. *Development* 134, 1259-68.

Tsien, R. Y. (2009). Constructing and exploiting the fluorescent protein paintbox (Nobel Lecture). *Angewandte Chemie* 48, 5612-26.

Tsukita, S., Furuse, M. and Itoh, M. (2001). Multifunctional strands in tight junctions. *Nat Rev Mol Cell Biol* 2, 285-93.

van Fürden, D., Johnson, K., Segbert, C. and Bossinger, O. (2004). The C. elegans ezrin-radixin-moesin protein ERM-1 is necessary for apical junction remodelling and tubulogenesis in the intestine. *Developmental biology* 272, 262-276.

Wang, Q. and Margolis, B. (2007). Apical junctional complexes and cell polarity. *Kidney Int* 72, 1448-58.

Weisz, O. A. and Rodriguez-Boulan, E. (2009). Apical trafficking in epithelial cells: signals, clusters and motors. *J Cell Sci* 122, 4253-66.

White, J. (1988). The anatomy. In *The nematode C. elegans*, (ed. W. B. Wood), pp. 81-122. New York: Cold Spring Harbor Laboratory Press.

White, J., Southgate, E., Thomson, J. and Brenner, S. (1986). The structure of the nervous system of Caenorhabditis elegans. *Philos Trans R Soc Lond B Biol Sci* 314, 1-340.

Wilkins, C., Dishongh, R., Moore, S. C., Whitt, M. A., Chow, M. and Machaca, K. (2005). RNA interference is an antiviral defence mechanism in Caenorhabditis elegans. *Nature* 436, 1044-7.

Wodarz, A., Ramrath, A., Grimm, A. and Knust, E. (2000). Drosophila atypical protein kinase C associates with Bazooka and controls polarity of epithelia and neuroblasts. *J Cell Biol* 150, 1361-74.

Wood, W. B. (1988). Preface/Front Matter. In *The nematode C. elegans*, (ed. W. B. Wood). Cold Spring Harbor, New York: Cold Spring Harbor Laboratory Press.

Yamanaka, T., Horikoshi, Y., Sugiyama, Y., Ishiyama, C., Suzuki, A., Hirose, T., Iwamatsu, A., Shinohara, A. and Ohno, S. (2003). Mammalian Lgl forms a protein complex with PAR-6 and aPKC independently of PAR-3 to regulate epithelial cell polarity. *Curr Biol* 13, 734-43.

Zhuang, J. J. and Hunter, C. P. (2011). Tissue-specificity of Caenorhabditis elegans Enhanced RNAi Mutants. *Genetics* 188, 235-7.

6

Intercellular Communication

Nuri Faruk Aykan
*Istanbul University, Institute of Oncology,
Turkey*

1. Introduction

Intercellular communication (transfert of information) is an essential issue for continuity of life in multicellular organisms. Several types of communication systems coordinate body functions to maintain homeostasis (Guyton & Hall, 2000). Until now, it has been accepted that two major organ systems control all physiologic processes within the human body: The **endocrine system** and the **nervous system** (Greenspan & Gardner, 2004). Beside them, a third organ system, **immune system,** is a super-system which provide recognition and destroy of foreign cells by specific coordination between their cells again within the body. The contact and communication between immun cells are used for the distinction between self and non-self. In recent years, considerable data supported the existence of dynamic interactions between these super-systems. For example, neuro-endocrine, neuro-immune, psycho-neuro-immuno-endocrinological cross communications have been identified (Downing & Miyan, 2000; Sternberg, 1997; Weihe et al., 1991). In addition, within organ systems, **autocrine, paracrine, juxtacrine, neurocrine, lumencrine (exocrine)** and finally **intracrine** communications have been defined (di Sant'Agnese, 1992; Greenspan & Gardner, 2004; Guyton & Hall, 2000; Hansson & Abrahamsson, 2001; Krantic et al., 2004; Miller, 2003; Patel et al., 1993; Re & Bryan, 1984; Re, 1989; Ruan & Lai, 2004; Sporn & Todaro, 1980; Sporn & Roberts, 1992; Zimmerman et al., 1993) (Figure 1). Intercellular communication in the organism is realized by specific molecules, except neural transmission exerted by action potentials (Despopoulos & Silbernagl, 2003; Faller & Schuenke, 2004; Guyton & Hall, 2000) and except information transfer by biophoton emission which has been reported very recently (Albrecht-Buehler, 1992; Cohen & Popp, 1997; Fels, 2009; Jaffe, 2005; Musumeci et al., 1999; Niggli, 1992). This review especially addresses chemical communication systems in the human body by simplified examples.

2. Modes of communication

This chapter will attempt to summarize the modes of communications in this order below; **autocrine** (including intracrine), **paracrine** (including juxtacrine, gap junctional, via Tunneling Nano-Tube like structures), **endocrine, neurocrine** (including neuro-endocrine) and **lumencrine** communications.

2.1 Autocrine communication

Autocrine communication (derived from *auto*: self and *krinein*: to secrete, Greek) is an activity of a hormone or growth factors (GFs) that binds to and affects the same cell that secreted it. These substances directly stimulate (or inhibit) the cell via their surface receptors. Autocrine secretion was described first by Sporn and Todaro in 1980. It explains self-regulation of cells. This concept is now not only important to explain malignant transformation, but is also mainstay of embryogenesis and morphogenesis. Autocrine regulation provides selective growth advantages during the earliest stages of embryogenesis before the development of a functioning circulatory system and endocrine function (Dockray, 1979; Sporn & Todaro 1980).

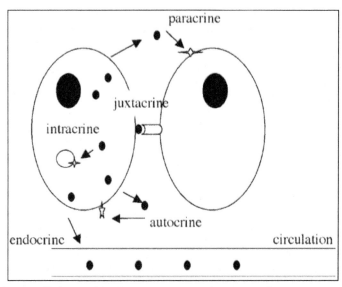

Fig. 1. Several intercellular modes of communication (courtesy of Mao-De Lai, corresponding author of the article Ruan & Lai, 2004)

Autocrine secretion is an important phenomenon in the regulation of the behavior of many normal cells such as macrophages, lymphocytes, fibroblasts and vascular smooth muscle cells. This regulation can be positive or negative manner. For example, oncogenes such as EGF, TGF-α and PDGF enhance autocrine pathways to increase cell replication during carcinogenesis. Same peptides also have an important autocrine role in tissue repair and wound healing in normal physiology. On the other hand TGF-β is a significant negative autocrine regulator in the adenoma-carcinoma sequence of human colon carcinogenesis (Sporn & Roberts, 1992). But, TGF-β is bifuntional like many other peptide growth factors and its stimulatory or inhibitory effects depends on many factors including cell type specificity, cell growth condition, and some other factors (Sporn & Roberts, 1988; Ruan & Lai 2004). In fact, cells are different than simple mechanical devices, they take new information from its environment and depending on conditions they give an appropriate response (Sporn & Roberts, 1992). For example, a specific autocrine cytokine such as TGF-β or interleukin-6 can act either positive or negative signal for growth in a given cell (Sporn & Roberts, 1988; Akira et al., 1990). Autocrine communication can be considered as a primitive mechanism of humoral regulation than endocrine secretion (Sporn & Todaro, 1980).

2.1.1 Internal autocrine (intracrine) communication

Another type of autocrine regulation is realized inside the cell. Internal autocrine or intracrine mode of action which is described first Re RN et al in 1984, indicates that some peptide hormones and growth factors (intracrines) bind and act in the cellular interior either after internalization by target cells or retention in their cells of synthesis (Re & Bryan, 1984; Re, 1989). Some endogenous cytokines such as interleukin-3 or PDGF can be modified and retained within the cell to ensure internal cellular action and they have high degree of intracellular biologic activity (Bejcek et al, 1989; Dunbar et al, 1989; Keating & Williams 1988). As shown in Figure 1, a chemical mediator (peptide growth factor or hormone) interacts with its specific receptor within the cell, bypassing the need of secretion outside, to exert functional activity. An intracrine system, in contrast to endocrine system, requires minimal amounts of biologically active hormones to exert their maximum hormonal effects. For this reason, the intracrine system plays an important role especially in the development of hormone-dependent neoplasms such as breast, prostate malignancies. As another example, locally produced bioactive androgens and/or estrogens exert their action in the cells where synthesis occurs without release in the extracellular space including circulation (Sasano et al., 2008). Labrie and colleagues described the formation of active androgens (such as DHT) from the inactive adrenal precursors in the some tissues or cells in adenocarcinoma of the prostate where biosynthesis takes place without release into the extracellular space as "intracrine activity" (Labrie et al., 1995, 2003, Sasano et al., 2008). On the other hand, estrogen-dependent breast carcinoma in which aromatase converts circulating androgens to estrogens (from androstenedione to estrone and from testosterone to estradiol, respectively) should also be considered as "intracrine tissue". One of the most studied example of intracrine function is about local renin-angiotensin system (RAS). As it is known, the proteolytic enzyme renin (an aspartyl protease) which cleaves angiotensinogen, is secreted mostly from juxtaglomerular cells in the renal afferent arteriole. Reduced renal arteriolar blood pressure and then the activation of local β2-adrenoreceptor stimulates the secretion of renin. The major source of plasma angiotensinogen is the liver, but it is also formed for local use in the heart and the brain. Cleavage of angiotensinogen by renin yields angiotensin I which has no biological activity. Further, angiotensin II is produced from angiotensin I by endothelial angiotensin-converting enzyme (ACE) and angiotensin III is produced from angiotensin II by aminopeptidase. Angiotensin II is a potent constrictor of vascular smooth muscle, and this action is mediated by the AT_1 receptor. Angiotensin II and III act on zona glomerulosa cells in the adrenal cortex and promote secretion of aldosterone. Today we know that renin is not simply a circulating enzyme but is a hormone and it is also an intracrine. Renin and angiotensin are also active within cells. A complete intracrine RAS exists in some cells (Re & Bryan, 1984). For example, an adrenal intracellular RAS has been reported (Peters et al., 1999). Prorenin and renin can bind to specific cellular receptors. Prorenin, and to a lesser extent, renin, can be internalized by cells where angiotensin II is produced (Re, 2003a). Internalized, activated prorenin causes both hypertension and cardiac injury. Nuclear angiotensin receptors were also reported (Re, 1999). There is a renin transcript in some cells (adrenal, brain) lacking the sequence encoding the secretory signal piece (renin exon 1A) (Clausmeyer et al., 2000; Peters et al.,

1999). Renin exon 1A generates angiotensin in mitochondria which have angiotensin receptors and stimulates aldosterone secretion in the adrenal cortex. This aldosterone secretion is inhibited by the angiotensin receptor blocker losartan (Peters et al., 1999). Renin exon 1A is upregulated by nephrectomy. Adrenal mitochondrial renin granules increase following nephrectomy (Peters et al, 1999). Release of angiotensin II in the intracellular space upregulates a series of genes including PDGF which stimulated proliferation (Re, 2003b). It has been reported that renin exon 1A upregulated in the ventricles of rats after myocardial infarction. Intracrine RAS may have a reasonable role in the processes like left ventricular hypertrophy, cardiac fibrosis and some forms of arrhytmia (Re, 2003b). So, the existence of intracrine RAS can be clinically important. Another important area related to intracrinology is the angiogenesis. Many intracrines (angiogenin, FGF-2, angiotensin) are angiogenic either directly or through the stimulation of vascular endothelial growth factor (VEGF) (Li & Keller, 2000; Re, 1999). Angiogenin is an RNase and needs nucleolar translocation to stimulate angiogenesis. VEGF also is an intracrine. Lee and colleagues, demonstrated that VEGFR1 expression was abundant in breast cancer cells (Lee et al., 2007). It was predominantly expressed internally in MDA-MB-231 and MCF-7 breast cancer cells and VEGFR1 antibody had no effects on the survival of these cells. Learning this intracrine concept has a practical significance because the usage of therapeutic antibodies against GFs have serious limitations in that an internal autocrine loop can not be accessible to antibody therapy (Sporn & Roberts, 1992).

2.2 Paracrine communication

Paracrine communication (derived from *para*: from beside by, Greek) is an activity of an agent (hormone or growth factor) that binds to and affects neighboring cells. The agent is directly released into the intercellular space and may involve many nearby cells that have receptors for this agent (Öberg, 1998; Raybould et al., 2003). One nice example of paracrine communication is the interaction between vascular endothelial cells and pericytes. The control of proliferation and migration of vascular endothelial cells can be mediated by neighboring cells; pericytes in a capillary (Antonelli-Orlidge et al, 1989), or smooth-muscle cells (Dennis & Rifkin, 1991) in an artery. A capillary endothelial cell synthesize latent TGF-β and a pericyte is required for activation of this latent molecule. This is a cooperative interaction via paracrine way. Loss of paracrine activation by pericytes may contribute diabetic proliferative vascular retinopathy (Antonelli-Orlidge et al, 1989). During recent years, interactions between endothelial cells and mural cells (pericytes and vascular smooth muscle cells) have gained increasing attention in physiological and pathological conditions including tumor angiogenesis, diabetic retinopathy, hereditary telangiectasia, lymphedema and hereditary stroke and dementia syndrome (Armulik et al, 2005). Some signaling pathways such as angiopoietin-Tie2, PDGF-B/PDGFR-β are described between endothelial cells and pericytes. Nitric oxide (NO) is also a paracrine agent; endothelial cells produce NO (and citrulline) from arginine as a substrate by endothelial nitric oxide synthase, and it diffuses into smooth muscle where induces relaxation and dilatation of blood vessels (Schechter & Gladwin, 2003). As another example, paracrine interactions between immune cells and fibroblasts are required for the normal repair of injured tissue (Sporn & Roberts, 1992). On the other hand, it has been demonstrated recently the multicellular autocrine and paracrine cross talk in the inflammatory tumor microenvironment; for example RAGE (receptor for advanced glycation end products) engagement in cancer cell surface with its ligands (AGEs, S100/calgranulins, amyloid A, amyloid-β and DNA-binding protein HMGB1) which are expressed and secreted by many cell types within the tumor microenvironment including fibroblasts, leukocytes and vascular cells, produces activation of multiple intracellular signalling mechanisms involved in several inflammation-associated clinical entities, such as cancer, diabetes, renal and heart failures and neurodegenerative diseases (Rojas et al, 2010). Intercellular bidirectional paracrine communication is essential also either in spermatogenesis or development of an egg competent to undergo fertilization and embryogenesis (Matzuk et al, 2002).

2.2.1 Juxtacrine communication

Another kind of paracrine communication between signaling and target cells is juxtacrine interactions. **Juxtacrine** mode of action (derived from *juxta*: nearby, Latin) is a direct and intimate contact between two cells such as macrophage-T lymphocyte, spermatogonia-Sertoli cell or endothelial cell (EC) and the leukocyte (Krantic et al., 2004; Patel et al., 1993; Zimmerman et al., 1993). This signaling form provides a mechanism for strict spatial control of activation of one cell by another and juxtacrine signaling is likely to be common in physiologic events that require tight regulation (Zimmerman et al., 1993). The term "juxtacrine" was coined by Anklesaria and Massagué and colleagues in 1990 (Anklesaria et al, 1990; Massagué, 1990). In juxtacrine systems the signaling factor acts while associated with the surface of signaling cells, rather than acting in the fluid phase. In the example of spermatogonia-Sertoli cell, spermatogonia produce somatostatin and Sertoli cells express sst2 receptors. Activation of sst2 receptor by somatostatin binding leads to a diminished expression of stem cell factor (SCF) expression by Sertoli cells. This inhibition of SCF is associated with a decrease in spermatogonia proliferation (Krantic et al, 2004). Immunologic synapse which involve multiple adhesion and regulatory molecules between antigen-presenting cell (APC) and T-cell can also be considered juxtacrine communication (Bromley et al. 2001; Biggs et al, 2011). Juxtacrine secretion provides a unique mechanism for preventing an undesirable diffuse action of a given cytokine on innocent bystander cells (Sporn and Roberts, 1992). For example tumor necrosis factor (TNF) is a cytokine that can act by a juxtacrine mechanism has been implicated as a critical mediator of cachexia, septic shock, rheumatoid artritis, autoimmune states, induction of HIV expression and the killing of tumor cells. Transmembrane form of TNF is highly active and cell-to-cell contact, without secretion into the intercellular space, is sufficient for TNF to kill a target tumor cell (Perez et al., 1990). Disruption of juxtacrine signaling may lead to pathologic outcomes, oxidant-injured endothelial cells is one example and this disruption may be a fundamental process in adult respiratory distress syndrome, shock and similar tissue injuries (Zimmerman et al., 1993).

2.2.2 Gap Junctional Intercellular Communication (GJIC)

GJIC is different than the other modes of communication where a ligand and its receptor interaction exists by diffuse (autocrine, paracrine, endocrine) or non-diffuse (juxtacrine) mechanism. This type of communication between adjacent cells is mediated via intercellular channels that cluster in specialized regions of the plasma membrane to form gap junctions (Robertson 1963, Revel and Karnovsky 1967, Wei et al 2004, as cited in Meşe et al, 2006). Gap junctional channels link the cytoplasm of two cells, and provide the exchange of ions (K+, Ca^{2+}), second messengers (cAMP, cGMP, IP3) and small metabolites like glucose (Kanno & Loewenstein, 1964; Lawrence et al., 1978, as cited in Meşe et al., 2006) . Valiunas et al (2005), recently showed that transfer of small interfering RNAs between neighboring cells trough gap junctions. GJIC is essential for many pysiological events such as cell synchronization, differentiation, cell growth, and metabolic coordination of avascular organs including epidermis and lens (Vinken et al., 2006; White and Paul, 1999). GJIC forms a close electrical and metabolic unit (syncytium). It is present in the epithelium, many smooth muscles, the myocardium, and the glia of the central nervous system (Despopoulos & Silbernagl, 2003). Electric coupling permits the transfer of excitation (*electrical synapses*); many examples can be given for this wave of excitation in the body such as atrium and ventricles of the heart, stomach, intestine, biliary tract, uterus and ureter. Gap junctions are formed by two unrelated protein families, the pannexins and connexins (Meşe et al, 2006). Connexins have four transmembrane domains and six connexins oligomerize to form hemichannels called "connexons". One connexon docks with another connexon on the adjacent cell, thereby forming a common channel which substances with molecular masses of up to around 1 kDa can pass (Despopoulos & Silbernagl, 2003). This organization requires the membranes of two adjacent cells leaving a 2-4 nm gap (Bruzzone et al., 1996; White & Paul 1999). Gap junction channels are selective permeable. There are at least 21 connexin isoforms and connexons can be formed either from a single type of connexin or from more than one type, leading to the formation of either homomeric or heteromeric hemichannels, respectively (Meşe et al, 2006) and this characteristic can explain selective permeability. For example connexin32 homomeric hemichannels were permeable to both cAMP and cGMP whereas connexin26/connexin32 hemichannels were permeable mainly to cGMP. The "Contact inhibition" process can be mediated in some cells by gap junctions (Trosko, 2007).

2.2.3 Intercellular communication via Tunneling Nano-Tube (TNT) like structures

Very recently, a novel mechanism for intercellular communication was discovered by which nanotubular structures, consisting of thin membrane bridges, mediate membrane continuity between mammalian cells (Rustom et al., 2004). These channels, referred to as tunneling nanotubes (TNT), were shown to actively traffic cytosolic content from cell to cell within the interior of their filaments (Rustom et al., 2004). TNTs were first described in cultured rat pheochromocytoma PC12 cells. Calcium ions, MHC class I proteins, prions, viral and bacterial pathogens, small organelles of the endosomal/lysosomal system and mitochondria are among identified TNT cargos until now (Eugenin et al., 2009; Gerdes, 2009; Gerdes et al., 2007; Gurke et al., 2008; Koyanagi et al., 2005). Intercellular exchange via TNT based cell-communication was reported in cells which have high motility and plasticity like progenitor cells, immune cells and tumor cells. The exchange of endosome-related organelles and other cellular components over long distances and the coordination of signaling between the connected cells are realized by this way (Rustom et al. 2004, Gerdes et al. 2007; Gerdes & Carvalho, 2008). Domhan et al (2011) reported also intercellular exchange by TNT, between human renal proximal epithelial cells; this may play an important role in renal physiology.

2.3 Endocrine communication

Endocrine system (derived from *endon*: inside, *krinein*: to secrete, Greek) is a radio-like communication system. It consists of endocrine glands and specialized groups of cells within organs of multicellular organism. The endocrine glands sends its hormonal messages like a radio broadcast to essentially all cells of human body by secretion into the circulation of blood. Hormones are chemical messengers of endocrine communication. They are transported through the bloodstream and cells which have a receiver (a *receptor*) take this message (Greenspan & Gardner, 2004). Hormones can be proteins (eg growth hormone, FSH, LH), peptides or peptide derivatives (eg ACTH), amino acid derivatives (eg catecholamines, thyroid hormones). Steroid hormones and vitamin D are derived from cholesterol. Retinoids are derived from carotenoids and eicosanoids are derived from fatty acids. Some hormones (eg insulin, growth hormone, prolactin, catecholamines) bind cell surface receptors, others (steroids, thyroid hormones) bind to intracellular receptors that act in the nucleus. Hormone binding alters receptor conformation and this alteration transmits the binding information into postreceptor events that influence cellular function (Greenspan & Gardner, 2004). Hormones serve as messenger substances that are mainly utilized for *slower, long-term transmission of signals*; they are carried by the blood to *target structures great distances away* (Despopoulos & Silbernagl, 2003). Endocrine system is essentially responsible for control and integration of multicellular organism. The principal functions of endocrine hormones at the target level, are to control and regulate enzyme activity, transport processes, growth, secretion of other hormones, exert negative or positive feedback control and coordinate cells of same type. Endocrinology is a great and expanding discipline of science (Table 1).

2.4 Neurocrine communication

Nervous communication is point-to-point through nerves and *electrical in nature* and *fast*. By this aspect, communication by nervous system is similar to sending messages by conventional telephone so it is a cable phone-like system. In neurocrine communication, neuronal cells release their products directly into the synaptic space; they act on another cell type (Öberg, 1998). A synapse is the site where the axon of a neuron communicates with effectors or other neurons (Despopoulos & Silbernagl, 2003). According to the termination of an axon, the synapse may be axo-dendritic, axo-axonic or axo-somatic (Faller and Schuenke, 2004). Chemical synapses utilize (neuro)transmitters for the transmission of information. The arrival of action potential to the synapse in the axon triggers the release of transmitter from the presynaptic terminals. The transmitter then diffuses across the narrow intercellular gap (synaptic cleft) which is approximately 10-50 nm, and it binds postsynaptic receptors in the membrane of a neuron or a glandular or muscle cell (Despopoulos & Silbernagl, 2003; Faller & Schuenke, 2004). Transmitters are released by exocytosis of synaptic cytosolic storage vesicles. Depending on the type of transmitter and receptor involved, the effect on the postsynaptic membrane may be excitatory or inhibitory. Neuroscience, like endocrinology is also another essential interdisciplinary science (Table 1).

2.4.1 Neuro-endocrine communication

In this system, neurocrine and endocrine communications exist together. Neuroendocrinology is a studying science the interactions between nervous and endocrine systems (Greenspan & Gardner, 2004). There are two major mechanism of neural regulation of endocrine function; the first is *neurosecretion* which refers to neurons that secrete hormones into the circulation. For example, hypothalamic neurons synthesize and secrete hormones (*releasing or release inhibiting hormones*) into blood vessels (a kind of portal venous system) that communicate with the anterior pituitary. Posterior pituitary hormones (oxytocin and ADH) are transported from hypothalamic neurons to the ends of the axons in the neurohypophysis where enter the systemic circulation directly (Despopoulos & Silbernagl, 2003). The second is the *direct autonomic innervation* of endocrine tissues (such as adrenal medulla, pancreatic islets and gut) which couples central nervous system signals to hormone release. Enterochromaffin cells (EC) in the gastrointestinal tract have close contact with nerve elements of both afferent and efferent type adjacent to the basal lamina of the mucosa and true synapses have been identified (Ahlman & Dahlström, 1983). One of the interesting example of neuro-endocrine communication is the stimulation of gastric secretion in the cephalic phase (Guyton & Hall, 2000). In this phase of gastric secretion, when we see or smell a nice food while we are hungry, gastric secretion increases even before food enters the stomach (Pavlov's sham-feeding assay). Neurogenic signals come by vagus to stomach. In this point the first chemical messengers are acetylcholine and gastrin-releasing peptide (GRP). It has been demonstrated that G cells have muscarinic receptors and muscarine-like action, particularly in the M3 receptor-mediated route, plays a significant role in acetylcholine-mediated gastrin secretion (Matsuno et al., 1997). And we know also that cephalic phase of acid secretion is augmented predominantly by acetylcholine and gastrin while histamine is of major importance during the gastric phase (Schusdziarra, 1993). Martinez and colleagues demonstrated similarly that atropine and gastrin antibody decrease basal acid secretion while gastrin antibody only did not block the rise in acid during sham feeding (Martinez et al., 2002). Gastrin stimulated acid secretion is through releasing histamine from ECL (enterochromaffine-like) cells (Waldum et al., 2002). Cholecystokinin 2 (gastrin) receptors in the stomach are only in the ECL cells (Waldum et al., 2002). Then, ECL cells release histamine and stimulate oxinthic cells via paracrine way (by binding H_2 receptors) to produce HCl. As a control of gastric acid secretion, somatostatin (SS14) secreted by D cells inhibits both G cell and the parietal cell.

2.5 Lumencrine (exocrine) communication

Another type of communication is lumencrine (exocrine) secretion in the open cell types like pancreas and prostate. Like the other modes of regulations, lumencrine mechanism can also play an important regulatory role both during growth and differentiation of the prostate as well as in the secretory process of the mature gland (di Sant'Agnese, 1992). Calcitonin, GRP (bombesin) and somatostatin have been reported in semen that they may be directly secreted into the ejaculate (Arver & Sjoberg, 1982; Bucht et al, 1986; Gnessi et al, 1989; Sasaki & Yoshinaga, 1989; Sjoberg et al, 1980; cited in di Sant'Agnese, 1992). It has also been shown that a decrease in seminal ionized calcium correlates with a decrease in motility of sperm

(Prien et al., 1990). On the other hand, lumencrine secretion of pancreatic enzymes, water and ions play a major role in the duodenal phase of the digestion (Raybould et al., 2003) (Figure 2).

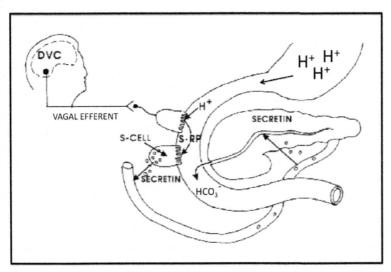

Fig. 2. Lumencrine, endocrine and neuro-endocrine communications during the regulation of pH in duodenum (modified from Li et al., and cited in Konturek et al, 2003). DVC: Dorsal vagal center, S-RP: Secretin releasing peptide.

2.6 Hypotheses about intercellular communication

Information theory was developed by Shannon, approximately more than 60 years ago (Shannon, 1948), to determine *quantitative* aspects of information exchange between a biologic source and a biologic receiver (Mayer and Baldi, 1991). According to this theory, peptidergic cell-to-cell communication between neurocrine, endocrine and growth factor-mediated messages require different encoding and decoding strategies. On the other hand *qualitative* component of the exchanged message is concerned with semantic information such as human language. The word "information" means "knowledge of order" in common language (Vincent, 1994). The laws of linguistics and semantics are valid not only at the organismic level, but also at the cellular and molecular level (Vincent, 1994). Today, we know that bacteria communicate by quorum sensing molecules (Miller & Bassler, 2001). Microbian language contains two component system which consists of a signal (input) and a response (output) (Pechère, 2007). Sensors receive the signal, effectors make the response. Cells use a molecule-based language called *cellese* which has the counterparts of sound- and visual-based human language (*humanese*) (Ji, 1999). What is transmitted is the meaning of the message (significance) which can be memorized by the cell, providing a possible following use (Vincent, 1994). It was suggested that cytokines can be viewed as symbols in an intracellular language (Sporn & Roberts, 1988). The participation of the extracellular matrix in the language of intercellular communication is a way that multicellular organisms can use past experience to determine the response to cytokines and interactions with matrix enable cytokines to elicit adaptive responses (Nathan & Sporn, 1991). Unique phenotype of cells based their carbohydrate determinants on their cell surfaces is another area of interest (glycobiology) in intercellular communication (Sporn & Roberts, 1992). Many peptide growth factors and cytokines are described and they are multifunctional ((Sporn & Roberts, 1988). It is apparent that they form part of complex cellular signaling language, in which the individual peptides are the equivalent of characters of an alphabet or code (Sporn & Roberts, 1988). Five years ago, we proposed an intercellular network model (message-adjusted network) in the physiology of gastro-entero-pancreatic (GEP) endocrine system, based on up-to-date information from medical publications (Aykan, 2007). In this network; **message** is an input which can affect the physiologic equilibrium, **mission** is an output to improve the disequilibrium, **aim** is always maintenance of homeostasis. Messages are picked up by biologic sensors or detectors. If we orientate to a transmission of a unique, physiologic, simple message we can design its proper network (Figure 3). In this model, different cells use different chemical messengers via different modes of regulations to transmit the same message. These **message-adjusted intercellular networks** may be most important (or unique) determinants in the formation of proper, environmentally adaptive multi-cellular organizations in the biology and it should be tested in the laboratory.

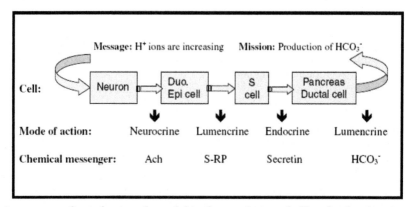

Fig. 3. A message-adjusted network model in the regulation of pH at the duodenum. Unbuffered hydrogen ions coming from stomach stimulate secretin release from S cells in the duodenum. Major effect of secretin is the secretion of bicarbonate ions from pancreatic ductal epithelium. Vagal stimulation results also small stimulation of pancreatic bicarbonate secretion.

Human body has more than 200 different cells (Alberts et al., 1994). The crucial question in the field of communication physiology is that "how they communicate each one with another?" Today, we know the words that are chemical messengers (bioactive peptides like hormones and neuromediators, cytokines, growth factors, some ions and other molecules), but we don't know languages specific to cell populations (tissue specific languages!). Although there is an expanding knowledge in molecular biology, some scientific disciplines which should be related to modes of communication are not developed yet in the literature (Table 1).

Intercellular Communication	Scientific Discipline
Intracrine	Intracrinology
Autocrine	?
Paracrine	?
Juxtacrine	?
Endocrine	Endocrinology
Neurocrine	Neuroscience
Neuro-endocrine	Neuroendocrinology
Lumencrine	?

Table 1. Modes of intercellular communication and related scientific disciplines.

On the other hand, we suggest that pheromones can play a role in lumencrine interindividual unconscious communication (Mayer & Baldi, 1991; Brennan & Keverne, 2004; Knecht et al., 2003) but we don't know if there is a wireless-like intercellular communication. This type of communication, if there is, should be between mobile cells, such as circulating cells, spermatozoa or mature oocyte and it should be bi-directional to provide cell to cell crosstalk. One potential example can be the induction of hypertrophy of draining lymph nodes by mast-cell derived tumour necrosis factor during infection (Mc Lachlan et al., 2003). The mode of this action of mast cells is defined as a remote control mechanism (Buckland, 2003). A second candidate for wireless communication can be again in the immune system; substance P and its receptors have been detected in granulocytes, monocytes and lymphocytes (Ferone et al, 2001). A third and most interesting example may be in the spermatozoa-egg communication; recent studies indicate that olfactory receptors might be a role in the chemotaxis of spermatozoa (Sliwa, 2003; Spehr et al, 2004; Fukuda et al, 2004; Eisenbach & Tur-Kaspa, 1999).

3. Conclusion

After human genome project (HU-GO) and protein organisations (HU-PO), it is time to resolve all parts of intercellular communication. Clarifying intercellular communication systems is as important as intracellular signal mechanisms. Finally, we believe that intercellular communication in our world becomes by specific molecules; these molecules are the words of the cell language.

4. Acknowledgment

Author thanks Ass. Prof. Barbaros Durgun, M.D. for his kind support in the preparation of figure 3 and Mao De Lai for kind permission of figure 1.

5. References

Ahlman, H. and Dahlström, A. (1983). Vagal mechanisms controlling serotonin release from the gastrointestinal tract and pyloric motor function. *J Auton Nerv Syst* 9:119-140.

Akira, S., Hirano, T., Taga, T., & Kishimoto, T. (1990). Biology of multifunctional cytokines: IL-6 and related molecules (IL-1 and TNF). *FASEB J.* 4:2860-2867.

Alberts B et al. (1994). Molecular biology of the cell. (3rd ed). Garland, p.1188.

Albrecht-Buehler, G. (1992). Rudimentary form of cellular "vision". *Proc. Natl. Acad. Sci. USA*, 89:8288-8292.

Anklesaria, P., J. Teixidó, M. Laiho, J. H. Pierce, J. S. Greenberger, & J. Massagué. (1990). Cell-cell adhesion mediated by binding of membrane-anchored transforming growth factor *a* to epidermal growth factor receptors promotes cell proliferation. *Proc. Natl. Acad. Sci. USA* 87:3289-3293.

Antonelli-Orlidge, A., Saunders, K.B., Smith, S.R. & D'Amore, P.A. (1989). An activated form of transforming growth factor β is produced by cocultures of endothelial cells and pericytes. *Proc. Natl. Acad. Sci. USA*. 86:4544-4548.

Armulik, A., Abramsson, A. & Betsholtz, C. (2005). Endothelial/pericyte interactions. *Circ Res.* 97:512-523.

Aykan, N.F. (2007). Message-adjusted network (MAN) hypothesis in gastro-entero-pancreatic (GEP) endocrine system. *Medical Hypotheses*, 69, 571-574.

Bejcek, B.E., Li, D.Y. & Deuel, T.F. (1989). Transformation by v-sis occurs by an internal autoactivation mechanism. *Science*, 245:1496-1498.

Biggs, M.J.P., Milone, M.C., Santos, L.C., Gondarenko, A. & Wind, S.J. (2011). High-resolution imaging of the immunological synapse and T-cell receptor microclustering through micrfabricated substrates. *J. R. Soc. Interface.* 8, 1462-1471.

Brennan, P.A., Keverne, E.B., (2004). Something in the air? New insights into mammalian pheromones. *Curr Biol.* 14(2):R81-9.

Bromley, S.K., Burack, W.R., Johnson, K.G., et al., (2001). The immunological synapse. *Annu. Rev. Immunol.* 19, 375-396.

Bruzzone, R., White, T.W., Paul, D.L. (1996) Connections with connexins: the molecular basis of direct intercellular signaling. *Eur J Biochem* 238:1-27.

Buckland, J. (2003). Mast cells act by remote control. *Nature Rev. Immunol.* 3, 927.

Clausmeyer, S., Reinecke A, Farrenkopf R, Unger T & Peters J. (2000). Tissue-specific expression of a rat renin transcript lacking the coding sequence for the prefragment and its stimulation by myocardial infarction. *Endocrinology* 141: 2963-2970.

Cohen S and Popp FA, (1997). Low-level luminescence of the human skin. *Skin Research and Technology* 3: 177-180.

Dennis, P.A. and Rifkin, D.B. (1991). Cellular activation of latent transforming growth factor β requires binding to the cation-independent mannose 6-phosphate/insulin-like growth factor type II receptor. *Proc Natl Acad Sci USA*. 88:580-584.

Despopoulos, A, Silbernagl, S. (Eds), (2003). *Color Atlas of Physiology*, 5th Edition, Georg Thieme Verlag, Stuttgart, New York.

di Sant'Agnese, P.A., (1992). Neuroendocrine differentiation in human prostatic carcinoma. *Hum. Pathol.* 23, 287-296.

Dockray, GJ. (1979). Evolutionary relationships of the gut hormones. *Fed Proc.* 38:2295-2301.

Domhan, S., Ma, L., Tai, A., Anaya, Z., Beheshti, A., Zeier, M., Hlatky, L. and Abdollahi, A. (2011). Intercellular communication by exchange of cytoplasmic material via tunneling nano-tube like structures in primary human renal epithelial cells. *PLoS One* 6(6): e21283.

Downing, J.E., Miyan, J.A., (2000). Neural immunoregulation: Emerging roles for nerves in immune homeostasis and disease. *Immunol. Today* 21, 281-289.

Dunbar, C.E., Browder, T.M., Abrams, J.S. & Nienhuis, A.W. (1989). COOH-terminal-modified interleukin-3 is retained intracellularly and stimulates autocrine growth. *Science*, 245:1493-1496.

Eisenbach, M. and Tur-Kaspa, I., (1999). Do human eggs attract spermatozoa? *Bioessays.* 21(3):203-210.

Eugenin, E.A., Gaskill, P.J., Berman, J.W. (2009) Tunneling nanotubes (TNT) are induced by HIV-infection of macrophages: a potential mechanism for intercellular HIV trafficking. *Cell Immunol* 254: 142-148.

Faller, A, Schuenke, M., (eds), (2004). *The Human Body. An introduction to structure and function.* Georg Thieme Verlag, Stuttgart, New York.

Fels, D. 2009. Cellular communication through light. *PloS One*, 4(4):e5086.

Ferone, D., Hofland, L.J., Colao, A. et al. (2001). Neuroendocrine aspects of immunolymphoproliferative diseases. *Ann. Oncol.* 12(Suppl. 2):S125-S130.

Fukuda, N., Yomogida, K., Okabe, M., Touhara, K. (2004). Functional characterization of a mouse testicular olfactory receptor and its role in chemosensing and in regulation of sperm motility. *J. Cell Sci.* 117(24):5835-45.

Gerdes, H.H. (2009) Prions tunnel between cells. *Nat Cell Biol* 11: 235-236.

Gerdes, H.H., Bukoreshtliev, N.V., Barroso, J.F. (2007) Tunneling nanotubes: a new route for the exchange of components between animal cells. *FEBS Lett* 581: 2194-2201.

Gerdes, H.H., Carvalho, R.N. (2008) Intercellular transfer mediated by tunneling nanotubes. *Curr Opin Cell Biol* 20: 470-475.

Greenspan, F.S., Gardner, D.G. (Eds), (2004). *Basic and Clinical Endocrinology*, Seventh Edition, pp. 2, McGraw-Hill, New York.

Gurke, S., Barroso, J.F., Hodneland, E., Bukoreshtliev, N.V., Schlicker, O., et al. (2008) Tunneling nanotube (TNT)-like structures facilitate a constitutive, actomyosin dependent exchange of endocytic organelles between normal rat kidney cells. *Exp Cell Res* 314: 3669-3683.

Guyton, A.C. and Hall, J.E. (Eds), (2000). *Textbook of Medical Physiology*, Tenth Edition, pp. 836, Saunders, Philadelphia.

Hansson, J., Abrahamsson, P.A., 2001. Neuroendocrine pathogenesis in adenocarcinoma of the prostate. *Ann. Oncol.* 12 (Suppl. 2): S145-S152.

Jaffe, L.F. (2005). Marine plants may polarize remote Fucus eggs via luminescence. *Luminescence* 20: 414-418.

Ji, S. (1999). The linguistics of DNA: words, sentences, grammar, phonetics, and semantics. *Ann NY Acad Sci* 870: 411-417.

Keating, M.T. & Williams, L.T. (1988). Autocrine stimulation of intracellular PDGF receptors in v-*sis*-transformed cells. *Science*, 239:914-916.

Knecht, M., Witt M, Abolmaali N, et al. (2003). The human vomeronasal organ. *Nervenarzt.* 74(10):858-862.

Konturek SJ, Pepera J, Zabielski K, et al. (2003): Brain-gut axis in pancreatic secretion and appetite control. *J Physiol Pharmacol* 54:3:293-317.

Koyanagi, M., Brandes, R.P., Haendeler, J., Zeiher, A.M. and Dimmeler, S. (2005) Cell-to-cell connection of endothelial progenitor cells with cardiac myocytes by nanotubes: a novel mechanism for cell fate changes? *Circ Res* 96: 1039-1041.

Krantic, S., Goddard, I., Saveanu, A., et al., (2004). Novel modalities of somatostatin actions. *Eur. J. Endocrinol.* 151, 643-655.

Labrie, F., Belanger, A., Simard, J., Luu-The, V. & Labrie, C. (1995). DHEA and peripheral androgen and estrogen formation: intracrinology, *Ann. N.Y. Acad. Sci.* 774 16–28.

Labrie, F., Luu-The, V., Labrie, C., Belanger, A., Simard, J., Lin, S.X. & Pelletier, G. (2003). Endocrine and intracrine sources of androgens in women: inhibition of breast cancer and other roles of androgens and their precursor dehydroepiandrosterone, *Endocr. Rev.* 24:152–182.

Lee TH, Seng S, Sekine M. et al. (2007). Vascular endothelial growth factor mediates intracrine survival in human breast carcinoma cells through internally expressed VEGFR1/FLT1. *PloS Med.* 4(6):e186.

Li, W. & Keller, G. (2000). VEGF nuclear accumulation correlates with phenotypical changes in endothelial cells. *J Cell Sci* 113: 1525–1534.

Martinez, V., Barrachina, M.D., Ohning, G. & Taché, Y. (2002). Cephalic phase of acid secretion involves activation of medullary TRH receptor subtype 1 in rats. *Am J Physiol Gastrointest Liver Physiol.* 283(6):G1310- G1319.

Massagué, J. (1990). Transforming growth factor-a. A model for membrane-anchored growth factors. *J. Biol. Chem.* 265:21393-21396.

Matsuno M, Matsui T, Iwasaki A, Arakawa Y. (1997). Role of acetylcholine and gastrin-releasing peptide (GRP) in gastrin secretion. *J Gastroenterol.* 32(5):579-586.

Matzuk, M.M., Burns, K.H., Viveiros, M.M. and Eppig, J.J. (2002). Intercellular communication in the mammalian ovary: oocytes carry the conversation. *Science.* Vol. 296, 2178-2180.

Mayer, E.A. and Baldi, J.P. (1991). Can regulatory peptides be regarded as words of a biological language? *Am J Physiol* 261: G171–G184.

McLachlan, J.B., Hart, J.P., Pizzo, S.V. et al., (2003). Mast cell-derived tumor necrosis factor induces hypertrophy of draining lymph nodes during infection. *Nat. Immunol.* 4(12):1199-205.

Meşe, G., Richard, G. and White, T.W. (2006). Gap junctions: basic structure and function. *Journal of Investigative Dermatology*, Vol.127, 2516-2524.

Miller, M.B. and Bassler, B.L. (2001). Quorum sensing in bacteria. *Annu Rev Microbiol.* 55:165-99.

Miller, L.J. (2003). Organization of the gut endocrine system. In: T. Yamada, D.H. Alpers, N. Kaplowitz, L Laine, C Owyang, D.W. Powell (Eds), *Textbook of Gastroenterology*, 4th ed, pp. 48-77, Lippincott Williams & Wilkins, Philadelphia.

Musumeci, F., Scordino, A., Triglia, A., Blandino, G., Milazzo, I. (1999). Intercellular communication during yeast cell growth. *Europhysics Letters* 47: 736–742.

Nathan, C. and Sporn, M. (1991). Cytokines in context. *J Cell Biol.* 113:5: 981-986.

Niggli H. (1992). Ultraweak photons emitted by cells: biophotons. *Journal of Photochemistry and Photobiology* B 14: 144–146.

Öberg, K. (Ed), (1998). *Neuroendocrine gut and pancreatic tumours; tumour biology, diagnosis and treatment.* Novartis Pharma AG, Basel.

Patel, K.D., Lorant, E., Jones, D.A., et al. (1993). Juxtacrine interactions of endothelial cells with leukocytes: tethering and signaling molecules. *Behring Inst. Mitt.* 92, 144-64.

Pechère, J-C. (ed), (2007). *The Intelligent Microbe.* First ed. Frison-Roche.

Perez, C., Albert, I., DeFay, K., Zachariades, N., Gooding, L., and Kriegler, M. (1990). A nonsecretable cell surface mutant of tumor necrosis factor (TNF) kills by cell-to-cell contact. *Cell.* 63, 251-258.

Peters J, Obermüller N, Woyth A, Peters B, Maser-Gluth C, Kranzlin B & Gretz N. (1999). Losartan and angiotensin II inhibit aldosterone production in anephric rats via different actions on the intraadrenal renin-angiotensin system. *Endocrinology* 140: 675-682.

Prien, S., Lox, C., Messer, R. and DeLeon, F. (1990). Seminal concentrations of total and ionized calcium from men with normal and decreased motility. *Fertil Steril* 54:171-172.

Raybould, H.E., Pandol, S.J., Yee, H., (2003). The integrated responses of the gastrointestinal tract and liver to a meal. In: T. Yamada, D.H. Alpers, N. Kaplowitz, L Laine, C Owyang, D.W. Powell (Eds), Textbook of Gastroenterology, 4th ed, pp. 2-12, Lippincott Williams & Wilkins, Philadelphia.

Re, R. (1999). The nature of intracrine peptide hormone action. *Hypertension* 34: 534-538.

Re, R.N. (1989). The cellular biology of angiotensin: paracrine, autocrine and intracrine action in cardiovascular tissues. *J Mol Cell Cardiol* 21, Suppl 5: 63-69.

Re, R.N. (2003a). Intracellular rennin and the nature of intracrine enzymes. *Hypertension*, 42:117-122.

Re, R.N. (2003b). Implications of intracrine hormone action for physiology and medicine. *Am J Physiol Heart Circ Physiol*, 284: H751-H757.

Re, R.N. and Bryan, S.E. (1984). Functional intracellular renin-angiotensin systems may exist in multiple tissues. *Clin Exp Hypertens* A6, Suppl 10-11: 1739-1742.

Rojas, A., Figueroa, H. and Morales, E. (2010). Fueling inflammation at tumor microenvironment: the role of multiligand/rage axis. *Carcinogenesis.* Vol.31, No.3 pp. 334-341.

Ruan, W.J. and Lai, M.D. (2004). Autocrine stimulation in colorectal carcinoma (CRC). *Medical Oncology*, vol. 21, no. 1, 1-7.

Rustom, A., Saffrich, R., Markovic, I., Walther, P., Gerdes, H.H. (2004) Nanotubular highways for intercellular organelle transport. *Science* 303: 1007-1010.

Sasano, H., Suzuki, T., Miki, Y. & Moriya, T. (2008). Intracrinology of estrogens and androgens in breast carcinoma. *J Steroid Biochem & Mol Biol.* 108: 181-185.

Schechter, A.N. and Gladwin, M.T. (2003). Hemoglobin and the paracrine and endocrine functions of nitric oxide. *N Engl J Med.* 348:15:1483-1485.

Schusdziarra, V. (1993). Physiologic regulation of gastric acid secretion. *Z Gastroenterol.* 31(3):210-5.

Shannon, C.E. (1948). A mathematical theory of communication. *Bell Syst Tech J* 27:379-423.

Sliwa, L. (2003). Chemotaxis of spermatozoa--an important and little known process accompanying fertilization. *Folia Med Cracov.* 44(1-2):153-8.

Spehr, M., Schwane, K., Riffell, J.A. et al. (2004). Particulate adenylate cyclase plays a key role in human sperm olfactory receptor-mediated chemotaxis. *J. Biol. Chem.* 279(38):40194-40203.

Sporn, M.B., Roberts, A.B. (1988). Peptide growth factors are multifunctional. *Nature* 332, 217-219.

Sporn, M.B., Roberts, A.B. (1992). Autocrine secretion-10 years later. *Ann Int Med.* 117:408-414.

Sporn, M.B., Todaro, G.J. (1980). Autocrine secretion and malignant transformation of cells. *New Engl. J. Med.* 303, 878-880.

Sternberg, E.M. (1997). Emotions and disease: From balance of humors to balance of molecules. *Nat. Med.* 3, 264-7.

Trosko, J.E. (2007). Gap junctional intercellular communication as a biological "Rosetta Stone" in understanding, in a systems biological manner, stem cell behavior, mechanisms of Epigenetic Toxicology, chemoprevention and chemotherapy. *J Membrane Biol.* 218: 93–100.

Valiunas, V., Polosina, Y.Y., Miller, H. et al. (2005). Connexin-specific cell-to-cell transfer of short interfering RNA by gap junctions. *J Physiol.* 568:459-468.

Vincent, L.M. (1994). Reflexions on the usage of information theory in biology. *Acta Biotheor.* 42(2-3): 167-179.

Vinken, M., Vanhaecke, T., Papeleu, P., Snykers, S., Henkens, T. and Rogiers V. (2006) Connexins and their channels in cell growth and cell death. *Cell Signal* 18:592–600.

Waldum, H.L., Kleveland, P.M., Sandvik, A.K., Brenna, E., Syversen, U., Bakke, I. and Tommeras, K. (2002). The cellular localization of the cholecystokinin 2 (gastrin) receptor in the stomach. *Pharmacol Toxicol.* 91(6):359-62.

Weihe, E., Nohr, D., Michel, S., Muller, S., Zentel, H.J., Fink, T. & Krekel, J. (1991). Molecular anatomy of the neuro-immune connection. *Int. J. Neurosci.* 59, 1-23.

White, T.W., Paul, D.L. (1999) Genetic diseases and gene knockouts reveal diverse connexin functions. *Annu Rev Physiol* 61:283–310.

Zimmerman, G.A., Lorant, D.E., McIntyre, T.M. & Prescott, S.M. (1993). Juxtacrine intercellular signaling: another way to do it. *Am. J. Respir. Cell Mol. Biol.* 9, 573-577.

Molecular and Sub-Cellular Gametogenic Machinery of Stem and Germline Cells Across Metazoa

Andrey I. Shukalyuk[1] and Valeria V. Isaeva[2]
[1]*Institute of Biomaterials and Biomedical Engineering,
University of Toronto, Toronto, Ontario,*
[2]*A. V. Zhirmunsky Institute of Marine Biology of The Far Eastern Branch of
The Russian Academy of Sciences, Vladivostok and A. N. Severtsov Institute of
Ecology and Evolution of The Russian Academy of Science, Moscow,*
[1]*Canada*
[2]*Russia*

1. Introduction

Metazoan life cycle and development include two main types of stem cells: the germline cells and the somatic stem cell lineages (Hogan, 2001; Rinkevich, 2009; Srouji & Extavour, 2011). In animals with asexual reproduction, the germ lineage is not segregated during embryogenesis, and the line of pluripotent stem cells is maintained continuously throughout the life of an individual or a colony, being predecessors of germ cells and a wide spectrum of somatic cells (Buss, 1987; Isaeva, 2011; Rinkevich, 2009; Sköld et al., 2009). Examples of such pluripotent gametogenic stem cells include sponge archaeocytes, cnidarian interstitial cells, planarian neoblasts, ascidian hemoblasts and stem cells of colonial rhizocephalans (reviews: Isaeva et al., 2008, 2009; Rinkevich et al., 2009; Sköld et al., 2009; Srouji & Extavour, 2011). In addition to the germline segregation by preformation, or mosaic developmental mode and epigenesis, or regulative mode (Extavour, 2008; Extavour & Akam, 2003; Gustafson & Wessel, 2010), somatic embryogenesis was also recognized (Buss, 1987; Blackstone & Jasker, 2003; Gustafson & Wessel, 2010; Rinckevich et al., 2009; Rosner et al., 2009). Earlier, somatic embryogenesis as natural cloning in animals was termed blastogenesis (Berrill, 1961; Ivanova-Kazas, 1996). In the life cycle of colonial animals, one generation of oozooid (an individual that has developed from an egg) alternates with numerous generations of blastozooids, with alternating morphogenetic processes: embryogenesis and blastogenesis (Ivanova-Kazas, 1996). Many animals, including placozoans, sponges, cnidarians, platyhelminths, nemerteans, entoproctans, ectoproctans, annelids, hemichordates and urochordates are capable of somatic embryogenesis (Buss 1987; Blackstone & Jasker, 2003; Gustafson & Wessel, 2010; Rosner et al., 2009). Among arthropods, many parasitic rhizocephalan crustaceans (Rhizocephala: Cirripedia: Crustacea) have asexual reproduction, somatic embryogenesis by budding without separation of blastozooids resulting in the emergence of colonial organization

(Høeg & Lützen, 1993, 1995; Høeg et al., 2005). We have found undifferentiated stem cells in stolons, buds and ovary rudiments of the colonial rhizocephalans *Polyascus polygenea* (Isaeva et al., 2001, 2004; Shukalyuk, 2002; Shukalyuk et al., 2005, 2007) and *Peltogasterella gracilis* (Isaeva et al., 2003; Shukalyuk et al., 2001, 2005). The rhizocephalan stem cells take part in the morphogenesis of the earliest buds, and later migrate to the developing ovary as primary germ cells. So, pluripotent gametogenic stem cells are a cellular source in the realization of reproductive strategy including both sexual and asexual reproduction in colonial rhizocephalan.

All stem cells are characterized by two common properties that extend across diverse taxa: first, the capacity for self-renewal, the ability to propagate without loss of stemness property; second, the ability to give rise to numerous progeny that are fated for further differentiation into specialized cells (Alié et al., 2011; Cox et al., 1998; Srouji & Extavour, 2011; Watanabe et al. 2009). Depending on the breadth of the potential range of cell differentiation, totipotent, pluripotent, multipotent, oligopotent, and unipotent stem cells are distinguished, but this terminology is not unified (see Isaeva, 2010; 2011).

Female germline cells can be qualified as unipotent, since they produce only one type of differentiated cells, and totipotent, taking into account their potential of developing into a whole organism. There is no doubt that differentiated and deeply specialized gametes are unipotent cells producing only oocytes or sperm under specific signaling control of their niche. However, the progenitors of germline cells are multipotent or even pluripotent, also capable of differentiating into somatic lineages *in vitro* or *in vivo*, causing various germline-based embryonic tumorogenesis.

Stem cells of invertebrates with asexual reproduction are capable of differentiation into both germline and all, most or many somatic cell types are traditionally referred to as totipotent, pluripotent or multipotent. We consider here these cells as gametogenic pluripotent stem cells. In asexually reproducing invertebrates no early segregation of the germ cell lineage is observed. The lineage of pluripotent, traditionally referred to as totipotent, stem cells ensures both sexual and asexual reproduction over the entire life span of an individual or a colony. These pluripotent stem cells can differentiate into gametes and somatic cells in adult organisms. We studied pluripotent gametogenic stem cells in asexually reproducing representatives of five animal types: archaeocytes in the sponge (Porifera) *Oscarella malakhovi* (Isaeva & Akhmadieva, 2011), interstitial cells in the colonial hydroids (Cnidaria) *Obelia longissima* and *Ectopleura crocea* (Isaeva et al., 2011), neoblasts in the planarian (Platyhelmintes) *Girardia tigrina* (Isaeva et al., 2005), stem cells in the colonial rhizocephalans (Arthropoda) *Peltogasterella gracilis* and *Polyascus polygenea* (Isaeva et al., 2003, 2004; Shukalyuk et al., 2005, 2007), hemoblasts, stem cells in the colonial ascidian (Chordata) *Botryllus tuberatus* (Akhmadieva et al., 2007), and also embryonic stem cells as a benchmark for pluripotency, using *in vitro* culture, electron microscopic, histological, histochemical and molecular methods.

Mammalian embryonic stem cells (ESCs) are considered as a standard cell culture model for studying pluripotency (Do & Schöler, 2009). In our studies, as well as in the present review, we compare our data on invertebrate pluripotent stem cells with the information on the molecular signature of pluripotent stem cells of various animals, including mouse ESCs (Isaeva et al., 2003; Shukalyuk, 2009; Shukalyuk et al., 2005, 2011; Shukalyuk & Stanford,

2008), taking into consideration that mammalian ESCs *in vitro* are in some sense artifacts of tissue culture (Shostak, 2006; Zwaka & Thomson, 2005).

In this review, we reveal some common principles in the sub-cellular and molecular machinery maintaining pluripotency and gametogenic potentiality. We hypothesize that evolutionary conserved molecular mechanisms underlie pluripotency, including gametogenic potentiality in germline, embryonic stem cells and other pluripotent stem cells of different metazoans.

2. The molecular and sub-cellular machinery of stem cell specifications

In all multicellular organisms, a stem cell system serves as a crucial cellular source during embryonic development building *de novo* the entire organism and during adulthood regenerating all types of cells and tissues of the individual. In colonial organisms with asexual reproduction in their life cycle, stem cells can be toti/pluripotent, producing not only all somatic types but germline as well. One of the critical features of the stem cell is self-renewal. Stem cells can divide indefinitely without losing their potent capacity and ability to differentiate. Typical stem cell morphology is characterized by a relatively organelle-free cytoplasm, large rounded nucleus and large prominent nucleolus (nucleoli) and diffuse euchromatin, presumably capable of genome-wide active transcription. They have also a significant proportion of inactive heterochromatin, which is silenced by histone methylation or siRNAs and appear as a compact electron-dense material at the ultrastructural level. Also, based on the chromatin organization state, stem cells commonly have the nuclear-to-cytoplasmic ratio shifted toward the nucleus of the cell. These morphological characteristics of stem cells are applicable for the germline cells as well. However, when germcells are specified they will stop actively proliferating until they reach the rudiment of the gonads. It is belived that germline cells will keep their toti-/pluripotent properties and self-renewal capacity while continuing to migrate and differentiate to gametes. Remarkable properties of germline cells, underlining their morpho-functional similarities and differences with other toti-/pluripotent cells, are our main focus in this section.

2.1 Germline cells

Germline cells across Metazoa are specialized cells which are usually specified in very early embryonic development, preserving their capacity to carry out important information about the entire organism, passing it down to the next generation. During preformation, maternally inherited factors of the egg cell, localized in specific areas of the cleaving zygote, leave a specific imprint in the blastomeres, which specialize into germline progenitors. It is believed that epigenesis gives an advantage for the multicellular organism in adaptation and species survival because it does not rely on the quality of a few blastomeres. Instead, during epiginesis, early blastomeres have equal developmental potential and have similar capacity contributing to the developing organism and germline specification (Fig. 1).

Germline cells can be indentified by their specific morpho-genetic signature. According to Extavour (Extavour, 2008; Extavour & Akam 2003), undifferentiated germ cells can be distinguished from somatic cells by several criteria, in addition to typical stem cell characteristics described above. The morphological features are mainly "default" characteristics of the undifferentiated state (Alié et al., 2011). An exception is the key organelles of germline cells referred to as germ (germinal) granules or nuage. They are considered to be a germline hallmark across the animal kingdom (Brown & Swalla, 2007; Eddy et al., 1975; Ikenishi, 1998; Lim & Kai, 2007; Mahowald, 2001; Matova & Cooley, 2001). Evolutionary conserved germ-cell-specific gene expression marks germline cells distinguishing them from somatic cells in all studied metazoans (Ewen-Camden et al., 2010; Extavour, 2008; Leatherman, Jongens, 2003; Matova & Cooley, 2001; Seydoux & Braun, 2006; Sroji & Extavour, 2011).

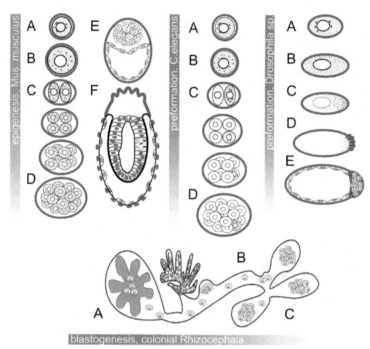

Fig. 1. Germline specification in Metazoa: epigenesis in mouse; preformation in *C. elegans* and *Drosophila* and blastogenesis in colonial Rhizocephala. Epigenesis and preformation: oocyte (A) with distinguished germ granules/germ plasm (red) by the nucleus; specific macromolecular complexes of cytoplasm (yellow), including proteins and transcripts of germline marker genes; mature egg or zygote (B), the germline granules presumably distribute evenly in the cell mass, in *Drosophila* the distinguished granules did not form yet; cleaving zygote (C–D), 2- to 16-cell blastomeres, maternal cytoplasm is distributing equally in mice and not equally in *C. elegans*, where only one blastomere inherited germ granules and the specific cytoplasmic factors, polarized cytoplasm in *Drosophila*, germ granules are located at the posterior pole; (E) in mouse, the group of pluripotent cells form the inner cell mass of the blastocyst, in *Drosophila*, germline cells specified by polarized germ granules; (F) germline specification occurs inside epiblast during mesoderm formation. Blastogenesis: in colonial Rhizocephala, gametogenic stem cells (SC, yellow) within stolon migrate to the rudiment of the gonad of the developing blastozooid (A), forming a germline lineage (blue colour, differentiating somatic tissues), totipotent SCs of the tubular-like stolon (B) forming buds (C).

2.1.1 Morphological evidence of "nuage"/germ granules/chromatoid bodies

"Germ plasm" (Keimplasma), Weismann's famous term, originally denoted the nuclear genetic material (Weismann, 1982, 1893) now is understood metaphorically as cytoplasmic compartment containing specific ultrastructural marker and a key organelle of metazoan germline cells (Amikura et al., 2001; Ikenishi, 1998; Lim & Kai, 2007; Matova & Cooley, 2001; Seydoux & Braun, 2006). Perinuclear germinal granules are almost universal specific organelles of germ cells. The ultrastructure of these organelles is similar, but they can be represented in cells of different organisms as either granules (bodies) or as a cloud (nuage) of fine-dispersed material. There are various terms for the specific electron-dense material: germ plasm granules, nuage, germ cell granules, polar, perinuclear, chromatoid, germinal, germ granules (bodies), dense bodies, etc. (Eddy, 1975; Flemr et al., 2010; Ikenishi, 1998; Isaeva, 2010, 2011; Mahowald, 2001; Seydoux & Braun, 2006; Lim & Kai, 2007). Despite the various terms for the germline granule in different model species, several common features are found (Fig. 2). It seems that the morphological appearance of the structure is linked to stage-specific function and macro-molecular composition, which will be discussed later in this review.

Nuage/germ granules/chromatoid bodies are morphologically and ultrastructurally identifiable and similar to each other. They are discrete, electron-dense organelles, composed of fibrillar and granular material, not bounded by a limiting membrane, often located in the perinuclear cytoplasm and usually associated with clusters of mitochondria. These bodies are found in germ cells in many stages of development, ranging from primordial germ cells in embryos to gametes in adult gonads. The observations suggest that germ granules/nuage/ chromatoid bodies represent different forms of the same material over time (Eddy, 1975; Ikenishi, 1998; Kloc et al., 2004; Mahowald, 2001; Parvinen, 2005). Germ granules have been called by a variety of names reflecting their different morphology at different developmental stages and in different organisms, for example, P granules in *Caenorhabditis elegans*, polar granules in primordial germ cells and nuage during later development in *Drosophila* and other insects, Balbiani body and germinal granules in *Xenopus*, chromatoid body in mammalian male germ cells (Eddy, 1975; Extavour, 2008; Kloc et al., 2004; Seydoux & Braun, 2006). Large complexes including other organelles as in *Xenopus* oocytes were called Balbiani body, mitochondrial cloud, intermitochondrial cement, yolk nucleus etc: the old nomenclature is "confusing and chaotic" (Kloc et al., 2004). The Balbiani body was also observed in oocytes of mouse (Pepling et al., 2007). Seydoux & Braun (2006) used the generic term "germ granules" to refer to all these structures considered as hubs for posttranscriptional regulation of gene expression. While the various names of these structures correspond to differences in morphology, composition, and animals in which they were first identified, it is believed that they are related entities (Eddy, 1975; Gustafson & Wessel, 2009; Parvinen, 2005). The exact relationship between all of these differently named structures has not been determined, but it is possible that they are all different morphological manifestations of the same germ line-specific body (Extavour & Akam, 2003). The chromatoid body of mammalian spermatocytes and spermatids is also suggested to be a mammalian counterpart of nuage on the basis of its structural features and protein composition (Parvinen, 2005; Pepling et al., 2007). Accumulating evidence indicates that the chromatoid body is involved in RNA storing and metabolism, being related to the RNA processing body (P-body: see below) of somatic cells (Kotaja et al., 2006; Nagamori et al., 2011). Here we will refer to these structures as germ plasm related bodies (GPRBs).

It was proposed that the determination of primordial germ cell fate in mammals is independent of germline-specific granules and occurs through an inductive process. Flemr and colleagues (2010) described the dynamics of the maternal stable untranslated transcripts (dormant maternal mRNAs) as components of P-bodies in mouse oocytes and reported that oocyte growth is accompanied by loss of P-bodies and a subcortical accumulation of several RNA-binding proteins, forming transient RNA-containing aggregates. The authors proposed that the cortex of growing oocytes contains a novel type of RNA granule related to P-bodies. Although early mouse oocytes contain granulo-fibrillar material reminiscent of germ cell granules in association with transiently appearing Balbiani bodies, later oocytes lack detectable germ granules (Flemr et al., 2010). Other authors argued that true P-bodies were not observed until the blastocyst stage of embryogenesis, providing evidence that mouse oocytes develop using molecular and developmental mechanisms widely conserved throughout the animal kingdom (Pepling, 2010; Pepling et al., 2007). Hubbard and Pera (2003) reasoned that basic germ-plasm machinery exists in mammalian germ cells as sub-microscopic complexes. In many organisms, GPRBs associate with nuclear pores (Eddy, 1975; Seydoux & Braun, 2006; Snee & Macdonald, 2004). The nuage is visible traversing the nuclear pores, so there is high probability that all or some of the nuage components originate in the nucleus or shuttle between the nucleus and nuage (Kloc et al., 2004). Continuity in electron-dense material between the nucleus and the chromatoid body through nuclear pore complexes has also been observed in male germ cells (Parvinen, 2005; Updike et al., 2011). Polysomes have been reported adjacent to nuage in *Drosophila* (Mahowald, 2001), and chromatoid bodies in rats (Parvinen, 2005). Perinuclear nuage clusters have remarkably dynamic composition, despite their relatively fixed positions around the nucleus (Snee & Macdonald, 2004). In the *Xenopus* oocyte and cleaving embryo, the germinal granules undergo constant transformation in size, number, and ultrastructure. Although the structure and behavior of germ line-specific structures show extraordinary variability, there are also striking similarities and common themes even among evolutionarily distant organisms (Kloc et al., 2004).

Thus, the presence of GPRBs with their specific organization and localization in the cell is an evolutionary conserved feature of metazoan germline cells. GPRBs have been found in more than 80 species of 8 animal types (Eddy, 1975). At least one new additional metazoan type can be added to Eddy's list – Porifera, because electron-dense bodies sometimes described as "nuclear extrusion" or "chromidia" were observed in oogonia and oocytes of several sponge species (see Harrison & De Vos, 1991; Isaeva & Akhmadieva, 2011). The germ granules, "the work horses" of germ cells, are thought to function as a specific cytoplasmic regulatory center, maintaining the genomic totipotency, preventing the expression of somatic differentiation genes, and protecting germline cells from somatic fate (Chuma et al 2006; Cinalli et al 2008; Extavour, 2008; Seydoux & Braun, 2006; Srouji & Extavour, 2011; Strome & Lehman, 2007), preventing somatic fate "by default" (Leatherman & Jongens, 2003).

2.1.2 Molecular signature

Germ granules and nuage contain products of marker germline genes, which are recognized as molecular signature of germline cells. GPRB's components include proteins, mRNAs, and noncoding RNAs. RNA-binding proteins in germinal granules are involved in mRNA localization, protection, and translation control. The molecular machinery and molecular signature of germline cell specification includes a set of evolutionary conserved proteins such as Vasa, Piwi/Aubergine, Nanos, Tudor, Pumilio, Staufen and some others whose homologues have been identified in all metazoans studied (Extavour & Akam, 2003; Leatherman & Jongens, 2003; Kloc et al., 2004; Parvinen, 2005; Chuma et al., 2006; Pepling et al., 2007; Lim & Kai, 2007; Extavour, 2008; Gustafson & Wessel, 2010; Flemr et al., 2010; Srouji & Extavour, 2011). It was shown that some proteins of the germinal granules determine germ cell fate, and their genes are evolutionary conservative in all studied metazoans (Ikenishi, 1998; Matova & Cooley, 2001; Mochizuki et al., 2001; Seydoux & Braun, 2006; Srouji & Extavour, 2011). Every known nuage component has a role in one or more types of posttranscriptional control of gene expression; the presence of shared components reinforces the notion that nuage and polar granules are closely related structures (Snee & Macdonald, 2004). Genes related to *vasa* (*vas*) and other genes of the DEAD family (Raz, 2000; Shukalyuk et al., 2007) and to *piwi/argonaute* family (Funayama al., 2010) were found in a diverse range of eukaryotes from yeast to plants and animals; molecular and functional similarities of these genes were found across the kingdoms (Mochizuki et al., 2001; Watanabe et al. 2009). Products of the *vasa*- and *piwi*-related genes are the most widely used molecular germline markers for Metazoa (Extavour & Akam, 2003; Ewen-Camden et al., 2010; Gustafson & Wessel, 2010; Alié et al., 2011)

2.1.2.1 Vasa, DEAD box family

vasa protein of the *Drosophila* (or its homologues), germline-specific RNA helicase is a key determinant of the fate of germline cells found in GPRBs of germline cells across animal kingdom (Alié et al., 2011; Cinalli et al., 2008; Extavour & Akam, 2003; Ewen-Camden et al., 2010; Gustafson & Wessel, 2010; Shibata et al., 1999, 2010; Sroji & Extavour, 2011; Sunanaga et al., 2006). Products of *vasa*-related genes are necessary for the formation and maintenance of the structural organization of GPRBs and, presumably, for the maintenance of pluri/totipotency of cells. Vasa and Pl10 are members of the DEAD-box family of RNA helicases, proteins known to function in all eukaryotes, from yeast up to plants and animals, in wide aspects of RNA metabolism, including unwinding double-stranded RNAs and controlling their export, splicing, editing, stability, and degradation. They are involved in ribosome biogenesis, translation initiation, and mediating both RNA-RNA and RNA-protein interactions, promoting expression of other germline genes (Cinalli et al., 2008; Gustafson & Wessel, 2010).

The Vasa-like protein and a set of RNA-binding proteins, as well as other translational regulators are common and invariable components of GPRBs in many organisms. The presence of Vasa-like proteins in the germ plasm of different animals indicates the conservation of molecular mechanisms underlying the formation and maintenance of the germ plasm across Metazoa (Extavour, 2008; Ewen-Campen et al. 2010; Gustafson & Wessel, 2010; Juliano et al., 2010; Kloc et al., 2004).

2.1.2.2 Piwi, Piwi/Argonaute family

Piwi/Argonaute family members serve as epigenetic regulators of stem cells in many systems. Piwi/Ago proteins are an animal germline-specific subclass, highly conserved across eukaryotes, specifically expressed in germ cells and playing a key role in germ cell maintenance and self-renewal, transposon silencing, and RNA silencing. These proteins are at the core of RNA-silencing machinery that uses small RNA molecules as guides to identify homologous sequences in RNA or DNA. The small RNAs regulate genes at the transcriptional or post transcriptional level affecting either chromatin structure or mRNA stability and mediating transcriptional gene silencing in germline maintenance (see Gustafson, Wessel, 2010; Peters, Meister, 2007; Thomson, Lin, 2009; Watanabe et al. 2009; Sroji & Extavour, 2011). Particularly, chromatoid bodies in male germ cells seem to operate as "intracellular nerve centers" of the microRNA pathway and function as subcellular concentration sites for components of the miRNA pathway, centralizing the miRNA posttranscriptional control system in the cytoplasm of haploid male germ cells (Kotaja et al., 2006; Nagamori et al., 2011). There are important interactions between Piwi and Vasa in the germline. PIWI-mediated microRNA pathways are evolutionarily conserved control mechanisms, found in bacteria, archaea and eukaryotes and are essential for stem cell division in both animal and plant kingdoms (see Ewen-Camden et al., 2010; Funayama et al., 2010; Watanabe et al., 2009).

2.1.2.3 Tudor-domain contained proteins

tudor (tud) gene products of *D. melanogaster* are key components of polar granules and nuage (see Anne, 2010; Arkov et al., 2006; Chuma et al 2006). Tudor motifs are found in many metazoan organisms and have been indentified to play a role in protein-protein interactions in which methylated protein substrates bind to these domains. Tudor protein interacts *in vitro* with Valois, which is a component of the methylosome in *Drosophila* (Mahowald, 2001; Anne, 2010). It also was shown to play a role in barrel-like folding, which creates the ability to bind and to recognize methylated histone H3-K4 and H4-K20 for a double Tudor-domain protein in human (Huang et al, 2006). The Tudor domain of the SMN (survival motor neuron) protein binds directly to spliceosomal SM proteins during spliceosome assembly (Selenko et al., 2001).

Thus, germline cells are relatively transcriptionally quiescent during most of embryonic development. Moreover, germ cells are typically mitotically quiescent from the time of their specification during embryogenesis, until the time that gametogenesis begins, usually during larval or adult life (Extavour, 2008). The transfer of most of the control of gene expression to the cytoplasm is an important evolutionary conservative acquisition ensuring plasticity for the germ cell genome (Seydoux & Braun, 2006).

2.2 Pluri/multipotent gametogenic stem cells of asexually reproducing animals

2.2.1 Germ granules/chromatoid bodies

In asexually reproducing invertebrates, stem cells capable of differentiating into germ and somatic cells can be identified by the presence of specific electron-dense cytoplasmic structures, morphologically similar or identical to germinal granules of germline cells (referring to GPRBs). In stem cells of asexually reproducing invertebrates, the germinal granules were revealed before our work in stem cells of cnidarians and flatworms. For example, the electron-dense bodies similar or identical to GPRBs of germline cells were found in interstitial cells of the hydra *Pelmatohydra robusta*. Bodies were associated with nuclear pores and mitochondria. The number and size of such dense bodies increased during early oogenesis and decreased during differentiation of somatic cells (cnidoblasts) from interstitial cells (Noda & Kanai, 1977). Studying many species across Metazoa, we also found significant similarity in the morphology of electron-dense bodies or granules (Fig. 2).

In planarians, electron-dense GPRBs were observed not only in germline cells but also in neoblasts. The chromatoid bodies in planarian neoblasts and germ cells are found near the nuclear envelope, in close proximity to mitochondria as well (Coward, 1974; Hori, 1982; Isaeva et al., 2005; Shibata et al., 1999). Nuage-like structures morphologically different from planarian chromatoid bodies were found in the flatworm *Macrostomum lignano* (Pfister et al., 2008). The chromatoid bodies in planarian neoblasts decrease in number and size during differentiation of somatic cells from neoblasts and disappear in completely differentiated cells, while in oogenic cells the chromatoid bodies were found during the entire life cycle (Hori, 1982; Shibata et al., 1999). These observations suggest that the chromatoid bodies are concerned with the cell totipotency maintenance (Shibata et al., 1999).

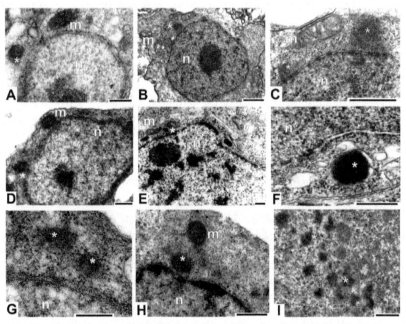

Fig. 2. Common morphofunctional feature of stem cells across Metazoa: transmition electron microscopic level of germ granules/nuage organization. Electron-dense granules (*) are usually localized near nucleus (n) pore and surrounded by mitochondira (m) in sponge *Oscarella malakhovi* (A, D and G), in planarian *Girardia tigrina* (E), in hydroids *Ectopleura crocea* (B and C) and *Obelia longissima* (F), as well as in mouse embryonic stem cells in culture (H, embryoid body, day 1) and in stem cells of inner cell mass of the mouse blastocyst (I). A, D and G, after Isaeva & Akhmadieva, 2011; E, after Isaeva et al., 2005; F, after Isaeva et al., 2011; H and I, Shukalyuk et al., unbubl. Scale bar is 0.1 μm.

The morpho-functional organization of pluripotent gametogenic stem and gonial cells in studied representatives of diverse metazoan phyla shares with germ and stem cells common properties as described above. Particularly, in the cytoplasm of archaeocytes in the sponge *Oscarella malakhovi* we have found germinal granules of a typical morphology located near the nuclear envelope and surrounded with polysomes (Isaeva & Akhmadieva, 2011). Electron-dense GPRBs were found earlier in the oogonia and oocytes of different sponges but have not been previously described in the archaeocytes or any other cells of sponges. We revealed electron-dense GPRBs in interstitial cells of the colonial hydroids *Obelia longissima* and *Ectopleura crocea* (Isaeva et al., 2011), similar to "dense bodies" of interstitial and germ cells in *Pelmatohydra robusta* (Noda & Kanai, 1977) and cnidarian oocytes. The GPRBs surrounded by mitochondria and in contact with nuclear pores have been found near the nuclear envelope in neoblasts and gonial cells of the planarian *Girardia tigrina* (Isaeva et al., 2005). We revealed typical GPRBs in the cytoplasm of embryonic stem cells and stem cells of the colonial rhizocephalans, *Peltogasterella gracilis* and *Polyascus polygenea* (Shukalyuk et al., 2005, 2007, 2011). In the cytoplasm of some stem cells in the early buds of colonial ascidian *Botryllus tuberatus* we have found small electron-dense bodies (Akhmadieva et al., 2007), similar to disperse material of nuage, often found in vertebrates. Perinuclear electron-dense germinal granules often associate with the nuclear pore membrane and bear signs of mitochondrial origin, in particular, cristae of the inner mitochondrial membrane.

Pluripotent or multipotent gametogenic stem cells in all studied asexually reproducing animals belong to 5 animal types: Porifera, Cnidaria, Platyhelminthes, Arthropoda, and Chordata. They all feature the presence of the germinal granules similarly to germline cells. Evidently, the electron-dense germ granules are ultrastructural markers and key organelles both of metazoan germline and potentially gametogenic pluripotent stem cells of asexually reproducing invertebrates.

2.2.2 Molecular signature in pluripotent gametogenic stem cells

In asexually reproducing animals, both germ and pluripotent stem cells express evolutionary conserved germ cell markers such as products of genes related to *vasa/pl10, piwi/argonaute, nanos, tudor* as well as high activity of alkaline phosphatase (AP, Fig. 3) and telomerase (Extavour, 2008; Ewen-Camden et al., 2010; Funayama et al., 2010; Gustafson & Wessel, 2010; Isaeva, 2010, 2011; Mochizuki et al., 2001; Shukalyuk et al., 2005, 2007; Sroji, Extavour, 2011). Specifically, *vasa*-related gene expression is characteristic not only of germline cells, but also pluripotent gametogenic stem cells involved in their determination and maintenance. Vasa expression as well as a high activity of AP and telomerase became the classic selective markers of these stem cells (see Isaeva, 2011; Mochizuki et al., 2001; Rinkevich et al., 2009; Shibata et al., 1999; Shukalyuk et al., 2007; Sköld et al., 2009; Sroji & Extavour, 20110). In many of the invertebrates, such as cnidarians, acoels, planarians, annelids and colonial urochordates, expression of Piwi and Vasa are not restricted to the germline but are expressed in multipotent stem cells (Alié et al., 2011). Piwi is considered to be an omnipresent stemness flag for self-renewal and maintenance of germ line and stem cells in diverse multicellular organisms (Rosner et al., 2009). Several studies indicate a functional relationship between Vasa and both the small interfering RNA and micro-RNA processing pathways (Gustafson & Wessel, 2010). In sponges, cnidarians, flatworms, and colonial botryllid ascidians, germ cells derive from adult pluripotent stem cells (Agata et al., 2006; Extavour, 2008; Isaeva, 2010; 2011; Rinkevich et al., 2009; Sköld et al., 2009; Srouji & Extavour, 2011). Gametogenic potentiality was observed also for stem cells in colonial rhizocephalan crustaceans *P. polygenea* and *P. gracilis*; the stem cells migrated into the developing ovary becoming oogonial cells (Isaeva et al., 2004; Shukalyuk et al., 2005).

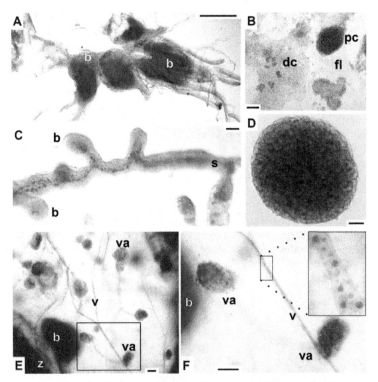

Fig. 3. Common histochemical feature of stem cells across Metazoa: selective expression of alkaline phosphatase (AP): A, buds (b) within a colony of *Peltogaster reticulatus* (Rhizocaphala); B, high AP-reaction in pluripotent *Mus musculus* ESCs and low expression in their differentiated colonies (dc) or fibroblasts (fl); C, AP-positive stem cells inside stolon (s) and buds (b) of *Peltogasterella gracilis* (Rhizocephala); D, AP reaction in neurosphere of *Mus musculus* ESCs in culture; E and F (selected E), AP-positive hemoblasts of *Botryllus tuberatus* inside zooid (z), bud (b), vessel (v) with AP-positive hemoblasts (enlarge in F), and vascular ampulla (va) during vascular budding. A, B & D, Shulalyuk, unpubl.; C, after Shukalyuk et al., 2005; D, after Akhmadieva et al., 2007. Scale bar is 20 μm (B, D−F) & 50 μm (A, C).

In the freshwater sponge Ephydatia fluviatilis (Porifera), expression of *piwi* orthologues was found in sponge archaeocytes and choanocytes, sponge pluripotent gametogenic stem cells (Funayama, 2008; Funayama et al., 2010). The expression of *nanos*-, *vasa*- and *PL10*-related genes (Mochizuki et al., 2000, 2001) was demonstrated for the adult interstitial and germline cells for the hydrozoan *Hydra magnipapillata* (**Cnidaria**); *piwi*-like expression was found in germline and stem cells of the jellyfish *Podocoryne carnea* (Seipel et al., 2004). *vasa*-like gene expresses in interstitial stem cells of *Hydractinia echinata* (Rebscher et al., 2008). Piwi/Ago, Pumilio, PCNA were revealed in the hydrozoan *Hydra magnipapillata* and the anthozoan *Nematostella vectensis*, whereas orthologues of Oct4 and Nanog were not found (Watanabe et al., 2009). In the acoel *Isodiametra pulchra* (**Acoelomorpha**) expression of *piwi* orthologue was shown in germ cells and neoblasts (De Mulder et al., 2009).

Planarian (**Plathyhelminthes**) neoblasts can differentiate into germ and somatic cells and express *vasa, piwi, nanos, pumilio, bruno, tudor* homologues (Agata et al., 2006; Pfister et al., 2008; Reddien et al., 2005; Rossi et al., 2007; Shibata et al., 1999, 2010; Solana et al., 2009). In *Dugesia japonica* two *vasa* homologues are expressed in the germ cells of the adult gonads. Only one of these homologues was expressed in neoblasts (Shibata et al. 1999). Flatworm *vasa* homologue of *Macrostomum lignano* was expressed in germ and stem cells (Pfister et al., 2008). Planarian homologues of *piwi* and *pumilio* genes were found specifically expressed in a neoblasts (Rossi et al., 2007) and the expression pattern of Piwi protein in planaria *Schmidtea mediterranea* coincides with the neoblasts (Reddien et al., 2005). The Tudor protein is a component of chromatoid bodies in germ cells and neoblasts in the planaria *Schmidtea polychroa* (Solana et al., 2009). Co-localization and co-expression of Piwi- and Tudor-related proteins also was detected in planarian neoblasts (see Shibata et al, 2010). In planarians, a high dose of irradiation significantly down-regulates neoblast's RNA metabolism, chromatin remodelling and transcription. However, a low dose of irradiation stimulates up-regulation of genes involved in signal transduction, cytoarchitecture organization, protein degradation, apoptosis, cell metabolism, intracellular trafficking and receptor/ligand activities (Rossi et al, 2007). Exposure to γ-irradiation demonstrates the presence of at least two irradiation-sensitive sub-populations of neoblasts in *Schmidtea mediterranea* (Eisenhoffer et al, 2008).

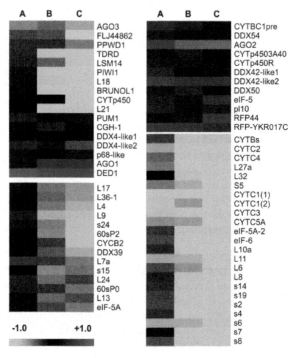

Fig. 4. Gene expression profile for neoblasts of planaria analyzed after irradiation at day 1 and day 7. Cluster analysis reveals set of genes which expression has changed slightly (right column) or dramatically (left column) from the level in wild type (A) to one day (B) or 7 days (C) post-irradiated animals. Heat map was generated using MultiExperiment Viewer v 4.3, emphasising gene expression in scale between -1.0 (bright green) and +1.0 (bright red).

A significant decrease of expression was observed in genes involved in translation, RNA processing, and chromatin transcription, synthesis and repair corresponding to the population of cells affected after 1 day or 7 days post-irradiation. However, the second neoblast population, presumably slower proliferating and more restricted, demonstrate significant down regulation of a specific genes only on day 7 post-irradiation, corresponding to the function of energy production, mitochondria maintenance, translation, metabolism and others. *piwi-* (smedwl-1) and *bruno*-like (brunol-1) genes were down-regulated in both days, marking a totipotent sub-population of neoblasts. Transcripts of mitochondrial carrier, cytochrome p450, ribosomal L21 and L18 proteins were specific to the 7-day irradiated neoblasts. We used publicly available MicroArray data (NCBI GEO #GSE11503; Eisenhoffer et al, 2008) to look at the expression profile of known components of germ plasm and nuage. In our analysis, we found it intriguing that as with Piwi1 and Brunol1 genes, expression of the tudor-like Tdrd gene was also down-regulated in both days post-irradiation (Fig. 4), as well as cytochrome C –family, L-family and S-family genes. However, high wild type expression of Ago2 up-regulated even more in both days post-irradiation, whereas Ddx42, Ddx50, and pl10 almost did not change in their expression (Fig. 4). In our opinion, mitochondrial cement might play an important role in gametogenic stem cell survival during specification by absorbing and utilizing some apoptotic factors that are released by mitochondria, such as cytochrome C, and typically found in the germline germ plasm functioning as stress relief granules. It is conceivable that Tudor, Piwi, and Bruno homologues form the core structure of the germ plasm and present all the time within the dynamic structure, playing an important role in self-renewal and stem cell maintenance.

Pluripotent stem cells of colonial parasitic rhizocephalan barnacles (Rhizocephala: Cirripedia: Crustacea: **Arthropoda**) are predecessors of somatic and germ cells, thereby ensuring the reproductive strategy with alternation of asexual and sexual reproduction. The earliest blastozooid primordia arise as epithelial buds of stolon-like structures filled with migrating stem cells; there is a cluster of undifferentiated stem cells within each bud; later stem cells migrate as germ cells into developing ovaries (Isaeva et al., 2001, 2003, 2004; Shulalyuk et al., 2005). Earlier we revealed the evolutionarily conserved sites of *vasa-* and *pl10-* related genes of the DEAD family, in DNA of the rhizocephalan crustaceans *Polyascus polygenea* and *Clistosaccus paguri* (Shukalyuk et al., 2007). Selective expression of RNA of the *vasa-* and *pl10-*related genes was observed in pluripotent stem cells, in oogenic and spermatogenic cells (Shukalyuk et al., 2007). We also found selectively high activity of AP histochemical marker in stem cells of *P. gracilis* along with expression of proliferating cell nuclear antigen (PCNA) in interna (Isaeva et al., 2003; Shukalyuk et al., 2005). Recently, we have shown for *P. gracilis* the presence of Piwi, Vasa and Nanog proteins in pluripotent stem cells, early blastozooids and early rudiments of the trophic system (Sukalyuk et al, 2011; Shukalyuk & Isaeva, unpubl. data).

Colonial ascidians (**Chordata**) can reproduce asexually, particularly, by palleal or vascular budding. In vascular budding of botryllid ascidians, pluripotent hemoblasts form buds generating a new individual. Hemoblasts are undifferentiated cells that can give rise to differentiated blood cells, somatic tissue cells of blastozooids during asexual reproduction, and evidently also to germline cells (see Rinkevich et al., 2009). *Vasa*-like gene expression was demonstrated in primary germline cells morphologically indistinguishable from hemoblasts in ascidian *Botryllus primigenus* (see Sunanaga et al., 2007). Brown and Swalla (2007) compared *vasa*-related gene expression in the solitary ascidian *Boltenia villosa* and the colonial ascidian *Botrylloides violaceus*. In *B. villosa*, mRNA of vasa-related gene was expressed in germ cells whereas mRNA of *vasa*-related gene of the ascidian *B. violaceus* in mature colonies was expressed in germ cells, in some circulating in the blood cells, in differentiating buds and zooids. Gustafson & Wessel (2010) reported Vasa mRNA expression in germ lines along with hemoblast aggregates in *Botryllus primigenus* and *Polyandrocarpa misakiensis*. In the colonial *Botryllus schlosseri*, mRNA and the proteins of *vasa-, Pl10-, piwi-* and *Oct4-* orthologues are not expressed exclusively in germ cell lineages, but emerging *de novo* also in circulating hemoblasts, thus indicating somatic embryogenesis (Rosner et al., 2009). The results strongly suggest that germline hemoblasts are recruited from undifferentiated hemoblasts in budding tunicates (Rosner et al., 2009; Sunanaga et al., 2007). Data on marker gene expression in gametogenic stem cells of colonial ascidians and their interpretation are rather contradictory (see below).

Pluripotent stem cells in various invertebrates with asexual reproduction as well as cells of the germ lineage display the expression of conserved genes related to *vasa, piwi* and others which function in the specification and maintenance of both cell types across different metazoan phyla (Agata et al., 2006; Gustafson & Wessel, 2010; Juliano & Wessel, 2010; Rinkevich et al., 2009; Sköld et al., 2009; Srouji & Extavour, 2011; Wu et al., 2011). Besides the default characteristics of undifferentiated cells, these stem cells contain electron-dense perinuclear germ granules and express germline marker genes demonstrating that pluripotent stem cells display all of the morphological and functional features commonly used to identify germ cells.

2.3 Mammalian embryonic stem cells via germline connection

Embryonic stem cells (ESCs) are derived from the inner cell mass (ICM) of the developing embryo. During their adaptation for culture conditions these cells or some population of the originally extracted cells will gain some new properties and will become some sort of an artificial system. First, this adaptation includes the cell's ability to attach to the supporting surface: tissue culture treated plastic, extracellular matrix or feeder layer. Second, highly proliferating cells will be selected over time with self-renewing capacity. Third, cells that are selected within the culture will be responsive to the cell culture medium signaling, for example, from serum supplements and LIF (Leukemia Inhibitory Factor) for mouse ESCs or basic FGF (Fibroblasts Growth Factor) for human ESCs. However, it is commonly accepted that the ICM cells of the embryo are pluripotent cells equal in their properties to ESCs in culture, and can recapitulate normal embryonic development *in vitro* when placed under specific conditions.

2.3.1 Nuage/germ granules

The inner mass cells of mammalian embryos *in vivo* contain P-bodies (Pepling et al., 2007; Pepling, 2010), but we do not know any data in the literature on P-bodies, nuage or germ granules for embryonic stem cells cultured *in vitro*. As Clock and coauthors (2004) wrote, it will be interesting to see whether the embryonic stem cells in mice contain chromatoid body similar to that present in totipotent cells in planarians. We were the first to report morphological evidence for electron-dense germinal granules and more dispersed nuage material located near the nuclear membrane in the cytoplasm of mouse EMCs *in vitro* using confocal and light microscopy, with localization of mouse vasa homologue DDX4/MVH in perinuclear granules or nuage (Shukalyuk, 2009).

2.3.2 Molecular signature

In pluripotent mammalian ESCs, molecular signature and a core transcriptional regulatory network dedicated to establishment and preservation of pluripotency include a set of marker genes overlapping with gene signature of germline cells (Kim et al., 2008). The transcription factors Nanog, Oct4 and Sox2 are considered to be the core of the transcriptional network involved in pluripotency and commitment in human or mouse ESCs and have been recognized to be essential *in vivo* and *in vitro* for early development and coordinately regulating the epigenetic network supporting ES cell pluripotency (see Chambers et al., 2003; Do & Schöler, 2009; Jaenisch & Young, 2008; Kim et al., 2008; Rosner et al., 2009; Seydoux, Braun, 2006; Stice et al., 2006; Walker et al., 2007). Moreover, Oct4, Sox2 and Nanog can also indirectly regulate gene transcription by affecting chromatin structure, DNA methylation, microRNA and X chromosome inactivation, changes in local and higher order conformation of DNA, and RNA interference; so, Oct4, Sox2 and Nanog are involved in the cellular machinery, which has an important role in cell fate determination (Atkinson & Armstrong, 2008; Do & Schöler, 2009).

Nanog is a transcription factor, homeodomain protein found in mammalian pluripotent ES and developing germ cells, essential for mammalian embryogenesis. Nanog is thought to be a key factor underlying pluripotency in early development and ESCs, maintaining self-renewal of ESCs and developing germ cells. Nanog is considered a core element of the transcription network and regulatory circuits underlying pluripotency and reprogramming, a hallmark of pluripotent cells *in vivo* and *in vitro* (Do & Schöler, 2009; Stice et al., 2006; Jaenisch & Young, 2008; Kim et al., 2008; Watanabe et al. 2009). Both Nanog and Oct4 are not expressed in mammalian somatic stem cells and loss of Nanog is an early marker of differentiation (Do & Schöler, 2009).

Previously we reported (Shukalyuk & Stanford, 2008) as others have mentioned (Lacham-Kaplan, 2006) that some germline related genes are spontaneously expressed in mouse ESCs and reprogrammed mouse induced pluripotent stem cells (iPS) (Shukalyuk, 2009), even when cultured under the pluripotent and self-renewing condition maintained with LIF. We also found that the mouse homologues of Vasa (Ddx4), Stella, Dazl, Piwi (Miwi) and p68 (Ddx5) can be found in a surprisingly similar proportion of the cells in various ESC lines. We also observed reorganization of proteins and their accumulation in granules visible under confocal microscope after 72 hrs of initiating spontaneous differentiation by withdrawing LIF. We showed the co-localization of DDX4 protein with mitochondrial cytochrome C oxidase IV (COX IV) and single strand binding protein (SSPB1) in germ-like perinuclear granules of mouse ESCs. We also demonstrated that among others, Stau1 and Stau2, mRNA of *staufen*-related genes, were significantly enriched in ESC's RNA using immune-precipitation with anti-mDDX4 antibody (Shukalyuk, 2009).

Using a publicly available on-line micro array data set (NCBI GEO #GSE7506, Walker et al, 2007) we focused on the expression profile of germ-plasm related genes in mouse ESCs under pluripotent culture condition (LIF plus) and during spontaneous differentiation without LIF (LIF minus) or direct differentiation under retinoic acid treatment (RA plus). It is known that embryonic stem cells in culture are heterogeneous in their level of pluripotency marker expression.

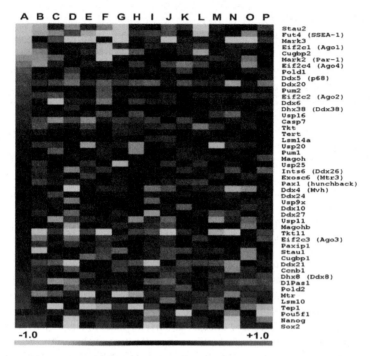

Fig. 5. Gene expression profile for OCT4-high (H), -medium (M) and -low (L) sub-populations of mouse ESCs differentiating over time under LIF- (5 days) or RA+ (2 days) conditions. Day0: A, LIF+(H); F, LIF+(M). Day1: B, LIF-(H); E, RA+(H); G, LIF-(M); J, RA+(M); L, LIF-(L); O, RA+(L). Day2: K, RA+(M), P, RA+(L). Day3: C, LIF-(M); H, LIF-(M), M, LIF-(L). Day5: D, LIF-(H); I, LIF-(M); N, LIF-(L). Heat map was generated using MultiExperiment Viewer v 4.3, emphasising the scale between -1.0 (bright green) and +1.0 (bright red).

In this particular data set, mouse ESCs were sorted based on their high, medium and low OCT4 (POU5F1) protein expression and each sub-population (high, medium and low) was differentiated under LIF-minus or RA-plus condition. As expected (Fig. 5), expression of Oct4, Sox2 and Nanog decreased overtime in each sub-population under both differentiating conditions. Expression of the Ddx4 gene, as compared to other DEAD-box contained genes (Ddx5, Ddx6, Ddx8, Ddx10, Ddx20, Ddx21, Ddx24, Ddx27), gradually increased, peaking on day 3, and dramatically decreased in each population by day 5 under LIFminus condition. An increase in Ddx4 expression was also observed for OCT4-low and OCT4-medium sub-populations of ESCs by day 2 under RAplus treatment (Fig. 5). Upregulation of Ddx4 gene expression by day 3 during spontaneous differentiation is directly correlated with the same trend of Stau2, Eif2c4 (Ago4), Eif2c2 (Ago2), Eif2c1 (Ago1), Lsm14a and p68 (Ddx5) gene expression for all sub-populations but not for Stau1, Eif2c3 (Ago3) or Lsm10. Remarkably, significant down-regulation of germ-plasm members such as Ddx4, Stau2, Ago-family, p68 and Lsm14a oppositely correlates with Casp7 gene expression which is down-regulated at day 3 but significantly up-regulated by day 5 in all ESC populations (Fig. 5). A quick switch over time between levels of expression was indentified for mictotubule affinity-regulating kinase Mark3, as well as for telomerase associated Tep1 gene and Tktl1, a known catalytic metal ion binding gene. These data are in line with existing *Drosophila* germ plasm composition studies revealing the presence, among transcriptional control regulators (Bruno, Nanos and Orb), of zinc ion binding and Ca^{2+} signaling element (Rangan et al., 2009). DDX4 is a DEAD-box family protein that, along with Oct4 and Nanog gene products, can express in the ICM, primordial germ cells (PGS) and ESCs (Chambers et al., 2003; Stice et al., 2006; Zwaka, Thomson, 2005). In undifferentiated mouse embryonic and induced pluripotent stem cells, we found expression of Ddx4/Vasa, Miwi/Piwi, Nanog and Oct4 (Shukalyuk, 2009). Futhermore, Oct4 expression appears to be crucial for the PGCs function and survival. Loss of Oct4 leads to the PGCs apoptosis in mammalians, instead promoting expected trophectodermal differentiation (Kehler et al., 2004).

Mammalian pluripoitent ESCs capable of differentiating into female and male germ cells *in vitro* are potentially gametogenic cells (Clark et al., 2004; Eguizabal et al., 2009; Mathews et al., 2009; Toyooka et al., 2003), along with invertebrate pluripotent stem cells. Because mammalian ESCs are capable of differentiating into germ cells, this suggests that these cells in mice contain all necessary components for the determination of germ cell fate and they are totipotent, despite their lack of differentiating towards extra-embryonic tissue. Observation of the common expression for protein and mRNA mammalian markers in both PGCs and ESCs has led to the hypothesis that embryonic stem cells are closely related, or even identical, to early germ cell precursors (Clark et al., 2004; Fox et al., 2007).

2.4 Evidence of *de novo* inducibility for germline markers

Generally, germline cells can be identified and retraced during development of an organism due to the availability of molecular markers. However, the molecular signature of pluripotent gametogenic stem cells becoming germline cells is not always and necessarily continuous during development and germline specification.

De novo epigenetical specification of hemoblasts into female germ cells was described in the colonial tunicates *Botryllus primigenus* and *Polyandrocarpa misakiensis* (Sunanaga et al., 2007; Rosner et al., 2009). In *B. primigenus*, *vasa* homologue expressing cells within the loose cell mass of the primary germline cells evidently arose from the *vasa*-negative cells at postembryonic stages. These results show that germ cell specification is inducible *de novo*. It has been suggested that germ cell formation in *B. primigenus* is a consequence of epigenetic induction during zooid differentiation. Similarly, in another budding ascidian, *P. misakiensis*, a *vasa* homolog was expressed strongly by loose cell aggregates and germ cells, indicating that germ cells arise *de novo* in developing zooids and suggesting that the *vasa* homologue plays a decisive role in switching the cell fate from coelomic stem cells to germ cells (Sunanaga et al., 2007). In *Botryllus schlosseri*, Vasa detected from the larva and the oozooid stages, repeatedly emerge *de novo* in the colony, independently of its sexual state (Rosner et al., 2009). The expression of *Pl10*-, -*piwi*- and *Oct4*- orthologues both in germline cells and also in circulating pluripotent stem cells, hemoblasts, in *B. schlosseri* in contrast to the observations in *B. primigenus* and *P. misakiensis* might reflect different modes of germ lineage sequestering between the species (Rosner et al., 2009). During development of the ascidian *Ciona intestinalis*, primary germ cells are localized to the tail of the tadpole and during metamorphosis migrate into the adult gonad rudiment. If the tail with primary germ cells is removed, adults still form mature germ cells, suggesting a compensatory mechanism that regulates ascidian germ line formation at a later ontogenetic stage (Takamura et al., 2002). In embryogenesis of the sea urchin *Strongylocentrotus purpuratus*, germ line determinants accumulate in the small micromere lineage. Vasa protein is enriched in the 16-cell stage micromeres and subsequent small micromeres. Experimental removal of Vasa-positive cells induces Vasa expression *de novo* in adjacent blastomeres (Voronina et al., 2008).

2.5 Transient expression of germline marker genes during development and cell differentiation

Among cnidarians, in *Hydra magnipapillata, pl10* mRNA is expressed not only in undifferentiated cells (multipotent interstitial stem cells and germline cells) but also in differentiating somatic cells of the interstitial cell lineage. One of two *vasa*-related genes appears to be expressed in all kinds of undifferentiated cells: multipotent stem cells, germline cells and the ectodermal epithelial cells in the body column. However, none of the *vasa*/*PL10* genes were expressed in fully differentiated somatic cells in *Hydra* (Mochizuki et al., 2001). Analyses of the *piwi*-related gene during embryogenesis and medusa formation in the hydrozoan *Podocoryne carnea* have shown this gene expression in somatic stem cells as well as the germ line cells (Seipel et al., 2004). In sea anemone *Nematostella vectensis* (Cnidaria) members of the *vasa* and *nanos* families are expressed not only in presumptive germline cells but also in broad somatic domains during early embryogenesis and later are restricted to primary germ cells (Extavour et al., 2005).

During embryonic development of the planaria *Schmidtea polychroa*, Tudor-related protein is expressed in differentiating cells rather than neoblasts (Solana et al., 2009).

In the larvae of polychaete annelid *Platynereis dumerilii*, *piwi*-, *vasa*-, *PL10*- and *nanos*-related genes are expressed altogether at the mesodermal posterior growth zone in highly proliferative stem cells providing the somatic mesoderm and the germ line. *vasa*-like gene expression was revealed in the germ line as well as in multiple somatic tissues, including the mesodermal bands, brain, foregut, and posterior growth zone (Rebscher et al., 2007). During embryonic development of the oligochaete annelid *Tubifex tubifex*, transient *vasa* homologue expression was observed in cells in nongenital segments (Oyama & Shimizu, 2007). In polychaete *Capitella sp.* during embryonic, larval, and juveniles stages, *vasa* and *nanos* orthologues are coexpressed in somatic and germ line tissue. Both these genes reveals expression in multiple somatic tissues with largely overlapping but not identical expression patterns; following gastrulation, expression is observed in the presumptive brain, mesodermal bands, and developing foregut (Dill & Seaver, 2008).

In various sea urchin species, Vasa, Nanos, and Piwi are expressed in descendants of the small micromeres and subsequently become restricted to the coelomic pouches, from which the entire adult rudiment will form, suggesting that these conserved molecular factors are involved in the formation of multipotent progenitor cells that contribute to the generation of the entire adult body, including both somatic and germ cells (Juliano et al., 2010; Voronina et al., 2008). In addition, echinoderm species lacking small micromeres, such as sea stars, also have Vasa protein and/or transcripts enriched in the larval coelomic pouches, suggesting a conserved mechanism for the formation of multipotent progenitor cells in the coelomic pouch to produce an adult rudiment within the echinoderms (Juliano & Wessel, 2010; Wu et al., 2011). In the colonial ascidian *Botryllus schlosseri*, *Pl10*, *piwi* and *Oct4* orthologues are highly expressed in differentiating soma cells (Rosner et al., 2009). In the cephalochordate amphioxus *Branchiostoma floridae* (Chordata), Vasa and Nanos, in addition to the early localization of their maternal transcripts in the primary germ cells, are also expressed zygotically in the tail bud, which is the posterior growth zone of highly proliferating somatic stem cells (Wu et al., 2011).

The data indicate a close relationship between presumptive germline cells and multipotent somatic stem cells during development (Wu et al., 2011) and suggest a common origin of germ cells and of somatic stem cells, which may constitute the ancestral mode of germ cell specification in Metazoa (Rebscher et al., 2007). A two-step model of germ cell specification was proposed as an ancestral mechanism involving co-specification of germ cells and stem cells: setting aside a population of undifferentiated pluripotent stem cells, which is excluded from somatic differentiation and has the potential to form both somatic and germ cells, from which the primary stem cells are segregated later (Rebscher et al., 2007).

2.6 Germline marker features beyond gametogenic stem cells

2.6.1 Processing bodies and cytoplasmic RNA granules in somatic cells

In eukaryotic somatic cells, mRNA metabolism is regulated by ribonucleoprotein (RNP) aggregates, RNP granules considered as possible equivalents of germ granules in germline cells. Post-transcriptional processes have a central role in the regulation of eukaryotic gene expression, and these processes are not only functionally linked, but are also physically connected by cytoplasmic granules (Eulalio et al., 2007). Cytoplasmic RNP granules function in determining mRNAs degradation, stabilization, intracellular localization, translational repression and RNA-mediated gene silencing. All RNA granules harbor translationally silenced mRNA. There are several classes of cytoplasmic granules in somatic cells named processing bodies (P-bodies, or P bodies), RNA or RNP granules, RNP particles, stress and neuronal granules (Anderson & Kedersha, 2006; Eulalio et al., 2007; Flemr et al., 2010; Lachke et al., 2011; Kiebler & Bassell, 2006; Kotaja et al., 2006; Seydoux, Brown, 2006).

In mammalian cells, P-bodies are the most common type of RNA granules and contain products of gene orthologues in germ cell granules (Flemr et al., 2010; Pepling, 2010). Processing bodies contain components of mRNA decay processes and microRNA-mediated silencing, serving as sites where mRNAs can be either stored or degraded (Kiebler & Bassell, 2006; Lachke et al., 2006). Argonaute proteins, and also miRNAs and miRNA-repressed mRNAs, were demonstrated to localize in P bodies in mammalian cells.

Unlike P bodies, stress and neuronal granules contain ribosomal subunits. Stress granules are dense aggregates accumulated in cells in response to environmental stress and regulated translational repression and mRNA recruitment to preserve cell integrity. Neuronal granules deliver mRNAs and inactive ribosomes to specific translation sites in dendrites (Anderson & Kedersha, 2006; Seydoux & Brown, 2006; Flemr et al., 2010). In mammalian neuronal cells, three classes of RNA granules were described: transport RNP particles, stress granules, and P bodies with potential functions in RNA localization, microRNA-mediated translational regulation, mRNA degradation, and localized translation of mRNAs involved in synapse formation or motility (Anderson & Kedersha, 2006; Kiebler et al., 2006). Electron dense perinuclear chromatoid body-like structures surrounding the nuclei of neurons were observed in the planaria *Dugesia japonica* (see Shibata et al., 2010).

Cytoplasmic RNP granules function in the posttranscriptional control of gene expression, but the extent of their involvement in developmental morphogenesis is unknown. Recently, a

Tudor domain-containing RNA binding protein (TDRD7) was identified as a component of a unique class of RNP granules with a conserved pattern of developmental expression in ocular lens fiber cells (Lachke et al., 2011). Furthermore, human TDRD7 mutations result in cataract formation via the misregulation of specific, developmentally critical lens transcripts. TDRD7 perturbation causes cataracts in chickens and mice. TDRD1, TDRD6, and TDRD7 have been associated with chromatoid bodies in mammalian male germ cells. Tdrd7 null mutant mice develop cataract and glaucoma; an arrest in spermatogenesis also was observed. Staining with the antibody of STAU1, a mammalian homologue of the *Drosophila* RNA-binding protein Staufen, revealed the presence of numerous STAU1-positive RNP particles in lens fiber cells co-localized to a high degree with TDRD7. The authors hypothesized that TDRD7 granules, either alone or through their interaction with STAU1- RNP granules and P bodies, might regulate the expression levels of specific lens transcripts (Lachke et al., 2011).

Several P body markers are highly concentrated in chromatoid bodies (Kotaja et al., 2006). These data suggest that the chromatoid bodies of male germ cells and P-bodies in somatic cells are functionally related, both acting as a site for mRNA decay and mRNA translational repression by the miRNA pathway (Anderson & Kedersha, 2006; Kotaja et al., 2006). P-bodies contain components of the RNA-dependent silencing machinery (Seydoux & Braun, 2006). P-body components are also present in two other classes of somatic RNP particles: stress granules and neuronal granules (Anderson & Kedersha, 2006). P-body components represent an ever-growing list of proteins involved in RNA metabolism, and the composition of P-bodies, stress granules in somatic cells and germ cell granules overlaps to some extent (Flemr et al., 2010). The data provide evidence of diversity of mammalian RNA granules. Although, they exhibit overlapping composition but different structures and functions (Flemr et al., 2010) sharing components and evolutionary conserved mechanism of post-transcriptional regulation with germ granules which function is distinguishably unique (Seydoux & Braun, 2006).

2.6.2 Germline marker gene expression in somatic stem pools and neurons

Among asexually reproducing animals, *vasa-* and *PL10*-related genes are expressed in somatic ectodermal epithelial cells (unipotent stem cells) in the hydrozoan *Hydra magnipapillata* (Mochizuki et al., 2001). In the colonial ascidians *Botryllus schlosseri*, *vasa*-related gene products are not exclusively expressed in germ lineages but also are strongly expressed in many embryonic and bud somatic cells. Expression of *vasa* and *piwi* orthologues were detected in somatic tissues and *Oct4*-related gene was also expressed in the somatic cells of the endostyle (Rosner et al., 2009).

Among animals without asexual reproduction, *piwi*, *vasa*, and *pl10* are expressed in somatic stem cells at the base of the tentacle bulb, giving rise to tentacle nematocytes in the hydromedusa (Cnidaria) *Clytia hemisphaerica* (Denker et al., 2008). Their expression, along with *bruno* orthologue, was found in germline cells in pluri/multipotent somatic stem cells in the tentacle root in *Pleurobrachia pileus* (Ctenophora), a species which reproduces only sexually (Alié et al., 2011). There is no experimental proof that ctenophore somatic stem cells are incapable of producing germ stem cells, but under normal conditions, this seems highly unlikely (Alié et al., 2011). There are also some other examples of canonic germline marker expression in somatic cells and tissues in bilaterian animals including vertebrates (see Alié et al., 2011; Gustafson & Wessels, 2010; Wu et al., 2011).

These data suggest two alternative hypotheses (see Alié et al., 2011). First, these genes are fundamentally associated with germinal potential, and when they are expressed in pluri/multipotent stem cells, they have the potential to generate germ cells. Second, these genes are components of an ancestral molecular toolkit of animal stem cells, whatever the fate of their progeny. Presumably, *piwi*, *vasa* and *pl10* belong to a gene network ancestrally acting in two contexts: germline and pluri/multipotent stem cells. Since the progeny of these multipotent stem cells includes both somatic and germinal derivatives, it remains unclear whether *vasa*, *piwi*, and *pl10* were ancestrally linked to stemness, or to germinal potential. Probably, total or partial restriction of these genes to the germline in some bilaterian groups (e.g., vertebrates and insects) is a derived, evolutionary secondary condition, and it is not appropriate to use these genes, including *vasa*, as germline markers (Alié et al., 2011). The fundamental reason why these genes are ancestrally linked to stemness, in addition to the germline, is probably the main function of the Piwi–piRNA pathway, i.e., genome protection through silencing. Genome protection is a crucial requirement not only for germ cells but also for somatic stem cells, and ancestral involvement of the same gene set in both the germline and somatic stem cells does not particularly imply their common origin in a genealogical sense, but the requirement of the same silencing pathway in two different contexts (Alié et al., 2011). Several conserved molecules are expressed in both germ cells and all types of stem cells (Sroji & Extavour, 2011).Tthe *piwi* gene family may represent the first class of genes with a common molecular mechanism shared by diverse stem cell types in diverse organisms (Cox et al., 1998).

There are some data on relationship of pluripotent stem cells and neuroblasts. Some proteins classically related to germ line development have been recently found to be involved in neuronal function and development. In the planaria *Schmidtea polychroa*, Tudor-related protein is expressed, beyond germline cells and neoblasts, in the central nervous system (Solana et al., 2009). *pumilio* and *bruno* planarian homologues are expressed similarly in neoblasts and in the central neural system, in perinuclear particles surrounding the nuclei of neurons (Salvetti et al., 2005; Guo et al., 2006). *nanos* and *pumilio* are involved in neuronal excitability, dendrite morphogenesis, and long-term memory in *D. melanogaster* (see Muraro et al., 2008; Solana et al., 2009). When mouse ESCs were cultured in serum- and feeder-deprived conditions colony-forming primitive neural stem cell populatiuons could be obtained (Stice et al., 2006). Tropepe and coauthors (2001) proposed neural fate specification from ESCs by a default choise.

All the data suggest that primordial germ cells can be segregated at almost any point during embryogenesis: before blastoderm formation; after embryonic rudiment formation but before germ layer separation; after germ layer separation but before gonadogenesis; or after gonadogenesis and continuously throughout adult life (Extavour & Akam, 2003).

2.7 Regulatory gene networks underlying gametogenic potential and pluripotency

The metazoan development program may be imagined as translation regulatory cascades. The regulatory transcriptional network to maintain stem cell function has been conserved during metazoan evolution (Watanabe et al., 2009). Genes *vasa/pl10*, *piwi/auberdine*, *nanos*, *tudor*, *pumilio*, and *staufen*, representing the core of the germline program, show striking evolutionary conservation (Alié et al., 2011; Chuma et al., 2006; Ewen-Camden et al., 2010; Extavour, 2008; Gustafson & Wessel, 2010; Leatherman & Jongens, 2003; Parvinen, 2005;

Sroji & Extavour, 2011). This gene network consists of gene modules whose interactions are highly stable and highly evolutionary conserved operating in similar ways both in different organisms, and in different places and/or times during the development of an animal organism. Interactions between *vasa* and other germ line genes have suggested a complex network of positive and negative regulation at multiple levels, including transcription, translation, and post-translational modification, epigenetic control of chromatin architecture mediated gene regulation crucial for the role in development (Cinalli et al., 2008; Ewen-Camden et al., 2010). Conserved germ cell-specific RNA networks repress transcriptional programs for somatic differentiation and promote germ cell maintenance (Cinalli et al., 2008). Maelstrom was identified as a nuage component that interacts with both mouse DDX4 and MIWI. It is required for spermatogenesis and also is involved in silencing transposable elements. Although still not definitive, the consistent association in multiple animals of Vasa and members of the RNAi pathway provides a strong argument that they have a functional relationship (Gustafson & Wessel, 2010).

We took adventage of the bioinformatic tool STRiNG (Snel at al, 2000) to construct and analyze germ plasm protein network. Using *Drosophila* known germ plasm components (Fig. 6A) and their homologues in mouse (Fig. 6B), we predicted interactions and indentified several pathways based on the protein functional domains, structure and sequence similarity (Fig. 6D). Similar to *Drosophila*, known germ plasm proteins (Ddx4, Tdrd1, Tdrd7, Tdrd9, Pum2, Nanos1, Nanos2, Nanos3 and others) formed a network (Fig. 6D) responsible for **germline differentiation** (module 3) upon RNA processing via sequence-specific RNA-binding, translation and mRNA-stabilisation (1, 4: Pum1, Nanos2, Pum2) as well as normal mRNA turnover and nonsense-mediated mRNA decay (15: Dcp1a, Dcp2, Edc3, Edc4). Some other important germline related functions are revealed: **transposable element repression** by piRNA machinery during spermatogenesis (5: Piwi-family, Mael, Tdrd6 and etc), **RNA-mediated gene silencing** (RNAi) by a RISC complex (2: Dicer1, Eif2-family/Ago2), **translational activation of mRNA** in the oocyte and early embryo (16, 17: Ddx3y, Ddx3x), and X-chromosome inactivation (8: Xiap). Several functions were related to **molecular metabolism** catalyzing the transsulfuration pathway from methionine to cysteine (20: Tdrd5, Cbs, Cth, Mthr-family), ribosomal protein complex (18: Mrps12, Mrps7, Mrpl11, Mrpl12, Rps20), proteasomal degradation and inhibition of the caspases (7, 8: Apaf1, Xiap, Casp7, Casp8ap2), mediation in activation of the stress-responsive elements upon DNA damage and also regulation of growth and apoptosis (9: Gadd-family, Casp7). A significant portion of the network is occupied by **cell cycle** pathways (10: Pcna, Cdk1) requiring the G2/M (mitosis) transition (7: Cyclin B1, Cdc20, Ccnb1, Ccnb2, Ccna1), DNA replication and polymerase function (11, 13: Fen1, Pold-family, Bub1b), progression from G1 to S phases (7: Lsm10, Lsm11) as well as the anaphase promoting complex/cyclosome regulation (12: Ube2c, Cdc-family, Mad-family). Presence of the cyclin B1 in the germ plasm of *Drosophila* was previously described by Dadly and Glover (1992). Other sub-sequential parts of the network are responsible for **tissue-specific regulation,** including alternative pre-RNA splicing (16: Ddx3y, Cugbp2, Usp9x, Ddx3x), endothelial cell motility and neurotrophic signalling for spinal and sensory neurons (19: Gpi1, Tkt-family), ventral **cell fate in the neural tube** and normal development of the vertebral column (14: Pax1, Pax9, Msx1, Shh), **survival of motor neurons** via spliceosomal Sm proteins function (6: Wdr77, Prmt5, Piwi-family, Lsm10, Lsm11) along with methylation of Piwi proteins required for interaction with Tudor-domain contained proteins.

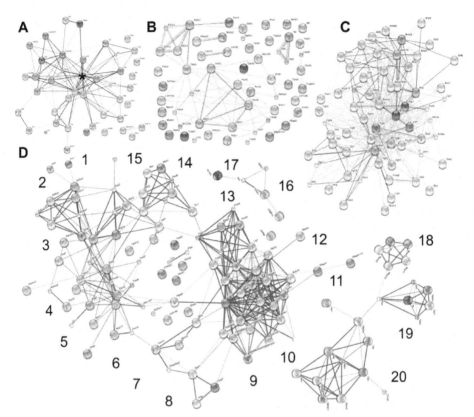

Fig. 6. Protein networks of *Drosophila melanogaster* germ-plasm (A, confidence mode), homologues (B, evidence mode) and expanded neighbouring interactions (D, confidence mode; 1–20 functional modules, see in the text) of *Mus musculus* germ-plasm; and interactions for *Mus musculus* BMP4 (C, action mode).

It was previously found that BMP-family proteins play an important role in primordial germline specification during normal mammalian development and are also required during germline differentiation of mouse and human ESCs *in vitro* (see Ying et al, 2001; Toyooka et al, 2003; Kee et al, 2006). The BMP4 protein, along with some nuage/germ granules components, is involved in a complicated network of transcriptional factors regulating early embryo development, stem cell fate and differentiation. Repression of the Shh gene, for example, prevents activation of the transcription for genes involved in ventral cell fate in the neural tube and also disrupts the polarizing signal for patterning of the anterior-posterior axis of the developing limb bud. It is not surprising that network control is looping from one gene to another (Wnt5a, Otx2, Shh, Bmp4, Runx2, Nog, and Fgf8) forming a "switch-off/switch-on" dynamic machinery in the cell cytoplasm (Fig. 6C). It is evident that the germ plasm or nuage complex is the sub-cellular localized organelle controlling this "on and off" loop depending on the cell fate. Indeed, the sub-cellular organization and localization of the nuclear pore would serve as a major station processing specific mRNA. Such a station can sort and degrade specific mRNA upon cell fate and differentiation via traditional nonsense-mediated mRNA decaying or by non-traditional Stau-based degeneration. Staufen is known to bind double-stranded RNA, along with other members of germ plasm (DEAD-box helicases), but Stau1 also binds tubulin cross-linking RNA macromolecular complexes (MMCs) to the cytoskeleton of the cell. Such organelle organization and function can quickly and reversibly mobilize MMCs to a specific compartment, preceding their post-transcriptional modification, and activate site-specific protein translation and localization. Furthermore, nuage can suppress or even put to silence transposable elements, preventing their mobilization via piRNA-based machinery. This can be done without significant structural changes in genome organization and without transcriptional de-activation of the crucial developmental loci and transcriptional factors that are involved in stem cell pluripotency (Pou5f1 and Sox2).

Besides the function in the germline, the germ plasm components control, or assist in the control, of tissue-specific pathways defining the pattern and fate in stem cell progenitors. They are also responsible for a wide range of gene transcription and translation regulation, as well as the cell cycle. It seems that the germ plasm acts as a switch between mesoderm and ectoderm fate. It stimulates endothelial cell motility, and possibly specifies or regulates the differentiation of cartilage and bone of notochord in chordates. Notochord induces neural plate formation and, by secreting SHH protein, signals differentiation of motoneurons in the neural tube.

The extensive molecular signatures and functional potential of germ cells and pluripotent stem cells suggest a shared evolutionary origin for these cell types and an ancestral pluripotency network including members of Vasa-like and Piwi-like class proteins, which are conserved components of both germ and stem cells across the metazoans (Alié et al., 2011; Ewen-Campen et al. 2010; Gustafson & Wessel, 2010; Sroji & Extavour, 2011). Based on the literature and our own data analysis, we support the idea that this regulatory gene network is not restricted to the germline cells but is expressed in stem cells that are capable of producing both somatic and germinal derivatives.

3. Conclusion

The data we have reviewed here suggest the existence of an evolutionary conserved basis of pluripotency and "stemness" of germ and gametogenic pluripotent stem cells. This mechanism is common for all studied metazoan representatives, from sponges to chordates, and operates at cellular, sub-cellular and molecular levels. In the studied asexually reproducing representatives of Porifera, Cnidaria, Platyhelminthes, Arthropoda and Chordata, stem cells serve as the predecessors of germ and somatic cells and are similar to cells of the germ lineage, displaying evolutionarily conserved features of the morphofunctional organization typical also of cells of the germ line (Ewen-Campen et al. 2010; Extavour, 2008; Funayama et al., 2010; Gustafson & Wessel, 2010; Isaeva et al., 2003, 2008, 2009; Rinkevich et al., 2009; Sköld et al., 2009; Shukalyuk et al., 2005, 2007, 2011; Sroji & Extavour, 2011). The reaction revealing the activity of alkaline phosphatase, earlier used for the identification of primary germ cells and embryonic stem cells in vertebrates, was successfully applied as a cytochemical marker of invertebrate stem cells (see Agata et al., 2006; Akhmadieva et al., 2007; Isaeva, 2011; Isaeva et al., 2003; Rinkevich et al., 2009; Shukalyuk et al., 2005; Sköld et al., 2009).

Since pluri/multipotent stem cells produce germline cells, they might be considered part of the germline (Mochizuki et al., 2001); such "primary" stem cells may be immortal contributing to the germ line, in contrast to somatic tissues (Weismann, 1893; Sköld et al 2009). It is gametogenesis that gives us an "afterlife," propelling our genome into future generations (Seydoux & Braun, 2006). Pluripotent stem cells of animals with asexual reproduction are predecessors of primary germ cells. Pluripotent gametogenic stem cells and germline cells share many morphological features and rely on the activity of related genes; their evolutionary and ontogenetic relationship has been proposed (Extavour, 2008; Extavour & Akam, 2003; Sköld et al., 2009; Strouji & Extavour, 2011).

Adult pluripotent stem cell systems are not restricted to primitive animals and probably evolved as components of asexual reproduction (Agata et al., 2006). The data on the asexual reproduction in some arthropods and chordates contradicts the dogma that asexual reproduction is common exclusively among the lower animals (Isaeva, 2010, 2011).

The term "somatic embryogenesis" (Buss, 1987; Blackstone & Jasker, 2003) suggests that stem cells, which ensure the asexual reproduction, are recognized as somatic ones; pluripotent stem cells in animals with asexual reproduction are often referred as somatic (Blackstone & Jasker, 2003; Extavour & Akam, 2003; Extavour, 2008; Rinkevich, 2009; Sköld et al., 2009; Funayama et al., 2010). However, pluriponent gametogenic stem cells of asexually reproducing invertebrates, like primary germ cells, do not belong to any germ layer, differentiated tissue, and population of specialized somatic cells or their somatic stem cells (Isaeva, 2010, 2011). Such pluripotent stem cells are dispersed in the organism, do not display contact inhibition of cell reproduction and movement and are similar to primary germ cells in their ability to perform amoeboid movements and large-scale migrations within the organism, directed to the localities of asexual reproduction and regeneration or to the gonads, respectively (Isaeva et al., 2008, 2009; Rinkevich et al., 2009; Sköld et al., 2009). We believe that the evolutionarily and ontogenetically related cells of early embryos, pluripotent gametogenic stem cells and germline cells belonging to cell populations capable of realizing the entire developmental program, including gametogenesis (and, potentially, subsequent embryogenesis) are not identical to somatic cells.

Pluripotent cells in invertebrates with asexual reproduction are similar in their potential and their molecular signature to mammalian embryonic stem cells, although the latter are artificial cell systems cultured *in vitro*. Thus, published and original data indicate the existence of evolutionary conserved, sub-cellular and molecular bases of toti/pluripotency and "immortality" and similarity of studied morphofunctional features and molecular signature of pluripotent stem cells in metazoans with asexual reproduction from sponges and cnidarians to chordates, germline and embryonic stem cells (Fig. 7).

Recent data indicate the broad and partially overlapping spectrum of gene expression in ECSs, germ, and pluri/multipotent potent stem cells, in particular, the possible inducibility of germline cells *de novo* without continuous expression of molecular markers of the germ line. The data also show a transient molecular signature typical of the germline in broad somatic domains during embryogenesis and the expression of germline marker genes in somatic stem pools. Embryonic stem, germ and pluripotent stem cells of various metazoans share the expression *piwi*, *vasa*-related and other germline marker genes. It is possible a functional diversification of paralogues of *vasa*, *piwi* and other marker "germline" genes fulfilling different functions in germ and other stem cells. In the animal kingdom, *vasa*-like genes are present in numbers from one to four (Shibata et al., 1999; Rebscher et al., 2007; Extavour et al., 2005; Pfister et al., 2008). In mammals, four Argonaute subfamily members have been shown to be involved in the miRNA pathway (Parvinen, 2005; Kotaja et al., 2006).

Although, the canonic, classical germline molecular markers remain reliable for germ cell identification within developing individual across Metazoa, more studies need to be done in order to understand molecular and cellular events underlying pluri/totipotency, stem cell self-renewal and self-preservation during germline specification.

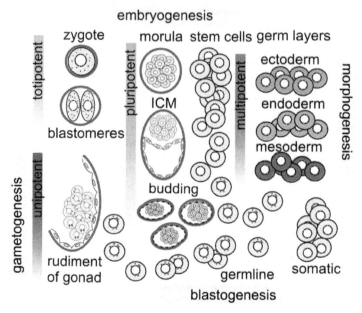

Fig. 7. Germ-plasm material on the sub-cellular level in animals with sexual or both sexual and asexual reproduction in their life span. In animals with epimorphosis, early blastomeres after first few divisions are believed to be fully totipotent along with zygote and capable of producing an entire organism. Macromolecular complexes (MMC, yellow) distributed in cytoplasm and germ-plasm granules around nucleus (in red) are present in zygote and early blastomers marking their totipotent ability. Through the embryonic development, the visual manifestations of the granules are disappeared in the pluripotent cells of morula, inner cell mass of blastocyst or embryonic stem cells in culture. However, specific MMCs are distributed in cytoplasm (yellow) marking their potency. Morphogenesis followed after specification of the three germ layers (ectoderm, endoderm and mesoderm) segregates true somatic cells lacking their potency (grey color) and the germline. The germline lineage expresses the specific MMCs and germ markers *de novo* during specification. The germ-plasm activates in granules (red) near nucleus pores. Germline specification occurs from the mesoderm (in mammalians) as well as spontaneously in culture of embryonic stem cells, and from the pool of the stem cells in asexually reproducing organisms. Presumably, re-activation of the germ-plasm in germline forms a unique sub-cellular niche for regulating and manipulating cellular pluripotency. Finally, germline cells with their specific molecular signature migrate and populate the rudiment of the developing gonad. A new signalling niche will significantly restrict germline abilities (to unipotency) during further gametogenesis. Eventually, cells will undergo dramatic morphological and molecular changes (oogenesis and spermatogenesis) ensuring protection and passing of undamaged information to the next generation.

Germ granules, chromatoid bodies, nuage, P-bodies and so on are the physical embodiment of overlapping but not identical gene networks. The patterns arose ancestrally as a gametogenic/stem program. Germ granules may be as diverse as P-granules of somatic cells, with functions ranging from RNA localization and decay to translational activation and repression (Anderson & Kedersha, 2006; Seydoux & Braun, 2006). It is also possible, that in some cases, this machinery exists in mammals as submicroscopic complexes that are rich in RNAs and RNA-binding proteins (Hubbard & Pera, 2003). The Oskar protein nucleates the formation of polar granules *de novo*, from cytoplasmic pools of the components shared with nuage. In this model, nuage could be an organelle that concentrates and thus potentiates the activity factors normally present in all cells, but that must be especially active in germline cells because of their intensive reliance on post-transcriptional controls of gene expression (Snee & Macdonald, 2004). Basic germ-plasm machinery may exist as discrete granules or bodies, large complexes as Balbiani bodies, dispersed nuage, small particles (for example, P-bodies) or submicroscopic RNP aggregates. It seems to us that similar networks in mammalians overlapped with the macromolecular frame of the germinal granules emerging earlier in germline cell evolution. Given that the factors that are associated with germ granules in non-mammalian species are also expressed in mammalian germ cells, we speculate that all multicellular animals share basic germ-plasm machinery, a nuage-like sub-cellular frame, which operates in a similar manner across Metazoa and might recruit other tissue-specific networks within.

Protection against apoptosis is very important for embryonic, germ and pluripotent stem cells as well as for long living neural precursors and neurons. Differentiation of toti-/pluripotent cells irrevocably drags their descendants into programmed death. Undifferentiated cells have only two choices: stay undifferentiated and immortal or start to differentiate and die. Breaking this rule leads cells to cancerogenesis. The germline cells and neuronal precursors evolutionarily obtained unique machinery for circumventing this rule, which allows them to continue their differentiation but keep their immortality over the life-span of the individual. However, cell death does occur during neurogenesis, matching the number of neurons to the number of target cells. It is also known that in zebrafish, loss of *piwi*-related gene function results in a progressive loss of germ cells due to apoptosis during larval development (Houwing et al., 2007). A mutation of mouse *vasa* homolog gene (*Mvh/Ddx4*) leads to restricted expression. In homozygotes, premeiotic germ cells cease differentiation and undergo apoptotic death (Tanaka et al., 2000). Studies of apoptosis have revealed the key role of the apoptosis induction factor of mitochondrial origin and apoptogenic functions of cytochrome C in this important, evolutionary conserved process (Green & Reed, 1998; Martinou, 1999). Contact with mitochondria is a typical property of germinal granules in diverse multicellular animals. The germinal granules in the cytoplasm of germline cells of *Drosophila* contain ribosomal RNAs of mitochondrial origin (Amikura et al., 2001). It is not coincidence that several cytochrome C oxidase and other mitochondrial products were found within the germ plasm. Moreover, we propose mitochondrial participation in biogenesis of germ granules (Isaeva et al., 2005; Isaeva, 2011). VASA protein homologue has been found both in germ granules and in the mitochondrial matrices in germ cells of *Xenopus* embryo (see Watanabe et al., 2009). Similarly, the protein encoded by the *tudor*-related gene, is present not only in polar granules, but also inside the mitochondria of early embryos of *Drosophila* (see Ding & Lipshitz, 1993). We also postulate the presence of molecular defence against apoptosis localized in germinal granules and related cytoplasmic structures in germline, embryonic, pluripotent stem and neural cells ensuring self-preservation against cell aging and death.

The hypothesis has been advanced that the germline originally evolved from primary stem cells (Sköld et al., 2009). Extavour (2008) also proposed that germ cells have their evolutionary origins in a pluripotent stem cell population. Taking a different position, Zwaka & Thomson (2005) hypothesized that embryonic stem and embryonic germ cells represent a family of related pluripotent cell lines, whose common properties reflect a common origin from germ cells. Rebscher et al. (2007) proposed a model in which *vasa*, *piwi*, and *pl10* ancestrally carried out a first step in germline determination by specifying a multipotent population of stem cells within which PGCs are sorted out later. An ancient association of "germline genes" with stemness (Watanabe et al. 2009) and "an ancestral gene fingerprint of stemness" (Alié et al. 2011) were proposed. The genes *vasa* and *piwi* are the most extensively studied of the genes, known as germline markers, which appear to be involved in the ancestral molecular signature of stemness and expressed in pluri-/multipotent stem cells across animal phyla. According to the hypothesis put forward by Alié et al. (2011), *piwi* and *vasa*-related genes belong to a gene network ancestrally associated with stemness. These genes determine gametogenic potential, but the main function of these genes is genome silencing. Agata et al. (2006) proposed that the pluripotent stem cell system supporting both asexual and sexual reproduction in many adult animals represents one type of origin of stem cell systems; the other system developed by separating multiple functions of primitive pluripotent stem cells into specialized cell lineages.

We suppose mainly common and partially overlapping molecular signatures in pluri-/multipotent stem cells of a wide range of animals from sponges to chordates. But we also suspect a wide spatial and temporal ontogenetic context as a continuum of toti-, pluri-, multipotent state within stem cells. Such a continuum ranges from pluripotent gametogenic stem cells to germline cells and up to multipotent somatic stem cells lacking gametogenic potential. All animals possess pluripotent gametogenic stem cells in different ontogenetic periods: as a transient state in early cleavage until segregation of germ line (preformation), as a longer state during embryogenesis (epigenesis), or as a continuous state during the entire life of asexually reproducing organisms. Pluripotent stem cells have the capacity to move away from pluripotency towards a special, restricted stem cell identity as germ cells (Sroji & Extavour, 2011) or to restricted identities as somatic multipotent stem cells, oligopotent stem cells and so on. A core regulatory gene network of pluripotent gametogenic stem cells, germline cells and multipotent stem cells evidently overlap to a large extent including also some specific distinctions and fluctuations of key gene expression. Genes of the molecular machinery of stem cells appeared to be interconnected in related pathways that are involved in post-transcriptional regulation and epigenetic modification acting in a coordinated manner, as part of a complex network of signal cascades that are known to regulate the balance between cell death and survival (Rossi et al., 2007). Since all animals have a common ancestor in single cell organisms it is possible to identify common principles in the regulatory mechanisms for the transcriptional and epigenetic machinery but at present there is no clear picture to what extent the regulatory transcriptional network to maintain stem cell function has been conserved during metazoan evolution (Watanabe et al. 2009), in germline cells or beyond the germ line (Gustafson & Wessel, 2010).

Taking into consideration our own data and the supporting literature, we see that there is enough evidence to suggest the existence of an ancient molecular basis of toti-, pluri-, or multipotency of germ and stem cells, common for all studied representatives of multicellular animals. However, the detailed molecular mechanisms and overlapping regulatory networks in different stem cell systems appeared to be more complex that was viewed before. Interconnecting regulatory networks of stem cells and germ cells still remain unclear in their capacity to decide cell fate transferring the control from the nuclear transcriptional networks to the cytoplasmic post-transcriptional machinery. Further comparative studies of stem cells in a wide variety of metazoans may provide significant and crucial data for our understanding of the common, evolutionary conservative basis of stemness, pluripotency and potential "immortality" of germ and stem cells across Metazoa.

4. Acknowledgment

This study was supported by a grant from the Russian Federation for Basic Research (no. 09-04-00019). The authors are grateful to Mr. Brian P. Ettkin for his help and support in manuscript preparation. This review was written in support of the Vladimir L. Kasyanov Foundation promoting developmental biology, stem cell studies and marine life ecology.

5. References

Agata, K., Nakajima E., Funayama N., Shibata N., Saito Y. & Umesono sY. (2006). Two different evolutionary origins of stem cell systems and their molecular basis, *Seminars in Cell & Devel. Biol.* Vol. 17: 503-509.

Akhmadieva, A.V., Shukalyuk, A.I., Aleksandrova, Y.N. & Isaeva, V.V. (2007). Stem cells in asexual reproduction of colonial ascidian *Botryllus tuberatus. Russian J. of Marine Biol.* Vol. 33 (No.2): 134-137.

Alié, A., Leclère, L., Jager, M., Dauraud, C., Chang, P., Guyader, H., Quéinnec, E., & Manuel, M. (2011). Somatic stem cells express *piwi* and *vasa* genes in an adult ctenophore: ancient association of "germline genes" with stemness. *Devel. Biol.* Vol. 350: 183-197.

Amikura, R., Hanyu, K., Kashikawa, M. & Kobayashi, S. (2001). Tudor protein is essential for the localization of mitochondrial RNAs in polar granules of *Drosophila* embryos. *Mechanisms of Development* Vol. 107, No. 1-2: 97-104.

Anderson, P. & Kedersha, N. (2006). RNA granules. *J. of Cell Biol.* Vol. 172: 803-808.

Anne, J. (2010). Targeting and anchoring Tudor in the pole plasm of the Drosophila oocyte. *PLoS ONE* Vol. 5(No. 12): e14362- e14362.

Arkov, A.L., Wang, J.Y.S., Ramos, A. & Lehmann, R. (2006). The role of Tudor domains in germline development and polar granule architecture. *Development* Vol. 133: 4053-4062.

Atkinson, S. & Armstrong, L. (2008). Epigenetics in embryonic stem cells: regulation of pluripotency and differentiation. *Cell Tissue Research* Vol. 331: 23-29.

Berrill, N.J. (1961) *Growth, Development, and Pattern*, San Francisco, London: Freeman and Company, 556 p.

Blackstone, N.W. & Jasker, B.D. (2003). Phylogenetic consideration of clonality, coloniality, and mode of germline development in animals. *J. of Experimental Zoology* Vol. 287B: 35-47.

Brown, F.D., & Swalla, B.J. (2007). *Vasa* expression in a colonial ascidian, *Botrylloides violaceus*. *Evolution & Development* Vol. 9: 165–177.
Buss, L.W. (1987). *The Evolution of Individuality*, Princeton University Press, Princeton, NJ.
Chambers, I., Colby, D., Robertson, M., Nichols, J., Lee, S., Tweedie, S. & Smith, A. (2003). Functional expression cloning of Nanog, a pluripotency sustaining factor in embryonic stem cells. Cell Vol. 113: 643–55.
Chuma, S., Hosokawa, M., Kitamura, K., Kasai, S., Fujioka, M., Hiyoshi, M., Takamune, K., Noce, T., & Nakatsuji N. (2006). Tdrd1/Mtr-1, a tudor-related gene, is essential for male germ-cell differentiation and nuage/germinal granule formation in mice. *PNAS USA* Vol. 103: 15894–15899.
Cinalli, R.M., Rangan, P. & Lehmann, R. (2008). Germ Cells Are Forever. *Cell* Vol. 132: 559-562.
Clark, A.T., Bodnar, M.S., Fox, M., Rodriquez, R.T., Abeyta, M.J., Firpo, M.T. & Pera R.A.R. (2004). Spontaneous Differentiation of Germ Cells from Human Embryonic Stem Cells in vitro, *Human Molecular Genetics*, Vol. 13, pp. 727-739.
Cox, D.N., Chao, A., Baker, J., Chang, L., Qiao, D. & Lin, H. (1998). A novel class of evolutionarily conserved genes defined by piwi are essential for stem cell self-renewal. *Genes & Devel.* Vol. 12: 3715-3727.
Coward, S.J. (1974). Chromatoid bodies in somatic cells of the planarian: Observation on their behavior during mitosis. *Anatomical Records* Vol. 180: 533-546.
De Mulder, K., Kuales, G., Pfister, D., Willems, M., Egger, B., Salvenmoser, W., Thaler, M., Gorny, A. K., Hrouda, M., Borgonie, G. & Ladurner P. (2009). Characterization of the stem cell system of the acoel *Isodiametra pulchra*. *BMC Dev. Biology* Vol. 9: 69
Denker, E., Manuel, M., Leclère, L., Le Guyader, H. & Rabet, N. (2008). Ordered progression of nematogenesis from stem cells through differentiation stages in the tentacle bulb of *Clytia hemisphaerica* (Hydrozoa, Cnidaria). *Developmental Biology* Vol. 315: 99–113.
Dill, K.K. & Seaver, E.C. (2008). Vasa and nanos are coexpressed in somatic and germ line tissue from early embryonic cleavage stages through adulthood in the *Polychaete capitella sp. Genes Devel. & Evol.* Vol. 218: 453–463.
Ding, D. & Lipshitz, H.D. (1993). A Molecular Screen for Polar-localized Maternal RNAs in the Early Embryo of *Drosophila*. *Zygote* Vol. 1: 257–271.
Do, J.T. & Schöler, H.R. (2009). Regulatory circuits underlying pluripotency and reprogramming. *Trends in Pharmacological Sciences* Vol. 30: 296-302.
Eddy, E.M. (1975). Germ Plasm and the Differentiation of the Germ Cell Line, *Int. Review of Cytology* Vol. 43: 229–280.
Eguizabal, C., Shovlin, T.C., Durcova-Hills, G., Surani, A. & McLaren, A. (2009). Generation of primordial germ cells from pluripotent stem cells. *Differentiation* Vol. 78: 116–123
Eisenhoffer, G.T., Kang, H. & Alvarado, A.S. (2008). Molecular analysis of stem cells and their descendants during cell turnover and regeneration in the planarian *Schmidtea mediterranea*. *Cell Stem Cell* Vol. 3(No. 3): 327-39.
Eulalio. A., Behm-Ansmant, I. & Izaurralde, E. (2007). P bodies: at the crossroads of post-transcriptional pathways. *Nat Rev Mol Cell Biol* Vol. 8: 9-22.
Extavour, C.G.M. (2008). Urbisexuality: the evolution of bilaterian germ cell specification and reproductive systems. In: *Evolving Pathways. Key Themes in Evolutionary Developmental Biology*, Eds. Minelli, A. & Fusco, G., Cambridge, New York,

Melbourne, Madrid, Cape Town, Singapore and São Paulo: Cambridge University Press, pp. 321–342.

Extavour, C.G. & Akam, M. (2003.) Mechanisms of Germ Cell Specification across the Metazoans: Epigenesis and Preformation. *Development* Vol. 130: 5869-5884, ISSN 1011-6370

Extavour, C.G., Pang, K., Matus, D.Q. & Martindale, M.Q. (2005). *vasa* and *nanos* expression patterns in a sea anemone and the evolution of bilaterian germ cell specification mechanisms. *Evol. & Devel.* Vol. 7: 201–215.

Ewen-Kampen, B., Schwager, E.E. & Extavour, C.G.M. (2010). The Molecular Machinery of Germ Line Specification. *Molecular Reproduction and Development* Vol. 77: 3-18.

Flemr, M., Ma, J., Schultz, R.M. & Svoboda, P. (2010). P-Body loss is concomitant with formation of a messenger RNA storage domain in mouse oocytes. *Biology of Reproduction* Vol. 82: 1008-1017.

Funayama, N. (2008). Stem cells of sponge, In: *Stem Cells. From Hydra to Man*, Ed. Bosch T.C.G., Springer, pp. 17-36.

Funayama, N., Nakatsukasa, M., Mohri, K., Masuda, Y., & Agata, K. (2010). Piwi expression in archeocytes and choanocytes in demosponges: insights into the stem cell system in Demosponges. *Evol. & Devel.* Vol. 12(No. 3): 275–287.

Green, D.R. & Reed, J.C. (1998). Mitochondria and Apoptosis. *Science* Vol. 281 (No. 5381): 1309-1312.

Gustafson, E.A. & Wessel, G.M. (2010). Vasa genes: emerging roles in the germ line and in multipotent cells. *Bioessays* Vol. 32: 626–637.

Harrison F.W., De Vos L. Porifera. In: Microscopic Anatomy of Invertebrates. Vol. 2: Placozoa, Porifera, Cnidaria, and Ctenophora. Eds. Harrison F.W., Westfall J.A. Wiley-Liss: New York e a. 1991. P. 28-89.

Hogan, B. (2001). Primordial germ cells as stem cells. In: *Stem cell biology*, Eds. Marshak, D.R., Gardner, R.L. & Gottlieb, D., Cold Spring Harbor Laboratory Press, New York, pp. 189-204.

Hori I. 1982. An ultrastructural study of the chromatoid body in planarian regenerative cells. *J. of Electron Microscopy* Vol. 31: 63-72.

Høeg, J.T. & Lützen, J. (1993). Comparative morphology and phylogeny of the family Thompsoniidae (Cirripedia, Rhizocephala, Akentrogonida), with descriptions of three new genera and seven new species. *Zoologica Scripta* Vol. 22: 363–386.

Høeg, J.T. & Lützen, J. (1995). Life cycle and reproduction in the Cirripedia Rhizocephala. *Oceanography and Marine Biology: Annual Review* Vol. 33: 427-485.

Høeg, J.T., Glenner, H. & Shields, J.D. (2005). Cirripedia, Thoracica and Rhizocephala (barnacles). In: *Marine Parasites*, Eds. Rohde, K. & Wallingford, U.K., CABI, and Collingwood, Victoria, Australia: CSIRO, pp. 154–165.

Huang, Y., Fang, J., Bedford, M.T., Zhang, Y. & Xu, R.M. (2006). Recognition of histone H3 lysine-4 methylation by the double tudor domain of JMJD2A. *Science* Vol. 5: 748-751.

Hubbard, E.J.A. & Pera, R.R.A. (2003). Germ-cell odyssey: fate, survival, migration, stem cells and differentiation. *EMBO reports* Vol 4: 352-357.

Ikenishi, K. (1998). Germ plasm in *Caenorhabditis elegans, Drosophila* and *Xenopus. Develop. Growth Differ.* Vol. 40: 1-10

Isaeva, V.V. (2010). The diversity of ontogeny in animals with asexual reproduction, and plasticity of early development. *Russian J. of Dev. Biology* Vol. 41(No. 5): 285–295.
Isaeva, V.V. (2011). Pluripotent Gametogenic Stem Cells of Asexually Reproducing Invertebrates. In: *Embryonic Stem Cells, Vol. 2*, InTech (in press).
Isaeva, V.V. & Akhmadieva, A.V. (2011). Germinal granules in archaeocytes of the sponge *Oscarella malakhovi* Ereskovsky, 2006. *Russian J. of Marine Biol.* Vol. 37: 209–216.
Isaeva, V., Akhmadieva, A.V., Alexandrova, Y.N. & Shukalyuk, A.I. (2009). Morphofunctional organization of reserve stem cells providing for asexual and sexual reproduction of Invertebrates. *Russian J. of Dev. Biology* Vol. 40(No. 2): 57–68.
Isaeva, V.V., Akhmadieva A., Aleksandrova, Y.N., Shukalyuk, A.I. & Chernyshev, A.V. (2011). Germinal Granules in Interstitial Cells of the Colonial Hydroids *Obelia longissima* Pallas, 1766 and *Ectopleura crocea* Agassiz, 1862. *Russian J. of Marine Biol.* Vol. 37(No. 4): 303–310.
Isaeva, V.V., Alexandrova, Y. & Reunov, A. (2005). Interaction between Chromatoid Bodies and Mitochondria in Neoblasts and Gonial Cells of the Asexual and Spontaneously Sexualized Planarian *Girardia (Dugesia) tigrina*. *Invertebrate Reproduction and Development* Vol. 48(No. 1-3): 119–128.
Isaeva, V.V., Shukalyuk, A.I. & Akhmadieva, A.V. (2008). Stem cells in reproductive strategy of asexually reproducing Invertebrates. *Russian J. of Marine Biol.* Vol. 34(No. 1): 1–8.
Isaeva, V.V., Shukalyuk, A.I. & Kizilova, E.A. (2003). Revealing of stem cells in colonial interna of the Rhizocephalans *Peltogasterella gracilis* and *Sacculina polygenea* at parasitic stage of their life cycle. *Tsitologiya* Vol. 45(No. 8): 758–763.
Isaeva, V.V., Shukalyuk, A.I., Korn, O.M. & Rybakov, A.V. (2004). Development of primordial externae in the colonial interna of *Polyascus polygenea* (Crustacea: Cirripedia: Rhizocephala). *Crustacean Research* No. 33: 61–71.
Isaeva, V.V., Shukalyuk, A.I., Trofimova, A.V., Korn, O.M. & Rybakov, A.V. (2001). The structure of colonial interna in *Sacculina polygenea* (Crtustacea: Cirripedia: Rhizocephala). *Crustacean Research* No. 30: 134–147.
Ivanova-Kazas, O.M. (1996). Blastogenesis, Cormogenesis, and Evolution. *Russian J. of Marine Biol.* Vol. 22(No. 5): 285–294.
Jaenisch, R. & Young, R. (2008). Stem cells, the molecular circuitry of pluripotency and nuclear reprogramming. *Cell* Vol. 132: 567–582.
Juliano, C.E., Yajima, M. & Wessel, G.M. (2010). Nanos functions to maintain the fate of the small micromere lineage in the sea urchin embryo. *Dev. Biol.* Vol. 337: 220–232.
Juliano, C. & Wessel, G. (2010). Versatile germline genes. *Science* Vol. 329: 640–641.
Kee, K., Gonsalves, J.M., Clark, A.M. & Pera R.R.A. (2006). Bone morphogenetic proteins induce germ cell differentiation from human embryonic stem cells. *Stem cells and Development* Vol. 15: 831–837.
Kiebler, M.A. & Bassell, G.J. (2006). Neuronal RNA Granules: Movers and Makers. *Neuron* Vol. 51: 685.
Kim, J., Chu, J., Shen, X., Wang, J. & Orkin, S.H. (2008). An extended transcriptional network for pluripotency of embryonic stem cells. *Cell* Vol. 132: 1049–1061.
Kloc, M., Bilinski, S. & Etkin, L.D. (2004). The Balbiani body and germ cell determinants: 150 Years Later. *Current Topics in Dev. Biol.* Vol. 59: 1–36.
Kotaja, N., Bhattacharyya, S.N., Jaskiewicz, L., Kimmins, S., Parvinen, M., Filipowicz, W. & Sassone-Corsi, P. (2006). The chromatoid body of male germ cells: Similarity with

processing bodies and presence of Dicer and microRNA pathway components. *PNAS* Vol. 103: 2647-2652.
Lacham-Kaplan, O., Chy, H., Trounson, A. (2006). Testicular cell conditioned medium supports differentiation of embryonic stem cells into ovarian structures containing oocytes. *Stem Cells* Vol. 24: 266-273.
Lachke, S.A., Alkuraya, F.S., Kneeland, S.C., Ohn, T., Aboukhalil, A., Howell, G.R., Saadi, I., Cavallesco, R., Yue, Y., Tsai, A.C.-H., Nair, K.S., Cosma, M.I., Smith, R.S., Hodges, E., AlFadhli, S.M., Al-Hajeri, A., Shamseldin, H.E., Behbehani, A.M., Hannon, G.J., Bulyk, M.L., Drack, A.V., Anderson, P.J., John, S.W.M. & Maas, R.L. (2011). Mutations in the RNA granule component TDRD7 cause cataract and glaucoma. *Science* Vol. 331: 1571-1576.
Leatherman, J.L., & Jongens, T.A. (2003). Transcriptional silencing and translational control: key features of early germline development. *Bioessays* Vol. 25: 326-335.
Lim, A.K. & Kai, T. (2007). Unique germ-line organelle, nuage, functions to repress selfish genetic elements in *Drosophila melanogaster*. *PNAS USA* Vol. 104: 6714-6719.
Mahowald, A.P. (2001). Assembly of the *Drosophila* germ plasm. *Int. Review of Cytology* Vol. 203: 187-213.
Martinou, J.C. (1999). Key to the mitochondrial gate. *Nature* Vol. 399(No. 6735): 411-412.
Matova, N. & Cooley, L. (2001). Comparative aspects of animal oogenesis. *Dev. Biol.* Vol. 16: 1-30.
Mochizuki, K., Sano, H., Kobayashi, S., Nishimiya-Fujisawa, C. & Fujisawa, T. (2000). Expression and evolutionary conservation of *nanos*-related genes in Hydra. *Development, Genes and Evolution* Vol. 210: 591-602.
Mochizuki, K., Nishimiya-Fujisawa, C. & Fujisawa, T. (2001). Universal occurrence of the *vasa*-related genes among Metazoans and their germline expression in Hydra. *Development, Genes and Evolution* Vol. 211: 299-308.
Muraro, N.I., Weston, A.J., Gerber, A.P., Luschnig, S., Moffat, K.G. & Baines, R.A. (2008). Pumilio binds para mRNA and requires Nanos and Brat to regulate sodium current in *Drosophila* motoneurons. *J. Neurosci.* Vol. 28: 2099-2109.
Nagamori, I., Cruickhank, A. & Sassone-Corsi, P. (2011). The chromatoid body: a specialized RNA granule of male germ cells. *Epigenetics and Human Health* Vol. 4: 311-328.
Noda, K. & Kanai, C. (1977). An ultrastructural observation of *Pelmatohydra robusta* at sexual and asexual stages, with a special reference to "germinal plasm". *J. Ultrastructural Research* Vol. 61: 284-294.
Oyama, A. & Shimizu, T. (2007). Transient occurrence of vasa-expressing cells in nongenital segments during embryonic development in the oligochaete annelid *Tubifex tubifex*. *Development, Genes and Evolution* Vol. 217: 675-690.
Parvinen, M. (2005). The chromatoid body in spermatogenesis. *Int. J. Androl.* Vol. 28: 189-201.
Pepling, M.E., Wilhelm, J.E., O'Hara, A.L., Gephardt, G.W. & Spradling, A.C. (2007). Mouse oocytes within germ cell cysts and primordial follicles contain a Balbiani body. *PNAS* Vol. 104: 187-192.
Pepling, M.E. (2010). A novel maternal mRNA storage compartment in mouse oocytes. *Biology of Reproduction* Vol. 82(No. 5): 807-808.
Peters, L. & Meister, G. (2007). Argonaute proteins: mediators of RNA silencing. *Mol. Cell* Vol. 26: 611-623.

Pfister, D., De Mulder, K., Hartenstein, V., Kuales, G., Borgonie, G., Marx, F., Morris, J. & Ladurner, P. (2008). Flatworm stem cells and the germ line: developmental and evolutionary implications of macvasa expression in *Macrostomum lignano*. *Dev. Biol.* Vol. 319: 146-159.
Rangan, P., DeGennaro, M., Jaime-Bustamante, K., Coux, R.X., Martinho, R.G. & Lehmann, R. (2009). Temporal and Spatial Control of Germ-Plasm RNAs. *Current Biology* Vol. 19: 72-77.
Reddien, P.W., Oviedo, N.J., Jennings, J.R., Jenkin, J.C. & Alvarado, S.A. (2005). SMEDWI-2 is a PIWI-like protein that regulates planarian stem cells. *Science* Vol. 310: 1327-1330.
Rebscher, N., Zelada-González, F., Banish, T., Raible, F. & Arendt, D. (2007). Vasa Unveils a Common Origin of Germ Cells and of Somatic Stem Cells from the Posterior Embryonic Stem Cells Growth Zone in the Polychaete *Platynereis dumerlii*. *Dev. Biol.* Vol. 306: 599-611.
Rebscher, N., Volk, C., Teo, R., & Plickert, G. (2008). The Germ Plasm Component vasa Allows Tracing of the Interstitial Stem Cells in the Cnidarian *Hydractinia echinata*. *Dev. Dyn.* Vol. 237: 1736-1745.
Rinkevich, B. (2009). Stem Cells: Autonomy Interactors that Emerge as Causal Agents and Legitimate Units of Selection, In: *Stem Cells in Marine Organisms*, Eds. Rinkevich, B. & Matranga, V., Springer: Dordrecht, Heidelberg, London, New York, pp. 1-20.
Rinkevich, Y., Matranga, V., Rinkevich, B. (2009). Stem cells in aquatic invertebrates: common premises and emerging unique themes. In: *Stem Cells in Marine Organisms*, Eds. Rinkevich, B., Matranga, V., Springer: Dordrecht, Heidelberg, London, New York, pp. 61-104.
Rosner, A., Moiseeva, E., Rinkevich, Y., Lapidot, Z. & Rinkevich, B. (2009). Vasa and the germ line lineage in a colonial urochordate. *Developmental Biology* Vol. 331: 113-128.
Rossi, L., Salvetti, A., Marincola, F.M., Lena, A., Deri, P., Mannini, L., Batistoni, R., Wang, E. & Gremigni, V. (2007). Deciphering the molecular machinery of stem cells: a look at the neoblast gene expression profile. *Genome Biology* Vol.8: R62.
Salvetti, A., Rossi, L., Lena, A., Batistoni, R., Deri, P., Rainaldi, G., Locci, M.T., Evangelista, M. & Gremigni, V. (2005). DjPum, a homologue of *Drosophila* Pumilio, is essential to planarian stem cell maintenance. *Development* Vol. 132: 1863-1874.
Seipel, K., Yanze, N., & Schmid, V. (2004). The germ line and somatic stem cell gene Cniwi in the jellyfish *Podocoryne carnea*. *Int. J. Dev. Biol.* Vol. 48: 1-7.
Selenko, P., Sprangers, R., Stier, G., Bühler, D., Fischer, U. & Sattler, M. (2001). SMN Tudor domain structure and its interaction with the Sm proteins. *Nature Structural Biology* Vol. 8(No. 1): 27-31.
Seydoux, G. & Braun, R.E. (2006). Pathway to totipotency: lessons from germ cells. *Cell* Vol. 127: 891-904.
Shibata, N., Rauhana, L. & Agata, K. (2010). Cellular and molecular dissection of pluripotent adult somatic stem cells in planarians. *Development, Growth and Diff.* Vol. 52: 27-41.
Shibata, N., Umesono, Y., Orii, H., Sakurai, T., Watanabe, K. & Agata, K. (1999). Expression of *vasa (vas)*-related genes in germline cells and totipotent somatic stem cells of planarians. *Develop. Biol.* V. 206: 73-87.
Shostak, S. (2006). (Re)defining stem cells. *BioEssays* Vol. 28: 301-308.

Shukalyuk, A.I., (2002). Organization of interna in *Sacculina polygenea* (Crustacea: Rhizocephala). *Russian Journal of Marine Biology* Vol. 28: 329-335.

Shukalyuk, A.I. (2009). Germline commitment in embryonic and induced pluripotent stem cells. *Proceeding of The American Society of Cell Biology: 49th Annual Meeting*, December 5-9, Regular Abstracts, pp. 585.

Shukalyuk, A.I., Baiborodin, S.I. & Isaeva, V.V. (2001). Organization of interna in the rhizocephalan barnacle *Peltogasterella gracilis*. *Russian J. of Marine Biol.* Vol. 27: 113-116.

Shukalyuk, A.I., Golovnina, K., Baiborodin, S., Gunbin, K., Blinov, A. & Isaeva, V. (2007). *vasa*-related genes and their expression in stem cells of colonial parasitic rhizocephalan barnacle *Polyascus polygenea* (Arthropoda: Crustacea: Cirripedia: Rhizocephala). *Cell Biology International* Vol. 31(No. 2): 97-108.

Shukalyuk, A.I., Isaeva, V.V., Akhmadieva, A.V., & Alexandrova, Y.N. (2011). Stem cells in asexually reproducing invertebrates, embryonic stem cells and germline cells share common, evolutionary conserved features. *Proceeding of the International Society for Stem Cell Research, 9th Annual Meeting*, June 15-18, Friday Abstracts, Toronto, p. 77.

Shukalyuk, A.I., Isaeva, V.V., Golovnina, K.A., Baiborodin, S.I. & Blinov, A.G. (2005). *vasa*-related genes and their expression in stem cells of colonial *Polyascus polygenea* (Crustacea: Cirripedia: Rhizocephala). *Proceeding of the International Society for Stem Cell Research: 3rd Annual Meeting*, San Francisco. Abstracts. p. 215.

Shukalyuk, A.I., Isaeva, V.V., Kizilova, E. & Baiborodin, S. (2005). Stem cells in reproductive strategy of colonial rhizocephalan crustaceans (Crustacea: Cirripedia: Rhizocephala). *Invertebrate Reproduction and Development* Vol. 48: 41-53.

Shukalyuk, A.I. & Stanford, W.L. (2008). Germ plasm signature in embryonic stem cells. *Proceeding of the International Society for Stem Cell Research, 6th Annual Meeting*, June 11-14, Philadelphia, USA, p. 353 N 238.

Sköld, H.N., Obst, M., Sköld, M. & Åkesson, B. (2009). Stem cells in asexual reproduction of marine invertebrates. In: *Stem Cells in Marine Organisms*. Eds. Rinkevich, B. & Matranga, V., Springer: Dordrecht, Heidelberg, London, New York, pp.105-137.

Snee, M.J. & Macdonald, P.M. (2004). Live imaging of nuage and polar granules: evidence against a precursor-product relationship and a novel role for Oskar in stabilization of polar granule components. *Journal of Cell Science* Vol. 117: 2109-2120.

Snel, B., Lehmann, G., Bork, P. & Huynen, M.A. (2000). STRING: a web-server to retrieve and display the repeatedly occurring neighbourhood of a gene. *Nucleic Acids Res.* Vol. 28(No. 18): 3442-3444.

Solana, J., Lasko, P. & Romero, R. (2009). Spoltud-1 is a chromatoid body component required for planarian long-term stem cell self-renewal. *Dev. Biol.* Vol. 328(No. 2): 410-421.

Srouji, J. & Extavour, C. (2011). Redefining stem cells and assembling germ plasm: key transition in the evolution of the germ plasm. In: *Key transition in animal evolution*, Eds. DeSalle, R. & Schierwater, B., Science Publishers New York, USA Abingdon, UK, pp. 360-397.

Stice, S.L., Boyd, N.L., Dhara, S.K., Gerwe, B.A., Machacek, D.W. & Shin, S. (2006). Human embryonic stem cells: challenges and opportunities. *Reproduction, Fertility & Development* Vol. 18: 839-846.

Sunanaga, T., Watanabe, A. & Kawamura, K. (2007). Involvement of *vasa* homolog in germline recruitment from coelomic stem cells in budding tunicates. *Dev. Genes Evol.* Vol. 217: 1-11.

Takamura, K., Fujimura, M. & Yamaguchi, Y. (2002). Primordial germ cells originate from the endodermal strand cells in the ascidian *Ciona intestinalis*. *Dev. Genes Evol.* Vol. 212: 11-18.

Tanaka, S.S., Toyooka, Y., Akasu, R., Katoh-Fukui, Y., Nakahara, Y., Suzuki, R., Yokoyama, M. & Noce, T. (2000). The mouse homolog of *Drosophila Vasa* is required for the development of male germ cells *Genes & Dev.* Vol. 14: 841-853.

Thomson, T. & Lin, H. (2009). The biogenesis and function of PIWI proteins and piRNAs: progress and prospect. *Annual Review of Cell & Dev. Biology* Vol. 25: 355-376.

Toyooka, Y., Tsunekawa, N., Akasu, R., & Noce, N. (2003). Embryonic stem cells can form germ gells *in vitro*. *PNAS USA* Vol. 100: 11457-11462.

Tropere, V., Hitoshi, S., Sigard, C., Mak, T.W., Rossant, J., & van der Kooy, D. (2001). Direct neural fate specification from embryonic stem cells: a primitive mammalian neural stem cell stage acquired through a default mechanism. *Neuron* Vol. 30: 65-78.

Updike, D.L., Hachey, S.J., Kreher, J. & Strome, S. (2011). P granules extend the nuclear pore complex environment in the *C. elegans* germ line. *J. Cell Biol.* Vol. 192: 939-948.

Voronina, E., Lopez, M., Juliano, C.E., Gustafson, E., Song, J.L., Extavour, C., George, S., Oliveri, P., McClay, D. & Wessel, G. (2008). Vasa protein expression is restricted to the small micromeres of the sea urchin, but is inducible in other lineages early in development. *Dev. Biol.* Vol. 314: 276-286.

Walker, E., Ohishi, M., Davey, R.E., Zhang, W., Cassar, P.A., Tanaka, T.S., Der, S.D., Morris, Q., Hughes, T.R., Zandstra, P.W. & Stanford W.L. (2007). Prediction and testing of novel transcriptional networks regulating embryonic stem cell self-renewal and commitment. *Cell Stem Cell* Vol. 1: 71-86.

Watanabe, H., Hoang, V.T., Mättner, R. & Holstein, T.W. (2009). Immortality and the base of multicellular life: Lessons from chidarian stem cells. Semin. *Cell Devel. Biol.* Vol. 20: 1114-1125.

Weismann, A. (1892). *Das Keimplasma. Eine Theorie der Vererbung*, Jena: Verlag von Gustav Fisher, 682 S.

Weismann. A. (1893). *The Germ Plasm. A Theory of Heredity*, New York: Charles Scriber's Sons, 468 p.

Wu, H.-R., Chen, Y.-T., Su, Y.-H., Luo, Y.J., Holland, L.Z. & Yu, K. (2011). Asymmetric localization of germline markers Vasa and Nanos during early development in the amphioxus *Branchiostoma floridae*. *Develop. Biol.* Vol. 353: 147-159.

Ying, Y., Qi, X. & Zhao, G.Q. (2001). Induction of primordial germ cells from murine epiblasts by synergistic action of BMP4 and BMP8B signaling pathways. *PNAS USA* Vol. 98: 7858-7862.

Zwaka, T.P. & Thomson, J.A. (2005). A germ cell origin of embryonic stem cells? *Development* Vol. 132: 227-233.

8

G Protein-Coupled Receptors-Induced Activation of Extracellular Signal-Regulated Protein Kinase (ERK) and Sodium-Proton Exchanger Type 1 (NHE1)

Maria N. Garnovskaya
Department of Medicine (Nephrology Division),
Medical University of South Carolina, Charleston,
USA

1. Introduction

G-protein-coupled receptors (GPCRs) comprise a large family of cell-surface molecules, involved in signal transmission, accounting for >2% of the total genes encoded by the human genome. GPCRs have been linked to key physiological functions, including immune responses, cardiac- and smooth-muscle contraction and blood pressure regulation, neurotransmission, hormone and enzyme release from endocrine and exocrine glands. Thus, GPCRs contribute to embryogenesis, tissue remodelling and repair, inflammation, angiogenesis and normal cell growth. Their dysfunction contributes to multiple human diseases, and GPCRs represent the target of over 50% of all current therapeutic agents (Reviewed by Pierce et al., 2002). In addition, recent studies indicate that many GPCRs are overexpressed in various cancer types, and contribute to tumor cell growth when activated by circulating or locally produced ligands, suggesting a crucial role of GPCRs in cancer progression and metastasis. For example, many potent mitogens such as thrombin, lysophosphatidic acid (LPA), endothelin and prostaglandins stimulate cell proliferation by acting on their cognate GPCRs in various cell types. (Reviewed by Dorsam & Gutkind, 2007). The mechanisms that control cellular proliferation are important in normal physiology and disease states. Multiple mitogens that activate GPCRs stimulate the extracellular signal-regulated protein kinase (ERK) and lead to proliferation of mammalian cells. Another extensively studied mitogenic effector pathway in addition to ERK that ultimately leads to cell proliferation, is the ubiquitous plasma membrane sodium-proton exchanger type 1 (NHE1). NHE1 and ERK have both been implicated as key mediators of growth signals (Noel & Pouyssegur, 1995; Rozengurt, 1986; Kapus et al, 1994; Krump et al, 1997), therefore the regulatory relationships between NHE1 and ERK have been the subject of a number of studies over the last decade. Because both proteins can serve mitogenic functions, and because both are activated by similar stimuli, it has been hypothesized that one may be a regulator for the other. Indeed, in some cell types ERK plays a clear role in either the short or long term activation of NHE1 (Aharonovitz & Granot, 1996; Bianchini et al., 1997; Wang et al., 1997; Sabri et al., 1998; Bouaboula et al., 1999; Gekle et al., 2001). However, several groups were unable to demonstrate any role of ERK in regulation of NHE1 in a number of cell types (Gillis et al., 2001; Kang et al., 1998; Pederson et al., 2002; Garnovskaya et al., 1998; Di Sario et al., 2003). In addition, a number of recent studies suggested that certain stimuli such as mechanical stretch, hypertrophy and inflammatory mediators require NHE1 to regulate ERK (Takewaki et al., 1995; Nemeth et al., 2002; Yamazaki et al., 1998; Javadov et al., 2006; Chen et al., 2007). At present, very little is known about GPCR-induced NHE1-dependent ERK regulation. One report suggests that NHE1 is not a regulator for LPA-induced ERK activation in C6 glioma cells (Cechin et al., 2005) and another paper demonstrates the lack of role of NHE1 in angiotensin II (Ang II)- and endothelin 1-induced ERK activation in cultured neonatal rat cardiomyocytes (Chen et al., 2007). At the same time, our group showed that NHE1 activation plays a necessary role in activation of ERK by AII AT_1 and serotonin 5-HT_{2A} receptors in vascular smooth muscle cells (VSMC) (Mukhin et al., 2004), and in bradykinin B_2 receptor-induced ERK activation in renal carcinoma A498 cells (Garnovskaya et al., 2008) thus suggesting that the critical role of NHE1 in GPCR-induced ERK activation is not restricted to one specific cell type and receptor. Studies on the involvement of NHE1 in ERK regulation may also have pathophysiological relevance. NHE1 is usually referred to as a "housekeeping" protein and is normally inactive, but it gets activated in response to multiple specific stimuli, and maintains homeostatic cell volume and pH through Na^+/H^+ transport. The role of NHE1 has been well established in the myocardial remodeling and heart failure process (reviewed by Karmazyn et al., 2008). NHE1 may play a key role in the maintenance of blood pressure because increased activity of NHE1 has been observed in cells and tissues from hypertensive animals and humans (Rosskopf, et al., 1993; Lucchesi et al., 1994). Northern blot analysis showed that cultured VSMC from Sprague-Dawley and Wistar-Kyoto rats express only the NHE1 isoform, and that steady-state mRNA levels are similar for normal and spontaneously hypertensive animals (Lucchesi et al., 1994; LaPointe et al., 1995). Because no mutations in the NHE1 DNA sequence have been found in hypertensive animals, this suggests that increased activity of the antiporter is caused by an alteration in the regulation of NHE1 (Lucchesi et al., 1994). In addition, NHE1-mediated intracellular alkalinization has been proposed to play role in cancer cells growth, and over-expression of NHE1 contributes to the transformed phenotype of multiple cancer cells (Cardone et al., 2005). The role of NHE1 in renal diseases is less known. Mice with a spontaneous point mutation that results in truncation between the 11th and 12th NHE1 trans-membrane domains and causes loss of NHE1 function (Cox et al., 1997) do not present visible renal phenotype, consistent with the concept that NHE1 "housekeeping" activity under normal conditions is not required. However, NHE1 activity was increased in cell lines derived from patients with diabetic nephropathy (Ng et al., 1994), suggesting that NHE1 activity may be important in the context of cellular stress. Further, it has been shown that genetic or pharmacological loss of NHE1 function causes renal tubule epithelial cell apoptosis and renal dysfunction in several models of kidney disease (ureteral obstruction, adriamycin-induced podocyte toxicity, and streptozotocin-induced diabetes), suggesting that NHE1 activity may be beneficial for chronic kidney disease (Schelling & Abu, 2008). Moreover both, ERK and NHE1, have been proposed as key therapeutic targets for vascular illnesses, such as congestive heart failure (Kusumoto et al., 2001), myocardial infraction and reperfusion injury (Avkiran & Marber, 2002), ventricular fibrillation (Gazmuri et al., 2001), and ventricular hypertrophy (Chen et al., 2001). Therefore, studies devoted to the regulatory relationships between NHE1 and ERK have a potential clinical relevance.

The purpose of this review is to describe the relationship between NHE1 and ERK when both pathways are activated by GPCRs, with a particular emphasis on the situations when NHE1 is regulating ERK activity leading to cell proliferation.

2. Mitogen-activated protein kinases and Sodium-Hydrogen Exchanger-1 as mediators of growth signals

2.1 Mitogen-activated protein kinases

Mitogen-activated protein kinases (MAPKs) are a family of highly conserved proline-directed serine/threonine kinases that are activated by a large variety of extracellular stimuli and play integral roles in controlling many cellular processes, from the cell surface to the nucleus (Widmann et al, 1999). The MAPK family in mammals includes four distinctly regulated groups of MAPKs: extracellular signal-regulated kinase 1/2 (ERK), p38, c-Jun N-terminal kinases/stress-activated protein kinases (JNK/SAPKs), and ERK5/Big MAPK (BMK1) (Chang & Karin, 2001; Bogoyevitch & Court, 2004; Johnson & Lapadat, 2002). MAPK cascades typically consist of three levels of protein kinases that are consecutively activated by phosphorylation events: MAPK kinase kinase (MAPKKK or MAP3K or MEKK) activates MAPK kinase (MAPKK (MKK or MEK) or MAP2K), which in turn activates MAPK (Figure 1). Even so, the different tiers are composed of many similar isoforms that can be activated by more than one MAPK, increasing the complexity and diversity of MAPK signaling. Regulation and function of different MAPKs as well as complexity of MAPK signaling have been recently discribed in several review articles (Pearson et al., 2001; Chang & Karin 2001; Shaul & Seger, 2007; Bodart; 2010). MAPKs are involved in transmitting signals from a wide variety of extracellular stimuli including those of growth factor receptors and GPCRs, as well as physical or mechanical stimuli. In fact, MAPKs are major components of signaling pathways regulating a large array of intracellular events, such as proliferation, differentiation, acute signaling in response to hormones, stress response, programmed cell death, and gene expression (Pearson et al., 2001; Chang & Karin 2001; Kim & Choi, 2010). ERK is one member of a family of kinases that participate in mitogenic signaling through complex phosphorylation cascades that convert cell surface signals into nuclear transcription programs. In the typical scenario, GTP-bound Ras, a small G protein, activates Raf-1 kinase. In an alternative scenario, protein kinase C (PKC) or other signaling molecules activate Raf-1. In either case, Raf-1 phosphorylates and activates mitogen and extracellular signal-regulated kinases kinase (MEK), which in turn phosphorylates and activates ERK (Cobb & Goldsmith, 1995). Activated ERK translocates to the nucleus, where it activates a number of transcription factors such as Elk-1.

Recently a number of scaffolding proteins that play important role in ERK regulation have been described. Examples of such proteins include the Kinase Suppressor of Ras (KSR), β-arrestins1/2, PEA15 (phosphoprotein enriched in astrocytes), paxillin, and Raf-1 (Kolch,

2005). Paxillin, a multi adaptor protein in focal adhesion assembly, serves as a connector between ERK and Focal Adhesion Kinase (FAK) signaling pathways binding Raf-1 and ERK in response to hepatocyte growth factor in epithelial cells (Ishibe et al., 2004).

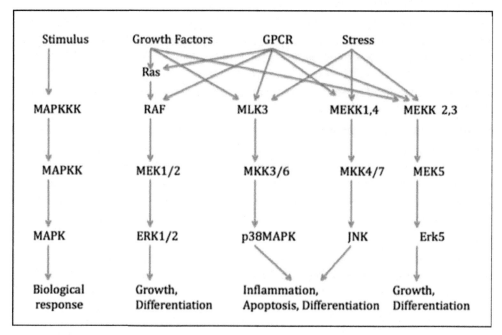

Fig. 1. Mitogen-Activated Protein Kinase Cascades.

2.1.1 Regulation of ERK by GPCRs

The ability of G-protein-coupled receptors (GPCRs) to generate signals that control cellular proliferation via activation of ERK pathways has been demonstrated in several studies (reviewed by Kranenburg & Moolenaar, 2001; and Luttrell, 2002). Although the mechanisms by which GPCRs control the activity of ERK vary between receptor and cell type, typically there are three categories of mechanisms: (1) signals that initiate classical G protein effectors, e.g., protein kinase A or protein kinase C causing the production of second messengers, (2) mechanisms that involve cross-talk between GPCRs and classical receptor tyrosine kinases, e.g., "transactivation" of epidermal growth factor (EGF) receptors, and (3) signals initiated by direct interaction between beta-arrestins and components of the MAP kinase cascade, e.g., beta-arrestin "scaffolds". Mitogenic pathways activated by different Gα families including activation of adenylyl cyclase/cAMP and phospholipase Cβ/protein kinase C second messenger pathways have been described in detail (reviewed by New & Wong, 2007). Angiotensin II promotes DNA synthesis and proliferation in many cell types by activating the G_q-coupled AT_1 receptor. AT_1 receptor activity in human adrenal cells induces Ras-dependent ERK activity, leading to increased levels of c-Fos and c-Jun transcription factors and to proliferation of the cells (Watanabe et al., 1996). Other mitogenic

GPCRs, including M_1 muscarinic and $α_{1B}$-adrenergic and purinergic receptors, induce ERK activity via the Ras-independent protein kinase C phosphorylation and activation of Raf-1 (Luttrell, 2002). G_s-coupled GPCRs utilize the adenylyl cyclase/ cAMP /Epac/Rap-1/B-Raf pathway to activate MAPK cascades and proliferation. In bone cells parathyroid hormone receptor promotes cAMP accumulation, which binds directly to the Rap-1 guanine nucleotide exchange factor Epac. Epac in turn activates Rap-1, a Ras family GTPase, which activates the kinase B-Raf, triggering ERK cascades (Fujita et al., 2002). Alternatively, PKA may directly activate Rap-1 (Luttrell, 2002).

However, activation of classical second messenger cascades cannot fully explain roles of GPCRs in stimulation of MAPK cascades. Additional signaling mechanisms including transactivation of the Receptor Tyrosine Kinases (RTKs) via the autocrine/paracrine release of epidermal growth factor (EGF)-like ligands at the cell surface and scaffolding of MAPK cascades, appear to contribute to GPCR-mediated MAPK activation. GPCR-mediated proliferation via the Gα or Gβγ subunit transactivation of RTKs has been described in several cell types (Ohtsu et al., 2006; Schafer et al., 2004). Thus, ligands for the LPA, endothelin-1 and thrombin receptors all stimulate cell proliferation in Rat-1 fibroblasts by transactivation of the epidermal growth factor receptor (EGFR, an RTK). Such transactivation requires the activation of matrix metalloproteases (MMPs) to release EGF from its membrane bound form, which then stimulates the EGFR and downstream ERK pathways (Schafer et al., 2004). Studies from our group demonstrated that bradykinin B_2 receptor activates ERK via EGFR transactivation in kidney cells (Mukhin et al., 2003; Mukhin et al., 2006; Kramarenko et al., 2010). The similar MMP/EGFR/ERK pathway have been also demonstrated in kidney cancer cells stimulated by LPA and angiotensin II (Schafer et al., 2004). A significant advance in the understanding of how GPCRs activate MAPK cascades is the discovery that beta-arrestin, a protein well known for its roles in both receptor desensitization and internalization, serves as a scaffolding protein for the GPCR-stimulated the extracellular signal regulated kinase ERK cascade. For example, agonist stimulation of the proteinase-activated receptor-2 (PAR2) leads to the formation of a large complex, which includes the receptor and beta-arrestin, MAPKKK, Raf-1, and activated ERK. Similarly, activation of neurokinin-1 receptor with the substance P, results in the formation of a complex, which includes the receptor, and beta-arrestin, c-Src and ERK. (Reviewed by Pierce et al., 2001).

ERK activation occurring via EGF receptor transactivation or via pathways employing second messengers (PKA- or PKC-dependent pathways) typically leads to sustained ERK activity and nuclear translocation of the kinase, thus contributing to regulation of cell cycle progression (Kranenburg & Moolenaar, 2001; Luttrell, 2002). In contrast, beta-arrestin/endocytotic pathway usually results in the retention of ERK in the cytoplasm and transient ERK activity, which is probably not sufficient to stimulate cell proliferation (Luttrell, 2002).

The intracellular pathways that mediate GPCR -induced ERK activation and regulation of cellular proliferation were recently reviewed by New & Wong (New & Wong, 2007).

2.2 Sodium-Hydrogen Exchanger-1 (NHE1)

The Na^+/H^+ exchange system was described in 1977 by Aickin and Thomas (Aickin & Thomas, et al., 1977), and the first Na^+/H^+ exchanger (NHE) gene was cloned in 1989 (Sarget et al., 1989). To date nine mammalian isoforms (NHE 1-9) have been identified in the family of Na^+/H^+ exchangers (Kemp et al, 2008). In this review we will focus only on the ubiquitously expressed, amiloride-sensitive integral plasma membrane protein NHE1 known as the "housekeeping enzyme", which is activated by various stimuli including growth factors, mitogens, and hyperosmolarity (Orlowski & Grinshtein, 2004; Wakabayashi et al., 1997). NHE1 is highly conserved across vertebrate species and is a major membrane transport mechanism, which plays an essential role in pH regulation, volume homeostasis, cell growth and differentiation (Bertrand et al., 1994). NHE1 is a phosphoglycoprotein of 815 amino acids that contains two functional domains: an NH2-terminal transmembrane ion translocation region with a proposed topology of 12 transmembrane domains, and a COOH-terminal cytoplasmic regulatory domain (Figure 2). The ion translocation domain catalyzes electroneutral exchange of extracellular sodium ion for intracellular hydrogen. Regulation of NHE1 activity in response to multiple stimuli including growth factors, hormones, and osmotic stress is mediated by a COOH-terminal cytoplasmic regulatory domain. The regulatory domain controls transport activity probably by altering affinity of a proton site in the transmembrane domain (Takahashi et al., 1999). This cytoplasmic domain includes a number of distinct subdomains modified either by phosphorylation or by the binding of regulatory proteins. The cytoplasmic domain contains high and low affinity Ca^{2+}/calmodulin-binding sites and several potential phosphorylation sites (Bertrand et al., 1994; Yan et al., 2001). Bertrand et al. first identified two calmodulin-binding sites on the

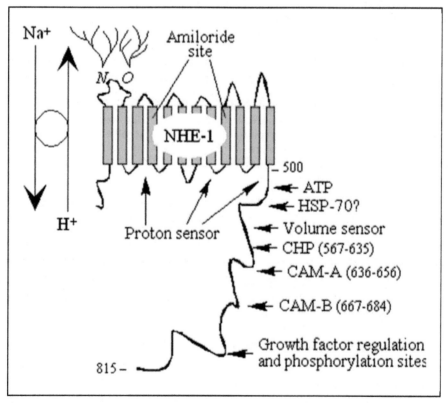

Fig. 2. Topographical model of NHE1.

cytoplasmic domain of NHE1 at amino acids 636-656 and 664-684, with high (Kd ~ 20 nM) and low (Kd~ 350 nM) affinities, respectively (Bertrand et al., 1994). In quiescent cells, the high-affinity calmodulin-binding domain may act as an autoinhibitory domain by interacting with the transmembrane domain, thus inhibiting ion translocation. Upon activation, NHE1 undergoes a conformational change that allows the Ca^{2+} -dependent binding of calmodulin (Wakabayashi et al., 1997). A phosphorylation domain at the distal COOH-terminus (amino acids 656-815 of human NHE1) contains a number of serine residues constitutively phosphorylated in quiescent cells that have increased phosphorylation levels in response to growth factors (Sarget et al., 1989). COOH-terminal serine residues on NHE1 molecule can be phosphorylated by the ERK-regulated kinase p90RSK (Tominaga et al., 1998) and by the Ste20-like Nck-interacting kinase (NIK) in response to growth factor receptors (Putney et al., 2002) and by Rho kinase 1 (ROCK1) in response to activation by GPCRs for thrombin and lysophosphatidic acid (Tominaga et al., 1998; Putney et al., 2002) and by integrin-induced cell adhesion (Tominaga & Barber, 1998). Because the cytoplasmic regulatory domain associates with multiple binding partners including the cytoskeleton-plasma membrane linker protein ezrin of the ezrin, moesin, radixin, (EMR) family (Denker et al., 2000), the calcineurin homolog protein CHP1 (Pang et al., 2001), calmodulin (CaM) (Yan et al., 2001), carbonic anhydrase II (Li et al., 2002), heat shock protein (Silva et al., 1995), and 14-3-3 protein (Lehoux et al., 2001), a novel function of NHE1 as a plasma membrane scaffold in the assembly of signaling complexes has been suggested (Baumgartner et al., 2004).

2.2.1 Regulation of NHE1 by GPCRs

While the activation of NHE1 and the kinetic alterations to the exchanger have been widely studied (Noel & Pouyssegur, 1995; Wakabayashi et al., 1997; Orlowski & Grinshtein, 1997) the signaling pathways that regulate NHE1 have not been fully elucidated. Because G protein-coupled receptors (GPCRs)-mediated regulation of sarcolemmal NHE activity is likely to play significant roles in modulating myocardial function in both physiological and pathophysiological conditions, most of the studies devoted to GPCR-induced NHE1 regulation were performed in cardiac myocytes (reviewed by Avkiran & Haworth, 2003). Sarcolemmal NHE activity is subject to exquisite regulation by a variety of extracellular stimuli, most of which act through GPCRs. Intriguingly, although the majority of the GPCR systems that have been studied to date have been shown to *stimulate* sarcolemmal NHE activity, there is also evidence that some may *inhibit* NHE activity or its stimulation through other pathways. A number of GPCRs, such as α_1-adrenergic receptors , angiotensin II AT_1 receptors, endothelin ET_A receptor, thrombin receptor, muscarinic receptors have been shown to increase sarcolemmal NHE activity through a change in the pH_i sensitivity of the exchanger. Interestingly, in contrast to the evidence that various G_q -coupled receptors (e.g. α_{1A}-ARs and angiotensin AT_1) mediate an increase in sarcolemmal NHE activity, GPCRs that signal through other G protein families (G_s and G_i) may attenuate NHE activity or its stimulation. Thus, β_1-AR stimulation *inhibits* sarcolemmal NHE activity, while adenosine A_1 and angiotensin AT_2 receptors attenuate stimulation of NHE1 by other ligands (Avkiran& Haworth, 2003). The mechanisms of GPCR-induced NHE1 activation are not fully understood. To date, several mechanisms of activation of NHE1 by G protein-coupled receptors have been proposed although not fully characterized : α_{1A}-adrenoceptor activates NHE1 through protein kinase C (Snabaitis et al., 2000; Avkiran & Haworth, 2003); lysophosphatidic acid stimulates NHE1 through RhoA and its effector ROCK (Tominaga et al., 1998), and angiotensin II AT_1 receptor regulates NHE1 activity through RSK (Takahashi et al., 1999). In addition, Wallert et al provided evidence that the specific α_1-adrenergic agonist, phenylephrine and the lysophosphatidic acid (LPA) activate NHE1 in CCL39 cells, and demonstrated a direct involvement of ERK in the α_1-adrenergic activation of NHE1 and a significant role for both ERK and RhoA in LPA stimulation of NHE1 in CCL39 fibroblasts (Wallert et al., 2004). Our group reported that a fibroblast NHE1 can be rapidly stimulated through the transfected human serotonin 5-HT_{1A} receptor via pertussis toxin-sensitive G protein α-subunits $G_{i\alpha 2}$ and $G_{i\alpha 3}$, in CHO cells (Garnovskaya et al., 1997), by endogenously expressed G_q-coupled bradykinin B_2 receptor in kidney cells (Mukhin et al., 2001), and by endogenously expressed G_q-coupled angiotensin II AT_1 and serotonin 5-HT_{2A} receptors in vascular smooth muscle cells (Garnovskaya et al., 2003). While studying the signaling pathway of bradykinin B_2 receptor-induced NHE1 activation in mIMCD-3 kidney cells, we found a new mechanism for the GPCR-induced regulation of Na^+/H^+ exchange (Mukhin et al., 2001). This novel pathway involved activation of phospholipase C, elevation of intracellular Ca^{2+}, activation of the non receptor tyrosine kinase, Janus kinase 2 (Jak2), tyrosine phosphorylation of Ca^{2+}/calmodulin (CaM), and binding of CaM to NHE1. Bradykinin rapidly stimulated the assembly of a signal transduction complex that includes CaM, Jak2, and NHE1. We suggested that Janus kinase 2 is involved in the activation of NHE1 by increasing the tyrosine phosphorylation of calmodulin, which appears to be a direct substrate for phosphorylation by Janus kinase 2. Further the same pathway has been demonstrated for the bradykinin B_2 receptor-mediated activation of Na^+/H^+ exchange in KNRK and CHO cells (Lefler et al., 2003), and for the G_q-coupled angiotensin II AT_1 and serotonin 5-HT_{2A} receptors, which stimulated NHE1 activation in vascular smooth muscle cells (Garnovskaya et al., 2003), suggesting that this pathway represent a fundamental mechanism for the rapid regulation of NHE1 by G_q-coupled receptors in multiple cell types. Further we have shown that the G_i – coupled serotonin 5-HT_{1A} receptor also rapidly stimulates NHE1 through a pathway that involves 1) activation of Janus kinase 2 downstream of the 5-HT_{1A} receptor; 2) formation of a complex that includes NHE1, Jak2, and CaM; 3) tyrosine phosphorylation of CaM through Jak2; and 4) increased binding of CaM to the carboxyl terminus of NHE1 (Turner et al., 2007).

2.3 Relationships between NHE1 and ERK

2.3.1 MAPK regulates NHE1

Whereas it has been known for some time that mitogens typically activate both NHE1 and ERK in concert (Noel & Pouyssegur, 1995; Rozengurt, 1986; Kapus et al., 1994; Krump et al., 1997) the exact relationships between NHE1 and ERK have only recently been explored in any great detail. Recent studies have shown that multiple stimuli that rapidly activate ERK pathways also rapidly increase NHE activity in many cell types, particularly in fibroblasts. Those stimuli include, but are not limited to: growth factors that modulate tyrosine phosphorylation cycles, integrins, hyperosmotic stress or cell shrinkage, protein kinase C (PKC), tyrosine phosphorylation cascades and heterotrimeric G proteins (Clark & Limbird, 1991; Barber, 1991; Rozengurt, 1986; Lowe et al., 1990). Those similarities provide evidence to suggest that ERK may be a direct proximal component of an NHE regulatory pathway (Noel & Pouyssegur, 1995; Aharonovitz & Granot, 1996). There is a growing awareness that tyrosine phosphorylation cycles are critical in regulating NHE activities in a number of cell types (Donowitz et al., 1994; Yamaji et al., 1995; Good, 1995; Fukushima et al., 1996) as has also been shown for ERK (Blumer & Johnson, 1994). Other studies have demonstrated that

NHE and ERK activities are modulated by overlapping upstream enzymes, including phosphoinositide 3′-kinase (PI-3K), phospholipase C, and PKC (Levine et al., 1993; Kapus et al., 1994; Voyno-Yasenetskaya et al., 1994; Bertrand et al., 1994; Ma et al., 1994; Dhanasekaran et al., 1994; Inglese et al., 1995). In aggregate, those studies implicate G proteins, lipid-recognizing enzymes, tyrosine kinases, and NHEs as playing interrelated roles along with ERK in cell growth (Barber, 1991; Noel & Pouyssegur, 1995; Aharonovitz & Granot, 1996; Blumer & Johnson, 1994; Lin et al., 1996). Relevant to the hypothesis that ERK regulates NHE1, are studies showing that microinjection of activated Ras (Hagag et al., 1987) or transfection of the Ha-Ras oncogene (Doppler et al., 1987; Maly et al., 1989; Kaplan & Boron, 1994) stimulates NHE activity in fibroblasts. The classical effect of GTP-bound Ras is the activation of the ERK1 and ERK2 (Blumer & Johnson, 1994). This is thought to occur primarily through a linear signalling pathway that flows as follows: Ras-GTP → Raf-1 kinase → MEK (MAPK/ERK kinase) → ERK. Thus, because Ras functions upstream of both NHE and ERK activities, ERK has been proposed as a logical funnel for signals from extracellular stimuli to the effector NHE. The effect of NHE activation due to the sustained activation of ERK is most likely secondary to the activation of transcription cascades that upregulate the NHE message/protein or modulate expression of key regulators of NHE activity. However, several studies suggest that ERK might regulate NHE activity in the short term, as well. The possibility that ERK rapidly regulates NHE activity was tested in platelets by Aharonovitz and Granot (Aharonovitz & Granot, 1996) who showed that arginine vasopressin (AVP) and PMA rapidly activated NHE by a pathway which was sensitive to PD98059, a specific inhibitor of MEK1. Moreover, the signal initiated by AVP was sensitive to genistein, a broad-spectrum inhibitor of tyrosine kinases (Aharonovitz & Granot, 1996). Bianchini et al. (Bianchini et al., 1997) went further to characterize the role of ERK in regulating NHE when cells were stimulated by combinations of growth factors or serum. Specifically, they showed that expression of a dominant negative p44 ERK or of the MAPK phosphatase MKP-1, or treatment with the MEK1 inhibitor PD98059 reduced activation of NHE-1 by mixtures of growth factors by about 50%. Further, it has been shown that short-term activation of ERK leads to rapid stimulation of NHE1 in multiple cell types (erythrocytes, fibroblasts, MDCK-11 cells, rabbit skeletal muscle, and cultured rat neonatal and adult ventricular cardiomyocytes) when activated by diverse stimuli including growth factors, angiotensin II, and aldosterone (Wang et al., 1997; Sabri et al., 1998; Bouboula et al., 1999; Gekle et al., 2001; Wei et al., 2001; Moor et al., 2001; Snabaitis et al., 2002). At least in some cases, the short term-stimulation of NHE1 by ERK is mediated by phosphorylation of NHE1 either by ERK itself, or by p90RSK, an ERK-regulated kinase (Takahashi et al., 1999). Cuello et al. demonstrated that ERK- dependent 90kDa ribosomal S6 kinase (RSK) is the principal regulator of cardiac sarcolemmal NHE1 phosphorylation and NHE activity after α_1-adrenergic stimulation in adult myocardium (Cuello et al., 2007). Thus, there is clear evidence that ERK can increase the activity of NHE1 by increasing its expression and/or by stimulating the activity of existing NHE1 molecules.

2.3.2 MAPK and NHE1 do not regulate each other

On the other hand, several groups have been unable to show any role for ERK in activating NHE1 in multiple cell types, including Xenopus oocytes (Kang et al., 1998), Ehrlich Ascites cells (Pederson et al., 2002), CHO cells (Garnovskaya et al., 1998), or hepatic stellate cells (Di Sario et al., 2003). Moreover, there is one report in which ERK was shown to mediate inhibition of NHE1 activity in MTAL cells (Watts & Good, 2002). Our group tested the hypothesis that ERK could mediate rapid, short-term activation of NHE activity in fibroblasts when both signals were initiated by a single G protein-coupled serotonin 5-HT$_{1A}$ receptor (Garnovskaya et al., 1998). These studies revealed a number of similarities between the regulation of ERK and NHE. Activation of the two processes shared similar concentration–response and time-course characteristics. Receptor-activated NHE and ERK also shared an overlapping sensitivity to some pharmacological inhibitors of tyrosine kinases (staurosporine and genistein), PI-3K (wortmannin and LY294002), and PC-PLC (D609), and neither pathway was sensitive inhibition of PKC. However, definitive studies designed to block signaling molecules possessing well-defined roles in activating ERK through the 5-HT$_{1A}$ receptor by transfecting cDNA constructs encoding inactive mutant PI-3K, Grb2, Sos, Ras, and Raf molecules were successful in attenuating ERK, but had essentially no effect upon NHE activation. Thus, our data do not support the hypothesis that ERK is a proximal short-term regulator of NHE in CHO cells when the signal is initiated by the G$_{i/o/z}$ protein-coupled 5-HT$_{1A}$ receptor. Therefore, the ability of ERK to stimulate NHE1 activity has not been a universal finding.

2.3.3 NHE1 as a regulator of MAPK

Despite the increasing interest in potential roles for ERK in the activation of NHE1, much less is known regarding the role of NHE1 in regulating ERK. There have been several reports that suggest that NHE1 might play a role in regulating ERK activation (reviewed by Pedersen et al., 2007). Mitsuka et al. had shown that specific inhibitors of NHE1 could reduce neointimal proliferation in a rat model of carotid artery injury (Mitsuka et al., 1993). However, in C6 glioma cells although lysophosphatidic acid (LPA) - increased proliferation was sensitive to NHE1 inhibitors, LPA-induced ERK activation was unaffected (Cechin et al., 2005). Takewaki et al. presented some evidence that a potent antagonist of NHE1 could partially inhibit stretch-induced activation of ERK in the cultured cardiomyocytes (Takewaki et al., 1995). Later the same group reported that in cultured neonatal rat cardiomyocytes NHE1 inhibition blocked the stretch-induced activation of Raf-1 and ERK, while angiotensin II (Ang II)- and endothelin 1-induced ERK activation remained unaffected (Yamazaki et al., 1998). On the other hand, our work in vascular smooth muscle cells (VSMC) demonstrated that activation of ERK by AII and serotonin was strongly dependent of NHE1 activity, and the effect of NHE1 occurs at or above the level of Ras (Mukhin et al., 2004). In human colon cancer epithelial cells, NHE1 inhibition suppressed activation of ERK and NF-κB and led to decreased production of interleukin-8 in response to inflammatory signals (Nemeth et al., 2002). Recently it has been also demonstrated that NHE1 inhibition prevented ERK activation during phenylephrine-induced hypertrophy in neonatal rat cardiomyocytes (Javadov et al., 2006), and prevented glucose-induced ERK activation in a high glucose model of cardiomyocyte hypertrophy (Chen et al., 2007). In Ehrlich Lettre Ascites cells under osmotic cell shrinkage NHE1 regulates ERK acting at or above the level of MEK (Pederson et al., 2002). Therefore, NHE1-dependent regulation of ERK in most cases has been described in cells stimulated by mechanical stretch, osmotic shrinkage, hypertrophy and inflammatory mediators (Takewaki et al., 1995; Nemeth et al., 2002; Yamazaki et al., 1998; Javadov et al., 2006; Chen et al., 2007; Mitsuka et al., 1993; Pederson et al., 2007). Very little is known about GPCR-induced NHE1-dependent ERK regulation. One report suggests that NHE1 is not a regulator for LPA-induced ERK activation in C6 glioma cells (Cechin et al., 2005) and another paper demonstrates the lack of role of NHE1 in AII- and endothelin 1-induced ERK activation in cultured neonatal rat cardiomyocytes (Chen et al., 2007).

2.3.3.1 NHE1 regulates ERK activity in GPCR-activated VSMC

Angiotensin II (Ang II), a potent hypertrophic factor for vascular smooth muscle cells, mediates its effects via specific plasma membrane AT_1 receptors that belong to GPCR family. Ang II stimulates multiple signaling pathways (reviewed by Touyz & Schiffrin, 2000) including MAPKs, Src family kinases, phospholipase D, and Janus kinase (Jak2). Ang II also has been shown to stimulate NHE1 activity in VSMC (Berk et al., 1987) but does not appear to increase the steady state levels of NHE1 mRNA. There also are reports on relationship between Ang II-induced NHE1 and ERK activities in VSMC suggesting that activation of the AT_1 receptor first leads to activation of the MEK-ERK-p90RSK pathway, and that activated p90RSK in turn directly phosphorylates and activates NHE1 in VSMC (Takewaki et al., 1995). However, this suggestion was based mainly on *in vitro* experiments in which p90RSK immunoprecipitated from Ang II-stimulated VSMC was able to phosphorylate recombinant NHE1, and it is still not clear whether Ang II-induced phosphorylation of NHE1 takes place in VSMC *in vivo* and if this phosphorylation is physiologically significant. Our group has described a novel pathway of the regulation of NHE1 activity in VSMC by two mitogens, Ang II and serotonin (5-HT) that involves the activation of Jak2, tyrosine phosphorylation of Ca^{2+}/calmodulin, and binding of calmodulin (CaM) to NHE-1 (Garnovskaya et al., 2003). In the same study we were not able to support any role for ERK in Ang II-induced NHE1 activation in VSMC (Garnovskaya et al., 2003). Further, we specifically investigated the roles of NHE and ERK (as stimulated by either 5-HT or Ang II) in the activation of each other in VSMC (Mukhin et al., 2004), and we have found evidence to support a novel role for NHE in the activation of ERK in VSMC. This evidence includes 1) dual stimulation of NHE and ERK by Ang II and 5-HT, with the activation of NHE preceding that of ERK, 2) similar concentration-response relationships for the stimulation of NHE and the phosphorylation of ERK by 5-HT and Ang-II, 3) blockade of the activation of ERK induced by 5-HT and Ang II by chemical inhibition of NHE, 4) blockade of the activation of ERK induced by 5-HT and Ang II by removal of sodium from incubation buffers, and 5) phosphorylation of ERK during recovery from an imposed acid load, a maneuver that induces receptor-independent activation of NHE.

Moreover, in the case of receptor-induced activation of ERK, NHE appears to be located upstream of MEK and ERK, and downstream of Ang II and 5-HT-mediated transactivation of the EGF receptor. NHE intersects the classical pathway of activation of ERK at or above the level of Ras. Figure 3 depicts one possible scheme that can account for our findings. Because it has been described that G_q-coupled receptors such as Ang II AT_1 and serotonin 5-HT_{2A} receptors activate ERK in VSMC through transactivation and phosphorylation of the epidermal growth factor (EGF) receptor (Eguchi et al., 1999), we wanted to establish whether NHE regulates ERK activation upstream of the EGF receptor. It appeared that inhibition of NHE activity by depriving the exchanger of extracellular sodium, or by blockade with the specific inhibitors, amiloride analog, methylisobutylamiloride (MIA), prevents activation of ERK by two GPCR ligands, Ang II and 5-HT. Those same maneuvers have no effect on EGF-stimulated ERK, suggesting that there are some differences in the pathways used by Ang II and 5-HT to activate ERK when compared with that used by EGF. Interestingly, the close connection between NHE and ERK activation is further underscored by the observation that receptor-independent activation of NHE also results in ERK phosphorylation only when the exchanger is allowed to mediate recovery from an imposed intracellular acid load. Thus, NHE activation is *necessary* for Ang II and 5-HT-induced activation of ERK, and is *sufficient* to activate ERK under conditions of an imposed acid load. In contrast, NHE activation is not *necessary* for EGF-mediated activation of ERK. The most likely explanation is that 5-HT or Ang II requires the parallel activation of NHE and the EGFR to activate ERK in VSMC. The two pathways intersect downstream of trans-phosphorylation of the EGFR, and upstream of ERK and MEK, most likely at or upstream of Ras (Figure 3). The precise mechanisms of NHE-dependent ERK activation by Ang II and 5-HT remain to be defined. One possibility is that NHE plays an accessory role in Ang II and 5-HT induced activation of ERK by facilitating cytoskeletal reorganization or by altering Na^+ or H^+ concentrations in cellular microdomains, thereby affecting enzyme activity or protein-protein interactions.

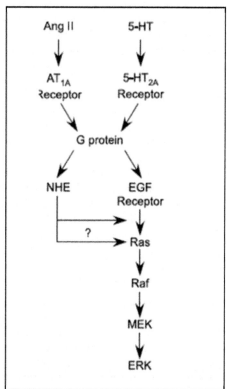

Fig. 3. Hypothetical scheme of NHE1-dependent ERK activation by GPCRs in VSMC.

The speculation regarding the cytoskeletal effects of NHE is particularly intriguing in light of work by Barber's group showing important functional links between NHE activity and the cytoskeleton (Denker et al., 2000). These findings have potential implications for the regulation of vascular tone, as well as for vascular pathobiology because Ang II and 5-HT are both potent vasoconstrictors, and Ang II has been shown to play major roles in various cardiovascular diseases including left ventricular hypertrophy and hypertension (Mitsuka et al., 1993).

2.3.3.2 NHE1 regulates ERK activity in Bradykinin-activated renal carcinoma cells

Because we have been able to detect the critical role of NHE in GPCR-mediated activation of ERK in cells of contractile phenotype, we thought that this relationship might be restricted to specific cell types and receptors. The cellular specificity of the relationship between NHE1 and ERK could be mediated by alternate accessory components of each signaling pathway, or by cell-specific compartmentalization of scaffolded signal transduction platforms (Luttrell & Lutrell, 2003). However, we have collected data that support a role of NHE1 in bradykinin B_2 receptor-induced ERK activation in renal carcinoma A498 cells, thus suggesting that the critical role of NHE1 in GPCR-induced ERK activation is not restricted to one specific cell type and receptor (Garnovskaya et al., 2008). In this study we investigated the endogenous intrarenal kinin hormone bradykinin (BK) that exerts its multiple pathophysiological functions via two known receptors, the bradykinin B_1 (BK B_1) and bradykinin B_2 (BK B_2) which belong to the superfamily of G protein-coupled receptors (GPCR) (Hess et al., 1992; Menke et al., 1994; Bagate et al., 2001). BK plays a significant role as a modulator of renal function such as electrolyte and water excretion (Mukai et al., 1996) as well as in renal cell growth and proliferation (El-Dahr et al., 1998; Jaffa et al., 1997). We have previously reported that BK activates NHE1 in a kidney cell line derived from the inner medullary collecting duct of mice (mIMCD-3 cells) (Mukhin et. al., 2001) via the similar pathway that Ang II and serotonin employ to activate NHE1 activity in VSMC, which involves the activation of Jak2, tyrosine phosphorylation of Ca^{2+}/calmodulin, and binding of calmodulin (CaM) to NHE-1 (Garnovskaya et al., 2003). We have also described that BK is a potent mitogenic factor for mIMCD-3 cells, and demonstrated that BK-induced cell proliferation was dependent on activation of epidermal growth factor receptor (EGFR) tyrosine kinase and subsequent activation of mitogen- and extracellular signal-regulated kinase kinase (MEK) and (Mukhin et al., 2003; Mukhin et al., 2006; Kramarenko et al., 2010). However, we were not able to establish the relationship between NHE1 and ERK in mIMCD-3 cells. Our data did not support either the hypothesis that ERK is a proximal short-term regulator of NHE or the hypothesis that NHE1 is necessary for the BK-induced ERK activation in normal kidney mIMCD-3 cells (Garnovskaya, unpublished data). Because there is evidence linking BK to the cancerogenic process (Bhoola et al., 2001; Chan et al., 2002), and because NHE1 has been proposed to play role in cancer cells growth (Cardone et al., 2005), we wanted to explore the possibility that BK exerts its mitogenic effects via activation of NHE in cancer cell lines. The expression of BK receptors has been demonstrated in clinical speciments of adenocarcinoma, squamous carcinoma, lymphoma, hepatoma and carcinoid, and in experimental mouse sarcoma 180 and colon adenocarcinoma 38 (Wu et al., 2002), in small cell and non-small cell carcinomas of the lung (Chee et al., 2008), and in oesophageal squamous cell carcinoma (Dlamini et al., 2005). The mitogenic effects of BK have been reported in primary cultured epithelial breast cancer cells and in MCF-7 breast cancer cell line, where BK stimulated cell proliferation through ERK activation (Greco et al., 2005; Greco et al., 2006). Because there were limited studies on the role of BK in renal cell carcinomas, we have chosen to use A498 cells, a transformed cell line derived from primary undifferentiated kidney carcinoma (Giard et al., 1973), which represents a widely used model for studying of renal carcinomas. Our results demonstrated that NHE1 is involved in BK-induced ERK activation and proliferation of A498 cells, and that BK B_2 receptor-induced ERK activation in A498 cells depends on NHE activity (Garnovskaya et al., 2008), suggesting that the critical role of NHE1 in GPCR-induced ERK activation is not restricted to one specific cell type and receptor. Previously, NHE1-mediated intracellular alkalinization has been proposed to play role in cancer cells growth because it has been shown that increased pH_i of tumor cells is associated with increased *in vivo* tumor growth, DNA synthesis, and cell-cycle progression, suggesting that over-expression of NHE1 contributes to the transformed phenotype of multiple cancer cells (reviewed by Cardone et al., 2005). The cellular alkalinization of tumor cells induced by hyperactivation of NHE1 has been shown to be directly related to increased protein synthesis and tumor cell growth (Cardone et al., 2005; Harguindey et al., 2005). It has been suggested that the mechanism of NHE1-mediated tumor cell growth and metastasis does not depend of its ion-transporting activities but rather employs NHE1 as a scaffolding protein to directly regulate cytoskeletal dynamics (Cardone et al., 2005). Further it has been shown that NHE1 antisense gene suppresses cell growth, induces cell apoptosis, and partially reverses the malignant phenotypes of human gastric carcinoma cells (Liu et al., 2008). Similarly, silencing of NHE1 gene by siRNA interference and /or inhibition of NHE1 activity by amiloride analogs effectively blocked the invasiveness of human hepatocellular carcinoma cells (Yang et al., 2011). Thus, inhibition of NHE1 might result in an antiproliferative effect, and NHE1 may be a potential target for chemotherapeutics to treatment of renal carcinoma.

3. Conclusion

The elucidation and understanding of the relationship between NHE1 and ERK cascade has been one of the most active areas in biological research over the past few years. As discussed above, experimental studies have strongly implicated a role for NHE1 in the regulation of ERK activity, although the precise pathway, which leads from the activation of NHE1 to ERK regulation still has to be defined. One possibility is that GPCR-induced NHE1-dependent ERK activation depends on NHE1-mediated Na^+/H^+ exchange. In that sense, Grinstein et al have demonstrated uneven distribution of NHE1 molecules throughout the cell surface with the focal accumulation at or near terminal edges of fibroblasts and CHO cells, and the areas of increased NHE1 density closely corresponded to sites of accumulation of cytoskeletal proteins (Grinshtein et al., 1993). It is possible that NHE1 regulates ERK by altering Na^+ or H^+ concentrations in cellular microdomains, thereby affecting enzyme activity or protein-protein interactions. Another possibility is that NHE1 acts as a plasma membrane scaffold (Baumgartner et al., 2004) in the assembly of signaling complexes independent of its ion exchange activity bringing together GPCRs, and the members of ERK-activation cascade. Regardless of the mechanisms, the critical role of NHE1 as an upstream molecule in GPCR-induced ERK activation could have significant physiological and pathophysiological relevance.

Because ERK-dependent cell proliferation is thought to be a critical component in many pathologic conditions, and NHE is involved in a variety of complex physiological and pathological events that include regulation of intracellular pH, cell movement, heart disease, and cancer, improved understanding of the molecular mechanisms that regulate NHE and ERK may allow alternative approaches to the therapeutic manipulation of ERK and NHE activity to be developed.

4. Acknowledgment

Studies from the author's laboratory were supported by grants from AHA (GIA 0655445U), the National Institutes of Health (DK52448 and GM63909), and Merit Awards from the Department of Veterans Affairs.

5. References

Aharonovitz, O. & Granot, Y. (1996). Stimulation of mitogen-activated protein kinase and Na^+/H^+ exchanger in human platelets. Differential effect of phorbol ester and vasopressin. *J. Biol. Chem.* 271, pp. 16494-99.

Aickin, C. & Thomas, R. (1977). An investigation of the ionic mechanism of intracellular pH regulation in mouse soleus muscle fibres. *J Physiol.* 273, pp. 295-316.

Avkiran, M. & Marber, M. (2002). Na^+/H^+ exchange inhibitors for cardioprotective therapy: progress, problems and prospects. *J. Am. Coll. Cardiol.,* 39, pp.747-753.

Avkiran, M. & Haworth, R. (2003). Regulatory effects of G protein-coupled receptors on cardiac sarcolemmal Na^+/H^+ exchanger activity: signalling and significance. *Cardiovascular Research,* 57, pp. 942-952.

Bagate, K.; Grima, M.; Imbs, J.; Jong, W.; Helwig, J. & Barthelmebs, M. (2001). Signal transduction pathways involved in kinin B(2) receptor-mediated vasodilation in the rat isolated perfused kidney. *Br J Pharmacol.,* 132, pp. 1735-1752.

Barber, D. (1991). Mechanisms of receptor-mediated regulation of Na-H exchange. *Cellular Signaling,* 3, pp. 387-397

Baumgartner, M.; Patel, H. & Barber, D. (2004). Na^+/H^+ exchanger NHE1 as plasma membrane scaffold in the assembly of signaling complexes. *Am. J. Physiol.,* 287, pp. C844-C850.

Bertrand, B.; Wakabayashi, S.; Ikeda, T.; Pouyssegur, J. & Shigekawa, M. (1994). The Na^+/H^+ exchanger isoform 1 (NHE1) is a novel member of the calmodulin-binding proteins. *J. Biol. Chem.,* 269, pp. 13703-13709.

Bhoola, K.; Ramsaroop, R.; Plendl, J.; Cassim, B.; Dlamini, Z. & Naicker, S. (2001). Kallikrein and kinin receptor expression in inflammation and cancer. Biological Chemistry, 382, pp. 77-90.

Bianchini, L.; L'Allemain, G. & Pouysségur, J. (1997). The p42/p44 MAP kinase cascade is determinant in medi-ating activation of the Na^+/H^+ exchanger in response to growth factors. *J. Biol. Chem.,* 272, pp. 271-279.

Blumer, K. & Johnson, G. (1994). Diversity in function and regulation of MAP kinase pathways. *Trends Biochem. Sci.* 19, pp. 236–240.

Bodart, J. (2010). Extracellular-regulated kinase-mitogen-activated protein kinase cascade: unsolved issues. *Journal of Cellular Biochemistry,* vol. 109, no. 5, pp. 850–857.

Bogoyevitch, M. & Court, N. (2004). Counting on mitogen-activated protein kinases—ERKs 3, 4, 5, 6, 7 and 8. *Cellular Signalling,* vol. 16, no. 12, pp. 1345–1354.

Bouaboula, M.; Bianchini, L.; McKenzie, F.; Pouyssegur, J. & Casellas, P. (1999). Cannabinoid receptor CB1 activates the Na^+/H^+ exchanger via G_i-mediated MAP kinase signaling transduction pathways. *FEBS Lett.,* 449, pp. 61-65.

Cardone, R.; Casavola, V. & Reshkin, S. (2005). The role of disturbed pH dynamics and the Na$^+$/H$^+$ exchanger in metastasis. *Nat Rev Cancer*, 5, pp. 786-795.

Cechin, S.; Dunkley, P. & Rodnight, R. (2005). Signal transduction mechanisms involved in the proliferation of C6 glioma cells induced by lysophosphatidic acid. *Neurochem. Res.*, 30, pp. 603-611.

Chan. D.; Gera, L.; Stewart, J.; Helfrich, B.; Verella-Garcia, M.; Johnson, G.; Baron, A.; Yang, J.; Puck, T. & Bunn, P. Jr. (2002). Bradykinin antagonist dimer, CU201, inhibits the growth of human lung cancer cell lines by a "biased agonist" mechanism. Proc Natl Acad Sci U S A, 99, pp. 4608-4613.

Chang, L. & Karin, M. (2001). Mammalian MAP kinase signalling cascades. *Nature*, vol. 410, no. 6824, pp. 37–40, 2001.

Chee, J.; Naran, A.; Misso, N.; Thompson P. & Bhoola, K. (2008). Expression of tissue and plasma kallikreins and kinin B1 and B2 receptors in lung cancer. *Biological Chemistry*, 389, pp. 1225-1233.

Chen, L.; Gan, X.; Haist, J.; Feng, Q.; Lu, X.; Chakrabarti, S. & Karmazyn, M. (2001). Attenuation of compensatory right ventricular hypertrophy and heart failure following monocrotaline-induced pulmonary vascular injury by the Na$^+$-H$^+$ exchange inhibitor cariporide. *J. Pharmacol. Exp. Ther.*, 298, pp. 469-476.

Chen, S.; Khan, Z.; Karmazyn, M. & Chakrabarti, S. (2007). Role of endothelin-1, Na$^+$-H$^+$ exchanger-1 and MAPK activation in glucose-induced cardiomyocyte hypertrophy. *Diabetes Metab. Res. Rev.*, 23 pp.356-67.

Clark, J. & Limbird, L. (1991). Na+/H+ exchanger subtypes: A predictive Review. *Am. J. Physiol.*, 261, pp. C945–C953.

Cobb, M. & Goldsmith, E. (1995). How MAP kinases are regulated. *J. Biol. Chem.*, 270, pp. 14843-14846.

Cox, G.; Lutz, C.; Yang, C.; Biemesderfer, D.; Bronson, R.; Fu, A.; Aronson, P.; Noebels, J. & Frankel , W. (1997). Sodium/hydrogen exchanger gene defect in slow-wave epilepsy mutant mice. *Cell*, 91, pp. 139-148.

Cuello, A.; Snabaitis, A.; Cohen, M.; Taunton J. & Avkiran, M. (2007). Evidence for direct regulation of myocardial Na+/H+ exchanger isoform 1 phosphorylation and activity by 90-kDa ribosomal S6 kinase (RSK): effects of the novel and specific RSK inhibitor fmk on responses to α_1-adrenergic stimulation. *Mol Pharmacol*, 71 (2007), pp. 799-806.

Denker, S.; Huang, D.; Orlowski, J.; Furthmayr, H. & Barber, D. (2000). Direct binding of Na$^+$/H$^+$ exchanger 1 to ERM proteins regulates the cortical cytoskeleton & cell shape independently of H$^+$ translocation. *Mol. Cell.*, 6, pp. 1425-1436.

Dhanasekaran, N.; Vara Prasad, M.; Wadsworth, S.; Dermott, J. & van Rossum, G. (1994). Protein kinase C-dependent and -independent activation of Na+/H+ exchanger by Ga12 class of G proteins. *J. Biol. Chem.*, 269, pp. 11802–11806

Di Sario, A.; Bendia, E.; Taffetani, S.; Marzioni, M.; Candelaresi, C.; Pigini, P.; Schindler, U.; Kleemann, H.; Trozzi, L.; Macarri, G. & Benedetti, A. (2003). Selective Na$^+$/H$^+$ exchange inhibition by cariporide reduces liver fibrosis in the rat. *Hepatology*, 37, pp. 256-266.

Dlamini, Z. & Bhoola, KD. (2005). Upregulation of tissue kallikrein, kinin B1 receptor, and kinin B2 receptor in mast and giant cells infiltrating oesophageal squamous cell carcinoma. *Journal of Clinical Pathology*, vol. 58, no. 9, pp. 915-922.

Donowitz, M.; Montgomery, J.; Walker, M. & Cohen, M. (1994). Brush-border tyrosine phosphorylation stimulates ileal neutral NaCl absorption and brush-border Na^+-H^+ exchange. *Am. J. Physiol.*, 266, pp. G647–G656

Doppler, W.; Jaggi, R. & Groner, B. (1987). Induction of v-mos and activated Ha-ras oncogene expression causes intracellular alkalinisation and cell cycle progression. *Gene*, 54, pp. 145-151.

Dorsam, R., & Gutkind, S. (2007). G-protein-coupled receptors and cancer. *Nature Reviews Cancer*, 7, pp. 79-94.

Eguchi, S.; Iwasaki, H.; Inagami, T.; Numaguchi, K.; Yamakawa, T.; Motley, E.; Qwada, K.; Marumo, F. & Hirata, Y. (1999). Activation of MAPKs by angiotensin II in vascular smooth muscle cells: metalloproteinase-dependent EGF receptor activation is required for activation of ERK and p38 MAP kinase, but not for JNK. *J. Biol. Chem.*, 276, pp. 7957-7962.

El-Dahr, S.; Dipp, S. & Baricos, W. (1998). Bradykinin stimulates the ERK Elk-1 Fos/AP-1 pathway in mesangial cells. *Am J Physiol.*, 275, pp. F343-F352.

Fujita T, Meguro T, Fukuyama R, Nakamuta H, Koida M. (2002). New signaling pathway for parathyroid hormone and cyclic AMP action on extracellular-regulated kinase and cell proliferation in bone cells. Checkpoint of modulation by cyclic AMP. *J Biol Chem.*, 277, pp. 22191-22200.

Fukushima, T.; Waddell, T.; Grinstein, S.; Goss, G.; Orlowski, J. & Downey, G. (1996). Na^+/H^+ exchange activity during phagocytosis in human neutrophils: role of Fcγ receptors and tyrosine kinases. *J. Cell Biol.*, 132, pp. 1037–1052.

Garnovskaya, M.; Gettys, T.; van Biesen, T.; Chuprin, J.; Prpic, V. & Raymond, J. (1997). G-protein-coupled 5-HT1A receptor activates Na^+/H^+ exchange in CHO-K1 cells through Gia2 and Gia3. *J. Biol. Chem.*, 272, pp. 7770-7776.

Garnovskaya, M.; Mukhin, Y. & Raymond, J. (1998). Rapid activation of Na^+/H^+-exchange and ERK in fibroblasts by G protein-coupled 5-HT_{1A} receptor involves distinct signaling cascades. *Biochem. J.*, 330, pp. 489-495.

Garnovskaya, M.; Mukhin, Y.; Turner, J.; Vlasova, T.; Ullian, M & Raymond, J. (2003). Mitogen-induced activation of Na^+/H^+ exchange in vascular smooth muscle cells involves Jak 2 and Ca^{2+}/calmodulin. *Biochemistry*, 42, pp. 7178-7187.

Garnovskaya, M.; Bunni, M.; Kramarenko, I.; Raymond, JR & Morinelli, T. (2008). Bradykinin B_2 Receptor Induces Multiple Cellular Responses Leading to Proliferation of Renal Carcinoma Cells. FASEB J. 22 (Meeting Abstract Supplement #297), San Diego, CA, April 2008.

Gazmuri, R.; Hoffner, E.; Kalcheim, J.; Ho, H.; Patel, M.; Ayoub, I.; Epstein, M.; Kingston, S. & Han, Y. (2001). Myocardial protection in ventricular fibrillation by reduction proton-driven sarcolemmal Na^+ influx. *J Lab Clin Med.*, 137, pp. 43-55.

Gekle, M.; Freudinger, R.; Mildenberger, S.; Schenk, K.; Marschitz, I. & Schramek H. (2001). Rapid activation of Na^+/H^+-exchange in MDCK cells by aldosterone involves ERK1/2. *Pflugers Arch.*, 441, pp. 781-786.

Giard, D.; Aaronson, S.; Todaro, G.; Arnstein, P.; Kersey, J.; Dosik, H.& Parks, W. (1973). In vitro cultivation of human tumors: establishment of cell lines derived from a series of solid tumors. *J Natl Cancer Inst.*, 51, pp. 1417-1423.

Gillis, D.; Shrode, L.; Krump, E.; Howard, C.; Rubie, E.; Tibbles, L.; Woodgett, J. & Grinstein, S. (2001). Osmotic stimulation of the Na^+/H^+ exchanger: relationship to activation of three MAPK pathways. *J. Membr. Biol* 181, pp. 205-214.

Good, D.W. (1995). Hyperosmolality inhibits bicarbonate absorption in rat medullary thick ascending limb via a protein-tyrosine kinase-dependent pathway. *J. Biol. Chem* 270, pp. 9883–9889

Greco, S.; Elia, M.; Muscella, A.; Romano, S.; Storelli, C. & Marsigliante, S. (2005). Bradykinin stimulates cell proliferation through an extracellular signal-regulated protein kinase 1- and 2- dependent mechanism in breast cancer cells in primary culture. *J Endocrin* 186, pp. 291-301.

Greco, S.; Storelli, C. & Marsigliante, S. (2006). Protein kinase C (PKC)-δ/-ε mediates the PKC/Akt-dependent phosphorylation of extracellular signal-regulated protein kinase 1 and 2 in MCF-7 cells stimulated by bradykinin. *J Endocrin* 188, pp. 79-89.

Grinstein, S.; Woodside, M.; Waddell, T.; Downey, G.; Pouyssegur, J.; Wong, D. & Foskett, J. (1993). Focal localization of the NHE-1 isoform of the Na+/H+ antiport: assessment of effects on intracellular pH. *EMBO J.*, 12, pp.5209-5218.

Hagag, N.; Lacal, J.; Graber, N.; Aaronson, S. & Viola, M. (1987). Microinjection of ras p21 induces a rapid rise in intracellular pH. *Mol. Cell. Biol* 7, pp. 1984–1988

Harguindey, S.; Orive, G.; Luis Pedraz, J.; Paradiso A. & Reshkin SJ. (2005). The role of pH dynamics and the Na^+/H^+ antiporter in the etiopathogenesis and treatment of cancer. Two faces of the same coin–one single nature. *Biochim Biophys Acta.* 1756, pp. 1–24.

Hess, J.; Borowski, J.; Young, G.; Strader, C. & Ransom, R. (1992). Cloning and pharmacological characterization of a human bradykinin (BK-2) receptor. *Biochem Biophys Res Commun.*, 184, pp. 260-268.

Huang, C.; Takenawa, T. & Ives, H. (1991). Platelet-derived growth factor-mediated Ca2+ entry is blocked by antibodies to phosphatidylinositol 4,5-bisphosphate but does not involve heparin-sensitive inositol 1,4,5-trisphosphate receptors. *J. Biol. Chem* 266, pp. 4045–4048

Inglese, J.; Koch, W.; Touhara, K. & Lefkowitz, R. (1995). Gbg interaction: factoring in pleckstrin homology domains and Ras-MAP kinase signaling pathways. *Trends Biochem. Sci.*, 20, pp. 151–156.

Ishibe, S.; Joly, D.; Liu, Z. & Cantley, L. (2004). Paxillin serves as an ERK-regulated scaffold for coordinating FAK and Rac activation in epithelial morphogenesis. *Mol. Cell* 16, pp. 257-267.

Jaffa, A.; Miller, B.; Rosenzweig, S.; Naidu, P.; Velarde, V. & Mayfield, R. (1997). Bradykinin induces tubulin phosphorylation and nuclear translocation of MAP kinase in mesangial cells. *Am J Physiol* 273, pp. F916-F924.

Javadov, S.; Baetz, D.; Rajapurohitam, V.; Zeidan, A.; Kirshenbaum, L. & Karmazyn, M. (2006). Antihypertrophic effect of Na^+/H^+-exchanger-1 inhibition is mediated by reduced MAPK activation secondary to improved mitochondrial integrity and decreased generation of mitochondrial-derived ROS. *J Pharmacol Expt Ther* 317, pp. 1036 -1043.

Johnson, G. & Lapadat, R. (2002). Mitogen-activated protein kinase pathways mediated by ERK, JNK, and p38 protein kinases. *Science*, vol. 298, no. 5600, pp. 1911–1912.

Kang, M.; Kulisz, A. & Wasserman, W. (1998). Raf-1 kinase, a potential regulator of intracellular pH in Xenopus oocytes. *Biol. Cell*, 90, pp. 477-485.

Kaplan, D. & Boron, W. (1994). Long-term expression of c-H-raps stimulates Na-H and Na$^+$-dependent Cl-HCO3 exchange in NIH-3T3 fibroblasts. *J. Biol. Chem* 269, pp. 4116-4124.

Kapus, A.; Grinstein, S.; Wasan, S.; Kandasamy, R. & Orlowski, J. (1994). Functional characterization of three isoforms of the Na$^+$/H$^+$ exchanger stably expressed in CHO cells. *J. Biol. Chem* 269, pp. 23544-23552.

Karmazyn, M.; Kilic, A. & Javadov, S. (2008). The role of NHE-1 in myocardial hypertrophy and remodeling. *J Mol Cell Cardiol.*, 44, pp. 647-653.

Kemp, G.; Young, H. & Fliegel, L. (2008). Structure and function of the human Na$^+$/H$^+$ exchanger isoform 1. *Channels*, 2, pp. 329-336.

Kim, E. & Choi, E. (2010). Pathological roles of MAPK signaling pathways in human diseases. *Biochimica et Biophysica Acta*, vol. 1802, no. 4, pp. 396–405.

Kolch, W. (2005). Coordinating MAPK signaling via scaffolds & inhibitors. *Nat Rev Mol Cell Biol* 6, pp. 827-37

Kranenburg, O. & Moolenaar, W. (2001). Ras-MAP kinase signaling by lysophosphatidic acid and other G protein-coupled receptor agonists. *Oncogene*, 20, pp. 1540-1546.

Krump, E.; Nikitas, K. & Grinstein, S. (1997). Induction of tyrosine phosphorylation and Na+/H+ exchanger activation during shrinkage of human neutrophils. *J. Biol. Chem* 272, pp. 17303-17311.

Kramarenko, I.; Bunni, M.; Raymond, J. & Garnovskaya, M. (2010). Bradykinin B$_2$ Receptor Interacts with Integrin α5/β1 to Transactivate Epidermal Growth Factor Receptor in Kidney Cells. *Mol. Pharmacol* 78, pp. 126-134.

Kusumoto, K.; Haist, J. & Karmazyn, M. (2001). Na$^+$/H$^+$ exchange inhibition reduces hypertrophy and heart failure after myocardial infarction in rats. *Am. J. Physiol.*, 280, pp. H738-H745.

LaPointe, M.; Ye, M.; Moe, O.; Alpern, R. & Battle, D. (1995). Na$^+$/H$^+$ antiporter (NHE-1 isoform) in cultured vascular smooth muscle from the spontaneously hypertensive rat. *Kidney Int* 47, pp. 78-87.

Lefler, D.; Mukhin, Y.; Pettus, T.; Leeb-Lundberd, L.M.F.; Garnovskaya, M. & Raymond, J. (2003). Jak2 and Ca2+/calmodulin are key intermediates for bradykinin B$_2$ receptor-mediated activation of Na$^+$/H$^+$ exchange in KNRK and CHO cells. *ASSAY and Drug Development Technologies*, 1, pp. 281-289.

Lehoux, S.; Abe, J.; Florian, J. & Berk, B. (2001). 14-3-3 binding to Na$^+$/H$^+$ exchanger isoform-1 is associated with serum-dependent activation of Na$^+$/H$^+$ exchange. *J. Biol. Chem.*, 276, pp. 15794-15800.

Levine, S.; Montrose, M.; Tse, M. & Donowitz, M. (1993). Kinetics and regulation of three cloned mammalian Na+/H+ exchangers stably expressed in a fibroblast cell line. *J. Biol. Chem.*, 268, pp. 25527–25535

Li, X.; Alvares, B.; Casey, J.; Reithmeier R. & Fliegel, L. (2002). Carbonic anhydrase II binds to and enhances activity of the Na^+/H^+ exchanger. *J. Biol. Chem.*, 277, pp. 36085-36091.

Lin, X.; Voyno-Yasanetskaya, T.; Hooley, R.; Lin, C.-Y.; Orlowski, J. & Barber, D. (1996) *J. Biol. Chem.* 271, pp. 22604-22610.

Liu, F. & Gesek, F. (2001). α_1-Adrenergic receptors activate NHE1 and NHE3 through distinct signaling pathways in epithelial cells. *Am. J. Physiol.*, 280, pp. F415-F425.

Liu, H-F.; Teng, X-C.; Zheng, J-C.; Chen, G. & Wang, X-W. (2008). Effect of NHE1 antisense gene transfection on the biological behavior of SGC-7901 human gastric carcinoma cells. *World Journal of Gastroenterology*, 14, pp. 2162-2167.

Lowe, J.; Huang, C. & Ives, H. (1990). Sphingosine differentially inhibits activation of the Na+/H+ exchanger by phorbol esters and growth factors. *J. Biol. Chem.*, 265, pp. 7188-7194.

Lucchesi, P.; DeRoux, N. & Berk, B. (1994). Na^+/H^+ exchanger expression in vascular smooth muscle of spontaneously hypertensive and Wistar-Kyoto rats. *Hypertension*, 24, pp. 734-738.

Luttrell, D. & Luttrell, L. (2004). Not so strange bedfellows: G-protein-coupled receptors and Src family kinases. *Oncogene*, vol. 23, no 48, pp. 7969-7978.

Luttrell, L. (2002). Activation and targeting of mitogen-activated protein kinases by G-protein-coupled receptors. *Can J Physiol Pharmacol.*, 80, pp. 375-382.

Ma, Y.-H.; Reusch, P.; Wilson, E.; Escobedo, J.; Fantl, W.; Williams, L. & Ives, H. (1994). Activation of Na^+/H^+ exchange by platelet-derived growth factor involves phosphatidylinositol 3'- kinase and phospholipase Cγ. *J. Biol. Chem.*, 269, pp. 30734-30739

Maly, K.; Uberall, F.; Loferer, H.; Doppler, W.; Oberhuber, H.; Groner, B. & Grunicke, H. (1989). Ha-ras activates the Na+/H+ antiporter by a protein kinase C-independent mechanism. *J. Biol. Chem* 264, pp. 11839-11842

Menke, J.; Borkowski, J.; Bierilo, K.; MacNeil, T.; Derrick, A.; Schneck, K. et al. (1994). Expression cloning of a human B1 bradykinin receptor. *J Biol Chem.*, 269, pp. 21583-21586.

Mitsuka, M.; Nagae, M. & Berk, B. (1993). Na^+-H^+ exchange inhibitors decrease neointimal formation after rat carotid injury. Effects on smooth muscle cell migration and proliferation. *Circ. Res.*, 73, pp. 269-275.

Moor, A.; Gan, X.; Karmazyn, M. & Fliegel, L. (2001). Activation of Na^+/H^+ exchanger-directed protein kinases in the ischemic and ischemic-reperfused rat myocardium. *J. Biol. Chem.*, 276, pp. 16113-16122.

Mukai, H.; Fitzgibbon, W.; Bozeman, G.; Margolius, H. & Ploth, D. (1996). Bradykinin B_2 receptor antagonist increases chloride and water absorption in rat medullary collecting duct. *Am J Physiol.*, 271, pp. R352-R360.

Mukhin, Y.; Vlasova, T.; Jaffa, A.; Collinsworth, G.; Bell, J.; Tholanikunnel, B.; Pettus, T.; Fitzgibbon, W.; Ploth, D.; Raymond, J. & Garnovskaya, M. (2001). Bradykinin B_2 receptors activate Na^+/H^+ exchange in mIMCD-3 cells *via* Janus kinase 2 and Ca^{2+}/calmodulin. *J. Biol. Chem.*, 276, pp. 17339-17346.

Mukhin, Y.; Garnovsky, E.; Ullian, M. & Garnovskaya, M. (2003). Bradykinin B2 receptor activates extracellular signal-regulated protein kinase in mIMCD-3 cells via epidermal growth factor receptor transactivation. *J Pharmacol Expt Ther.*, 304, pp. 968-977.

Mukhin, Y.; Garnovskaya, M.; Ullian, M. & Raymond, J. (2004). ERK is regulated by sodium-proton exchanger in rat aortic vascular smooth muscle cells. *J. Biol. Chem.*, 279, pp. 1845-1852.

Mukhin, Y.; Gooz, M.; Raymond, J. & Garnovskaya, M. (2006). Collageneses 2 and 3 mediate epidermal growth factor receptor transactivation by Bradykinin B2 receptor in kidney cells. *J Pharmacol Expt Ther.*, 318, pp. 1033-1043.

Nemeth, Z.; Deitch, E.; Szabo, C.; Mabley, J.; Pacher, P.; Fekete, Z.; Hauser, C. & Hasko, G. (2002). Na^+/H^+ exchanger blockade inhibits enterocyte inflammatory response and protects against colitis. *Am. J. Physiol.*, 283, pp. G122-G132.

New, D. & Wong, Y. (2007). Molecular mechanisms mediating the G protein-coupled receptor regulation of cell cycle progression. *Journal of Molecular Signaling*, vol. 2, 2, doi:10.1186/1750-2187-2-2.

Ng, L.; Davies, J.; Sickowski, M.; Sweeney, F.; Quinn, P.; Krolewski, B. & Krolewski, A. (1994). Abnormal Na^+/H^+ antiporter phenotype and turnover of immortalized lymphoblasts from type 1 diabetic patients with nephrophathy. *J Clil Invest.*, 93, pp. 2750-2757.

Noel, J. & Pouysségur, J. (1995). Hormonal regulation, pharmacology, and membrane sorting of vertebrate Na^+/H^+ exchanger isoforms. *Am. J. Physiol.*, 268, pp. C283-C296.

Orlowski, J. & Grinstein, S. (1997). Na+/H+ exchangers in mammalian cells. *J. Biol. Chem.*, 272, pp. 22373-22376.

Orlowski, J. & Grinstein, S. (2004). Diversity of the mammalian NHE SLC9 gene family. *Pflugers Arch.*, 447, pp. 549-565.

Ohtsu, H.; Dempsey, P. & Eguchi, S. (2006). ADAMs as mediators of EGF receptor transactivation by G protein-coupled receptors. *Am J Physiol.*, 291, pp. C1-10.

Pang, T.; Su, X.; Wakabayashi, S. & Shigekawa, M. (2001). Calcineurin homologous protein as an essential cofactor for Na^+/H^+ exchangers. *J. Biol. Chem.* 276, pp. 17367-17372.

Pearson, G.; Robinson, F.; Gibson, T.B et al. (2001). Mitogen-activated protein (MAP) kinase pathways: regulation and physiological functions. *Endocrine Reviews*, vol. 22, no. 2, pp. 153-183.

Pederson, S.; Varming, C.; Christensen, S. & Hoffmann, E. (2002). Mechanisms of activation of NHE by cell shrinkage and by calyculin A in Ehrlich ascites tumor cells. *J. Membr. Biol.*, 189, pp. 67-81.

Pedersen, S.; Darborg, B.; Rentsch, M. & Rasmussen, M. (2007). Regulation of mitogen-activated protein kinase pathways by plasma membrane Na^+/H^+ exchanger, NHE1. *Arch. Biochem. Biophys.*, 462, pp. 195-201.

Pedersen, S.; Darborg, B.; Rasmussen, M.; Nylandsted, J. & Hoffmann, E. (2007). The Na^+/H^+ exchanger 1 differentially regulates MAPK subfamilies after osmotic shrinkage in Elrlich Letter Ascites cells. *Cell. Physiol. Biochem.*, 20, pp. 735-750.

Pierce, K.; Luttrell, L. & Lefkowitz, R., (2001). New mechanisms in heptahelical receptor signaling to mitogen activated protein kinase cascades *Oncogene*, 20, pp. 1532-1539.

Pierce, K.; Premont, R. & Lefkowitz, R. (2002). Seven-transmembrane receptors. Nature Reviews. Mol. Cell Biol. 3, pp. 639–650.

Putney, L.; Denker, S. & Barber, D. (2002). The changing face of the Na^+/H^+ exchanger, NHE1: structure, regulation, and cellular actions. *Annu. Rev. Pharmacol. Toxicol.*, 42, pp. 527-552.

Rosskopf, D.; Dusing, R. & Siffert, W. (1993). Membrane Na^+/H^+ exchange & hypertension. *Hypertension*, 21, pp. 607-617.

Rozengurt, E. (1986). Early signals in the mitogenic response. *Science*, 234: pp. 161-166.

Sabri, A.; Byron, K.; Samarel, A.; Bell J. & Lucchesi, P. (1998). Hydrogen peroxide activates MAP kinases and Na^+-H^+ exchange in neonatal rat cardiac myocytes. *Circ. Res.*, 82, pp. 1053-1062

Sarget, C.; Franchi, A. & Pouyssegur, J. (1989). Molecular cloning, primary structure and expression of the human growth f actor-activatable Na^+/H^+ antiporter. *Cell*, 56, pp. 271-280.

Schafer, B.; Marg, B.; Gschwind, A. & Ullrich, A. (2004). Distinct ADAM metalloproteinases regulate G protein-coupled receptor-induced cell proliferation and survival. *J Biol Chem.*, 279, pp. 47929-47938.

Schelling, J. & Abu Jawdeh, B. (2008). Regulation of cell survival by NHE-1. *Am J Physiol.*, 295, pp. F625-F632.

Shaul, Y. & Seger, R. (2007). The MEK/ERK cascade: from signaling specificity to diverse functions. *Biochimica et Biophysica Acta*, vol. 1773, no. 8, pp. 1213–1226.

Silva, N.; Haworth, R.; Singh, D. & Fligel, L. (1995). The carboxy-terminal region of the Na^+/H^+ exchanger interacts with mammalian heat shock protein. *Biochemistry*, 34, pp. 10412-10420.

Snabaitis, A.; Yokoyama, H. & Avkiran M. (2000). Roles of mitogen-activated protein kinases and protein kinase C in α(1A)-adrenoceptor-mediated stimulation of the sarcolemmal Na^+-H^+ exchanger. *Circ. Res.*, 86, 2, pp. 214–220.

Snabaitis, A.; Hearse, D. & Avkiran, M. (2002). Regulation of sarcolemmal Na^+/H^+ exchanger by hydrogen peroxide in adult rat ventricular myocytes. *Cardiovasc. Res.*, 53, pp. 470-480.

Takahashi, E.; Abe, J.; Gallis, B.; Aebersold, R.; Spring, D.; Krebs, E. & Berk B. (1999). p90(RSK) is a serum-stimulated Na^+/H^+ exchanger isoform-1 kinase. *J. Biol. Chem.*, 274, pp. 20206-20214.

Takewaki, S.; Kuro-o, M.; Hiroi, Y.; Yamazaki, T.; Noguchi, T.; Miyagishi, A.; Nakahara, K.; Aikawa, M.; Manabe, I.; Yazaki Y, *et al.* (1995). Activation of Na^+-H^+ antiporter (NHE-1) gene expression during growth, hypertrophy and proliferation of the rabbit cardiovascular system. *J. Mol. Cell. Cardiol.*, 27, pp. 729-742.

Tominaga, T.; Ishizaki, T.; Narumiya, S. & Barber, D. (1998). p160ROCK mediates RhoA activation of Na-H exchange. *EMBO J.*, 17, pp. 4712-4722.

Tominaga, T. & Barber, D. (1998). Na-H exchange acts downstream of RhoA to regulate integrin-induced cell adhesion and spreading. *Mol. Biol Cell.*, 9, pp. 2287-2303

Touyz, R. & Schiffrin, E. (2000). Signal transduction mechanisms mediating the physiological and pathophysiological actions of Ang II in vascular smooth muscle cells. *Pharmacol Rev.*, 52, pp. 639-672.

Turner, J.; Garnovskaya, M.; Coaxum, S.; Vlasova, T.; Yakutovich, M.; Lefler, D. & Raymond, J. Sr. (2007). Ca^{2+}-calmodulin and Jak2 are Required for Activation of Sodium-proton Exchange by the Gi-coupled 5-hydroxytryptamine$_{1A}$ Receptor. *J Pharmacol Expt Ther.*, 320, pp. 314-322.

Voyno-Yasenetskaya, T.; Conklin, B.; Gilbert, R.; Hooley, R.; Bourne, H. & Barber, D. (1994). G alpha 13 stimulates Na-H exchange. *J. Biol. Chem.* 269, pp. 4721–4724.

Wakabayashi, S.; Shigekawa, M. & Pouyssegur, J. (1997). Molecular physiology of vertebrate Na^+/H^+ exchangers. *Physiol. Rev.* 77, pp. 51-74.

Wallert, M.; Thronson, H.; Korpi, N.; Olmschenk, S.; McCoy, A.; Funfar, M. & Provost, J. (2005). Two G protein-coupled receptors activate Na+/H+ exchanger isoform 1 in Chinese hamster lung fibroblasts through an ERK-dependent pathway. *Cell Signal.* 17(2), pp. 231-242.

Wang, H.; Silva, N.; Lucchesi, P.; Haworth, R.; Wang, K.; Michalak, M.; Pelech, S. & Fliegel, L. (1997). Phosphorylation and regulation of the Na^+/H^+ exchanger through MAPK. *Biochemistry*, 36, pp. 9151-58.

Watanabe G, Lee RJ, Albanese C, Rainey WE, Batlle D, Pestell RG. (1996). Angiotensin II activation of cyclin D1-dependent kinase activity. *J Biol Chem* 1996, 271:22570-22577.

Watts, B. III & Good, D. (2002). ERK mediates inhibition of Na^+/H^+ exchange and HCO_3^- absorption by nerve growth factor in MTAL. *Am. J. Physiol.* 282, pp. F1056-1063.

Wei, S.; Rothstein, E.; Fliegel, L.; Dell'Italia, L. & Lucchesi P. (2001). Differential MAPK activation and Na^+/H^+ exchanger phosphorylation by H_2O_2 in rat cardiac myocytes. *Am. J.Physiol.* 281, pp. C1542-C1550.

Widmann, C.; Gibson, S.; Jarpe, M. & Johnson, G. (1999). Mitogen-activated protein kinase: conservation of a three-kinase module from yeast to human. *Physiol. Rev.* 79, pp. 143-180.

Wu, J.; Akaike, T.; Hayashida, K.; Miyamoto, Y.; Nakagawa, T.; Miyakawa, K.; Müller-Esterl, W. & Maeda, H. (2002). Identification of bradykinin receptors in clinical cancer specimens and murine tumor tissues. *International Journal of Cancer*, 98, pp. 29-35.

Yamaji, Y.; Amemiya, M.; Cano, A.; Preisig, P.; Miller, R.; Moe, O. and Alpern, R. 1995). Overexpression of csk inhibits acid-induced activation of NHE-3. *Proc. Natl. Acad. Sci. U.S.A.*, 92, pp. 6274–6278

Yamazaki, T.; Komuro, I.; Kudoh, S.; Zou, Y.; Nagai, R.; Aikawa, R.; Uozumi, H. & Yazaki, Y. (1998). Role of ion channels and exchangers in mechanical stretch-induced cardiomyocyte hypertrophy. *Circ. Res.*, 82, pp. 430-37.

Yan, W.; Nehrke, K.; Choi, J. & Barber, D. (2001). The Nck-interacting kinase (NIK) phosphorylates the Na^+/H^+ exchanger NHE1 and regulates NHE1 activation by PDGF. *J. Biol. Chem.*, 276, pp. 31349-31356.

Yang, X.; Wang, D.; Dong, W.; Song, Z. & Dou, K. (2011). Suppression of Na+/H+ exchanger 1 by RNA interference or amiloride inhibits human hepatoma cell line SMMC-7721 cell invasion. *Medical Oncology*, 28, pp. 385-390.

Section 2

Cellular Basis of Disease and Therapy

9

Biology of Cilia and Ciliopathies

David Alejandro Silva, Elizabeth Richey and Hongmin Qin
*Department of Biology, Texas A&M University, College Station,
USA*

1. Introduction

Cilia and flagella are microtubule-based appendages extending from the basal body of most eukaryotic cells, and are classified as either motile or primary. Motile cilia or flagella can be found on many cells such as *Chlamydomonas*, sperm, and respiratory tract epithelial cells. This type of cilia is responsible for movement of the cell itself or generation of fluid flow. In contrast, primary cilia are non-motile organelles that are critically involved in visual, olfactory and auditory signal transduction and play key roles in the regulation of gene expression, development, and behavior. This chapter reviews the current understanding of the various mechanisms involved in cilia and flagellar assembly and maintenance. Consistent with the nearly ubiquitous cellular distribution, cilia have been implicated in numerous human diseases collectively known as ciliopathies. This chapter also discusses several major ciliopathies including primary ciliary dyskinesia, hydrocephalus, polycystic kidney disease, Bardet-Biedl syndrome, and cancer.

2. Mechanics of ciliogenesis

2.1 Basic biology of cilia

Cilia and flagella are long, slender structures protruding from the body of ciliated cells and are composed of a microtubule-based core known as the axoneme. The main structural element is an array of nine doublet microtubule pairs. The a-subfiber of these pairs exists as a fully enclosed filament and is fused to the incomplete b-subfiber containing fewer tubulin subunits. The individual pairs of subfibers are linked together by nexin proteins, forming an enclosed cylinder around a central pair of singlet subfibers; this layout is known as the "9+2" arrangement. The axoneme itself originates from the basal body, a modified form of the centriole consisting of nine triplet microtubules which anchors the cilia to the plasma membrane. The area between the triplet microtubules of the basal body and doublet pairs of the axoneme is referred to as the transition zone. Proteinaceous extensions from this area called transition fibers serve to mark the enclosure of the flagellar compartment and create a barrier from the cytoplasm. The outer and inner dyneins, located on the a-subfiber of the axoneme, are motor proteins that produce the bending and sliding of the microtubules by exerting force on the b-subfiber via ATP hydrolysis. Defects in the inner and outer motor complexes can yield paralyzed or uncoordinated sliding of the axoneme, resulting in inefficient movement. The outer doublet ring and the central pair of microtubules are connected by structures known as radial spokes [1].

Cilia can be distinguished into two types: primary (nonmotile) and motile cilia. While they have the same basic structure, the biological functions of these two types can be very different. Primary cilia typically do not contain the central pair of microtubules (having a "9+0" structure), and also lack other accessory proteins important for generating the ciliary waveform stroke. This form of cilia is considered a sensory antenna for the cell due to a highly specialized membrane protein profile and ability to extend in the luminal space of various tissues. Historically, motile cilia/flagella are known to be important for locomotion in single-celled organisms. In humans, however, motile cilia are important for various physiological processes, ranging from mucous clearing in the trachea to aiding in establishing proper left-right symmetry in developing organisms [2]. Because of their structural similarities, the terms cilia and flagella will be used interchangeably.

2.2 Intraflagellar transport at a glance

Intraflagellar transport (IFT) is a term used to describe the bi-directional movement of non-membrane protein particles that move along the microtubule doublet core (or axoneme), between the space of the ciliary membrane and the axoneme. IFT was originally discovered by Kozminksi and colleagues in 1993, using digital interference contrast (DIC) microscopy to visualize the continuous movement of "bulges" beneath the membrane of a *Chlamydomonas reinhardtii* mutant with paralyzed-cilia [3]. Anterograde movement, towards the ciliary tip or plus end of microtubules, is powered by heterotrimeric kinesin-2, and retrograde movement is driven by cytoplasmic dynein1b [4-14]. A multi-meric protein complex known as the IFT particle attaches to the motor complex and is itself comprised of two large protein sub-complexes [5]. The axoneme is undergoing constant turnover at its tip, meaning tubulin and other accessory proteins must be constantly replenished at the distal tip [15]. The well-conserved IFT motors and particles are tasked with assembling and maintaining the whole cilia structure by serving as adaptors for the transport of axonemal precursors and the recycling of turnover products [1]. A secondary, though equally important, function of IFT is to ferry in ciliary membrane proteins through a secondary adapter complex known as the BBSome [16, 17].

2.2.1 IFT Motors

2.2.1.1 Anterograde motor

First isolated in sea-urchin, the anterograde IFT motor is a heterotrimeric kinesin-2, comprised of three individual subunits [18]. Homologs are found in a variety of ciliated organisms, including *Tetrahymena, Caenorhabditis elegans*, and humans. In *Chlamydomonas*, FLA10 and FLA8 comprise the motor portion of the complex and together they interact with FLA3, a kinesin-associated protein [19]. First evidence for the role of the kinesin-2 in anterograde movement came from the characterization of a temperature sensitive mutant in FLA10. While incubated at the permissive temperature (22°C), the biflagellated green algae possess two, full-length flagella. However, following a shift to the non-permissive temperature (32°C), FLA10 subunit denatures and levels of IFT proteins significantly reduce within the first hour [6]. Cessation of IFT results in the dismantling of the axoneme and the entire ciliary structure is retracted back into the cell body because of the normal turnover [5]. In addition, an isolated null mutant for the FLA10 subunit produces no flagella [20]. Taken together, these studies demonstrate the importance of kinesin-2 to ciliogenesis.

These observations, however, are not entirely consistent among all ciliated organisms. Mutations in the kinesin-2 motor subunits of different species do not result in a cilia-less cell phenotype because of a secondary, homodimeric kinesin known as OSM-3 in *C. elegans* and KIF17 in *Homo sapiens*[13]. Studies investigating the function of OSM-3 conclude that the canonical kinesin-2 motor and OSM-3 work in a concerted effort to build sensory cilia in *C. elegans* [13]. Single mutants in KLP-11 (FLA8) and KAP-1 (FLA3) in *C. elegans* appear to form intact sensory cilia due to the redundancy of OSM-3 function in ciliogenesis (Signor et al, 1999). However, perturbations of OSM-3 results in loss of the ciliary distal segment comprised of singlet microtubule extensions beyond the doublet axoneme core. In these mutants, the heterotrimeric anterograde motor still allows formation of the middle segment.

It could be possible the transferring of the IFT particle from the canonical kinesin-II to OSM-3 may insure proper, sequential construction of the cilia. However, it has been well documented that OSM-3 speed actually increases in disrupted kinesin-II mutants [21], suggesting that kinesin-II may in fact be negatively regulating OSM-3. If so, kinesin-II would ultimately be involved in determining the re-supply rate of axonemal precursors to the flagella compartment. Defects in retrograde IFT clearly demonstrate the negative impact that excess precursors and turnover products have on proper ciliary function. Therefore, accumulation of axonemal components, due to a faster influx of proteins by OSM-3, could also unbalance the natural turnover vs. assembly in favor of creating longer cilia, which is a phenotype that has been observed in kinesin-II mutants. Recently, a null mutant for a relatively new kinesin, KLP-6 in *C. elegans* males, demonstrated a slower procession of OSM-3/KAP-1-associated IFT particle within the ciliary compartment [22]. Although it was observed moving independently of the canonical IFT particle/motor complex, KLP-6 function may have a positive influence on ciliary length. This conclusion is supported by a reduction of *klp-11/klp-6* double mutant cilia compared to the single *klp-11* mutant; *klp-11* mutant has comparatively longer cilia than wild-type.

2.2.1.2 Retrograde motor

IFT-dynein, cytoplasmic dynein 2 (previously known as dynein 1b), powers the retrograde movement of IFT [12]. To date, four proteins are confirmed members of the dynein 2 complex: heavy chain DHC1b, light chain LC8, light intermediate chain D1bLIC, and an intermediate chain FAP133 [7-9, 14, 23-25]. *C. elegans* null mutants defective in dynein components undergo normal anterograde movement but accumulate large amounts of IFT proteins and turnover products within the ciliary compartment [26]. Retrograde-defective cilia are severely truncated and contain protein aggregates that appear as noticeable large, electron-dense clots. These results suggest IFT dynein is responsible for the retrograde movement of the IFT; this result has been seen in *Chlamyomonas*, where defects in IFT dynein lead to protein accumulations in the flagella compartment [27]. Anterograde movement remains active in these mutants; however, the characteristic bulbous cilia are present as a result of axonemal turnover outpacing the dysfunctional retrograde IFT. It has become fairly evident that IFT particles do not passively diffuse out of the flagella compartment and turnover products must be actively removed by dynein 2 in order to allow unhindered trafficking of the IFT trains.

The current model for retrograde activation is fragmented at best. IFT-dynein is carried into the compartment in an inactivated form as part of the IFT cargo. Once it reaches the tip, a poorly understood remodeling occurs, initiating the dynein-powered return of the IFT train back to the cell body [27, 28]. During retrograde movement, the kinesin is inactivated, but it is unknown whether kinesin-2 is removed by the IFT particle as part of the turnover cargo or simply diffused out. IFT-dynein has been historically shown to associate with complex A in *Chlamydomonas*, primarily due to IFT-A temperature sensitive mutants exhibiting similar phenotypes as retrograde mutants [29, 30]. During remodeling at the distal tip, IFT-A likely facilitates the activation of dynein-2 in order to initiate retrograde movement, although a detailed mechanistic overview is lacking [28].

The newest addition to the retrograde movement model suggests OSM-3 and kinesin-II may directly transport IFT dynein, independently of the IFT particle in *C. elegans* [11]. The conclusion is derived from IFT-dynein undergoing normal IFT transportation speeds despite the uncoupling of IFT complex A/kinesin-2, and complex B/OSM-3. Another new concept suggests IFT172 may in fact mediate the interaction between inactivated dynein and the IFT particle during anterograde movement [28]. A new study in *C. elegans* has revealed the presence of a new retrograde dynein motor, specific to outer labial quadrant neurons, which are able to form full functional cilia in canonical IFT dynein mutants [11].

2.2.1.3 Regulation of the motors

IFT motor regulation continues to be of high interest in the cilia field due to the motors' important functions. Defects in FLA3, the kinesin associated protein, lead to mislocalized kinesin-II and subsequently produce a bald or flagella-less phenotype in *Chlamydomonas* [31]. Isolation of DYF-11 null mutant, a homolog of human microtubule-interacting protein (MIP)-T3 and IFT54 in *C. elegans*, reveals that this protein may serve as an anchoring protein for the priming/loading of the entire IFT motor/particle complex onto the transition zone of cilia [32]. KIF17, OSM-3 homolog of *C. elegans*, in human primary cilia was discovered to be under the regulation of a RAN gradient between the cell body and flagellar compartment. This mechanism operates in similar fashion to the RAN gradient active in regulating the trafficking of proteins across the nuclear pore complex [33]. A ciliary localization signal (CLS) at the tail end of KIF17 was shown to contribute to the interaction with another accessory protein known as importin-β2, a nuclear import protein; this interaction was inhibited by RAN-GTP. Similar CLS signals may exist for additional proteins in other model organisms due to the conserved nature of IFT. Recent study into the regulation of IFT-dynein flagellar entry has suggested that *Chlamydomonas* IFT172, a peripheral protein of IFT-B subcomplex, may be directly involved in the transport of IFT-dynein into the flagellar compartment [28].

2.3 IFT particle

By comparing the flagellar proteome of a *fla10^{ts}* mutant after incubation at the non-permissive temperature to a wild-type proteome, Cole and colleagues biochemically observed the depletion of certain proteins from the flagellar compartment [5]. After further analysis, members of the IFT particle were discovered. Results from this study demonstrated that the IFT particle was actually comprised of two sub-complexes, IFT-A and IFT-B, which to date consist of 6 and 12 polypeptides, respectively [19]. A recent study shed new light on the organization of the IFT-B subcomplex, demonstrating it can be separated further into two tetrameric subdomains: IFT25/27/74/81/72 and IFT52/46/88/70 [34-36].

IFT52 serves as the interface between the IFT74/81 and IFT52/46/88/70 [36]; IFT74/81 functions as the intermediate complex between IFT25/27 and IFT52/46/88/70. Currently IFT-A complex is understood to be composed of IFT144/121/140/121/139/43 [30], however its structural organization remains unclear.

A majority of the IFT members are enriched in WD40 and tetracopeptide repeats (TPR), multi-protein binding domains that possibly form a circularized beta-propeller structure and alpha helical solenoid, respectively, to behave as scaffolding elements [19]. WAA is another binding motif present in these proteins, although it is poorly understood. Most of the IFT proteins contribute to the overall integrity of their respective complexes, evident by the subsequent instability and depletion of complex-mates following disruption of certain IFT proteins. Depending on which complex is disrupted, flagella morphology is typically affected in one of two ways: structurally sound but severely truncated cilia (IFT-B mutants) or short bulbous flagella (IFT-A) [1, 13, 27]. The resulting phenotype reveals the different nature of the two complexes; short flagella convey IFT-B's importance in anterograde movement and protein buildup in the flagella compartment suggest IFT-A is involved in retrograde IFT. Nonetheless, the many parts of IFT machinery must work in a concerted effort to strike an efficient balance between retrograde and anterograde transportation dynamics.

Defects in other IFT-B proteins typically lead to a bald phenotype, making any biochemical analysis a challenge to determine individual function. A null mutant of *ift88*, the first IFT protein to be implicated in disease [37], displays a bald phenotype in *Chlamydomonas*. The absence of IFT54/MIP-T3 causes the entire IFT motor/particle complex, with the exception of OSM-3, to mislocalize at the ciliary base in *C. elegans* [32]. In IFT46 mutants, IFT-B complex still assembles the complex B core proteins but stability is severely affected, evident by the presence of structural sound yet short flagella in *Chlamydomonas* [38]. A suppressor IFT46 mutant sufficiently stabilizes the IFT complex to produced full length flagella. However, upon closer analysis, the axoneme lack outer dynein arms [38]. Thus, IFT46 serves as an adaptor protein for the specific transport of ODA16, a component of the outer dynein arms [39]. It is unclear whether IFT46 functions as a structural protein, since it appears not to be an essential contributor to complex B structural integrity (Richey and Qin, unpublished). It could possess a secondary function as a molecular indicator or chaperone for the IFT-B complex assembly, since the IFT-B can still assemble on a sucrose density gradient (Richey and Qin, unpublished). In addition to the known IFT core proteins, there are a few peripheral proteins associated with complex B: IFT57, IFT20, IFT172, and IFT80 [34]. IFT20 is a particularly interesting protein, because it is the only IFT protein that can localize in the Golgi apparatus, the central hub for the sorting and packaging of macromolecules for secretion. The current model suggests IFT20 is involved in directing vesicles transporting ciliary-specific proteins near the basal body, and participating in the trafficking of membrane proteins into the flagellar compartment [40]. IFT57 can target to the transition fibers of the axoneme, and serves as an anchoring protein for IFT20 to IFT-B in zebrafish [41]. IFT172 is also an interesting protein, since it readily dissociates from the IFT particle and has been shown to be important for retrograde movement. Using temperature sensitive mutant *fla11^{ts}* (IFT172), Pederson et al. 2006 concluded that IFT172 directly interacts with CrEB1, a protein exclusively located at the flagella tip, and accumulates IFT-B but not IFT-A nor IFT dynein proteins in the flagella [42, 43]. Recently, evidence of IFT172 involvement in the flagellar entry of IFT-dynein was detected in *Chlamydomonas*. Upon incubation at the non-permissive temperature, IFT-dynein is depleted from the flagella compartment of the temperature sensitive IFT172 mutant while the rest of IFT particle remains at wild-type levels [28].

The function of individual IFT-A sub-complex proteins are even more enigmatic. Much like IFT-B proteins, disruption or depletion of a single IFT-A protein leads to the instability of the complex and subsequent depletion from the cell body [28]. Mouse IFT122 was shown to regulate members of the sonic hedgehog pathway in a number of ways by uniquely affecting the localization of certain proteins differently than IFT-A and IFT dynein mutants [44]. In *Drosophila*, an IFT140 mutant does not have detectable levels of ciliary TRPV ion channels; while the mRNA levels were unchanged, IFT140 may instead be important for the post-translational stability of the ion channels [45]. The more predominant understanding of the IFT-A function comes from its importance to retrograde movement. At the permissive temperature, electron-dense bulges are present within the cilia of temperature-sensitive *Chlamydomonas* mutants in IFT139 (*fla17*) and IFT144 (*fla15*). Following a shift to the non-permissive temperature leads to the complete breakdown of retrograde IFT and retraction of the axoneme, resulting in lolli-pop shaped bulbs filled with IFT-B proteins [29, 30]. This phenotype is also observed in IFT dynein mutants, thought to arise from the possible hindrance of retrograde IFT activation, and ultimately leading to the buildup of turnover products and IFT particles within the flagellar compartment [43]. In *C. elegans*, IFT-A directly interacts with kinesin-II while IFT-B is transported by OSM-3. IFT-A and IFT-B are linked together by the BBSome, a secondary adaptor complex important for ciliary membrane biogenesis [21, 46].

2.4 BBSome role in membrane biogenesis

Originally discovered during genetic disease screens, the BBSome protein complex functions as an adaptor complex for the IFT particle and facilitates the transport of ciliary membrane proteins. Interaction assays using BBS4 led to the discovery of the seven conserved proteins, BBS1/2/4/5/7/8/9, that comprise the BBSome complex [47]. In the same study, BBS5 was found to interact with phosphoinositides, phospholipids important for recruitment of trafficking proteins to the plasma membrane, implicating a role for the BBSome in vesicle trafficking. In addition, the BBS1 was shown to interact with Rabin8, a guanine nucleotide exchange factor (GEF) for Rab8, two proteins important for ciliary protein trafficking [16]. BBS1 direct association with Rabin8 stimulates the protein's GEF-activity to promote Rab8 activation. Rab8 and Rabin8 contribution to ciliogenesis will be discussed below. Arl6 (BBS3), although not important for BBSome assembly, is important for the recruitment of BBSome to primary cilia and purified liposomes [48]. In *Chlamydomonas*, perturbation of BBSome proteins does not lead to any morphological defects; however, cells are unable to undergo phototaxis, suggesting a role in signal transduction. The ciliary membrane in BBS1, BBS4, and BBS7 mutants accumulate several proteins which are thought to hinder Ca^{2+} signaling pathways involved in the phototaxic response [17]. In *C. elegans*, the BBSome serves as a linking bridge between IFT-A/kinesin-II and IFT-B/OSM-3; disruption of the complex results in the uncoupling of the IFT machinery and leads to cilia morphology defects in some cases [21, 46]. Protein models predict the BBSome functions as a vesicle coat, much like clathrin and COPI/II coat, directing post trans-Golgi network (TGN) vesicles to the ciliary compartment and accompanying them as a mediator between the IFT machinery [48]. The BBSome also sporadically "falls off" the IFT train, possibly in the event of cargo unloading [17]. However, only a few ciliary membrane proteins have been confirmed to be BBSome-dependent for proper localization, most notably Somatostatin receptor 3 (SSTR30) and some G-coupled receptors. Thus, it is unclear whether the BBSome-dependent ciliary transport is a general mechanism or a protein-specific system [17].

2.5 Role of GTPases in ciliogenesis

Research in small GTPases involved in ciliogenesis is a growing branch of the field and the results have been quite interesting. ADP-ribosylating factor–like (ARL) 13, BBS3 and the BBSome are involved in the targeting and entry of flagellar membrane proteins into the compartment [16]. ARL-13 and ARL-3 are small G-proteins antagonistically operating to maintain the stability of IFT particles during middle segment transport in *C. elegans* (IFT A and B) [49, 50]. ARL-13 may also have roles involved in maintaining axonemal integrity since null mutant animals have a variety of gross cilia abnormalities [49-51]. It has been suggested that ARL13 may in fact regulate the coupling of IFT-A and IFT-B, while ARL3 regulates IFT-B interaction with OSM-3; together they regulate the integrity of the IFT machinery in *C. elegans* [50]. Rab8 is recruited to the transition zone by Rabin8 (Rab8GEF), following stimulation from BBS1, a core member of the BBSome; this ultimately results in the fusion of post-Golgi vesicles shuttling ciliary membrane proteins near the basal body [52, 53]. Dominant negative and constitutive active constructs demonstrate the impact that the nucleotide state of Rab8 has on its entry into the ciliary compartment and its role in ciliogenesis [47]. Arf4 and Rab11 form a complex with Arf GTPase activating protein ASAP1 and FIP3 to package and transport rhodopsin from the trans-golgi-network to photoreceptor cilia [54]. This interaction between rhodopsin and Arf4 is dependent on a VxPx motif, a ciliary localization signal also found in other ciliary membrane proteins [16]. Recently, the VxPx motif has been shown to be essential for the trafficking of polycistin-1 protein, and to be involved in the recruitment of Rab8, thereby promoting fusion of ciliary membrane protein-containing vesicles [55]. Another small GTPase, Rab23, was found to be responsible for the turnover of sonic hedgehog signaling protein, Smoothened, from the ciliary compartment [56]. As mentioned in IFT motor regulation section, a RAN-GTP ciliary/cytoplasmic gradient regulates the entry of kinesin motor KIF17 in the primary cilia of cultured cells [33].

Two mysterious members of the IFT-B complex, IFT27/RABL4 and IFT22/RABL5, are the only small GTPases known to directly interact with the IFT particle. IFT25 is a phosphoprotein of unknown function, though it is known to interact with the small GTPase-like IFT27 [57]. Recent work on IFT27 confirmed its GTP binding and GTPase activity along with solving the crystal structure of the sub-complex IFT25/27. However, the exact function of IFT25/27 remains unknown [58]. IFT22 has been the more controversial of the two, since in recent studies with *C. elegans* and *Trypanosome* IFT22 homologs produced conflicting results. In *C. elegans*, a putative constitutive active form (GTP-locked) of the IFTA-2 (IFT22 homolog) can enter the ciliary compartment while dominant negative (GDP-locked) diffusely localizes throughout the neuronal cell body and is notably excluded from the ciliary compartment [59]. IFTA-2 null mutants exhibited extended lifespans, reminiscent of insulin IGF-1-like signaling pathway defects, and a failure to enter dauer formation, a type of survival mode. The null IFTA-2 (IFT22) mutant had intact sensory cilia, effectively suggesting IFTA-2 is not essential to ciliogenesis. However, RNAi knockdown experiments of *Trypanosome* RABL5 lead to the buildup of IFT particles in the flagella compartment and subsequent shortening of the flagella [60]. This phenotype is similar to mutants with defective retrograde IFT, suggesting RABL5 is important for ciliogenesis.

2.6 Gating the ciliary compartment

As mentioned above, the transition fibers mark the entrance of the flagella compartment by tethering the plasma membrane to the base of the flagella. The ciliary proteome contains various proteins not found at such concentrated levels in cytoplasm, implying an inherent selectivity to the transition zone barrier [61]. Although the complete regulatory pathway remains poorly understood, various studies have begun to demonstrate the complexity of the flagella gating mechanism. Recent biochemical characterization of *cep290* mutant in *Chlamydomonas* revealed that the protein CEP290 functions as an intricate member of the transition zone barrier proteins [62]. CEP290 is part of the MKS/MKSR/NPHP proteins (Meckel-Gruber syndrome/related and nephronophthisis), shown to localize at the base of the flagella. Together these proteins form the transition zone and function as the ciliary selective barrier, evident by the accumulation of non-ciliary proteins in the cilia of various TZ mutants [63, 64]. Although IFT anterograde movement was normal and retrograde slightly slower, the *cep290* flagella accumulated IFT-B proteins and BBS4, yet had a reduction of IFT-A, some membrane proteins and axonemal precursors. This phenotype suggests CEP290 plays a role in the mechanical selectivity of the transition zone; it could be possible that IFT-B binding/priming at the transition zone requires CEP290, which could explain the mostly unhindered movement of IFT and the buildup of IFT-B and not IFT-A.

Additional selectivity mechanisms have become more apparent, such as the requirement of ciliary transport signal (CTS) for access to the flagella compartment [16]. The VxPx motif, a CTS, is important for the targeting and entry of ciliary membrane proteins polycistin-1 and rhodopsin. In contrast, a recent study discovered a mechanism for molecular retention, whereby passive diffusion into the ciliary membrane is inhibited by a transferable retention signal [65]. Podoclayxin was shown to contain a four- amino acid PDZ binding motif that facilitated its interaction with NA+/H+ exchanger 3 regulatory factor NHERF1, a protein attached to the apical actin cytoskeleton [66]. The conserved four- amino acid sequence in the PDZ binding motif was shown to be sufficient to prevent passive diffusion into the cilary membrane domain [65]. Although the ciliary entry is passive, the ciliary membrane protein retention appears to require the protein to be firmly attached to the axoneme. Thus, a ciliary retention signal is likely to be necessary for membrane protein accumulation in the ciliary compartment. Additionally, much like the gating system for the nuclear pore complex, a RAN-GTP has been found to exist between the cilia and cytosol that is important for import of KIF17, via its Ran-GTP dependent association with importin-β2 [33].

2.7 Microtubule post translational modifications

The microtubule component of the axoneme undergoes various post-translational modifications (PTMs) that play a vital role in promoting the mechanical movement of the organelle. The dramatic impact of microtubule PTMs has been well characterized, but recent studies have begun to deepen our understanding of how PTMs affect ciliary assembly and maintenance. For an extensive review on the impact of PTMs on ciliogenesis and cell motility, please see a recent review by Wloga and Gaertig [67, 68].

2.7.1 Acetylation of tubulin

N-Acetylation is the only PTM that occurs within the microtubule core, at the highly conserved K40 residue on α-tubulin [69-71]. Mammalian microtubules undergo acetylation of lysine residues on multiple sites located on both α- and β- tubulin [72]. Recently MEC-17, a previously uncharacterized protein now known as αTAT1 (α-tubulin acetyltransferase), may be the sole enzyme responsible for the acetylation of K40 in mammalian cilia [73]. Knockdown of αTAT1 does not produce any severe morphological defects nor does it affect microtubule polymerization. However, as a BBSome-associated protein, it was suspected to be involved in cilia assembly. The depletion of αTAT1 leads to a delayed assembly of primary cilia; taken together with recent information, K40 acetylation may be involved in the dynamics of axonemal assembly and disassembly [74]. Recent studies demonstrate that acetylation of α-tubulin may target these subunits for degradation, since they are preferentially selected to be ubiquitinated over β-tubulin during disassembly [75]. A newly discovered BBSome subunit, BBIP10, functions as a positive regulator of microtubule stability, as a reduction of cytoplasmic microtubules and increase in free tubulin is seen in BBIP10-depleted cells. In addition, overexpression of BBIP10, microtubule acetylation was dramatically increased. The function of BBIP10 on microtubule stability appears to be either dependent or independent of the BBSome [76]

2.7.2 Glutamylation and glycylation

Glutamylation has been observed on microtubules in general, while glycylation has been found to be restricted to the ciliary microtubules of flagellated cell types [77]. These side chains are synthesized in distinct steps of initiation and elongation, typically carried out by two types of enzymes known as tyroslytubulin ligase-like proteins (TTLLs) [78, 79]. Recent studies have begun to investigate the function of these types of PTMs for ciliogenesis, most notably by the impact of polyglutamylation on inner dynein dynamics [80, 81].

3. Ciliopathies

Ciliated cells can be found in various tissues throughout the human body. These include the eye, the trachea, the kidney, the reproductive tract, the intestines, the heart, and many others. In each of these tissues, the cilia perform a significant role in allowing proper function of the tissues. Since ciliated cells are in most important organ tissues, malfunctioning cilia contribute substantially to human disease. Diseases caused or related to faulty cilia are called ciliopathies. The list of ciliopathies is just as diverse as the variety of tissues in which cilia are found. These diseases support the fact that cilia, once thought to be unimportant cellular appendages, are essential for sustaining health in the human body. There are too many ciliopathies to mention, and the list continues to grow. Some major ciliopathies include Primary Ciliary Dyskinesia (PCD), Hydrocephalus, Polycystic Kidney Disease (PKD), Bardet-Biedl Syndrome (BBS), and even cancer [82].

3.1 Primary Ciliary Dyskinesia (PCD)

The relationship between Primary Ciliary Dyskinesia (PCD) and ciliary defects was first discovered in the 1970's [83], making PCD the first human disorder found to be linked to cilia function [84]. PCD is a multi-symptomatic ciliopathy that is present in all major ethnic groups and occurs 1 in every 20,000 live births, although this is likely an underestimation due to a failure to properly diagnose the disorder [85, 86]. PCD was first called "immotile cilia syndrome," but was renamed because it was later found that the cilia were not always immotile, but often had abnormal motility [87].

PCD is characterized by many symptoms that are expressed to various degrees [84]. It affects mainly the respiratory system beginning at birth or within the first month of life. Early signs typically involve a persistent cough and chronic nasal congestion. Other symptoms often include other respiratory problems such as sinusitis and bronchiectasis [87]. The respiratory symptoms of PCD are caused by the lack of uniform ciliary movement to transport particles, or mucous in or out of the organs or the cells themselves. There are about 200 motile cilia in the respiratory tract of a healthy individual. Beating coordinately, these cilia function to remove mucous and debris from the airway in a process called mucocilliary clearance [88]. When the cilia malfunction, there is buildup of mucous and debris in the tract, which leads to respiratory difficulties.

However, this disease is not only associated with the respiratory system. It also has an impact in development, fertility, and aural health. Fifty percent of patients with PCD have total situs inversus, with organs developing on the opposite side of the body. This is thought to be due to the importance of cilia in producing correct direction of nodal flow during embryo development. This was seen in a 2002 study which showed that when direction of nodal flow was artificially reversed to right instead of left, mice developed organs that were a mirror image to normal orientation [89]. There are three known genes involved in left-right axis (LRA) determination, including lrd (left-right dynein), hfh-4 (hepatocyte nuclear factor/forkhead homologue 4), and kif3B (kinesin member 3B) [90-92]. PCD also affects fertility. Males with PCD are typically infertile, and females have a higher rate of ectopic pregnancies. Significant hearing impairment is seen in about fifty percent of children with PCD [93]. This is likely due to the condition known as chronic secretory otitis, which is found almost universally in PCD patients [94]. This is a condition that causes collection of fluid in the middle ear and can cause serious hearing loss and pain. Studies have been done to show that people with chronic secretory otitis have a slower ciliary beat frequency, resulting in failure to move fluid out of the ear canal, leading to infection [95].

PCD is a genetically heterogeneous disease typically caused by autosomal recessive mutations in ciliary protein genes. The most common mutations occur in DNAH5 (dynein heavy chain) and DNAI1 (intermediate chain dynein) genes. Studies show that the disease is caused by abnormalities in the axonemal structure of the cilia. In most patients, the outer dynein arms are missing, of which fifty percent of these patients have mutations in DNAH5 and DNAI1 [96-98]. However, it is thought that lacking the inner dynein arms, central pair of microtubules, or radial spokes can also lead to PCD. Seventy to eighty percent of patients have outer or inner dynein arm defects, while only five to ten percent are missing radial spokes. PCD can also be caused by a disorientation of cilia and transposed microtubules [85]. Transposed microtubules occur when a cilium lacks the central microtubule pair and a peripheral doublet with dynein arms transposes to the center of the axoneme. This microtubule defect, coined "central microtubule agenesis," causes a circular rotation of the cilia rather than the normal back and forth motion [99]. In addition to genetic causation, PCD can also be a result of acquired defects caused by epithelial damage by chronic infection or irritant exposure [100].

Beat frequency, beat patterns, and protein localization are all parameters measured in diagnosis of the disease. Nasal brush biopsies are most commonly used to collect a patient's cells, but bronchoscopic brush biopsies can also be used [101]. Typically, a diagnosis is made by examining fixed cells under a transmission electron microscope to identify ciliary structural defects in addition to measuring ciliary beat frequency in live cells by video microscopy. However, in the case of central microtubule agenesis, it is important to also look at ciliary beat pattern, since the cilia typically maintain a normal beat frequency, with aberrant movement. These patients also tend to have normally functioning nodal cilia, which allow normal situs, accounting for the fact that not all PCD patients have situs inversus [99]. This may pose a problem for diagnosis if doctors are using organ orientation as a diagnostic factor. Protein localization is a relatively new and uncommon means of diagnosis. In respiratory cells, certain ciliary proteins have been shown to mislocalize in the case of PCD. These include DNAH5 and DNAI1, which under normal circumstances, colocalize throughout the axoneme. In patients with PCD caused by a mutation in the DNAI1 gene, the protein fails to localize to the distal tip, leading to abnormal motility. In patients with a mutation in DNAH5, the protein is entirely absent from the axoneme, causing complete paralysis. This is useful for diagnosis since immunofluorescence can show if this mislocalization is present [102]. Genetic testing can also be done, but is not very reliable since the disease is very genetically heterogeneous, with the discovery of at least ten related genes from various loci on multiple chromosomes [103]. False positive diagnoses can occur during or after a respiratory infection or inflammation, since these conditions show impaired ciliary function. Therefore diagnosis is only accurate at least four to six weeks post-infection [101].

Screening for the disease is also very important to help rule PCD out for patients with similar symptoms. Among the screening techniques, measuring levels of nasal nitric oxide (nNO) is relatively new but fairly promising. Very low levels of nNO are exhaled from patients with PCD. Therefore, if the patient has high or normal levels, they likely do not have PCD. Levels are not diagnostic of PCD, however, since low levels can also be found in accordance with other diseases such as cystic fibrosis, chronic sinusitis, and others. One downfall of this screening technique is that it is ineffective for children under five years old, since younger children will not be able to blow into the apparatus [104]. Saccharin testing is also used for PCD screening. This is a measurement of the time it takes for a patient to be able to taste saccharin, which is related to the function of the cilia in the taste bud cells. However, this is not a very useful test since it is unreliable in children under twelve. Radioaerosol mucociliary clearance testing, which measures how well mucous is removed from the respiratory tract, is useful in infants for screening, but again is not diagnostic of PCD [101].

Current treatment for PCD is mainly focused on treating and preventing the symptoms of the disease. Studies are being done to test the efficacy of antibiotics, airway clearance, and anti-inflammatory treatments [101]. Nebulized DNase has also shown promise in a 1999 case study with a PCD patient. The patient's symptoms were not relieved from treatment with antibiotic and bronchodilator treatments alone. However, when DNase was used in addition to these treatments, there was a significant improvement in symptoms overnight. This showed the therapeutic potential of DNase, but more studies need to be done to further test the efficacy [105]. Gene therapy is also being considered. A 2010 pilot study showed that a lentiviral vector can incorporate ciliary protein into the axoneme to restore ciliary function [96]. Further studies need to be done to make these treatments more mainstream. In addition to treatments, preventative measures such as avoiding respiratory irritations and exercising, can be very helpful. If treatments and preventative measures are unsuccessful, it is often necessary to undergo surgery.

PCD research has taught us a lot about cilia. It has shown the importance of cilia in development, in respiratory function, in aural health, and in fertility. It has also taught us a lot about the structure of cilia and the function of various ciliary proteins. In fact, a 2010 study revealed 208 potential ciliary genes based on PCD research [106]. PCD will continue to give us new insights on ciliary function and will continue to be an important area of ciliopathy research.

3.2 Hydrocephalus

Hydrocephalus is a disease in which cerebral spinal fluid (CSF) accumulates in the brain. CSF is mostly produced by the choroid plexuses of the lateral, third, and fourth ventricles [107]. To maintain equilibrium it is important for the excess fluid to drain into the subarachnoid space where it is resorbed into the venous system [108]. When this process malfunctions, the fluid builds up, causing swelling in the brain that leads to many complications. The disease dates back to the time of Hippocrates when he described a "liquefaction of the brain" that showed symptoms such as headache, vomiting, and visual impairment [109]. Other symptoms can include high intracranial pressure, in addition to impairments in gait, cognition, alertness, and continence [110]. Although this disease can have various causes, it has been shown that there is a higher prevalence of hydrocephalus in patients that are known to have ciliary defects, and the disease can also be linked to genes known to impact cilia function or structure [111]. Hydrocephalus is often seen in conjunction with other ciliopathies such as PKD as seen in the Tg737 mouse [112] and more rarely in PCD [113].

Although hydrocephalus has a rich history in research, the disease as a ciliopathy is a relatively new area of study, and much is yet to be learned about its link with cilia. In animal models, 43 mutations from 9 different genes are known to be related to hydrocephalus. In contrast, there is only one gene that has been identified in humans [114]. X-linked hydrocephalus is caused by a mutation in the neural cell adhesion molecule L1 (L1CAM) gene [115]. This form of the disease was first discovered in 1949 [116] and occurs about 1 in every 30,000 male births [117]. No research has been done up to this point to determine if there is a link between L1CAM and cilia.

Many mouse models, however, have allowed identification of multiple ciliary genes that, when mutated, can lead to hydrocephalus. Accumulation of CSF can be caused by malfunctions in CSF production, CSF flow, or CSF absorption. Tg737 mice are missing intraflagellar transport protein IFT88/Polaris. This protein is important for ion transport and the mutation shows an overproduction of CSF, leading to the development of hydrocephalus [107]. This suggests the importance of cilia in signaling and regulation of CSF levels. Mice with this mutation also have abnormal beating of the motile cilia of the ependymal cells, disrupting the CSF flow. However, this is not causal in this case since development of the disease occurs before the motile cilia form [107]. In contrast, CSF flow disruption seems to be the cause of hydrocephalus in Mdnah5 (axonemal dynein heavy chain) and Hydin mutants. These mutants cause structural defects in the axoneme of the cilia, leading to impaired motility [111, 118-120]. In these two mutants, the cilia are unable to create sufficient flow to remove CSF from the ventricles. Other hydrocephalus mutants are thought to be linked to important development pathways that may disrupt the CSF equilibrium. Inactivation of Pten and β-catenin leads to hydrocephalus [121]. These proteins are key players in Wnt signaling pathway and important for proper midbrain development, in which the function of primary cilia could be involved [122]. Polycystin-1, a ciliary membrane protein, is also shown to be important in development and regulating fluid in the brain [112]. Ptch1 and parkin-qk1 mutations also lead to hydrocephalus [123] most likely due to a disruption in the Hedgehog pathway, which also involves the primary cilia [124]. All these findings show the importance of cilia in signaling, development, and movement of fluid, and their roles in maintaining a healthy CSF balance in the brain.

An early method of diagnosis required ventricular puncture to test for dilation and occlusive lesions. This procedure was risky, so it was eventually replaced by computed tomography (CT) and magnetic resonance imaging (MRI) [109]. Prenatal diagnosis can be done for x-linked recessive hydrocephalus by doing serial ultrasound scans to test for abnormal growth of the baby's head [125]. Since there is only one known human hydrocephalus gene, and it is for the rare x-linked type, genetic testing is not yet useful for diagnosis.

The oldest treatment known for hydrocephalus was to tightly bandage the baby's deformed swollen head to decrease the size and swelling. This method was abandoned since it increased the intracranial pressure. In the 18th and 19th centuries, special diets were recommended and dehydration was induced with laxatives, diuretics, potassium iodide, etc. In 1957 acetazolamide was first used in practice and is still used to reduce production of CSF. Other treatments were abandoned such as isosorbide and irradiation of the choroid plexus [109]. Vasoconstrictors such as dihydroergotamine have shown promise in allowing better arterial pulsation and reducing ventricular dilation [126]. Various surgical treatments have been used including external CSF drainage, serial lumbar puncturing, and implantation of an internal shunt, which is a catheter that allows drainage of CSF out of the ventricles. These surgical procedures brought about complications such as infections, improper placement of shunts, and hydraulic mismanagement due to body positioning. These complications necessitated further research and improvements. One of the most pivotal improvements is the development and modification of shunts with adjustable, autoregulating, antisiphon, and gravitational valves. There are currently at least 127 designs of valves and more than 20 shunting procedures that have been suggested, with ventriculoperitoneal shunts being the most commonly utilized. Valved shunts are now standard treatment for hydrocephalus, being the choice treatment for about 80% of cases [109, 127]. In elderly patients shunt implantation is very risky. Repeated lumbar puncturing in patients with communicating hydrocephalus can potentially prevent the need for shunt surgery in these patients [110]. The newest addition to shunt technology is antibiotic-impregnated shunts that help prevent post-surgical infections [128].

The relationship between cilia and hydrocephalus and the genes that are involved are poorly understood in humans. Advancements in this knowledge may eventually lead to better, less invasive forms of treatment and a better understanding of how cilia function in the brain.

3.3 Polycystic Kidney Disease (PKD)

There are two types of PKD, autosomal dominant polycystic kidney disease (ADPKD) and autosomal recessive polycystic kidney disease (ARPKD). ADPKD is the most common of all the potentially lethal autosomal dominant diseases, with an incidence of 1 in 1000 [129]. Symptoms typically do not present themselves until between the ages of 35-50 years. These symptoms include acute abdominal and lower back pain, hypertension, palpable kidneys, recurrent urinary tract infections, shortness of breath, early satiety, hydrocephalus, kidney stones and cysts in the kidney, liver, thyroid, subarachnoid space, and seminal vesicles [130]. ADPKD most often ends in end stage renal disease (ESRD) between 55 and 75 years old [82]. ARPKD, on the other hand presents itself immediately. Although delayed presentation is possible, ARPKD can often be seen in utero, and leads to ESRD in the neonatal stage and infants often die due to respiratory complications. Also unlike ADPKD, there is usually no cyst formation in other organs in the case of ARPKD [130]. It is also much more rare, with an incidence of 1 in 20000, and has a very high infant mortality rate [131].

Although other factors may be involved, strong correlations have been made between PKD and cilia function. Primary cilia are now known to act as mechanosensors that regulate Ca^{2+} influx. Fluid flow in the kidney causes the primary cilia to bend, which allows the Ca^{2+} channels to open, allowing increased intracellular Ca^{2+}. This process is disrupted in some forms of PKD. However, primary cilia have also been implicated in pathways such as Hedgehog, Wnt, cAMP, and Planar Cell Polarity. Disruptions of these pathways can lead to abnormal polarity, differentiation, and proliferation, which can lead to cyst formation [130]. Therefore the role of primary cilia in cystogenesis is not only important, but is very multidimensional.

In human, ADPKD is caused by a mutation in PKD1 (85 % of the time) or PKD2 (15% of the time). There are more than 500 mutations known in PKD1 and 120 in PKD2 [132, 133]. These genes code for proteins polycystin-1 and polycystin-2, which both localize to renal cilia [134]. These proteins prove to be important in renal tube development and cell differentiation in the kidney. They are involved in the calcium signal transduction cascade that regulates proliferation and differentiation [135, 136]. ARPKD is caused by a mutation in PKHD1, which encodes fibrocystin, a receptor-like protein associated with the membrane and colocalizes with polycystin-2 in primary cilia [82, 137]. It also plays a role in collecting duct and biliary cell differentiation [138]. Studies in other species have made other connections between cilia and PKD. In addition to hydrocephalus, Tg737 mice have cysts in kidney and pancreas, hepatic fibrosis, and polydactyly. Mice with this mutation have elevated polycystin-2 levels and have cilia that are much shorter than normal [135, 139]. These factors result in cyst formation. A mutation in Kif3A (a subunit of kinesin-II in kidney epithelium) causes increased canonical Wnt activity. Deletion of this gene leads to absence of cilia, and after the cilia are lost, cystogenesis occurs [140]. Seahorse is another mutation that shows a disruption in Wnt signaling in zebrafish and may function downstream from cilia [141]. Out of the 11 genes identified in zebrafish that relate to PKD, 6 were found to be ciliary genes [142]. These mutants and more, show the importance of cilia and pathways in regulating renal cells.

Although CT and MRI techniques can be used, diagnosis by ultrasound and positive family history are the choice methods to test for PKD [143]. ARPKD can even be detected in utero by ultrasound which reveals large kidneys that take up most of the fetal abdominal cavity and the lack of urine in the bladder [130]. Nuclear magnetic resonance (NMR) spectroscopy also allows discrimination between PKD and other kidney diseases with an accuracy of over 80%, by creating a fingerprint of urinary protein biomarkers with key PKD features [143]. Since most PKD patients have mutations in known PKD genes, genetic testing may be plausible. However, since there are hundreds of possible mutations, the only commonly used genetics testing is direct sequencing to screen for the disease [144].

There is currently no effective treatment that is widely used for PKD. Transplantation and dialysis are often the only options. Apart from that, pain control, antibiotics for urinary tract infections, increase in fluid intake, and refraining from smoking and caffeine, are ways of lessening the symptoms [130]. However, many potential drugs for treating cystogenesis are under investigation. Many drugs have shown to slow cyst growth in animal models. Rapamycin helps regulate cell proliferation by inhibiting the mTOR pathway, which is often overexpressed in PKD patients. This drug increases apoptosis and shedding of cystic cells, which decreases kidney size and restored kidney function [145]. Roscovitine, currently used in cancer treatment because it inhibits cell cycle, show positive effects in PKD models [146]. Lisinopril, an inhibitor of angiotensin converting enzyme (ACE) also alters proliferative and apoptotic pathways, reducing cyst development [147]. Patients with PKD show high levels of circulating vasopressin. Tolvaptin is a vasopressin-2 receptor antagonist and shows reduction in kidney and cyst volumes [148]. Ocreatide, an analogue of the hormone somatostatin, inhibits cAMP production. In doing so, it inhibits secretion and reduces liver cysts and kidney volume. However, it does not seem to improve renal function [149, 150]. Epidermal growth factor receptor (EGFR) tyrosine kinase inhibition using EKI-785 also showed promise in PKD animal studies [151]. Anti-inflammatory drug, colchicine, is a microtubule inhibitor that has been shown to delay formation of cysts and is a candidate for prolonged clinical use [152].

PKD mutants are giving us a better understanding of the importance of cilia in many regulatory pathways and sensory functions. The disease further emphasizes how complex these seemingly simple cellular organelles are, and how necessary they are in maintaining systemic health and prevention of cystic growth.

3.4 Bardet-Biedl Syndrome (BBS)

Many syndromes are related to cilia, Bardet-Biedl syndrome being the most well-known. Bardet-Biedl Syndrome (BBS) is a multi-symptomatic disorder with symptoms including obesity, retinitis pigmentosa, genital hypoplasia, polydactyly, and mental retardation [153]. Typically at age 8, night blindness occurs which can eventually lead to complete blindness between ages 15 and 20. Kidney cysts are also common, making end-stage renal failure the most common cause of premature death in BBS patients [154]. Mid-facial deformities are often seen in humans with BBS and also in mice mutants for BBS4 and BBS6 [131]. Although this is a very multi-symptomatic disorder, obesity is what it is most known for. Ninety-eight percent of BBS patients become obese with a body mass index greater than 30% [155]. Diabetes occurs secondary to obesity, and patients often show lower locomotor activity than normal [156, 157].

Phototransduction proteins and others necessary for vision are produced in the inner segment of photoreceptor cells. To maintain the outer segments, the phototransduction proteins have to be transported to the outer segments of the photoreceptors through the connecting cilia, the only bridge between outer and inner segments. BBS proteins form a BBSome and are believed to be responsible for recruiting membrane vesicles to the cilia [16, 47]. BBS proteins are also involved in important pathways such as Wnt and Hedgehog pathways which are very important for proper development and function [153]. Hedgehog and Wnt signaling are anti-adipogenic, preventing obesity [158-160]. It is now believed that obesity in BBS patients is linked to the impairment of ciliated cells in the hypothalamus to sense satiety, inducing hyperphagia [161]. BBS proteins may be important in transport of leptin in and out of the cells, causing loss of leptin signaling ability when cilia are lost. Moreover, melanin concentrating hormone receptor-1 (Mchr1) is involved in regulation of feeding behavior, and fails to localize to cilia in BBS mutants [161, 162]. These are just a few of the possible factors that may be involved in causing obesity in BBS patients.

BBS is an autosomal recessive disease showing pleiotropy, with multiple traits being affected. Development of the disorder often requires more than three mutations in at least 2 BBS genes [163]. Apart from the causative mutations, other BBS mutations often serve as disease modifiers. So far there are 12 known BBS genes, BBS1-12, which code for proteins important in trafficking cargo to the basal body and along the cilia. Other ciliary proteins have also been linked to BBS [154]. Among these are Kif3A and Tg737 mutants, which show hyperphagic activity and obesity, in addition to elevated plasma levels of glucose, insulin, and leptin [164].

Diagnosis of BBS typically requires the presence of at least four primary symptoms or three primary in addition to two secondary symptoms. Primary symptoms include cone-rod dystrophy, polydactyly, obesity, learning disability, genital defects, and renal anomalies. Secondary symptoms include speech impairment, brachydactyly (short digits), syndachtyly (fused digits), developmental delay, polyuria (excessive urination), ploydypsia (excessive thirst), ataxia (lack of muscle coordination), diabetes, heart and liver problems, olfactory deficits, and defects in pain and temperature sensation [154].

Since BBS is syndromic and affects multiple systems, treatment can be multi-faceted. Gene therapy has shown promise in treating vision impairments. Mice, in which Bbs4 has been deleted, show an inability of rhodopsin to localize to rod cilia and cone opsins to localize to cone cilia. This failure to localize leads to photoreceptor apoptosis and the deterioration of electroretinogram (ERG) a- and b-waves, causing serious vision impairment. Recently, adeno-associated viral (AAV) vectors have shown to be successful in incorporating Bbs4 into these mutants and restoring the localization of rhodopsin. In doing so, photoreceptor death was prevented, function of the retina was restored, and mice showed recovery of visual behavioral responses [165]. Surgical techniques are used for other symptoms of BBS. For instance renal transplantation is used in the case of cystic kidneys. Obesity is important to treat, since it has been shown to cause a fifty percent increase in mortality rates most likely caused by complications from secondary diseases. These include diabetes, cardiovascular diseases, cerebrovascular diseases, digestive disorders, gall bladder cancer, breast cancer, endometrium cancer, prostate cancer, etc [166, 167]. Bariatric surgeries such as gastric bypass and gastric band operations greatly improve weight loss in patients with obesity. These operations also show a dramatic improvement in reducing the risk of secondary diseases that are known to be associated with obesity. Because of the reduction in the incidence of these diseases, the mortality rate decreases substantially [168]. In addition to surgical procedures, drug treatments have also been under review for the treatment of obesity. There are only two drugs currently accepted by the U.S. Food and Drug Administration for long term obesity treatment. This is due to the high prevalence of severe side effects correlated with the use of weight loss drugs. These side effects include heart attack, gastrointestinal distress, liver damage, anxiety, memory problems, suicide, and are often habit-forming. Because of these serious issues, it is important to be skeptical in determining which drugs are safe to be used to treat obesity [169]. These are only a few of the current treatments being used for BBS-related symptoms, and many others are yet to be discovered.

Again, BBS adds further support to the idea that cilia are important in many areas throughout the body, and they are very important in maintaining the overall health of the individual.

3.5 Cancer

Relating cancer to cilia is one of the newest areas of ciliary research in the field today. The cilia assembly-disassembly cycle is closely linked with the cell cycle. Cilia assemble upon exit of mitosis to the stationary phase, and resorb when the cell exits the S-phase and enters mitosis [170]. When the ciliogenesis is disrupted, it may have adverse effects on cell cycle and lead to cancer. This may be due to centrosomal amplification and genetic instability [171]. Proteins necessary for ciliogenesis colocalize to the centrosome, also supporting the link between cilia and cell cycle [172]. Cilia have also been found to play roles in important pathways such as Sonic Hedgehog, signal transduction pathways, and ligand-induced signaling. These pathways have recently been shown to relate to cancer. New insights are being made to connect ciliary dysfunction to carcinogenesis [172].

Primary cilia appear to have dual opposing functions in development of different types of cancer, so that some refer to them as being an On/Off switch, regulating tumorigenesis [173, 174]. One of the most important pathways related to ciliogenesis and cancer is the Hedgehog (Hh) pathway. Hh is normally suppressed by Ptch1 which prevents the trafficking of Smo in primary cilia. When Smo fails to localize to cilia, it prevents Hh signaling. Under cancerous conditions, however, Smo is able to localize to the cilia and Hh signaling is overexpressed leading to oncogenesis [175]. SmoM2 leads to brain tumors only if primary cilia are present. However, Gli2ΔN induces brain tumors only in the absence of cilia, since absence of cilia causes a disruption in Gli3, a repressor of Gli2ΔN [174, 176]. It appears that primary cilia are required both for the suppression and expression of oncogenisis. Primary cilia suppress oncogenic mutations that act downstream of cilia, but allow expression of oncogenic mutations upstream of cilia [173, 174, 177]. Therefore, cilia can play opposite roles depending on the causal mutation of the cancer development [173, 174].

Disruptions in Hh, Wnt, and PDGF pathways all have been linked to tumor formation. Cilia-associated genes such as Gli1, RPGRIP1, and DNAH9 are often mutated in breast cancer [177]. Nek8 is also localized to cilia and upregulated in breast cancer [178, 179]. Pancreatic ductal adenocarcinoma (PDAC) is the fourth leading cause of cancer deaths in the United States [180]. In this case, excessive Hh signaling is caused by a mutation in a Kras gene[181], which codes for a protein known to be important in ciliogenesis. Therefore affected cells lack primary cilia, and are unable to regulate proliferation [182].

German pathologist, Rudolf Virchow, first defined cancer in the mid-1800s when he realized that Leukemia was caused by a rapid duplication of healthy cells that had mutated and in response, multiplied. Before the advent of modern-day technology, cancer was not diagnosed or treated until the tumors became visible and palpable, in which case they were surgically removed. Since then, however, an emphasis has been placed on early detection so that doctors can treat the cancer before it turns into large fatal tumors. Most doctors recommend certain regular screening for some types of cancer. Among these are mammograms for breast cancer, colonoscopies for colon cancer, Pap smears for cervical cancer, and prostate exams for prostate cancer. Other diagnostic tests include blood tests, ultrasounds, computed tomography (CT), X-ray, magnetic resonance imaging (MRI), and fine-needle biopsies. These techniques are used to detect early signs of cancer, allowing an earlier treatment and an attempt to prevent fatality [183].

Cancer treatment is one of the most prioritized areas of research today. Many treatments are currently being used and even more are being tested. These treatments include chemotherapy, radiation therapy, surgery, and gene therapy. Perhaps the implications in the relationship with cilia and cell cycle will help lead to the development of new treatments.

Primary ciliary dyskinesia, hydrocephalus, polycystic kidney disease, Bardet-Biedl syndrome, and cancer are only a few of the known ciliopathies, and more are still being discovered. These diseases are not only important to study for diagnostic and treatment purposes, but also give us a clearer understanding about cilia and their role in most critical bodily functions. Further ciliopathy studies will continue to shed new light on these important cellular structures.

The research in the Qin lab is supported by the NSF grant MCB-0923835.

4. References

[1] H. Ishikawa and W. F. Marshall, "Ciliogenesis: building the cell's antenna," *Nat Rev Mol Cell Biol*, vol. 12, pp. 222-34, Apr 2011.
[2] G. J. Pazour and G. B. Witman, "The vertebrate primary cilium is a sensory organelle," *Curr Opin Cell Biol*, vol. 15, pp. 105-10, Feb 2003.
[3] K. G. Kozminski, K. A. Johnson, P. Forscher, and J. L. Rosenbaum, "A motility in the eukaryotic flagellum unrelated to flagellar beating," *Proc Natl Acad Sci U S A*, vol. 90, pp. 5519-23, Jun 15 1993.
[4] J. Rosenbaum, "Intraflagellar transport," *Curr Biol*, vol. 12, p. R125, Feb 19 2002.
[5] D. G. Cole, D. R. Diener, A. L. Himelblau, P. L. Beech, J. C. Fuster, and J. L. Rosenbaum, "Chlamydomonas kinesin-II-dependent intraflagellar transport (IFT): IFT particles contain proteins required for ciliary assembly in Caenorhabditis elegans sensory neurons," *J Cell Biol*, vol. 141, pp. 993-1008, May 18 1998.
[6] K. G. Kozminski, P. L. Beech, and J. L. Rosenbaum, "The Chlamydomonas kinesin-like protein FLA10 is involved in motility associated with the flagellar membrane," *J Cell Biol*, vol. 131, pp. 1517-27, Dec 1995.
[7] G. J. Pazour, C. G. Wilkerson, and G. B. Witman, "A dynein light chain is essential for the retrograde particle movement of intraflagellar transport (IFT)," *J Cell Biol*, vol. 141, pp. 979-92, May 18 1998.

[8] G. J. Pazour, B. L. Dickert, and G. B. Witman, "The DHC1b (DHC2) isoform of cytoplasmic dynein is required for flagellar assembly," *J Cell Biol*, vol. 144, pp. 473-81, Feb 8 1999.
[9] P. Rompolas, L. B. Pedersen, R. S. Patel-King, and S. M. King, "Chlamydomonas FAP133 is a dynein intermediate chain associated with the retrograde intraflagellar transport motor," *J Cell Sci*, vol. 120, pp. 3653-65, Oct 15 2007.
[10] D. J. Asai, V. Rajagopalan, and D. E. Wilkes, "Dynein-2 and ciliogenesis in Tetrahymena," *Cell Motil Cytoskeleton*, vol. 66, pp. 673-7, Aug 2009.
[11] L. Hao, E. Efimenko, P. Swoboda, and J. M. Scholey, "The Retrograde IFT Machinery of C. elegans Cilia: Two IFT Dynein Complexes?," *PLoS One*, vol. 6, p. e20995, 2011.
[12] J. M. Scholey, "Intraflagellar transport motors in cilia: moving along the cell's antenna," *J Cell Biol*, vol. 180, pp. 23-9, Jan 14 2008.
[13] J. M. Scholey, "Intraflagellar transport," *Annu Rev Cell Dev Biol*, vol. 19, pp. 423-43, 2003.
[14] M. E. Porter, R. Bower, J. A. Knott, P. Byrd, and W. Dentler, "Cytoplasmic dynein heavy chain 1b is required for flagellar assembly in Chlamydomonas," *Mol Biol Cell*, vol. 10, pp. 693-712, Mar 1999.
[15] W. F. Marshall and J. L. Rosenbaum, "Intraflagellar transport balances continuous turnover of outer doublet microtubules: implications for flagellar length control," *J Cell Biol*, vol. 155, pp. 405-14, Oct 29 2001.
[16] M. V. Nachury, E. S. Seeley, and H. Jin, "Trafficking to the ciliary membrane: how to get across the periciliary diffusion barrier?," *Annu Rev Cell Dev Biol*, vol. 26, pp. 59-87, Nov 10 2010.
[17] K. F. Lechtreck, E. C. Johnson, T. Sakai, D. Cochran, B. A. Ballif, J. Rush, G. J. Pazour, M. Ikebe, and G. B. Witman, "The Chlamydomonas reinhardtii BBSome is an IFT cargo required for export of specific signaling proteins from flagella," *J Cell Biol*, vol. 187, pp. 1117-32, Dec 28 2009.
[18] D. G. Cole, S. W. Chinn, K. P. Wedaman, K. Hall, T. Vuong, and J. M. Scholey, "Novel heterotrimeric kinesin-related protein purified from sea urchin eggs," *Nature*, vol. 366, pp. 268-70, Nov 18 1993.
[19] D. G. Cole, "The intraflagellar transport machinery of Chlamydomonas reinhardtii," *Traffic*, vol. 4, pp. 435-42, Jul 2003.
[20] K. Matsuura, P. A. Lefebvre, R. Kamiya, and M. Hirono, "Kinesin-II is not essential for mitosis and cell growth in Chlamydomonas," *Cell Motil Cytoskeleton*, vol. 52, pp. 195-201, Aug 2002.
[21] J. J. Snow, G. Ou, A. L. Gunnarson, M. R. Walker, H. M. Zhou, I. Brust-Mascher, and J. M. Scholey, "Two anterograde intraflagellar transport motors cooperate to build sensory cilia on C. elegans neurons," *Nat Cell Biol*, vol. 6, pp. 1109-13, Nov 2004.
[22] N. S. Morsci and M. M. Barr, "Kinesin-3 KLP-6 Regulates Intraflagellar Transport in Male-Specific Cilia of Caenorhabditis elegans," *Curr Biol*, vol. 21, pp. 1239-44, Jul 26 2011.
[23] Y. Hou, G. J. Pazour, and G. B. Witman, "A dynein light intermediate chain, D1bLIC, is required for retrograde intraflagellar transport," *Mol Biol Cell*, vol. 15, pp. 4382-94, Oct 2004.
[24] C. A. Perrone, D. Tritschler, P. Taulman, R. Bower, B. K. Yoder, and M. E. Porter, "A novel dynein light intermediate chain colocalizes with the retrograde motor for

intraflagellar transport at sites of axoneme assembly in chlamydomonas and Mammalian cells," *Mol Biol Cell*, vol. 14, pp. 2041-56, May 2003.

[25] J. C. Schafer, C. J. Haycraft, J. H. Thomas, B. K. Yoder, and P. Swoboda, "XBX-1 encodes a dynein light intermediate chain required for retrograde intraflagellar transport and cilia assembly in Caenorhabditis elegans," *Mol Biol Cell*, vol. 14, pp. 2057-70, May 2003.

[26] D. Signor, K. P. Wedaman, J. T. Orozco, N. D. Dwyer, C. I. Bargmann, L. S. Rose, and J. M. Scholey, "Role of a class DHC1b dynein in retrograde transport of IFT motors and IFT raft particles along cilia, but not dendrites, in chemosensory neurons of living Caenorhabditis elegans," *J Cell Biol*, vol. 147, pp. 519-30, Nov 1 1999.

[27] L. B. Pedersen and J. L. Rosenbaum, "Intraflagellar transport (IFT) role in ciliary assembly, resorption and signalling," *Curr Top Dev Biol*, vol. 85, pp. 23-61, 2008.

[28] S. M. Williamson, D. A. Silva, E. Richey, and H. Qin, "Probing the role of IFT particle complex A and B in flagellar entry and exit of IFT-dynein in *Chlamydomonas*" *Protoplasma*, 2011.

[29] G. Piperno, E. Siuda, S. Henderson, M. Segil, H. Vaananen, and M. Sassaroli, "Distinct mutants of retrograde intraflagellar transport (IFT) share similar morphological and molecular defects," *J Cell Biol*, vol. 143, pp. 1591-601, Dec 14 1998.

[30] C. Iomini, L. Li, J. M. Esparza, and S. K. Dutcher, "Retrograde intraflagellar transport mutants identify complex A proteins with multiple genetic interactions in Chlamydomonas reinhardtii," *Genetics*, vol. 183, pp. 885-96, Nov 2009.

[31] J. Mueller, C. A. Perrone, R. Bower, D. G. Cole, and M. E. Porter, "The FLA3 KAP subunit is required for localization of kinesin-2 to the site of flagellar assembly and processive anterograde intraflagellar transport," *Mol Biol Cell*, vol. 16, pp. 1341-54, Mar 2005.

[32] C. Li, P. N. Inglis, C. C. Leitch, E. Efimenko, N. A. Zaghloul, C. A. Mok, E. E. Davis, N. J. Bialas, M. P. Healey, E. Heon, M. Zhen, P. Swoboda, N. Katsanis, and M. R. Leroux, "An essential role for DYF-11/MIP-T3 in assembling functional intraflagellar transport complexes," *PLoS Genet*, vol. 4, p. e1000044, Mar 2008.

[33] J. F. Dishinger, H. L. Kee, P. M. Jenkins, S. Fan, T. W. Hurd, J. W. Hammond, Y. N. Truong, B. Margolis, J. R. Martens, and K. J. Verhey, "Ciliary entry of the kinesin-2 motor KIF17 is regulated by importin-beta2 and RanGTP," *Nat Cell Biol*, vol. 12, pp. 703-10, Jul 2010.

[34] B. F. Lucker, R. H. Behal, H. Qin, L. C. Siron, W. D. Taggart, J. L. Rosenbaum, and D. G. Cole, "Characterization of the intraflagellar transport complex B core: direct interaction of the IFT81 and IFT74/72 subunits," *J Biol Chem*, vol. 280, pp. 27688-96, Jul 29 2005.

[35] B. F. Lucker, M. S. Miller, S. A. Dziedzic, P. T. Blackmarr, and D. G. Cole, "Direct interactions of intraflagellar transport complex B proteins IFT88, IFT52 and IFT46," *J Biol Chem*, vol. 2010, Apr 30 2010.

[36] M. Taschner, S. Bhogaraju, M. Vetter, M. Morawetz, and E. Lorentzen, "Biochemical Mapping of Interactions within the Intraflagellar Transport (IFT) B Core Complex: IFT52 BINDS DIRECTLY TO FOUR OTHER IFT-B SUBUNITS," *J Biol Chem*, vol. 286, pp. 26344-52, Jul 29 2011.

[37] G. J. Pazour, B. L. Dickert, Y. Vucica, E. S. Seeley, J. L. Rosenbaum, G. B. Witman, and D. G. Cole, "Chlamydomonas IFT88 and its mouse homologue, polycystic kidney

disease gene tg737, are required for assembly of cilia and flagella," *J Cell Biol,* vol. 151, pp. 709-18, Oct 30 2000.

[38] Y. Hou, H. Qin, J. A. Follit, G. J. Pazour, J. L. Rosenbaum, and G. B. Witman, "Functional analysis of an individual IFT protein: IFT46 is required for transport of outer dynein arms into flagella," *J Cell Biol,* vol. 176, pp. 653-65, Feb 26 2007.

[39] N. T. Ahmed, C. Gao, B. F. Lucker, D. G. Cole, and D. R. Mitchell, "ODA16 aids axonemal outer row dynein assembly through an interaction with the intraflagellar transport machinery," *J Cell Biol,* vol. 183, pp. 313-22, Oct 20 2008.

[40] J. A. Follit, R. A. Tuft, K. E. Fogarty, and G. J. Pazour, "The intraflagellar transport protein IFT20 is associated with the Golgi complex and is required for cilia assembly," *Mol Biol Cell,* vol. 17, pp. 3781-92, Sep 2006.

[41] B. L. Krock and B. D. Perkins, "The intraflagellar transport protein IFT57 is required for cilia maintenance and regulates IFT-particle-kinesin-II dissociation in vertebrate photoreceptors," *J Cell Sci,* vol. 121, pp. 1907-15, Jun 1 2008.

[42] L. B. Pedersen, M. S. Miller, S. Geimer, J. M. Leitch, J. L. Rosenbaum, and D. G. Cole, "Chlamydomonas IFT172 is encoded by FLA11, interacts with CrEB1, and regulates IFT at the flagellar tip," *Curr Biol,* vol. 15, pp. 262-6, Feb 8 2005.

[43] L. B. Pedersen, S. Geimer, and J. L. Rosenbaum, "Dissecting the molecular mechanisms of intraflagellar transport in chlamydomonas," *Curr Biol,* vol. 16, pp. 450-9, Mar 7 2006.

[44] J. Qin, Y. Lin, R. X. Norman, H. W. Ko, and J. T. Eggenschwiler, "Intraflagellar transport protein 122 antagonizes Sonic Hedgehog signaling and controls ciliary localization of pathway components," *Proc Natl Acad Sci U S A,* vol. 108, pp. 1456-61, Jan 25 2011.

[45] E. Lee, E. Sivan-Loukianova, D. F. Eberl, and M. J. Kernan, "An IFT-A protein is required to delimit functionally distinct zones in mechanosensory cilia," *Curr Biol,* vol. 18, pp. 1899-906, Dec 23 2008.

[46] G. Ou, O. E. Blacque, J. J. Snow, M. R. Leroux, and J. M. Scholey, "Functional coordination of intraflagellar transport motors," *Nature,* vol. 436, pp. 583-7, Jul 28 2005.

[47] M. V. Nachury, A. V. Loktev, Q. Zhang, C. J. Westlake, J. Peranen, A. Merdes, D. C. Slusarski, R. H. Scheller, J. F. Bazan, V. C. Sheffield, and P. K. Jackson, "A core complex of BBS proteins cooperates with the GTPase Rab8 to promote ciliary membrane biogenesis," *Cell,* vol. 129, pp. 1201-13, Jun 15 2007.

[48] H. Jin, S. R. White, T. Shida, S. Schulz, M. Aguiar, S. P. Gygi, J. F. Bazan, and M. V. Nachury, "The conserved Bardet-Biedl syndrome proteins assemble a coat that traffics membrane proteins to cilia," *Cell,* vol. 141, pp. 1208-19, Jun 25 2010.

[49] S. Cevik, Y. Hori, O. I. Kaplan, K. Kida, T. Toivenon, C. Foley-Fisher, D. Cottell, T. Katada, K. Kontani, and O. E. Blacque, "Joubert syndrome Arl13b functions at ciliary membranes and stabilizes protein transport in Caenorhabditis elegans," *J Cell Biol,* vol. 188, pp. 953-69, Mar 22 2010.

[50] Y. Li, Q. Wei, Y. Zhang, K. Ling, and J. Hu, "The small GTPases ARL-13 and ARL-3 coordinate intraflagellar transport and ciliogenesis," *J Cell Biol,* vol. 189, pp. 1039-51, Jun 14 2010.

[51] N. A. Duldulao, S. Lee, and Z. Sun, "Cilia localization is essential for in vivo functions of the Joubert syndrome protein Arl13b/Scorpion," *Development*, vol. 136, pp. 4033-42, Dec 2009.
[52] C. J. Westlake, L. M. Baye, M. V. Nachury, K. J. Wright, K. E. Ervin, L. Phu, C. Chalouni, J. S. Beck, D. S. Kirkpatrick, D. C. Slusarski, V. C. Sheffield, R. H. Scheller, and P. K. Jackson, "Primary cilia membrane assembly is initiated by Rab11 and transport protein particle II (TRAPPII) complex-dependent trafficking of Rabin8 to the centrosome," *Proc Natl Acad Sci U S A*, vol. 108, pp. 2759-64, Feb 15 2011.
[53] A. Knodler, S. Feng, J. Zhang, X. Zhang, A. Das, J. Peranen, and W. Guo, "Coordination of Rab8 and Rab11 in primary ciliogenesis," *Proc Natl Acad Sci U S A*, vol. 107, pp. 6346-51, Apr 6 2010.
[54] J. Mazelova, L. Astuto-Gribble, H. Inoue, B. M. Tam, E. Schonteich, R. Prekeris, O. L. Moritz, P. A. Randazzo, and D. Deretic, "Ciliary targeting motif VxPx directs assembly of a trafficking module through Arf4," *Embo J*, vol. 28, pp. 183-92, Feb 4 2009.
[55] H. H. Ward, U. Brown-Glaberman, J. Wang, Y. Morita, S. L. Alper, E. J. Bedrick, V. H. Gattone, 2nd, D. Deretic, and A. Wandinger-Ness, "A conserved signal and GTPase complex are required for the ciliary transport of polycystin-1," *Mol Biol Cell*, vol. 2011, Jul 20 2011.
[56] C. Boehlke, M. Bashkurov, A. Buescher, T. Krick, A. K. John, R. Nitschke, G. Walz, and E. W. Kuehn, "Differential role of Rab proteins in ciliary trafficking: Rab23 regulates smoothened levels," *J Cell Sci*, vol. 123, pp. 1460-7, May 1 2010.
[57] Z. Wang, Z. C. Fan, S. M. Williamson, and H. Qin, "Intraflagellar transport (IFT) protein IFT25 is a phosphoprotein component of IFT complex B and physically interacts with IFT27 in Chlamydomonas," *PLoS One*, vol. 4, p. e5384, 2009.
[58] S. Bhogaraju, M. Taschner, M. Morawetz, C. Basquin, and E. Lorentzen, "Crystal structure of the intraflagellar transport complex 25/27," *Embo J*, Apr 19 2011.
[59] J. C. Schafer, M. E. Winkelbauer, C. L. Williams, C. J. Haycraft, R. A. Desmond, and B. K. Yoder, "IFTA-2 is a conserved cilia protein involved in pathways regulating longevity and dauer formation in Caenorhabditis elegans," *J Cell Sci*, vol. 119, pp. 4088-100, Oct 1 2006.
[60] C. Adhiambo, T. Blisnick, G. Toutirais, E. Delannoy, and P. Bastin, "A novel function for the atypical small G protein Rab-like 5 in the assembly of the trypanosome flagellum," *J Cell Sci*, vol. 122, pp. 834-41, Mar 15 2009.
[61] J. L. Rosenbaum and G. B. Witman, "Intraflagellar transport," *Nat Rev Mol Cell Biol*, vol. 3, pp. 813-25, Nov 2002.
[62] B. Craige, C. C. Tsao, D. R. Diener, Y. Hou, K. F. Lechtreck, J. L. Rosenbaum, and G. B. Witman, "CEP290 tethers flagellar transition zone microtubules to the membrane and regulates flagellar protein content," *J Cell Biol*, vol. 190, pp. 927-40, Sep 6 2010.
[63] H. Omran, "NPHP proteins: gatekeepers of the ciliary compartment," *J Cell Biol*, vol. 190, pp. 715-7, Sep 6 2010.
[64] C. L. Williams, C. Li, K. Kida, P. N. Inglis, S. Mohan, L. Semenec, N. J. Bialas, R. M. Stupay, N. Chen, O. E. Blacque, B. K. Yoder, and M. R. Leroux, "MKS and NPHP modules cooperate to establish basal body/transition zone membrane associations and ciliary gate function during ciliogenesis," *J Cell Biol*, vol. 192, pp. 1023-41, Mar 21 2011.

[65] S. S. Francis, J. Sfakianos, B. Lo, and I. Mellman, "A hierarchy of signals regulates entry of membrane proteins into the ciliary membrane domain in epithelial cells," *J Cell Biol*, vol. 193, pp. 219-33, Apr 4 2011.

[66] T. Takeda, T. McQuistan, R. A. Orlando, and M. G. Farquhar, "Loss of glomerular foot processes is associated with uncoupling of podocalyxin from the actin cytoskeleton," *J Clin Invest*, vol. 108, pp. 289-301, Jul 2001.

[67] J. Gaertig and D. Wloga, "Ciliary tubulin and its post-translational modifications," *Curr Top Dev Biol*, vol. 85, pp. 83-113, 2008.

[68] D. Wloga and J. Gaertig, "Post-translational modifications of microtubules," *J Cell Sci*, vol. 123, pp. 3447-55, Oct 15 2010.

[69] K. Greer, H. Maruta, S. W. L'Hernault, and J. L. Rosenbaum, "Alpha-tubulin acetylase activity in isolated Chlamydomonas flagella," *J Cell Biol*, vol. 101, pp. 2081-4, Dec 1985.

[70] S. W. L'Hernault and J. L. Rosenbaum, "Reversal of the posttranslational modification on Chlamydomonas flagellar alpha-tubulin occurs during flagellar resorption," *J Cell Biol*, vol. 100, pp. 457-62, Feb 1985.

[71] S. W. L'Hernault and J. L. Rosenbaum, "Chlamydomonas alpha-tubulin is posttranslationally modified by acetylation on the epsilon-amino group of a lysine," *Biochemistry*, vol. 24, pp. 473-8, Jan 15 1985.

[72] C. Choudhary, C. Kumar, F. Gnad, M. L. Nielsen, M. Rehman, T. C. Walther, J. V. Olsen, and M. Mann, "Lysine acetylation targets protein complexes and co-regulates major cellular functions," *Science*, vol. 325, pp. 834-40, Aug 14 2009.

[73] J. S. Akella, D. Wloga, J. Kim, N. G. Starostina, S. Lyons-Abbott, N. S. Morrissette, S. T. Dougan, E. T. Kipreos, and J. Gaertig, "MEC-17 is an alpha-tubulin acetyltransferase," *Nature*, vol. 467, pp. 218-22, Sep 9 2010.

[74] T. Shida, J. G. Cueva, Z. Xu, M. B. Goodman, and M. V. Nachury, "The major alpha-tubulin K40 acetyltransferase alphaTAT1 promotes rapid ciliogenesis and efficient mechanosensation," *Proc Natl Acad Sci U S A*, vol. 107, pp. 21517-22, Dec 14 2010.

[75] K. Huang, D. R. Diener, and J. L. Rosenbaum, "The ubiquitin conjugation system is involved in the disassembly of cilia and flagella," *J Cell Biol*, vol. 186, pp. 601-13, Aug 24 2009.

[76] A. V. Loktev, Q. Zhang, J. S. Beck, C. C. Searby, T. E. Scheetz, J. F. Bazan, D. C. Slusarski, V. C. Sheffield, P. K. Jackson, and M. V. Nachury, "A BBSome subunit links ciliogenesis, microtubule stability, and acetylation," *Dev Cell*, vol. 15, pp. 854-65, Dec 2008.

[77] J. C. Bulinski, "Tubulin posttranslational modifications: a Pushmi-Pullyu at work?," *Dev Cell*, vol. 16, pp. 773-4, Jun 2009.

[78] K. Rogowski, F. Juge, J. van Dijk, D. Wloga, J. M. Strub, N. Levilliers, D. Thomas, M. H. Bre, A. Van Dorsselaer, J. Gaertig, and C. Janke, "Evolutionary divergence of enzymatic mechanisms for posttranslational polyglycylation," *Cell*, vol. 137, pp. 1076-87, Jun 12 2009.

[79] D. Wloga, D. M. Webster, K. Rogowski, M. H. Bre, N. Levilliers, M. Jerka-Dziadosz, C. Janke, S. T. Dougan, and J. Gaertig, "TTLL3 Is a tubulin glycine ligase that regulates the assembly of cilia," *Dev Cell*, vol. 16, pp. 867-76, Jun 2009.

[80] S. Suryavanshi, B. Edde, L. A. Fox, S. Guerrero, R. Hard, T. Hennessey, A. Kabi, D. Malison, D. Pennock, W. S. Sale, D. Wloga, and J. Gaertig, "Tubulin glutamylation

regulates ciliary motility by altering inner dynein arm activity," *Curr Biol*, vol. 20, pp. 435-40, Mar 9 2010.

[81] T. Kubo, H. A. Yanagisawa, T. Yagi, M. Hirono, and R. Kamiya, "Tubulin polyglutamylation regulates axonemal motility by modulating activities of inner-arm dyneins," *Curr Biol*, vol. 20, pp. 441-5, Mar 9 2010.

[82] F. Hildebrandt, T. Benzing, and N. Katsanis, "Ciliopathies," *N Engl J Med*, vol. 364, pp. 1533-43, Apr 21 2011.

[83] B. A. Afzelius, "A human syndrome caused by immotile cilia," *Science*, vol. 193, pp. 317-9, Jul 23 1976.

[84] M. A. Zariwala, M. R. Knowles, and M. W. Leigh, "Primary Ciliary Dyskinesia," 1993.

[85] M. Meeks, A. Walne, S. Spiden, H. Simpson, H. Mussaffi-Georgy, H. D. Hamam, E. L. Fehaid, M. Cheehab, M. Al-Dabbagh, S. Polak-Charcon, H. Blau, A. O'Rawe, H. M. Mitchison, R. M. Gardiner, and E. Chung, "A locus for primary ciliary dyskinesia maps to chromosome 19q," *J Med Genet*, vol. 37, pp. 241-4, Apr 2000.

[86] M. Meeks and A. Bush, "Primary ciliary dyskinesia (PCD)," *Pediatr Pulmonol*, vol. 29, pp. 307-16, Apr 2000.

[87] A. Bush and C. O'Callaghan, "Primary ciliary dyskinesia," *Arch Dis Child*, vol. 87, pp. 363-5; discussion 363-5, Nov 2002.

[88] A. Wanner, M. Salathe, and T. G. O'Riordan, "Mucociliary clearance in the airways," *Am J Respir Crit Care Med*, vol. 154, pp. 1868-902, Dec 1996.

[89] S. Nonaka, H. Shiratori, Y. Saijoh, and H. Hamada, "Determination of left-right patterning of the mouse embryo by artificial nodal flow," *Nature*, vol. 418, pp. 96-9, Jul 4 2002.

[90] D. M. Supp, D. P. Witte, S. S. Potter, and M. Brueckner, "Mutation of an axonemal dynein affects left-right asymmetry in inversus viscerum mice," *Nature*, vol. 389, pp. 963-6, Oct 30 1997.

[91] J. Chen, H. J. Knowles, J. L. Hebert, and B. P. Hackett, "Mutation of the mouse hepatocyte nuclear factor/forkhead homologue 4 gene results in an absence of cilia and random left-right asymmetry," *J Clin Invest*, vol. 102, pp. 1077-82, Sep 15 1998.

[92] S. Nonaka, Y. Tanaka, Y. Okada, S. Takeda, A. Harada, Y. Kanai, M. Kido, and N. Hirokawa, "Randomization of left-right asymmetry due to loss of nodal cilia generating leftward flow of extraembryonic fluid in mice lacking KIF3B motor protein," *Cell*, vol. 95, pp. 829-37, Dec 11 1998.

[93] B. A. Afzelius, B. Mossberg, and S. E. Bergstrom, "Immotile cilia syndrome (primary ciliary dyskinesia), including Kartagener syndrome. ," in *The metabolic and molecular bases of inherited disease.* , C. R. Scriver, A. L. Beaudet, W. S. Sly, and D. Valle, Eds. New York: McGraw-Hill Medical Publishing Division, 2001, pp. 4817-4827.

[94] P. J. Hadfield, J. M. Rowe-Jones, A. Bush, and I. S. Mackay, "Treatment of otitis media with effusion in children with primary ciliary dyskinesia," *Clin Otolaryngol Allied Sci*, vol. 22, pp. 302-6, Aug 1997.

[95] A. Gurr, T. Stark, M. Pearson, G. Borkowski, and S. Dazert, "The ciliary beat frequency of middle ear mucosa in children with chronic secretory otitis media," *Eur Arch Otorhinolaryngol*, vol. 266, pp. 1865-70, Dec 2009.

[96] L. E. Ostrowski, W. Yin, K. E. Thonmpson, M. Patel, and J. C. Olsen, "Pilot studies of gene therapy for primary ciliary dyskinesia," *Am J Respir Crit Care Med*, 2010.

[97] N. Hornef, H. Olbrich, J. Horvath, M. A. Zariwala, M. Fliegauf, N. T. Loges, J. Wildhaber, P. G. Noone, M. Kennedy, S. E. Antonarakis, J. L. Blouin, L. Bartoloni, T. Nusslein, P. Ahrens, M. Griese, H. Kuhl, R. Sudbrak, M. R. Knowles, R. Reinhardt, and H. Omran, "DNAH5 mutations are a common cause of primary ciliary dyskinesia with outer dynein arm defects," *Am J Respir Crit Care Med*, vol. 174, pp. 120-6, Jul 15 2006.

[98] G. Pennarun, E. Escudier, C. Chapelin, A. M. Bridoux, V. Cacheux, G. Roger, A. Clement, M. Goossens, S. Amselem, and B. Duriez, "Loss-of-function mutations in a human gene related to Chlamydomonas reinhardtii dynein IC78 result in primary ciliary dyskinesia," *Am J Hum Genet*, vol. 65, pp. 1508-19, Dec 1999.

[99] W. Stannard, A. Rutman, C. Wallis, and C. O'Callaghan, "Central microtubular agenesis causing primary ciliary dyskinesia," *Am J Respir Crit Care Med*, vol. 169, pp. 634-7, Mar 1 2004.

[100] G. B. Harris, J. C. R. Bermejo, and M. C. Suarez, "Different frequency of cilia with transposition in human nasal and bronchial mucosa. A case of acquired ciliary dyskinesia," *Virchows Archiv-an International Journal of Pathology*, vol. 437, pp. 325-330, Sep 2000.

[101] A. Barbato, T. Frischer, C. E. Kuehni, D. Snijders, I. Azevedo, G. Baktai, L. Bartoloni, E. Eber, A. Escribano, E. Haarman, B. Hesselmar, C. Hogg, M. Jorissen, J. Lucas, K. G. Nielsen, C. O'Callaghan, H. Omran, P. Pohunek, M. P. Strippoli, and A. Bush, "Primary ciliary dyskinesia: a consensus statement on diagnostic and treatment approaches in children," *Eur Respir J*, vol. 34, pp. 1264-76, Dec 2009.

[102] M. Fliegauf, H. Olbrich, J. Horvath, J. H. Wildhaber, M. A. Zariwala, M. Kennedy, M. R. Knowles, and H. Omran, "Mislocalization of DNAH5 and DNAH9 in respiratory cells from patients with primary ciliary dyskinesia," *Am J Respir Crit Care Med*, vol. 171, pp. 1343-9, Jun 15 2005.

[103] H. N. Morillas, M. Zariwala, and M. R. Knowles, "Genetic causes of bronchiectasis: primary ciliary dyskinesia," *Respiration*, vol. 74, pp. 252-63, 2007.

[104] S. D. Sagel, "Nasal nitric oxide: diagnostic value and physiological significance in primary ciliary dyskinesia," *J Pediatr*, vol. 159, pp. 363-5, Sep 2011.

[105] M. ten Berge, G. Brinkhorst, A. A. Kroon, and J. C. de Jongste, "DNase treatment in primary ciliary dyskinesia--assessment by nocturnal pulse oximetry," *Pediatr Pulmonol*, vol. 27, pp. 59-61, Jan 1999.

[106] M. Geremek, M. Bruinenberg, E. Zietkiewicz, A. Pogorzelski, M. Witt, and C. Wijmenga, "Gene expression studies in cells from primary ciliary dyskinesia patients identify 208 potential ciliary genes," *Hum Genet*, vol. 129, pp. 283-93, Mar 2011.

[107] B. Banizs, M. M. Pike, C. L. Millican, W. B. Ferguson, P. Komlosi, J. Sheetz, P. D. Bell, E. M. Schwiebert, and B. K. Yoder, "Dysfunctional cilia lead to altered ependyma and choroid plexus function, and result in the formation of hydrocephalus," *Development*, vol. 132, pp. 5329-39, Dec 2005.

[108] H. J. Garton and J. H. Piatt, Jr., "Hydrocephalus," *Pediatr Clin North Am*, vol. 51, pp. 305-25, Apr 2004.

[109] A. Aschoff, P. Kremer, B. Hashemi, and S. Kunze, "The scientific history of hydrocephalus and its treatment," *Neurosurg Rev*, vol. 22, pp. 67-93; discussion 94-5, Oct 1999.

[110] T. S. Lim, S. W. Yong, and S. Y. Moon, "Repetitive lumbar punctures as treatment for normal pressure hydrocephalus," *Eur Neurol*, vol. 62, pp. 293-7, 2009.
[111] I. Ibanez-Tallon, A. Pagenstecher, M. Fliegauf, H. Olbrich, A. Kispert, U. P. Ketelsen, A. North, N. Heintz, and H. Omran, "Dysfunction of axonemal dynein heavy chain Mdnah5 inhibits ependymal flow and reveals a novel mechanism for hydrocephalus formation," *Hum Mol Genet*, vol. 13, pp. 2133-41, Sep 15 2004.
[112] C. Wodarczyk, I. Rowe, M. Chiaravalli, M. Pema, F. Qian, and A. Boletta, "A novel mouse model reveals that polycystin-1 deficiency in ependyma and choroid plexus results in dysfunctional cilia and hydrocephalus," *PLoS One*, vol. 4, p. e7137, 2009.
[113] M. A. Greenstone, R. W. Jones, A. Dewar, B. G. Neville, and P. J. Cole, "Hydrocephalus and primary ciliary dyskinesia," *Arch Dis Child*, vol. 59, pp. 481-2, May 1984.
[114] J. Zhang, M. A. Williams, and D. Rigamonti, "Genetics of human hydrocephalus," *J Neurol*, vol. 253, pp. 1255-66, Oct 2006.
[115] S. Weller and J. Gartner, "Genetic and clinical aspects of X-linked hydrocephalus (L1 disease): Mutations in the L1CAM gene," *Hum Mutat*, vol. 18, pp. 1-12, 2001.
[116] D. S. Bickers and R. D. Adams, "Hereditary stenosis of the aqueduct of Sylvius as a cause of congenital hydrocephalus," *Brain*, vol. 72, pp. 246-62, Jun 1949.
[117] S. Kenwrick, M. Jouet, and D. Donnai, "X linked hydrocephalus and MASA syndrome," *J Med Genet*, vol. 33, pp. 59-65, Jan 1996.
[118] I. Ibanez-Tallon, S. Gorokhova, and N. Heintz, "Loss of function of axonemal dynein Mdnah5 causes primary ciliary dyskinesia and hydrocephalus," *Hum Mol Genet*, vol. 11, pp. 715-21, Mar 15 2002.
[119] K. F. Lechtreck, P. Delmotte, M. L. Robinson, M. J. Sanderson, and G. B. Witman, "Mutations in Hydin impair ciliary motility in mice," *J Cell Biol*, vol. 180, pp. 633-43, Feb 11 2008.
[120] K. F. Lechtreck and G. B. Witman, "Chlamydomonas reinhardtii hydin is a central pair protein required for flagellar motility," *J Cell Biol*, vol. 176, pp. 473-82, Feb 12 2007.
[121] A. Ohtoshi, "Hydrocephalus caused by conditional ablation of the Pten or beta-catenin gene," *Cerebrospinal Fluid Res*, vol. 5, p. 16, 2008.
[122] J. B. Wallingford and B. Mitchell, "Strange as it may seem: the many links between Wnt signaling, planar cell polarity, and cilia," *Genes Dev*, vol. 25, pp. 201-13, Feb 1 2011.
[123] C. Gavino and S. Richard, "Patched1 haploinsufficiency impairs ependymal cilia function of the quaking viable mice, leading to fatal hydrocephalus," *Mol Cell Neurosci*, vol. 47, pp. 100-7, Jun 2011.
[124] L. Jacob and L. Lum, "Deconstructing the hedgehog pathway in development and disease," *Science*, vol. 318, pp. 66-8, Oct 5 2007.
[125] V. Varadi, K. Csecsei, G. T. Szeifert, Z. Toth, and Z. Papp, "Prenatal diagnosis of X linked hydrocephalus without aqueductal stenosis," *J Med Genet*, vol. 24, pp. 207-9, Apr 1987.
[126] D. Greitz, T. Greitz, and T. Hindmarsh, "A new view on the CSF-circulation with the potential for pharmacological treatment of childhood hydrocephalus," *Acta Paediatr*, vol. 86, pp. 125-32, Feb 1997.
[127] L. Jia, Z. X. Zhao, C. You, J. G. Liu, S. Q. Huang, M. He, P. G. Ji, J. Duan, Y. J. Zeng, and G. P. Li, "Minimally-invasive treatment of communicating hydrocephalus using a percutaneous lumboperitoneal shunt," *J Zhejiang Univ Sci B*, vol. 12, pp. 293-7, Apr 2011.

[128] D. M. Sciubba, R. M. Stuart, M. J. McGirt, G. F. Woodworth, A. Samdani, B. Carson, and G. I. Jallo, "Effect of antibiotic-impregnated shunt catheters in decreasing the incidence of shunt infection in the treatment of hydrocephalus," *J Neurosurg*, vol. 103, pp. 131-6, Aug 2005.

[129] V. E. Torres, P. C. Harris, and Y. Pirson, "Autosomal dominant polycystic kidney disease," *Lancet*, vol. 369, pp. 1287-301, Apr 14 2007.

[130] C. R. Halvorson, M. S. Bremmer, and S. C. Jacobs, "Polycystic kidney disease: inheritance, pathophysiology, prognosis, and treatment," *Int J Nephrol Renovasc Dis*, vol. 3, pp. 69-83, 2010.

[131] R. J. Quinlan, J. L. Tobin, and P. L. Beales, "Modeling ciliopathies: Primary cilia in development and disease," *Curr Top Dev Biol*, vol. 84, pp. 249-310, 2008.

[132] J. Hoefele, K. Mayer, M. Scholz, and H. G. Klein, "Novel PKD1 and PKD2 mutations in autosomal dominant polycystic kidney disease (ADPKD)," *Nephrol Dial Transplant*, vol. 26, pp. 2181-8, Jul 2011.

[133] Y. C. Tan, J. D. Blumenfeld, R. Anghel, S. Donahue, R. Belenkaya, M. Balina, T. Parker, D. Levine, D. G. Leonard, and H. Rennert, "Novel method for genomic analysis of PKD1 and PKD2 mutations in autosomal dominant polycystic kidney disease," *Hum Mutat*, vol. 30, pp. 264-73, Feb 2009.

[134] B. K. Yoder, X. Hou, and L. M. Guay-Woodford, "The polycystic kidney disease proteins, polycystin-1, polycystin-2, polaris, and cystin, are co-localized in renal cilia," *J Am Soc Nephrol*, vol. 13, pp. 2508-16, Oct 2002.

[135] G. J. Pazour, J. T. San Agustin, J. A. Follit, J. L. Rosenbaum, and G. B. Witman, "Polycystin-2 localizes to kidney cilia and the ciliary level is elevated in orpk mice with polycystic kidney disease," *Curr Biol*, vol. 12, pp. R378-80, Jun 4 2002.

[136] A. C. Ong and D. N. Wheatley, "Polycystic kidney disease--the ciliary connection," *Lancet*, vol. 361, pp. 774-6, Mar 1 2003.

[137] S. Wang, J. Zhang, S. M. Nauli, X. Li, P. G. Starremans, Y. Luo, K. A. Roberts, and J. Zhou, "Fibrocystin/polyductin, found in the same protein complex with polycystin-2, regulates calcium responses in kidney epithelia," *Mol Cell Biol*, vol. 27, pp. 3241-52, Apr 2007.

[138] I. Kim, Y. Fu, K. Hui, G. Moeckel, W. Mai, C. Li, D. Liang, P. Zhao, J. Ma, X. Z. Chen, A. L. George, Jr., R. J. Coffey, Z. P. Feng, and G. Wu, "Fibrocystin/polyductin modulates renal tubular formation by regulating polycystin-2 expression and function," *J Am Soc Nephrol*, vol. 19, pp. 455-68, Mar 2008.

[139] G. J. Pazour, "Intraflagellar transport and cilia-dependent renal disease: the ciliary hypothesis of polycystic kidney disease," *J Am Soc Nephrol*, vol. 15, pp. 2528-36, Oct 2004.

[140] F. Lin, T. Hiesberger, K. Cordes, A. M. Sinclair, L. S. Goldstein, S. Somlo, and P. Igarashi, "Kidney-specific inactivation of the KIF3A subunit of kinesin-II inhibits renal ciliogenesis and produces polycystic kidney disease," *Proc Natl Acad Sci U S A*, vol. 100, pp. 5286-91, Apr 29 2003.

[141] N. Kishimoto, Y. Cao, A. Park, and Z. Sun, "Cystic kidney gene seahorse regulates cilia-mediated processes and Wnt pathways," *Dev Cell*, vol. 14, pp. 954-61, Jun 2008.

[142] Z. Sun, A. Amsterdam, G. J. Pazour, D. G. Cole, M. S. Miller, and N. Hopkins, "A genetic screen in zebrafish identifies cilia genes as a principal cause of cystic kidney," *Development*, vol. 131, pp. 4085-93, Aug 2004.

[143] W. Gronwald, M. S. Klein, R. Zeltner, B. D. Schulze, S. W. Reinhold, M. Deutschmann, A. K. Immervoll, C. A. Boger, B. Banas, K. U. Eckardt, and P. J. Oefner, "Detection of autosomal dominant polycystic kidney disease by NMR spectroscopic fingerprinting of urine," *Kidney Int*, vol. 79, pp. 1244-53, Jun 2011.

[144] M. A. Garcia-Gonzalez, J. G. Jones, S. K. Allen, C. M. Palatucci, S. D. Batish, W. K. Seltzer, Z. Lan, E. Allen, F. Qian, X. M. Lens, Y. Pei, G. G. Germino, and T. J. Watnick, "Evaluating the clinical utility of a molecular genetic test for polycystic kidney disease," *Mol Genet Metab*, vol. 92, pp. 160-7, Sep-Oct 2007.

[145] J. M. Shillingford, N. S. Murcia, C. H. Larson, S. H. Low, R. Hedgepeth, N. Brown, C. A. Flask, A. C. Novick, D. A. Goldfarb, A. Kramer-Zucker, G. Walz, K. B. Piontek, G. G. Germino, and T. Weimbs, "The mTOR pathway is regulated by polycystin-1, and its inhibition reverses renal cystogenesis in polycystic kidney disease," *Proc Natl Acad Sci U S A*, vol. 103, pp. 5466-71, Apr 4 2006.

[146] N. O. Bukanov, L. A. Smith, K. W. Klinger, S. R. Ledbetter, and O. Ibraghimov-Beskrovnaya, "Long-lasting arrest of murine polycystic kidney disease with CDK inhibitor roscovitine," *Nature*, vol. 444, pp. 949-52, Dec 14 2006.

[147] G. Jia, M. Kwon, H. L. Liang, J. Mortensen, V. Nilakantan, W. E. Sweeney, and F. Park, "Chronic treatment with lisinopril decreases proliferative and apoptotic pathways in autosomal recessive polycystic kidney disease," *Pediatr Nephrol*, vol. 25, pp. 1139-46, Jun 2010.

[148] X. Wang, V. Gattone, 2nd, P. C. Harris, and V. E. Torres, "Effectiveness of vasopressin V2 receptor antagonists OPC-31260 and OPC-41061 on polycystic kidney disease development in the PCK rat," *J Am Soc Nephrol*, vol. 16, pp. 846-51, Apr 2005.

[149] T. V. Masyuk, A. I. Masyuk, V. E. Torres, P. C. Harris, and N. F. Larusso, "Octreotide inhibits hepatic cystogenesis in a rodent model of polycystic liver disease by reducing cholangiocyte adenosine 3',5'-cyclic monophosphate," *Gastroenterology*, vol. 132, pp. 1104-16, Mar 2007.

[150] P. Ruggenenti, A. Remuzzi, P. Ondei, G. Fasolini, L. Antiga, B. Ene-Iordache, G. Remuzzi, and F. H. Epstein, "Safety and efficacy of long-acting somatostatin treatment in autosomal-dominant polycystic kidney disease," *Kidney Int*, vol. 68, pp. 206-16, Jul 2005.

[151] W. E. Sweeney, Y. Chen, K. Nakanishi, P. Frost, and E. D. Avner, "Treatment of polycystic kidney disease with a novel tyrosine kinase inhibitor," *Kidney Int*, vol. 57, pp. 33-40, Jan 2000.

[152] Y. Solak, H. Atalay, I. Polat, and Z. Biyik, "Colchicine treatment in autosomal dominant polycystic kidney disease: many points in common," *Med Hypotheses*, vol. 74, pp. 314-7, Feb 2010.

[153] J. L. Tobin and P. L. Beales, "Bardet-Biedl syndrome: beyond the cilium," *Pediatr Nephrol*, vol. 22, pp. 926-36, Jul 2007.

[154] K. Baker and P. L. Beales, "Making sense of cilia in disease: the human ciliopathies," *Am J Med Genet C Semin Med Genet*, vol. 151C, pp. 281-95, Nov 15 2009.

[155] P. L. Beales, N. Elcioglu, A. S. Woolf, D. Parker, and F. A. Flinter, "New criteria for improved diagnosis of Bardet-Biedl syndrome: results of a population survey," *J Med Genet*, vol. 36, pp. 437-46, Jun 1999.

[156] P. Sen Gupta, N. V. Prodromou, and J. P. Chapple, "Can faulty antennae increase adiposity? The link between cilia proteins and obesity," *J Endocrinol*, vol. 203, pp. 327-36, Dec 2009.

[157] K. Rahmouni, M. A. Fath, S. Seo, D. R. Thedens, C. J. Berry, R. Weiss, D. Y. Nishimura, and V. C. Sheffield, "Leptin resistance contributes to obesity and hypertension in mouse models of Bardet-Biedl syndrome," *J Clin Invest*, vol. 118, pp. 1458-67, Apr 2008.

[158] C. Christodoulides, C. Lagathu, J. K. Sethi, and A. Vidal-Puig, "Adipogenesis and WNT signalling," *Trends Endocrinol Metab*, vol. 20, pp. 16-24, Jan 2009.

[159] P. J. King, L. Guasti, and E. Laufer, "Hedgehog signalling in endocrine development and disease," *J Endocrinol*, vol. 198, pp. 439-50, Sep 2008.

[160] W. Cousin, C. Fontaine, C. Dani, and P. Peraldi, "Hedgehog and adipogenesis: fat and fiction," *Biochimie*, vol. 89, pp. 1447-53, Dec 2007.

[161] N. F. Berbari, J. S. Lewis, G. A. Bishop, C. C. Askwith, and K. Mykytyn, "Bardet-Biedl syndrome proteins are required for the localization of G protein-coupled receptors to primary cilia," *Proc Natl Acad Sci U S A*, vol. 105, pp. 4242-6, Mar 18 2008.

[162] N. F. Berbari, A. D. Johnson, J. S. Lewis, C. C. Askwith, and K. Mykytyn, "Identification of ciliary localization sequences within the third intracellular loop of G protein-coupled receptors," *Mol Biol Cell*, vol. 19, pp. 1540-7, Apr 2008.

[163] J. R. Lupski, N. Katsanis, S. J. Ansley, J. L. Badano, E. R. Eichers, R. A. Lewis, B. E. Hoskins, P. J. Scambler, W. S. Davidson, and P. L. Beales, "Triallelic inheritance in Bardet-Biedl syndrome, a Mendelian recessive disorder," *Science*, vol. 293, pp. 2256-2259, Sep 21 2001.

[164] J. R. Davenport, A. J. Watts, V. C. Roper, M. J. Croyle, T. van Groen, J. M. Wyss, T. R. Nagy, R. A. Kesterson, and B. K. Yoder, "Disruption of intraflagellar transport in adult mice leads to obesity and slow-onset cystic kidney disease," *Curr Biol*, vol. 17, pp. 1586-94, Sep 18 2007.

[165] D. L. Simons, S. L. Boye, W. W. Hauswirth, and S. M. Wu, "Gene therapy prevents photoreceptor death and preserves retinal function in a Bardet-Biedl syndrome mouse model," *Proc Natl Acad Sci U S A*, vol. 108, pp. 6276-81, Apr 12 2011.

[166] A. L. Negri, F. R. Spivacow, E. E. Del Valle, M. Forrester, G. Rosende, and I. Pinduli, "Role of overweight and obesity on the urinary excretion of promoters and inhibitors of stone formation in stone formers," *Urol Res*, vol. 36, pp. 303-7, Dec 2008.

[167] L. Garfinkel, "Overweight and mortality," *Cancer*, vol. 58, pp. 1826-9, Oct 15 1986.

[168] N. V. Christou, J. S. Sampalis, M. Liberman, D. Look, S. Auger, A. P. McLean, and L. D. MacLean, "Surgery decreases long-term mortality, morbidity, and health care use in morbidly obese patients," *Ann Surg*, vol. 240, pp. 416-23; discussion 423-4, Sep 2004.

[169] E. Westly, "Fat attack," *Sci Am*, vol. 303, pp. 20-2, Sep 2010.

[170] L. M. Quarmby and J. D. Parker, "Cilia and the cell cycle?," *J Cell Biol*, vol. 169, pp. 707-10, Jun 6 2005.

[171] M. Fliegauf, T. Benzing, and H. Omran, "When cilia go bad: cilia defects and ciliopathies," *Nat Rev Mol Cell Biol*, vol. 8, pp. 880-93, Nov 2007.

[172] E. J. Michaud and B. K. Yoder, "The primary cilium in cell signaling and cancer," *Cancer Res*, vol. 66, pp. 6463-7, Jul 1 2006.

[173] S. Y. Wong, A. D. Seol, P. L. So, A. N. Ermilov, C. K. Bichakjian, E. H. Epstein, Jr., A. A. Dlugosz, and J. F. Reiter, "Primary cilia can both mediate and suppress Hedgehog pathway-dependent tumorigenesis," *Nat Med,* vol. 15, pp. 1055-61, Sep 2009.

[174] Y. G. Han and A. Alvarez-Buylla, "Role of primary cilia in brain development and cancer," *Curr Opin Neurobiol,* vol. 20, pp. 58-67, Feb 2010.

[175] R. Rohatgi, L. Milenkovic, and M. P. Scott, "Patched1 regulates hedgehog signaling at the primary cilium," *Science,* vol. 317, pp. 372-6, Jul 20 2007.

[176] Y. G. Han, H. J. Kim, A. A. Dlugosz, D. W. Ellison, R. J. Gilbertson, and A. Alvarez-Buylla, "Dual and opposing roles of primary cilia in medulloblastoma development," *Nat Med,* vol. 15, pp. 1062-5, Sep 2009.

[177] K. Yuan, N. Frolova, Y. Xie, D. Wang, L. Cook, Y. J. Kwon, A. D. Steg, R. Serra, and A. R. Frost, "Primary cilia are decreased in breast cancer: analysis of a collection of human breast cancer cell lines and tissues," *J Histochem Cytochem,* vol. 58, pp. 857-70, Oct 2010.

[178] A. J. Bowers and J. F. Boylan, "Nek8, a NIMA family kinase member, is overexpressed in primary human breast tumors," *Gene,* vol. 328, pp. 135-42, Mar 17 2004.

[179] M. R. Mahjoub, M. L. Trapp, and L. M. Quarmby, "NIMA-related kinases defective in murine models of polycystic kidney diseases localize to primary cilia and centrosomes," *J Am Soc Nephrol,* vol. 16, pp. 3485-9, Dec 2005.

[180] A. Jemal, R. Siegel, E. Ward, Y. Hao, J. Xu, T. Murray, and M. J. Thun, "Cancer statistics, 2008," *CA Cancer J Clin,* vol. 58, pp. 71-96, Mar-Apr 2008.

[181] C. Almoguera, D. Shibata, K. Forrester, J. Martin, N. Arnheim, and M. Perucho, "Most human carcinomas of the exocrine pancreas contain mutant c-K-ras genes," *Cell,* vol. 53, pp. 549-54, May 20 1988.

[182] E. S. Seeley, C. Carriere, T. Goetze, D. S. Longnecker, and M. Korc, "Pancreatic cancer and precursor pancreatic intraepithelial neoplasia lesions are devoid of primary cilia," *Cancer Res,* vol. 69, pp. 422-30, Jan 15 2009.

[183] K. Pickert, "Screening dilemma. Are some cancers better left undiscovered?," *Time,* vol. 177, pp. 60-4, 67, Jun 13 2011.

10

Adult Stem Cells in Tissue Homeostasis and Disease

Elena Lazzeri, Anna Peired, Lara Ballerini and Laura Lasagni
University of Florence,
Italy

1. Introduction

Stem cells (SCs) are a rare population of cells characterized by the ability to self-renew in order to preserve the SC pool and to differentiate in different lineage to produce progeny needed for the physiological functions of tissues and organs. SC can be classified as embryonic SC (ESC) and adult or somatic SC (ASC): ESC have been isolated from the inner cell mass of the blastocyst and are pluripotent cells, that is cells able to differentiate into all the cell types required to form an entire organism (Smith, 2001); ASC are tissue-resident SC that, based on their differentiation potency, can be classified as multipotent, oligopotent or even unipotent. It is still controversial whether every mammalian tissue and organ possesses an ASC, but many tissue-specific ASC have been successfully identified and isolated e.g., hematopoietic SCs (HSCs), mammary SCs, muscle SCs (satellite cells), intestinal SCs, and mesenchymal SCs. All these tissues need to constantly replace damaged or dead cells throughout the life of the animal. This process of continual cell replacement critical for the maintenance of adult tissues, is called tissue homeostasis, and is maintained through the presence of ASC (Fig. 1). The homeostatic replacement of cells varies substantially among different tissues. The epithelium of the intestine is one of the most rapidly self-renewing tissue in adult mammals and it completely self-renews in around 5 days (van der Flier & Clevers, 2009). By contrast, interfollicular epidermis takes 4 weeks to renew (Blanpain & Fuchs, 2009), whereas the lung epithelium can take as long as 6 months to be replaced (Rawlins & Hogan, 2006). Moreover, apart from the maintenance of tissue homeostasis, ASC are devoted to the regeneration and repair of highly specialized tissues. Regeneration refers to the proliferation of cells to replace lost structures, such as the growth of an amputated limb in amphibians. In mammals, whole organs and complex tissues rarely regenerate after injury, but tissues with high proliferative capacity, such as the hematopoietic system and the epithelia of the skin and gastrointestinal tract, renew themselves continuously and can regenerate after injury, as long as the SC of these tissues are not destroyed (Fig. 1). Repair most often consist of a combination of regeneration and scar formation by the deposition of collagen which relative contribution depends on the ability of the tissue to regenerate and the extent of the injury. For instance, in superficial injury of the skin, wound can heal through the regeneration of the surface epithelium. However, scar formation is the predominant healing process that occurs when the extracellular matrix framework is damaged by severe injury (Fig. 1). This last mechanism results in restoration of tissue continuity but with or without function (Gurtner et al., 2008).

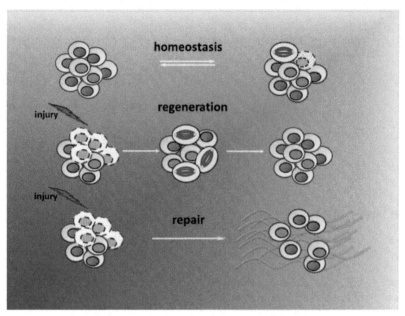

Fig. 1. Normal homeostasis and healing responses. In normal homeostasis a balance between proliferation and cell death maintains the tissue structure and function. Healing after acute injury can occur by regeneration, that restores normal tissue structure, or repair with deposition of collagen fibers and scar formation.

2. SCs and their niches

Self-renewal and differentiation of ASC are supported by two types of cell division known as symmetric and asymmetric (Morrison & Kimble, 2006). With symmetric division both the daughter cells acquire similar fates, while the asymmetric division, a fundamental and nearly universal mechanism for the generation of cellular diversity and pattern, gives rise to daughter cells with dissimilar fates. Divergent fates in daughter cells may be recognized by various characteristics: (i) morphological, such as cell size and shape; (ii) molecular, such as the segregation of proteins into only one daughter cell; or (iii) behavioural, such as the subsequent descendant types produced by either of the daughter cells. One mechanism for fate determination of daughter cells following symmetric and asymmetric cell divisions is the partitioning of fate-determining molecules during mitosis of the mother cell (Tajbakhsh et al., 2009). The idea that specific molecules can be partitioned unequally to daughter cells and behave as fate determinants had been hypothesized over a century earlier, following observations of cell divisions in simple organisms. When an intrinsic mechanism is used, cells establish an axis of polarity, orient the mitotic spindle along this axis and localize cell fate determinants to one side of the cell. During cytokinesis, determinants are then segregated into one of the two daughter cells where they direct cell fate (Betschinger & Knoblich, 2004). However, this hypothesis was only experimentally validated a little under two decades ago, with the identification of the first asymmetrically segregated cell fate determinant – Numb (Rhyu et al., 1994).

Alternatively, the SC depends on the contact with the surrounding microenvironment (the SC "niche") for maintaining the potential to self-renew (Li & Xie, 2005). By orienting its mitotic spindle perpendicularly to the niche surface, the SC will placed the two daughters in distinct cellular environments either inside or outside the SC niche, leading to asymmetric fate choice. However, when SC divides parallel to the niche it may also generate two identical SC in order to increase SC number or to compensate for occasional SC loss (Yamashita et al, 2010). The concept of the "niche" was proposed first by Schofield (Schofield, 1978) who hypothesized that proliferative, hematopoietic cells derived from the spleen displayed decreased proliferative potential when compared to HSC obtained from the bone marrow because they were no longer in association with a complement of cells, the "niche", which supports long term SC activity. This concept subsequently has proven relevant to many different SC systems, and the definition of the niche has been expanded further to include functional regulation of SC by both cellular and acellular (extracellular matrix) component of the niche. Thus the niche comprise all the microenvironment surrounding SCs, which provides diverse external cues to instruct SCs activities, preserve their proliferative potential and block maturation (Jones & Wagers, 2008).

3. Signaling pathways regulating SC function

Despite morphological and functional differences among different ASC, common signaling pathways appear to control SC self-renewal, activation, and differentiation, including Notch and Wingless-type (Wnt).

3.1 Notch signaling pathway

The Notch signaling pathway was discovered in flies more than 90 years ago (Morgan, 1917), and it is among the most well-conserved signaling pathways in animals. It arose with the evolution of multicellular organisms and the concomitant need for juxtacrine cell-to-cell communication to coordinate development. In mammals, four Notch transmembrane receptors (Notch1-4) have been described. Notch ligands are also transmembrane proteins comprising two different subtypes (Delta, Jagged), each containing several members (Jagged1-2, Delta-like1, 3, and 4) (Kopan & Ilagan, 2009). In Notch signaling, a 'signal-sending cell' presents the Notch ligand to the 'signal-receiving cell', which expresses the Notch receptor. Triggering of Notch receptor by ligand binding promotes two proteolytic cleavage events at the Notch receptor (Fig. 2) (Kopan & Ilagan, 2009). The first cleavage is catalyzed by the ADAM-family of metalloproteases, whereas the second cleavage is mediated by γ-secretase, an enzyme complex that contains presenilin, nicastrin, PEN2 and APH1. The second cleavage releases the Notch intracellular domain (NICD), which is free to translocate to the nucleus where it engages CSL, converting it from a transcriptional repressor to an activator and activates transcription of genes containing CSL binding sites (Kopan & Ilagan, 2009). In the absence of a Notch signal, CSL represses transcription of Notch target genes by interacting with the basal transcription machinery and recruiting ubiquitous corepressor proteins to form multiprotein transcriptional repressor complexes (Lai, 2002). In the presence of a Notch signal, NICD binding to CSL displaces corepressors from CSL. The best characterized Notch target genes belong to the hairy enhancer of split (Hes) complex and consist of the b-HLH transcription factors Hes (1-7) and Hey (1-3) (Bray & Bernard, 2010).

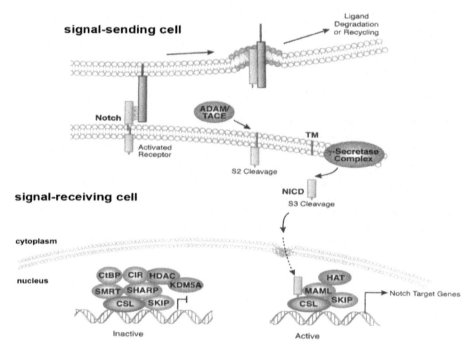

Fig. 2. Model of Notch signaling pathway. See the text for detail.

3.2 Wnt signaling pathway

The Wnt signaling pathway is a highly conserved developmental pathway, and orchestrates development and morphogenesis in many different tissues. Wnt proteins are secreted proteins, that bind to receptors of the Frizzled family (FZD) (Wodarz & Nusse, 1998), of which 10 members were found, and several coreceptors such as lipoprotein receptor-related protein (LRP)-5/6, (Pinson et al., 2000) Ryk, or Ror2 (Logan & Nusse, 2004). Wnt signals can be transduced to the canonical, or Wnt/β-catenin, pathway and to the noncanonical, or β-catenin independent, pathway.

3.2.1 Canonical Wnt signaling pathway

The canonical Wnt pathway involves the multifunctional protein β-catenin (MacDonald et al., 2009). In the absence of Wnt, β-catenin is targeted to a multimeric destruction complex with adenomatous polyposis coli (APC) and Axin and is phosphorylated by casein kinase 1α, followed by phosphorylation by glycogen synthase kinase (GSK)3β (Fig.34) (Ikeda et al., 1998). This phosphorylation targets β-catenin for ubiquitination and degradation by the proteasome. The binding of Wnt ligands to the FZD receptors results in the disassembly of the destruction complex and the stabilization of β-catenin. This process also involves the protein dishevelled (DVL). Cytoplasmic β-catenin accumulates and is eventually imported into the nucleus, where it serves as a transcriptional coactivator of transcription factors of the TCF/LEF family (Arce et al., 2006). TCF/LEF target genes are then involved in regulating cell proliferation, SC maintenance, or differentiation.

3.2.2 Noncanonical Wnt signaling pathway

Different noncanonical Wnt signals are transduced through FZD receptors and coreceptors. Depending on the major intracellular mediators used, those are called the Wnt/JNK (Veeman et al., 2003) or Wnt/calcium pathway (Fig. 3). The core element of the Wnt/JNK pathway (or planar cell polarity –PCP- pathway) includes the activation of small GTPases of the rho family, such as rac, cdc42, and rhoA. The GTPases can activate more downstream mediators like JNK or rho kinase (ROK). In this branch, Dvl is also recruited by a FZD receptor and promotes the asymmetrical localization of the PCP core proteins within the cell (Montcouquiol, et al. 2006). The asymmetrical subcellular localization of these elements in an epithelial sheet directs cytoskeletal reorganization. The same mechanism is used in mesenchymal cells to direct cell movement and migration during gastrulation (convergent and extension movements) (Roszko, et al., 2009).

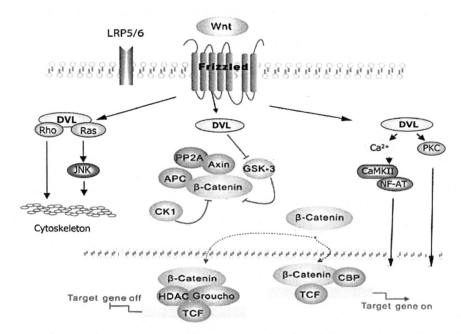

Fig. 3. Model of canonical and noncanonical Wnt signaling pathway. See the text for detail.

The existence of the Wnt/calcium pathway was hypothesized because injection of RNA coding for certain Wnts or FZD into early zebrafish embryos triggered intracellular calcium release (Slusarski et al., 1997) and loss of Wnt-11 or Wnt-5A function resulted in reduced intracellular calcium signaling (Eisenberg & Eisenberg, 1999; Westfall et al., 2003). This finding was subsequently expanded by the observation that the Wnt-induced release of intracellular calcium is sufficient to activate different intracellular calcium-sensitive enzymes such as protein kinase C, PKC (Sheldahl et al., 1999), calcium–calmodulin-dependent kinase II, CamKII (Kuhl et al., 2000) and the calcium-sensitive phosphatase calcineurin (Saneyoshi et al., 2002). Through calcineurin the Wnt/calcium pathway connects to NFAT (nuclear factor of activated T cells) transcription factor and gene expression.

Presently, a series of recent findings clearly indicate that different Wnt signaling pathways are simultaneously active within the same cell type, supporting the idea that Wnt pathways are highly connected to form a Wnt signaling network. This network seems to be activated by either one or more ligands acting on a certain cell type (Kestler & Kuhl, 2008).

3.3 Wnt signaling inhibitors

Secreted frizzled-related proteins (SFRP1, 2, 3, 4, 5), WIF1, DKK1, -2, -3, and -4 are secreted-type Wnt signaling inhibitors. WIFs and SFRPs can directly bind to Wnt proteins in the extracellular space, thereby affecting receptor occupancy and, ultimately, the cellular response (Bovolenta et al., 2008). DKK1 is among the best-characterized inhibitors of the canonical Wnt pathway. DKK1 itself is a target gene of Wnt/β-catenin signaling, thereby establishing a negative-feedback loop (Niida et al., 2004). There are two possible mechanisms by which DKK1 inhibits β-catenin signaling. One possible mechanism is that DKK1 prevents the formation of Wnt–FZD–LRP6 complexes on the cell surface by binding to LRP6 (Seto et al., 2006). Another possibility, which is related to the internalization of LRP6, is that DKK1 binds to another class of receptor, Kremen (Krm). In this model, the binding of DKK1 to LRP6 and Krm results in the formation of a ternary structure and induces rapid endocytosis and the removal of LRP6 from the plasma membrane, and thereby attenuates β-catenin signaling (Mao et al., 2002).

4. Hematopoietic SCs

In adult mammals, HSCs form a rare population of multipotent SCs that reside primarily in the bone marrow (BM). They have the capability to both self-renew and constantly give rise to lineage-specific progenitor cells and effector blood cells that perform the physiological functions of the hematopoietic system. Blood cells can be classified into various cell types, from the myeloid (monocytes and macrophages, neutrophils, basophils, eosinophils, erythrocytes, megakaryocytes/platelets, dendritic cells), and lymphoid lineages (T-cells, B-cells, NK-cells) (Liu et al., 2010).

HSCs are functionally defined by their capacity to reconstitute the hematopoietic system of immunodeficient animals such as NOD/SCID mice or contribute to functional reconstitution in human transplant settings. HSCs can be identified and isolated by a combination of presence and absence of cell surface markers. The most commonly used combination is characterized by the positive expression of the tyrosine kinase receptor c-Kit (CD117) and the membrane glycoprotein Sca-1 (Okada et al., 1992), together with the lack of markers of terminal differentiation (Ter119, Gr-1, Mac-1, B220, CD4 and CD8), collectively known as Lineage markers. The resulting c-Kit+ Sca-1+ Lin- population, is commonly referred to as KSL cells. More recently, an alternative method was described, using a signature of SLAM (Signaling lymphocyte activation molecule) family of cell surface molecules, CD150+ CD244- CD48- (Kiel et al, 2005). This is the first family of receptors whose combinatorial expression precisely distinguishes HSCs from hematopoietic progenitor cells (HPC).

The BM microenvironment –also called niche- plays an important role in the regulation of self-renewal and differentiation of HSCs. It is composed of different types of cells and structures surrounding the bone, which regulates the fate of hematopoietic cells through direct or indirect means, facilitating a stable generation of all the blood cells needed in a steady state situation. But the niche also adapts in times of hematopoietic stress. A failure to maintain a strict regulation of the hematopoietic cells can lead to a variety of malignancies such as leukemia, the most common form of cancer in humans (Renstrom et al., 2010).

4.1 Notch pathway as a regulator of HSC behavior

All Notch receptors and ligands are expressed on HSCs (Singh et al., 2000) and it is now well established that Notch signaling is essential for the production of HSCs during embryogenesis. However, its role in subsequent stages of mammalian HSC development is still controversial (Liu et al, 2010; Radtke et al., 2010).

In adult hematopoiesis, activation of Notch signaling has been reported to promote HSCs self-renewal, proliferation and differentiation *in vitro* and *in vivo*, and in both mice and humans. Constitutive expression of NICD by HSCs, leading to the constitutive activation of the Notch pathway, enhances proliferation and consequently delays hematopoiesis. Conversely, it inhibits differentiation in response to various cytokines, mostly under myeloid promoting conditions (Carlesso et al, 1999). Several reports show that HSCs stimulated with soluble or membrane-bound Notch ligand Delta 1 (Karanu et al, 2001) or Jagged1 (Karanu et al. 2000) increase in expansion potential *in vitro* and in reconstitution capacity *in vivo*. Although these gain-of-function studies show an important role for Notch in expanding the HSC pool, they do not prove that Notch is essential for post-natal hematopoiesis. The controversy arises from several loss-of-function studies in mice that did not fully support the previous conclusions. In particular, inactivation of Notch receptors (Notch1, Notch2), ligands (Jagged1) or downstream effectors (CSL/RBPJ, Mastermind-like1) does not impair HSC function (Cerdan & Bhatia, 2010). Additional studies failed to identify a protective role for Notch when HSCs were exposed to oxidative stress. Taken together, these results show that Notch signaling is not a major regulator of adult HSC maintenance *in vivo*. Downstream of HSCs, Notch signaling plays a critical role in cell fate decision of a variety of oligopotent progenitor cells in the hematopoietic system, such as in T-cell development. Inactivation of Notch signaling in HPCs results in early blockade of T-cell lymphopoiesis, due to a failure in commitment to the T-cell lineage. Transgenic mice with a conditional deletion of Notch1 do not develop T-cells but develop ectopic B-cells in the thymus, while immunodeficient mice expressing a constitutively active form of Notch1 develop ectopic T-cells in the bone marrow (BM) but no B-cell (Tanigaki & Honjo, 2007). Additionally, Notch1 signaling is necessary at various stages of T-cell development, such as progression through thymocyte maturation, regulation of T-cell Receptor β (TCR-β) gene rearrangement, regulation of lineage decisions between αβ and γδ lineages (Tanigaki & Honjo, 2007).

4.2 Role of Notch in T-cell leukemia

The pathological role for a deregulated Notch signaling was first described in a rare human T-cell acute lymphoblastic leukaemia/lymphoma (T-ALL), in which a t(7;9) chromosomal translocation results in the generation of a constitutively active, but truncated form of the Notch1 receptor named TAN1 (Translocation Associated Notch homolog) (Ellisen et al., 1991). Evidence that constitutively active Notch1 is responsible for disease development was provided by murine BM reconstitution experiments. Irradiated mice transplanted with BM progenitors expressing activated forms of Notch1 developed clonal hematopoietic tumors characterized as T-ALL. Experiments performed using other truncated Notch isoforms, including Notch2 and Notch3, showed similar results. However, mice having a defect in T-cell development failed to produce tumors. These results reveal that Notch1 has a special oncotropism for T-cell progenitors (Radtke et al., 2010). These findings became extremely relevant when a study of a large number of T-ALL patients revealed in more than 50% of them the presence of at least one gain-of-function mutation in the Notch1 receptor, emphasizing the oncogenic role of Notch (Weng et al., 2004). Notch1 mutations found in T-ALL affect critical domains responsible for preventing the spontaneous activation of the receptor in the absence of ligand or for terminating Notch1 signaling in the nucleus.

Studies of the genes and pathways controlled by Notch in T-ALL identified Notch1 as a central regulator, promoting leukemia cell growth by multiple direct and indirect mechanisms (Fig. 4) (Paganin & Ferrando, 2011). Analysis of Notch1 expression in T-ALL showed that it acts as a direct transcriptional activator of multiple genes. Notch1 also promotes the expression of the MYC oncogene, which in turn further enhances its direct effect on anabolic genes and facilitates cell growth. Indeed, many of the anabolic genes directly controlled by Notch1 are also direct targets of MYC, creating a feed-forward-loop transcriptional network that promotes leukemic cell growth (Palomero et al., 2006). Additionally, Notch1 facilitates the activation of the PI3K-AKT-mTOR signaling pathway, a critical regulator of cell growth and metabolism, via transcriptional downregulation of the PTEN tumor suppressor gene by Hes1, a transcriptional repressor directly downstream of Notch1 signaling (Palomero et al., 2007). The mTOR signaling was suppressed in T-ALL cells upon inhibition of Notch signaling, illustrating the importance of this indirect mechanism of regulation. The transcriptional program activated by oncogenic Notch1 also has a direct effect on cell cycle progression, promoting of G1/S cell cycle progression in T-ALL. This effect is mediated in part by transcriptional upregulation of CCND3, CDK4 and CDK6. Moreover, Notch1 induces the transcription of the S phase kinase-associated protein 2 (SKP2), which mediate the proteasomal degradation of CDKN1B (p27/Kip1) and CDKN1A (p21/Cip1), promoting premature entry of the cells into S phase (Sarmento et al, 2005). Notch1 can also modulate the survival of T-ALL cells by interacting with NF-κB, upregulating its activity by increasing expression of IκB kinase and upregulating both the expression and the nuclear localization of NF-κB. Inhibition of NF-κB in T-ALL can efficiently restrict tumor growth both *in vitro* and *in vivo* (Vilimas et al., 2007).

In addition, Notch1 modulates the NFAT cascade through the activation of calcineurin, which is a calcium-activated phosphatase that is important for the activation and translocation of NFAT factors to the nucleus. Calcineurin inhibition resulted in T-ALL cell death, as well as tumor regression and prolonged survival of leukemic mice (Medyouf et al., 2007). Finally, Notch1 regulates the activity of p53, lowering its expression through repression of the ARF-mdm2-p53 surveillance network. Attenuation of Notch signaling led to increase p53 expression and to tumor regression by inducing apoptosis (Beverly et al., 2005). A strong body of evidence supports a central role of Notch1 in promoting cell metabolism, growth and proliferation, as well as in enhancing the activity of signaling pathways that reinforce these functions and also promote cell survival. These results suggest that blocking Notch1 signaling may reduce the self-renewal capacity of T-ALL cells and/or selectively affect the leukemia initiating cell population.

Only few Notch mutations have been reported in myelogenous leukemias, but it is unclear whether Notch aberrant expression is responsible for the disease.

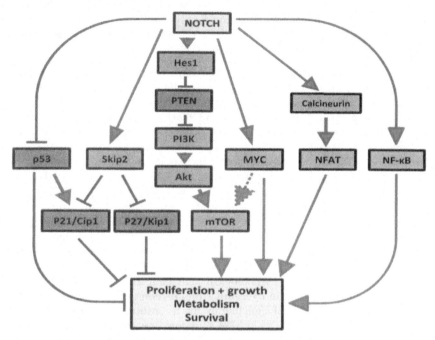

Fig. 4. Genes and pathways controlled by Notch in T-ALL

4.3 Wnt pathway and HSC

In hematopoiesis, Wnt pathway activity is required in the BM niche to regulate HSC proliferation and preserve self-renewal capacity (Malhotra & Kincade, 2009). Even though the role of canonical signaling on the regulation of adult hematopoiesis has been studied in great detail, controversy remains, possibly explained by differences in strength and duration of Wnt signaling or redundancy with other pathways. A role for Wnt signaling in hematopoiesis is supported by observations that Wnt ligands enhance proliferation of HSCs ex vivo (Van Den Berg et al, 1998) and that Wnt antagonists inhibit HSC proliferation and reconstitution. In particular, only short-term repopulation was reported using HSCs from normal mice cultured with Wnt3A (Reya et al., 2003; Willert et al., 2003). Subsequent studies reported that noncanonical Wnt5a inhibited canonical Wnt3a-mediated signaling to promote the maintenance of quiescent, functionally transplantable HSCs. In addition constitutively active nuclear β-catenin signaling reduces HSC quiescence and blocks HSC differentiation (Kirstetter et al., 2006). On the other hand, osteoblast-specific expression of Dkk1 results in increased HSC cycling and reduced regenerative capacity (Fleming et al, 2008). These findings suggest that Wnt pathway activation in the niche limits HSC proliferation and preserves self-renewal. These observations suggest that fine-tuning of Wnt/β-catenin activity in the microenvironment is crucial for maintaining SC quiescence.

The canonical Wnt pathway has also been shown to be necessary for appropriate HSC development (Zhao et al., 2008). In this model, Ctnnb1-/- bone marrow cells are deficient in long-term HSC maintenance and compete poorly against wild-type cells. However, experiments in adult HSC revealed that Ctnnb1 is dispensable for HSC maintenance in fully developed HSC (Koch et al., 2008). This indicates differential requirements for self-renewal pathways in development versus maintenance of HSC.

In the context of development, genetic studies have demonstrated the requirement for canonical signaling in the formation of mesoderm (Kelly et al., 2004; Liu et al., 1999). Recent advances have provided insights into the uniqueness of the biological functions of canonical and noncanonical pathways. It has been found that non-canonical and canonical Wnts affected different target populations and stages of hematopoietic development (Vijayaragavan et al., 2009). Consistent with its previously defined role in human adult cells (Van Den Berg et al., 1998), canonical signaling increased proliferation of blood committed progenitors when administered during the proper window of time during EB development. However, a short pulse of non-canonical signaling was necessary and sufficient to control exit of hESCs from the pluripotent state and subsequent entry into the mesendoderm/mesoderm lineages (Vijayaragavan et al., 2009). Taken together, these findings provide the first evidence of a unique role for non-canonical signaling in early specification of hematopoiesis from hESCs, whereas canonical signaling affects the proliferation of cells already fated to blood. These studies provide a valuable model system for examining the possibility of chronological activation and interaction between non-canonical and canonical signaling in the cellular progression from mesoderm to blood. The controversial function of canonical signaling on the reconstituting capacity of adult HSCs, combined with these present findings in hESCs, underscores the importance of fine tuning the strength and duration of Wnt signaling towards therapeutically exploiting the balance between self-renewal and lineage commitment of HSCs.

However, there are conflicting reports on the requirement for Wnt/β-catenin signaling in basal hematopoiesis: conditional disruption of β-catenin in adult HSCs does not affect their ability to self-renew and reconstitute hematopoietic lineages (Huang et al, 2009). In addition, although overexpression of stabilized β-catenin increases immunophenotypic HSCs, this is associated with a loss of repopulating activity and hematopoietic failure *in vivo* (Kirstetter et al., 2006), findings that appear incompatible with a positive role for β-catenin in hematopoiesis. A general conclusion from these apparently conflicting reports is that the role of Wnt signaling in hematopoiesis is complex and context dependent (Staal & Sen,. 2008). However, although the β-catenin loss-of-function studies suggest that canonical Wnt signaling is not essential for basal hematopoiesis in adults, they do not rule out a possible role for the Wnt/β-catenin pathway under nonbasal conditions and are still compatible with gain-of-function experiments in which the pathway is activated.

4.4 Wnt signaling and malignant HSC

Stem cell quiescence is closely associated with protection from myelotoxic insults (Cheshier et al, 1999). Similar to the role of tissue SCs in normal tissues, several cancers are also propagated by small populations of quiescent cancer stem cells (CSCs) that are resistant to both conventional chemotherapy and targeted therapies, and are retained and contribute to relapse following discontinuation of therapy (Dick, 2008).

When Ctnnb1 was deleted contemporaneously with activation of BCR-ABL using retroviral infection and transformation of HSC, chronic myeloid leukemia stem cell (CML-LSC) failed to engraft in secondary recipient mice (Hu Y et al., 2009). These experiments clearly indicate a pivotal role of Wnt signaling in CML-LSC development. More recently, Ctnnb1 has been investigated in the maintenance of already engrafted CML-LSC. In this clinically relevant setting, pharmacologic or genetic inactivation of Ctnnb1 after onset of the myeloproliferative disease acted synergistically with imatinib, reduced LSC numbers, and improved survival in a BM transplant model (Abrahamsson et al., 2009). Thus, despite its dispensability for adult HSC, CML-LSCs seem to retain dependency on canonical Ctnnb1 to maintain self-renewal capacity. In human disease, Ctnnb1 activation via the canonical Wnt pathway has been shown to occur in CML-blast crisis LSCs. Aberrant splicing of GSK3 appears to contribute to this hyperactivation in blast crisis samples (Abrahamsson et al., 2009). Thus, there is growing evidence that canonical Wnt signaling is an attractive target pathway in the treatment of CML-LSC. Moreover, cell extrinsic inhibition of Wnt signaling through ectopic DKK1 expression impairs leukemia cell proliferation in vitro (Zhu et al., 2009).

5. Intestinal SCs

Homeostasis of the intestinal epithelium is maintained by an intestinal SC (ISC) compartment that resides at the bottom of the crypt, safely far from the shear stresses and potentially toxic agents. These ISC are at the top of a cellular hierarchy and are crucial for the renewal of the differentiated progeny within the intestinal layer (Medema & Vermeulen, 2011). Indeed, as they migrate out of their niche, they cease to proliferate and initiate differentiation into the different cell lineages of the mature villi: absorptive enterocytes, mucin-secreting-goblet cells, peptide hormone-secreting neuroendocrine cells, and microbicide-secreting Paneth cells. Until relatively recently, ISCs were a rather elusive entity at the bottom of the intestinal crypt, and the discovery of ISC markers has only partly detailed the organization of the intestinal crypt and villi. Briefly, the marker LGR5 identifies crypt base columnar cells (CBCC) located in between the Paneth cells at the crypt bottom (Barker et al., 2007), whereas the markers BMI1 and TERT identify the +4 position in the crypt, just above the Paneth cells (Montgomery et al., 2011; Sangiorgi & Capecchi, 2008). Knock-in constructs that allow expression of GFP and Cre from the Lgr5 locus show that LGR5 expression is confined to CBCCs, and that these cells give rise to the variety of epithelial cells present in crypts, proving that CBCCs function as ISCs as well (Barker et al., 2007, Sato et al., 2009). The existence of these different types of ISC remains a matter of debate and notably, remains to be determined whether and how BMI1+ +4 cells ISCs and LGR5+ ISCs relate to each other. Interestingly, recent data indicate that TERT-expressing ISCs can generate LGR5+ISCs (Montgomery et al., 2011) suggesting that these different ISC types may act in a hierarchical fashion. Regardless of this dispute about ISC identity, there is a consensus that ISCs reside in a niche that provides the cells with essentials signals such as Wnt, Notch and Hedgehog. Under normal circumstances, the Paneth cell signals dictate the size of the SC pool to maintain the total number of SCs within the niche constant. SCs may divide asymmetrically, so that one SC remains within the niche, resulting in self-renewal, whilst the other daughter cell gives rise to progenitor cells that can migrate up the crypt and become more differentiated as they reach the top. Alternatively, two recent studies (Lopez-Garcia et al., 2010; Snippert et al., 2010) support that SCs may divide symmetrically either forming two daughter SCs (leading to expansion) or two daughter non-stem progenitor cells (leading to extinction). Several pathways play a role in maintaining and regulating stem ISCs, including Wnt and Notch.

5.1 Notch signaling in intestinal epithelium

In the intestine, Notch activity determines lineage decisions between enterocyte and secretory cell differentiation. Several components of the Notch pathway are expressed in adult intestinal crypt cells, suggesting a role for Notch signaling in gene expression programs in immature proliferating compartment cells (Sander & Powell, 2004; Schroder & Gossler, 2002). The first evidence that Notch signaling plays a role in cell-type specification in the intestine was reported in Hes1 knockout mice (Jensen et al., 2000). The deletion of the Hes1 gene resulted in the generation of excessive numbers of goblet cells, enteroendocrine cells, and Paneth cells. Subsequently, it was shown that Math1 (mouse atonal homolog1), one of the genes repressed by Hes1, is required for the differentiation into the three secretory lineages, because the intestinal epithelium of Math1-mutant mice is populated only by absorptive cells (Yang et al., 2001). These data suggest that the choice between the absorptive or secretory fate might be the first decision made by each progenitor cells, and that Hes1 and Math1 activated by Notch signal play opposite roles in this decision making. Recently, using the villin promoter to drive the expression of a constitutively active form of mouse Notch1 receptor, it was noticed an expansion of proliferating intestinal progenitor cells (Fre et al., 2005). Moreover, Notch activation inhibited the differentiation of secretory cells in the mouse intestine, as there was a complete depletion of goblet cells, a marked reduction in enteroendocrine cells, and a low expression of early marker for Paneth cells. These results clearly suggest that Notch signaling is required for maintaining crypt cells in a proliferative state, at least in part, through its negative regulation of Math1. Conversely, conditional removal of the Notch pathway transcription factor CSL/RBP-J increases the proportion of goblet cells in the murine intestine, and a similar phenotype was observed using a γ-secretase inhibitor (van Es et al., 2005). These results suggest that Notch pathway is not only a gatekeeper for proliferating crypt progenitor cells, but is also involved in controlling the balance between secretory and absorptive cell types. Data suggest that the ISC microenvironment delivers Notch-activating signals to maintain stemness, which is consistent with the observation that Paneth cells express Notch ligands (Sato et al., 2011). In particular, recent papers identified Dll1 and Dll4 as the physiologically relevant Notch1 and Notch2 ligands within the small intestine of the mouse. These ligands cooperate and exhibit a partial functional redundancy to maintain the crypt progenitor compartment (Pellegrinet et al., 2011). However, Notch seems to have dual functions in the crypt, as it acts together with Wnt to affect significantly crypt homeostasis (Fre et al., 2005; van Es et al., 2005).

5.2 Canonical Wnt signaling in intestinal epithelium

The Wnt pathway proteins regulate cellular fate along the crypt-villus axis in normal gut epithelium and have been implicated in ISC self-renewal. The nuclear accumulation of β-catenin is preferentially observed in cells located at the base of crypts and decreases as cells move toward the top of the crypts (van der Wetering et al., 2002). Wnt target genes EphB2 and EphB3 control crypt cellular segregation (Batlle et al., 2002), Sox9 regulates Paneth cell differentiation (Mori-Akiyama et al., 2007), and Lgr5 (Barker et al., 2007). TCF4 null mice died shortly after birth and showed an embryonic epithelium made entirely of differentiated cells without proliferative compartments in the crypts (Korinek et al., 1998) suggesting that TCF4 maintains the proliferation of SCs in the murine small intestine. Notably, deletion of the Wnt/TCF4 target gene c-Myc led to a loss of intestinal crypts in a murine model (Muncan et al., 2006). The importance of the Wnt signaling pathway in maintaining the architecture and homeostasis of the adult intestinal epithelium was also shown in a murine model through adenoviral expression of Dkk1. This induced Wnt inhibition in fully adult mice, resulted in inhibition of proliferation in the small intestine and colon, with progressive loss of crypts, villi and glandular structure (Kuhnert et al., 2004). By contrast, when the Wnt pathway is overactivated by mutations in APC or β-catenin, many of the epithelial cells enter into the proliferative state and display a failure of the differentiation programs (Andreu et al., 2005; Sansom et al., 2004). According with these data, recent papers demonstrated that injection of R-spondin1 (R-Spo1), a potent activator of the Wnt signaling pathways, induced rapid onset of crypt cell proliferation displaying epithelial hyperplasia in the intestine of normal mice through β-catenin stabilization and subsequent transcriptional activation of target genes such as murine Axin2, Ascl2, and Lgr5 (Kim et al., 2005; Takashima et al., 2011). The effects of R-Spo1 administration determine protection against radiation-induced colitis by stimulating proliferation of intestinal SCs and protect them against a damage after allogeneic bone-marrow transplantation, suppressing inflammatory cytokine cascades and donor T cell activation (Takashima et al., 2011). These, *in vivo*, data suggest that Wnt signaling is directly linked to the promotion of cellular proliferation and, more specifically, the regulation of progression through cell cycle. In this regard, previous papers pointed to the downregulation of $p21^{cip1waf1}$, a cyclin-dependent kinase inhibitor (CKI), as an important mechanism that might mediate Wnt-dependent growth promotion. A microarray analysis showed that $p21^{cip1waf1}$ was one of the genes whose expression was increased by inhibition of Wnt signaling in human colorectal cancer-derived LS174T cells (van der Wetering et al., 2002). Furthermore, the TCF4 target gene c-Myc has been shown to play a central role in Wnt-mediated repression of $p21^{cip1waf1}$ expression at the transcriptional level through its direct binding to the $p21^{cip1waf1}$ gene promoter (van der Wetering et al., 2002). These data suggest that the repression of $p21^{cip1waf1}$ by c-Myc might be the intracellular mechanism by which Wnt signaling regulates the G1/S transition and cell cycle progression. This signaling cascade has been shown to be functional *in vivo*, because abnormal features of proliferation/differentiation in the adult murine intestine, which occur with the single deletion of APC, are mostly rescued when c-Myc gene is simultaneously deleted (Sansom et al., 2007). Furthermore, this restoration of the morphologically normal phenotype in double mutant mice for APC and c-Myc is accompanied by restoration of p21 expression within the crypts, suggesting the involvement of p21 in the Wnt-c-Myc pathway-mediated growth control of progenitor cells. Indeed, raises the possibility that p21 is an intracellular molecular switch between proliferation and differentiation. Moreover, it has been shown that conditional expression of $p21^{cip1waf1}$ alone allow cells to differentiate (van der Wetering et al., 2002) suggesting that the cell fate choice between proliferation and differentiation is regulated by modulation of the expression of $p21^{cip1waf1}$ via the direct induction of c-Myc by Wnt signaling.

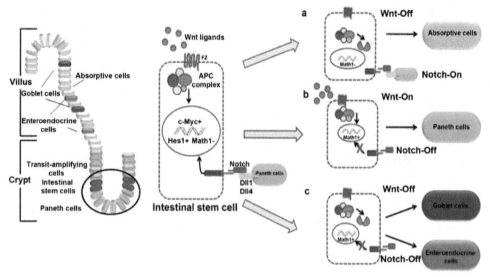

Fig. 5. The role for Notch and Wnt pathways in intestinal epithelial proliferation and differentiation. The ISC can give rise to four lineages of terminally differentiated cells: a is absorptive cells, b and c (Paneth, goblet and enteroendocrine cells) have secretory phenotypes. See the text for detail.

In general, the data strongly support a model in which Notch directs proliferation when Wnt signal activity is high, and directs enterocyte differentiation when Wnt activity levels drop towards the top of the crypt. The multipotent progenitors require both Wnt and Notch signals to be activated for fulfilling continuous proliferation without differentiation. Once some cells in this Wnt and Notch-activated population escape from the Notch signal, they stop proliferating and acquire the Math1 function. These cells raise the terminally differentiation in secretory cells in areas where the Wnt signal is not active (Pinto et al., 2003), whereas they differentiate in Paneth cells if they remain at the bottom of the crypt where Wnt ligands are abundant. By contrast, if cells in this Wnt and Notch-active population lose the Wnt signal, for example, because of their positional changes along the vertical axis, they differentiate as absorptive cells (Fig. 5).

5.3 SCs and the origin of intestinal cancer

Despite stringent homeostatic maintenance in the intestine, the high number of patients with colorectal cancer (CRC) indicates that these regulatory mechanisms often fall short in protecting against malignant transformation. Both environmental and genetic risk factors have been defined for CRC, and deregulation of morphogenetic pathways plays a key part in cancer development. Notably, the vast majority of sporadic CRC cases carry Wnt pathway mutations, highlighting the importance of this pathway in CRC. The hit that induces transition from normal to polypoid tissue is accompanied by several changes in crypt appearance and behavior, cells show a more immature phenotype and a higher proliferative index which results in expansion of the pre-malignant clone. Although mutation of APC or β-catenin is an early event in the transformation of colonic epithelial cells, studies have revealed that colon carcinomas do not contain nuclear β-catenin homogeneously (Fodde & Brabletz, 2007). This so-called β-catenin paradox indicate that Wnt signaling has a preponderant role only for a subset of tumour cells, cancer SCs (CSCs), which are endowed with tumorigenic capacity (Vermeulen et al., 2008). Indeed, the past decade has seen a shift in the way tumours are perceived, and the now widely accepted model is that tumours contain a small population of self-renewing CSCs, as well as a large compartment of more differentiated tumour cells (Vermeulen et al., 2008). Cellular hierarchy within CRC is maintained, at least in part, by microenvironmental factors regulating stemness and differentiation. In agreement, tumour cells located next to myofibroblast-rich regions, have a much higher incidence of nuclear-localized β-catenin, suggesting for microenvironment-modulated Wnt signaling (Fodde & Brabletz, 2007). A recent paper point to hepatocyte growth factor (HGF) as the myofibroblast-derived signal that, at least in part, orchestrates this intimate relationship and enhances Wnt activity in more differentiated tumour cells, thereby reinstalling CSCs features (dedifferentiation) (Vermeulen et al., 2010). Indeed, using a TCF/LEF reporter that directs the expression of enhanced green fluorescent protein, authors provided evidence that Wnt signaling activity is a marker for colon CSCs and is regulated by the microenvironment. Moreover, they show that differentiated cancer cells can be reprogrammed to express CSC markers and regain their tumorigenic capacity when stimulated with myofibroblast-derived factors (Vermeulen et al., 2010). Although, these data clearly ascertain a role for the Wnt pathway in CRC stemness, Notch inhibition with an antibody against the Notch ligand Dll4 results in human colon CSCs differentiation, reduction of CRC growth in a xenotransplantation model and chemosensitization (Hoey et al., 2009).

6. Identification of Renal SCs

The mammalian kidney shares with the majority of organs the ability to repopulate and at least partially repair structures that have sustained some degree of injury. Indeed, tubular integrity can be rescued after acute damage, and even severe glomerular disorders sometimes may undergo regression and remission, suggesting that glomerular injury is also reparable (Imai & Iwatani, 2007; Remuzzi, et al., 2006). However, the existence of renal SC (RSC) has been a matter of long debate. Recently, converging data definitively demonstrated the existence of a population of stem/progenitor cells in the parietal epithelium of the Bowman's capsule of adult human kidney (Sagrinati, et al., 2006) (Fig.6). These SC coexpress both CD24, a surface molecule that has been used to identify different types of human SC, and CD133, a marker of several types of adult tissue SC, lack lineage-specific markers, express transcription factors that are characteristic of multipotent SC, and exhibit self-renewal, high clonogenic efficiency and multidifferentiation potential. When injected intravenously in SCID mice that had acute kidney injury, RSC regenerated tubular structures from different portions of the nephron and also reduced the morphological and functional kidney damage (Sagrinati, et al., 2006).

In addition, it was demonstrated that RSC are arranged in a precise sequence within Bowman's capsule of adult human kidneys (Ronconi, et al., 2009) (Fig. 6).

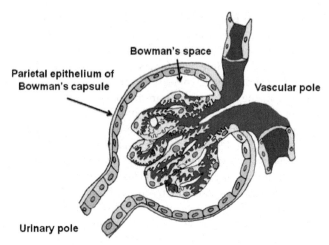

Fig. 6. Localization of RPC in the glomerulus. RPC (green) are localized in the Bowman's capsule epithelium. A transitional cell population (podocyte progenitors, green/yellow) displays features of either RPC or podocyte (yellow) and localize between the urinary pole and the vascular stalk. Cells that express only podocyte markers and the phenotypic features of differentiated podocytes (yellow) localize at the vascular stalk of the glomerulus.

These findings obtained in human kidneys were confirmed in a parallel study performed in murine kidney by Appel (Appel, et al., 2009), who also demonstrated the existence of transitional cells with morphological and immunohistochemical features of both parietal epithelial cells and podocyte in proximity of the glomerular vascular stalk and that podocytes are recruited from parietal epithelial cells, which proliferate and differentiate from the urinary to the vascular stalk, then generating novel podocytes (Fig. 6). This occurs as the kidney grows, during childhood and adolescence, and may also take place following an injury which allows a slow, regulated generation of novel podocytes, such as uninephrectomy. Recently, a rare subpopulation of CD133+CD24+ cells has also been describe in renal tubules (Lindgren, et al., 2011). These cells are able to proliferate and differentiate after tubular injury. Accordingly, tubular epithelium regenerating on acute tubular necrosis displayed long stretches of CD133+CD24+ cells, further substantiating that the cells that are repairing tubular epithelium may simply represent the result of proliferation and differentiation of CD133+CD24+ tubular progenitors.

6.1 Involvement of RSC in glomerular disorders and cancer

It has been widely recognized that a disruption in the strictly regulated balance of SC self-renewal and differentiation not only impairs regenerative mechanisms but can even generate disorders. In the glomerulus, the response to podocyte injury may cause aberrant epithelial cell proliferation, hypercellular lesions formation and Bowman's space obliteration, as seen in collapsing glomerulopathy and in crescentic glomerulonephritis (Albaqumi & Barisoni, 2008; Thorner, et al., 2008). Until now, theories explaining the origin of aberrant epithelial cells in collapsing glomerulopathy and crescentic glomerulonephritis have been controversial. One possibility is that these cells are exclusively of parietal epithelial origin (Thorner et al., 2008), while another is that some dedifferentiated podocytes acquire markers of parietal epithelial cells (Moeller et al., 2004). It was recently demonstrated that the majority of cells present in the hyperplastic lesions in collapsing glomerulopathy or crescentic glomerulonephritis exhibits the RSC markers CD133 and CD24, with or without coexpression of podocyte markers (Smeets et al., 2009). Therefore, it is suggested that the glomerular hyperplastic lesions are generated by RSC of Bowman's capsule at different stages of their differentiation towards mature podocytes. Support for this hypothesis came from lineage tracing experiments performed in transgenic mice with genetically labeled parietal epithelial cells in a model of inflammatory crescentic glomerulonephritis, and of collapsing glomerulopathy (Smeets et al., 2009).

Finally, a close relationship between the transcriptome of CD133+ tubular progenitors and the one derived by papillary renal cell carcinomas was demonstrated (Lindgren et al. 2011). Moreover, a strong CD133 expression was observed in the papillary renal cell carcinomas analysed. Thus, these observations raise the provocative hypothesis that papillary renal cell carcinomas may directly derive from $CD133^+CD24^+$ renal tubular progenitors, whereas clear renal cell carcinomas may derive from other more differentiated proximal tubular cells.

6.2 Signaling pathway regulating the RSC niche

The molecular mechanisms regulating the proliferation of RSC, as well as the cell fate determination in the podocyte lineage are unknown. We recently demonstrate the role of the Notch signaling pathway in both these processes (Lasagni et al., 2010). Notch activation triggers the expansion of renal progenitors by promoting their entry into the S-phase of the cell cycle and mitotic division. Moreover, Notch downregulation is required for differentiation toward the podocyte lineage. However, Notch downregulation was neither sufficient nor necessary for the acquisition of a podocyte phenotype, but an impaired downregulation of the Notch pathway led to podocyte death. Indeed, renal progenitor differentiation into podocytes was associated with cell cycle checkpoint activation and G_2/M arrest, reflecting an intrinsic barrier to replication of mature podocytes. Persistent activation of the Notch pathway induced podocytes to cross the G_2/M checkpoint, resulting in cytoskeleton disruption and cell death (Lasagni et al., 2010). Notch expression was virtually absent in the glomeruli of healthy adult kidneys, while a strong upregulation was observed in renal progenitors and podocytes in patients affected by glomerular disorders. Accordingly, inhibition of the Notch pathway in mouse models of focal segmental glomerulosclerosis ameliorated proteinuria and reduced podocyte loss during the initial phases of glomerular injury, while inducing reduction of progenitor proliferation during the regenerative phases of glomerular injury with worsening of proteinuria and glomerulosclerosis. Taken altogether, these results suggest that the severity of glomerular disorders depends on the Notch-regulated balance between podocyte death and regeneration provided by renal progenitors (Lasagni et al., 2010).

7. References

Abrahamsson AE, Geron, I., Gotlib, J., Dao, KH., Barroga, CF., Newton, IG., Giles, FJ., Durocher, J., Creusot, RS., Karimi, M., Jones, C., Zehnder, JL., Keating, A., Negrin, RS., Weissman, IL.& Jamieson, CH. (2009). Glycogen synthase kinase 3beta missplicing contributes to leukemia stem cell generation. *Proc Natl Acad Sci USA*, Vol.106, No.10, (March 2009), pp.3925-9, ISSN1091-6490.

Albaqumi, M. & Barisoni, L. (2008) Current views on collapsing glomerulopath,. *J Am Soc Nephrol.*, Vol.19, No.7, (July 2008), pp. 1276-1281, ISSN 1046-6673

Andreu, P.; Colnot, S.; Godard, C.; Gad, S.; Chafey, P.; Niwa-Kawakita, M.; Laurent-Puig, P.; Kahn, A.; Robine, S.; Perret, C. & Romagnolo, B. (2005). Crypt-restricted proliferation and commitment to the Paneth cell lineage following Apc loss in the mouse intestine. *Development*, Vol.132, No.6, (March 2005), pp. 1443-1451, ISSN 1011-6370

Appel, D.; Kershaw, D.B.; Smeets, B.; Yuan, G.; Fuss, A.; Frye, B.; Elger, M.; Kriz, W.; Floege, J. & Moeller, M.J. (2009) Recruitment of podocytes from glomerular parietal epithelial cells, *J Am Soc Nephrol*, Vol.20, No.2, (February 2009), pp. 333-343, ISSN 1046-6673

Arce, L.; Yokoyama, N.N. & Waterman, M.L. (2006) Diversity of LEF/TCF action in development and disease. *Oncogene*. Vol.25, No.57, (December 2006), pp. 7492-504, ISSN 0950-9232

Barker, N.; van Es, JH.; Kuipers, J.; Kujala, P.; van den Born, M.; Cozijnsen, M.; Haegebarth, A.; Korving, J.; Begthel, H.; Peters, PJ. & Clevers, H. (2007) Identification of stem cells in small intestine and colon by marker gene Lgr5. *Nature*, Vol.449, No.7165, (October 2007), pp. 1003-7, ISSN 00280836

Batlle, E.; Henderson, J.T.; Beghtel, H.; van den Born, M.M.; Sancho, E.; Huls, G.; Meeldijk, J.; Robertson, J.; van de Wetering,; M. Pawson, T. & Clevers, H. (2002). Beta-catenin and TCF mediate cell positioning in the intestinal epithelium by controlling the expression of EphB/ephrinB. *Cell*, Vol.111, No.2, (October 2002), pp. 251-263, ISSN 0092-8674

Betschinger, J. & Knoblich, J.A. (2004) Dare to be different: asymmetric cell division in Drosophila, C. elegans and vertebrates, *Curr Biol*, Vol.14, No.16, (August 2004), pp. R674-685, ISSN 0960-9822

Beverly, L.J.; Felsher, D.W. & Capobianco, A.J. (2005). Suppression of p53 by Notch in lymphomagenesis: Implications for initiation and regression. *Cancer Research*, Vol.65, No.16, (August 2005), pp. 7159-7168, ISSN 1538-7445.

Blanpain C, Fuchs E. (2009) Epidermal homeostasis: a balancing act of stem cells in the skin, *Nat Rev Mol Cell Biol*, Vol.10, No.3, (March 2009), pp. 207-217, ISSN 1471-0072

Bovolenta, P.; Esteve, P.; Ruiz, J.M.; Cisneros, E. & Lopez-Rios, J.(2008) Beyond Wnt inhibition: new functions of secreted Frizzled-related proteins in development and disease. *J Cell Sci*, Vol.121, No.6, (March 2008), pp. 737–746, ISSN 0021-9533

Bray, S. & Bernard, F. (2010) Notch targets and their regulation. *Curr. Top. Dev. Biol.*, Vol.92, 253–275 ISSN: 0070-2153

Carlesso, N.; Aster, J.C.; Sklar, J. & Scadden, D.T., (1999). Notch1-induced delay of human hematopoietic progenitor cell differentiation is associated with altered cell cycle kinetics. *Blood*, Vol.93, No.3, (February 1999), pp. 838-848, ISSN 0006-4971.

Cedarn, C. & Bhatia, M. (2010) Novel roles for Notch, Wnt and Hh in hematopoiesis derived from human pluripotent stem cells. *Int J Dev Biol*, Vol.54, No.6-7, (2010), pp. 955-963, ISSN 02146282

Cheshier, SH., Morrison, SJ., Liao, X. & Weissman IL. (1999). In vivo proliferation and cell cycle kinetics of long-term self-renewing hematopoietic stem cells. *Proc Natl Acad Sci USA*, Vol.96, No.6, (March 1999), pp.3120–5, ISSN 1091-6490.

Dick, JE. (2008). Stem cell concepts renew cancer research. *Blood*, Vol.112, No.13, (December 2008), pp.4793–807, ISSN 0006-4971.

Eisenberg, C.A. & Eisenberg, L.M. (1999) WNT11 promotes cardiac tissue formation of early mesoderm, *Dev. Dyn.*, Vol.216, No.1, (September 1999), pp. 45–58, ISSN 1058-8388

Ellisen, L.W.; Bird, J.; West, D.C.; Soreng, A.L.; Reynolds, T.C.; Smith, S.D. & Sklar, J. (1991). TAN-1, the human homolog of the Drosophila Notch gene, is broken by chromosomal translocations in T lymphoblastic neoplasms. *Cell*, Vol.66, No.4, (August 1991), pp. 649-661, ISSN 0092-8674.

Fleming, HE., Janzen, V., Lo Celso, C., Guo, J., Leahy, KM., Kronenberg, HM. & Scadden DT. (2008). Wnt signaling in the niche enforces hematopoietic stem cell quiescence and is necessary to preserve self-renewal in vivo. *Cell Stem Cell*, Vol.2, No.3, (March 2008), pp.274–83, ISSN 19345909

Fodde, R. & Brabletz, T. (2007). Wnt/beta-catenin signaling in cancer stemness and malignant behavior. *Curr Opin Cell Biol*, Vol.19, No. , (April 2007), pp. 150-158, ISSN 09550674

Fre, S.; Huyghe, M.; Mourikis, P.; Robine, S.; Louvard, D. & Artavanis-Tsakonas, S. (2005). Notch signals control the fate of immature progenitor cells in the intestine. *Nature*, Vol. 435, No. 7044, (June 2005), pp. 964-968, ISSN 0028-0836

Gurtner, G.C.; Werner, S.; Barrandon, Y. & Longaker, M.T. (2008) Wound repair and regeneration, *Nature*, Vol.453, No.7193, (May 2008), pp. 314-321, ISSN 0028-0836

Hoey, T.; Yen, W.C.; Axelrod, F.; Basi, J.; Donigian, L.; Dylla, S.; Fitch-Bruhns, M.; Lazetic, S.; Park, I.K.; Sato, A.; Satyal, S.; Wang, X.; Clarke, M.F.; Lewicki, J. & Gurney, A. (2009). DLL4 blockade inhibits tumor growth and reduces tumor-initiating cell frequency. *Cell Stem Cell*, Vol.5, No.2, (August 2009), pp. 168-177, ISSN 19345909

Hu Y, Chen, Y., Douglas, L. & Li, S. (2009). beta-Catenin is essential for survival of leukemic stem cells insensitive to kinase inhibition in mice with BCR-ABLinduced chronic myeloid leukemia. *Leukemia*, Vol.23, No.1, (January 2009), pp.109–16, ISSN 08876924.

Huang, J., Zhang, Y., Bersenev, A., O'Brien, WT., Wei Tong, W., Emerson, SG. & Klein, PS. (2009). Pivotal role for glycogen synthase kinase-3 in hematopoietic stem cell homeostasis in mice *J. Clin. Invest*, Vol.119, No.12, (December 2009), pp.3519–3529, ISSN 00219738.

Ikeda, S.; Kishida, S.; Yamamoto, H.; Murai, H.; Koyama, S. & Kikuchi, A. (1998) Axin, a negative regulator of the Wnt signaling pathway, forms a complex with GSK-3beta and beta-catenin and promotes GSK-3beta-dependent phosphorylation of beta-catenin, *EMBO J.*, Vol.17, No.5, (March 1998), pp. 1371-84, ISSN 0261-4189

Imai, E. & Iwatani, H. (2007) The continuing story of renal repair with stem cells. *J Am Soc Nephrol.*, Vol.18, No.9, (September 2007), pp. 2423-2424, ISSN 1046-6673

Jensen, J.; Pedersen, E.E.; Galante, P.; Hald, J.; Heller, R.S.; Ishibashi, M.; Kageyama, R.; Guillemot, F.; Serup, P. & Madsen, O.D. (2000). Control of endodermal endocrine development by Hes-1. *Nat Genet*, Vol.24, No.1, (January 2000), pp. 36-44, ISSN 1061-4036

Jones, D.L. & Wagers, A.J. No place like home: anatomy and function of the stem cell niche, *Nature Review Molecular Cell Biology*, Vol.9, No.1, (January 2008), pp. 11-21, ISSN: 1471-0072

Karanu, F.N.; Murdoch, B.; Gallacher, L.; Wu, D.M.; Koremoto, M.; Sakano, S. & Bathia, M. (2000). The Notch ligand Jagged-1 represents a novel growth factor of human hematopoietic stem cells. *J Exp Med*, Vol.192, No.9, (November 2000), pp. 1365-1372, ISSN 1540-9538.

Karanu, F.N.; Murdoch, B.; Miyabayashi, T.; Ohno, M.; Koremoto, M.; Gallacher, L.; Wu, D.; Itoh, A.; Sakano, S. & Bathia, M. (2001). Human homologues of Delta-1 and Delta-4 function as mitogenic regulators of primitive human hematopoietic cells. *Blood*, Vol.97, pp. 1960-1967, ISSN 0006-4971.

Kelly, OG., Pinson, KI. & Skarnes, W.C. (2004) The Wnt co-receptors Lrp5 and Lrp6 are essential for gastrulation in mice. *Development*, Vol.131, No.12, (June 2004), pp. 2803-2815, ISSN 09501991.

Kestler, H.A. & Kühl, M. (2008) From individual Wnt pathways towards a Wnt signaling network. *Philos Trans R Soc Lond B Biol Sci*, Vol.363, No.1495, (April 2008), pp. 1333-47, ISSN 0080-4622

Kiel, M.J.; Yilmaz, O.H.; Iwashita, T.; Yilmaz, O.H.; Terhorst, C. & Morrison S.J. (2005). SLAM family receptors distinguish hematopoietic stem and progenitor cells and reveal endothelial niches for stem cells. *Cell*, Vol.121, No.7, (July 2005), pp. 1109-1121, ISSN 0092-8674.

Kim, K.A.; Kakitani, M.; Zhao, J.; Oshima, T.; Tang, T.; Binnerts, M.; Liu, Y.; Boyle, B.; Park, E.; Emtage, P.; Funk, W.D. & Tomizuka K. (2005). Mitogenic influence of human R-spondin1 on the intestinal epithelium. *Science*, Vol. 309, No. 5738, (August 2005), pp. 1256-1259, ISSN 0036-8075

Kirstetter, P., Anderson, K., Porse, BT., Jacobsen, SE. & Nerlov, C. (2006) Activation of the canonical Wnt pathway leads to loss of hematopoietic stem cell repopulation and multilineage differentiation block. *Nat Immunol*, Vol.7, No.10, (October 2006), pp.1048-1056, ISSN 15292916.

Koch, U., Wilson, A., Cobas, M., Kemler, R., Macdonald, H.R. & Radtke, F. (2008) Simultaneous loss of beta- and gamma-catenin does not perturb hematopoiesis or lymphopoiesis. *Blood*, Vol.111, No.1, (January 2008), pp.160-164, ISSN 15280020.

Kopan, R. & Ilagan, M. X. (2009) The canonical Notch signaling pathway: unfolding the activation mechanism. *Cell*, Vol.137, No.2, (April 2009), pp. 216–233, ISSN 0092-8674

Korinek, V.; Barker, N.; Moerer, P.; van Donselaar, E.; Huls, G.; Peters, P.J. & Clevers, H. (1998). Depletion of epithelial stem-cell compartments in the small intestine of mice lacking Tcf-4. *Nat Genet.*, Vol.19, No.4, (August 1998), pp. 379-383, ISSN 1061-4036

Kuhl, M.; Sheldahl, L.C.; Malbon, C.C. & Moon, R.T. (2000) Ca^{2+}/calmodulin-dependent protein kinase II is stimulated by Wnt and Frizzled homologs and promotes ventral cell fates in Xenopus, *J. Biol. Chem.*, Vol.275, No.17, (April 2000), pp. 12□701-12□711, ISSN 0021-9258

Kuhnert, F.; Davis, C.R.; Wang, H.T.; Chu, P.; Lee, M.; Yuan, J.; Nusse, R. & Kuo, C.J. (2004). Essential requirement for Wnt signaling in proliferation of adult small intestine and colon revealed by adenoviral expression of Dickkopf-1. *Proc Natl Acad Sci U S A*, Vol.101, No.1, (January 2004), pp. 266-271, ISSN 1091-6490

Lai, E.C.(2002) Keeping a good pathway down: transcriptional repression of Notch pathway target genes by CSL proteins, *EMBO Rep*, Vol.3, No.9, (September 2002), pp. 840-845, ISSN 1469-221X.

Lasagni, L.; Ballerini, L.; Angelotti, M.L.; Parente, E.; Sagrinati, C.; Mazzinghi, B.; Peired, A.; Ronconi, E.; Becherucci, F.; Bani, D.; Gacci, M.; Carini, M.; Lazzeri, E. & Romagnani, P. (2010) Notch activation differentially regulates renal progenitors proliferation and differentiation toward the podocyte lineage in glomerular disorders. *Stem Cells*, Vol.28, No.9, (September 2010), pp. 1674-1685, ISSN 066-5099

Li, L. & Xie, T. (2005) Stem cell niche: structure and function, *Annu Rev Cell Dev Biol.*, Vol.21, (November 2005), pp. 605-31, ISSN 1081-0706

Lindgren, D.; Boström, A.K.; Nilsson, K.; Hansson, J.; Sjölund, J.; Möller, C.; Jirström, K.; Nilsson, E.; Landberg, G.; Axelson, H. & Johansson, M.E. (2011) Isolation and characterization of progenitor-like cells from human renal proximal tubules, *Am J Pathol*, Vol.178, No.2, (February 2011), pp. 828-837, ISSN 0002-9440

Liu, J.; Sato, C.; Cerletti, M. & Wagers, A. (2010).Notch signaling in the regulation of stem cell self-renewal and differentiation. *Curr Top Dev Biol*, Vol.92, No.7, (April 2001), pp. 367-409, ISSN 0070-2153.

Liu, P., Wakamiya, M., Shea, MJ., Albrecht, U., Behringer, RR. & Bradley, A. (1999). Requirement for Wnt3 in vertebrate axis formation. *Nat Genet* , Vol.22, No.4, (August 1999), pp. 361-365, ISSN 10614036.

Logan, C.Y. & Nusse, R. (2004) The Wnt signaling pathway in development and disease. *Annu Rev Cell Dev Biol.*, Vol.20, (July 2004), pp. 781–810, ISSN 1081-0706

Lopez-Garcia, C. Klein, A.M. Simons, B.D. Winton, D.J. (2010). Intestinal stem cell replacement follows a pattern of neutral drift. *Science*, Vol.330, No.6005, (November 2010), pp. (822-825), ISSN 0036-8075

MacDonald, B.T.; Tamai, K. & He, X. (2009) Wnt/beta-catenin signaling: components, mechanisms, and diseases. *Dev Cell*, Vol.17, No.1, (July 2009), pp. 9-26, ISSN. 1534-5807

Malhotra, S. & Kincade, P.W. (2009). Wnt-related molecules and signaling pathway equilibrium in hematopoiesis. *Cell Stem Cell*, Vol.4, No.1, (January 2009), pp.27–36, ISSN, 19345909.

Mao, B.; Wu, W.; Davidson, G.; Marhold, J.; Li, M.; Mechler, B.M.; Delius, H.; Hoppe, D.; Stannek, P.; Walter, C.; Glinka, A. & Niehrs, C. (2002) Kremen proteins are Dickkopf receptors that regulate Wnt/β-catenin signaling, *Nature*, Vol. 417, No.6889, (June 2002), pp. 664–667, ISSN 0028-0836

Medema, JP. Vermeulen, L. (2011). Microenvironmental regulation of stem cells in intestinal homeostasis and cancer. *Nature*, Vol.474, No.7351, (June 2011), pp. 318-326, ISSN 0028-0836

Medyouf, H.; Alcade, H.; Berthier, C.; Guillemain, M.C.; dos Santos, N.R.; Janin, A.; Decaudin, D.; de Thé, H. & Ghysdael, J. (2007). Targeting calcineurin activation as a therapeutic strategy for T-cell acute lymphoblastic leukemia. *Nature Medicine*, Vol.13, No.6, (June 2007), pp. 736-741, ISSN 1078-8956.

Moeller, M.J.; Soofi, A.; Hartmann, I.; Le Hir, M.; Wiggins, R.; Kriz, W. & Holzman, L.B. (2004) Podocytes populate cellular crescents in a murine model of inflammatory glomerulonephritis. *J Am Soc Nephrol*, Vol.15, No.1, (January 2004), pp. 61-67, ISSN 1046-6673

Montcouquiol, M.; Crenshaw, E.B. 3r &, Kelley MW (2006) Noncanonical Wnt signaling and neural polarity. *Annu Rev Neurosci.*, Vol.29, (July 2006), pp. 363-86, ISSN 0147-006X

Montgomery, R.K. Carlone, D.L. Richmond, C.A. Farilla, L. Kranendonk, M.E. Henderson, D.E. Baffour-Awuah, N.Y. Ambruzs, D.M. Fogli, L.K. Algra, S. Breault, D.T. (2011). Mouse telomerase reverse transcriptase (mTert) expression marks slowly cycling intestinal stem cells. *Proc Natl Acad Sci U S A*, Vol. 108, No. 1, (January 2011), pp. (179-184), ISSN 1091-6490

Morgan, T. (1917) The theory of the gene, *Am. Nat.*, Vol.51, No.609, (September 1917), pp. 513-544. ISSN 00030147

Mori-Akiyama, Y.; van den Born, M.; van Es, J.H.; Hamilton, S.R.; Adams, H.P.; Zhang, J.; Clevers, H. & de Crombrugghe, B. (2007). SOX9 is required for the differentiation of paneth cells in the intestinal epithelium. *Gastroenterology*, Vol.133, No.2, (August 2007), pp. 539-546, ISSN 0016-5085

Morrison, S.J. & Kimble J. (2006) Asymmetric and symmetric stem-cell division in development and cancer, *Nature*, Vol.441, No.7097, (June 2006), pp. 1068-1074, ISSN 0028-0836

Muncan, V.; Sansom, O.J.; Tertoolen, L.; Phesse, T.J.; Begthel, H.; Sancho, E.; Cole, A.M.; Gregorieff, A.; de Alboran, I.M.; Clevers, H.; Clarke, A.R. (2006). Rapid loss of intestinal crypts upon conditional deletion of the Wnt/Tcf-4 target gene c-Myc. *Mol Cell Biol*, Vol.26, No.22, (November 2006), pp. 8418-8426, ISSN 1098-5549

Niida, A.; Hiroko, T.; Kasai, M.; Furukawa, Y.; Nakamura, Y.; Suzuki, Y.; Sugano, S. & Akiyama, T. (2004) DKK1, a negative regulator of Wnt signaling, is a target of the beta-catenin/TCF pathway, *Oncogene*, Vol.23, No.52, (November 2004), pp. 8520-6, ISSN 0950-9232

Okada, S.; Nakauchi, H.; Nagayoshi, K.; Nishikawa, S.; Miura, Y. & Suda, T. (1992). In vivo and in vitro stem cell function of c-kit- and Sca-1-positive murine hematopoietic cells. *Blood*, Vol.80, No.12, (December 1992), pp. 3044-3050, ISSN 1528-0020.

Paganin, M. & Ferrando, A. (2011). Molecular pathogenesis and targeted therapies for NOTCH1-induced T-cell acute lymphoblastic leukemia. *Blood*, Vol.25, No.2, (March 2011), pp. 83-90, ISSN 1528-0020.

Palomero, T.; Lim, W.K.; Odom, D.T.; Sulis, M.L.; Real, P.J.; Margolin, A.; Barnes, K.C.; O'Neil, J.; Neuberg, D.; Weng, A.P.; Aster, J.C.; Sigaux, F.; Soulier, J.; Look, A.T.; Young, R.A.; Califano, A. & Ferrando, A.A. (2006). NOTCH1 directly regulates c-MYC and activates a fee-forward-loop transcriptional network promoting leukemic cell growth. *Proc Natl Acad Sci, USA* , Vol.103, No.48, (November 2008), pp. 18261-18266, ISSN 1091-6490.

Palomero, T.; Sulis, M.L.; Cortina, M.; Real, P.J.; Barnes, K.; Ciofani, M.; Caparros, E; Buteau, J.; Brown, K.; Perkins, S.L.; Bhagat, G.; Mishra, A.; Basso, G.; Parsons, R.; Zúñiga-Pflücker, J.C.; Dominguez, M. & Ferrando A.A. (2007). Mutational loss of PTEN induces resistance to NOTCH1 inhibition in T-cell leukemia. *Nature medicine*, Vol.13, No.10, (October 2007), pp. 1203-1210, ISSN 1074-7613.

Pellegrinet, L.; Rodilla, V.; Liu, Z.; Chen, S.; Koch, U.; Espinosa, L.; Kaestner, K.H.; Kopan, R.; Lewis, J. & Radtke, F. (2011). Dll1- and dll4-mediated notch signaling are required for homeostasis of intestinal stem cells. *Gastroenterology*, Vol.140, No.4, (April 2011), pp. 1230-1240, ISSN 0016-5085

Pinson K.I, Brennan J, Monkley S, Avery B.J, Skarnes W.C (2000) An LDL-receptor-related protein mediates Wnt signaling in mice. *Nature.* Vol.407, No.6803, (September 2000), pp. 535–538, ISSN 0028-0836

Pinto, D.; Gregorieff, A.; Begthel, H. & Clevers, H. (2003). Canonical Wnt signals are essential for homeostasis of the intestinal epithelium. *Genes Dev*, Vol.17, No.14, (July 2003), pp. 1709-1713, ISSN 1549-5477

Radtke, F.; Fasnacht, N. & MacDonald, H.R. (2010). Notch Signaling in the Immune System. *Immunity*, Vol.32, No.1, (January 2010), pp. 14-27, ISSN 1074-7613.

Rawlins, E.L. & Hogan, B.L. (2006) Epithelial stem cells of the lung: privileged few or opportunities for many? *Development*, Vol.133, No.13, (July 2006), pp. 2455-2465, ISSN 1011-6370

Remuzzi, G.; Benigni, A. & Remuzzi A. (2006) Mechanisms of progression and regression of renal lesions of chronic nephropathies and diabetes. *J Clin Invest.*, Vol.116, No.2, (February 2006), pp. 288-296, ISSN 0021-9738

Renstrom, J.; Kroger, M., Peschel, C. & Oostendorp, R.A.J. (2010). How the niche regulates hematopoietic stem cells. *Chemico-Biological Interactions*, Vol.184, No.1-2, (March 2010), pp. 7-15, ISSN 0009-2797.

Reya, T., Duncan, AW., Ailles, L., Domen, J., Scherer, D.C., Willert, K., Hintz, L., Nusse, R. & Weissman, IL. (2003) A role for Wnt signaling in self-renewal of haematopoietic stem cells. *Nature*, Vol.423, No.6938, (May 2003), pp.409-414 ISSN 0028-0836.

Rhyu, M.S.; Jan, L.Y. & Jan, Y.N. (1994) Asymmetric distribution of numb protein during division of the sensory organ precursor cell confers distinct fates to daughter cells, *Cell*, Vol.76, No.3, (February 1994), pp. 477-491, ISSN 0092-8674

Ronconi, E.; Sagrinati, C.; Angelotti, M.L.; Lazzeri, E.; Mazzinghi, B.; Ballerini, L.; Parente, E.; Becherucci, F.; Gacci, M.; Carini, M.; Maggi, E.; Serio, M.; Vannelli, G.B.; Lasagni, L.; Romagnani, S. & Romagnani, P. (2009) Regeneration of glomerular podocytes by human renal progenitors. *J Am Soc Nephrol.*, Vol.20, No.2, (February 2009), pp. 322-332, ISSN 1046-6673

Roszko, I.; Sawada, A. & Solnica-Krezel, L. (2009) Regulation of convergence and extension movements during vertebrate gastrulation by the Wnt/PCP pathway, *Semin Cell Dev Biol.*, Vol.20, No.8, (October 2009), pp. 986-97, ISSN 1084-9521

Sagrinati, C.; Netti, G.S.; Mazzinghi, B.; Lazzeri, E.; Liotta, F.; Frosali, F.; Ronconi, E.; Meini, C.; Gacci, M.; Squecco, R.; Carini, M.; Gesualdo, L.; Francini, F.; Maggi, E.; Annunziato, F.; Lasagni, L.; Serio, M.; Romagnani, S. & Romagnani, P. (2006) Isolation and characterization of multipotent progenitor cells from the Bowman's capsule of adult human kidneys. *J Am Soc Nephrol.*, Vol.17, No.9, (September 2006), pp. 2443-2456, ISSN 1046-6673.

Sander, G.R. & Powell, B.C. (2004). Expression of notch receptors and ligands in the adult gut. *J Histochem Cytochem*, Vol. 52, No. 4, (April 2004), pp. 509-516, ISSN 0022-1554

Saneyoshi, T.; Kume, S.; Amasaki, Y. & Mikoshiba, K. (2002) The Wnt/calcium pathway activates NF-AT and promotes ventral cell fate in Xenopus embryos. *Nature*, Vol.417, No.6886, (May 2002), pp. 295-299, ISSN 0028-0836

Sangiorgi, E. Capecchi, M.R. (2008). Bmi1 is expressed in vivo in intestinal stem cells. *Nat Genet.* Vol.40, No.7, (July 2008), pp. 915-920, ISSN 1061-4036

Sansom, O.J.; Meniel, V.S.; Muncan, V.; Phesse, T.J.; Wilkins, J.A.; Reed, K.R.; Vass, J.K.; Athineos, D.; Clevers, H. & Clarke, A.R. (2007). Myc deletion rescues Apc deficiency in the small intestine. *Nature*, Vol.446, No.7136, (April 2007), pp. 676-679, ISSN 0028-0836

Sansom, O.J.; Reed, K.R.; Hayes, A.J.; Ireland, H.; Brinkmann, H.; Newton, I.P.; Batlle, E.; Simon-Assmann, P.; Clevers, H.; Nathke, I.S.; Clarke, A.R. & Winton D.J. (2004). Loss of Apc in vivo immediately perturbs Wnt signaling, differentiation, and migration. *Genes Dev.* Vol.18, No.2, (June 2004), pp. 1385-1390, ISSN 1549-5477

Sarmento, L.M.; Huang, H.; Limon, A.; Gordon, W.; Fernandes, J.; Tavares, M.J.; Miele, L.; Cardoso, A.A.; Classon, M. & Carlesso, N. (2005). Notch1 modulates timing of G1-S progression by inducing SKP2 transcription and p27 Kip1 degradation. Journal of Experimental Medicine, Vol.202, No.1, (July 2005), pp. 157-168, ISSN 1540-9538.

Sato, T. van Es, J.H. Snippert, H.J. Stange, D.E. Vries, R.G. van den Born, M. Barker, N. Shroyer, N.F. van de Wetering, M. Clevers, H. (2011). Paneth cells constitute the niche for Lgr5 stem cells in intestinal crypts. *Nature*, Vol.469, No. 330, (January 2011), pp. 415-418, ISSN 0028-0836

Sato, T. Vries, R.G. Snippert, H.J. van de Wetering, M. Barker, N. Stange, D.E. van Es, J.H. Abo, A. Kujala, P. Peters, P.J. Clevers, H. (2009). Single Lgr5 stem cells build crypt-villus structures in vitro without a mesenchymal niche. *Nature*, Vol.459, No.7244, (May 2009), pp. 262-265, ISSN 0028-0836

Schofield, R. (1978) The relationship between the spleen colony-forming cell and the haematopoietic stem cell, *Blood Cells*, Vol.4, No.1-2, pp. 7-25, ISSN 0340-4684

Schröder, N. & Gossler, A. (2002). Expression of Notch pathway components in fetal and adult mouse small intestine. *Gene Expr Patterns*, Vol.2, No.3-4, (December 2002), pp. 247-250, ISSN 1567-133X

Seto. E.S. & Bellen, H.J.(2006) Internalization is required for proper Wingless signaling in Drosophila melanogaster, *J. Cell Biol*, Vol.173, No.1, (April 2006), pp. 95–106, ISSN 0021-9525

Sheldahl, L.C.; Park, M.; Malbon, C.C. & Moon, R.T. (1999) Protein kinase C is differentially stimulated by Wnt and Frizzled homologs in a G-protein-dependent manner, *Curr. Biol*, Vol.9, No.13, (July 1999), pp. 695-698, ISSN 0960-9822

Singh, N.; Phillips, R.A.; Iscove, N.N. & Egan, S.E. (2000). Expression of notch receptors, notch ligands, and fringe genes in hematopoiesis. *Exp hematol*, Vol.28, No.5, (May 2000), pp. 527-534, ISSN 0301-472X.

Slusarski, D.C.; Yang-Snyder, J.; Busa, W.B. & Moon, R.T. (1997) Modulation of embryonic intracellular Ca^{2+} signaling by Wnt-5A, *Dev. Biol.*, Vol.182, No.1, (February 1997), pp. 114-120, ISSN 0012-1606

Smeets, B.; Angelotti, M.L.; Rizzo, P.; Dijkman, H.; Lazzeri, E.; Mooren, F.; Ballerini, L.; Parente, E.; Sagrinati, C.; Mazzinghi, B.; Ronconi, E.; Becherucci, F.; Benigni, A.; Steenbergen, E.; Lasagni, L.; Remuzzi, G.; Wetzels, J. & Romagnani, P. (2009) Renal progenitor cells contribute to hyperplastic lesions of podocytopathies and crescentic glomerulonephritis. *J Am Soc Nephrol*, Vol.20, No.12, (December 2009), pp. 2593-2603, ISSN 1046-6673

Smith, A.G. (2001). Embryo-derived stem cells: of mice and men, *Annu Rev Cell Dev Biol.*, Vol.17, (November 2001), pp. 435-462, ISSN 1081-0706

Snippert, H.J. van der Flier, L.G. Sato, T. van Es, J.H. van den Born, M. Kroon-Veenboer, C. Barker, N. Klein, A.M. van Rheenen, J. Simons, B.D. Clevers, H. (2010). Intestinal crypt homeostasis results from neutral competition between symmetrically dividing Lgr5 stem cells. *Cell*, Vol.143, No.1, (October 2010), pp. 134-144, ISSN 0092-8674

Staal, F.J. & Sen, J.M. (2008). The canonical Wnt signaling pathway plays an important role in lymphopoiesis and hematopoiesis. *Eur. J. Immunol,* Vol.38, No.7, (Jul 2008), pp.1788-1794, ISSN 00142980.

Tajbakhsh, S.; Rocheteau, P. & Le Roux, I. (2009) Asymmetric cell divisions and asymmetric cell fates, *Annu Rev Cell Dev Biol,* Vol.25, (November 2009), pp. 671-699, ISSN 1081-0706

Takashima, S.; Kadowaki, M.; Aoyama, K.; Koyama, M.; Oshima, T.; Tomizuka, K.; Akashi, K. & Teshima T. (2011). The Wnt agonist R-spondin1 regulates systemic graft-versus-host disease by protecting intestinal stem cells. *J Exp Med,* Vol.208, No.2, (February 2011), pp. 285-294, ISSN 0022-1007

Tanigaki, K. & Honjo, T. (2007). Regulation of lymphocyte development by Notch signaling. *Nature Immunology,* Vol.8, No.5, (May 2007), pp. 451-456, ISSN 1529-2916.

Thorner, P.S.; Ho, M.; Eremina, V.; Sado, Y. & Quaggin, S. (2008) Podocytes contribute to the formation of glomerular crescents. *J Am Soc Nephrol,* Vol.19, No.3, (March 2008), pp. 495-502, ISSN 1046-6673.

van de Wetering, M. Sancho, E. Verweij, C. de Lau, W. Oving, I. Hurlstone, A. van der Horn, K. Batlle, E. Coudreuse, D. Haramis, A.P. Tjon-Pon-Fong, M. Moerer, P. van den Born, M. Soete, G. Pals, S. Eilers, M. Medema, R. Clevers, H. (2002). The beta-catenin/TCF-4 complex imposes a crypt progenitor phenotype on colorectal cancer cells. *Cell,* Vol.111, No.2, (October 2002), pp. (241-250), ISSN 0092-8674

Van Den Berg, DJ., Sharma, AK., Bruno, E. & Hoffman, R. (1998). Role of members of the Wnt gene family in human hematopoiesis. *Blood,* Vol.92, No.9, (November 1998), pp. 3189-3202, ISSN 15280020.

van Es, J.H.; van Gijn, M.E.; Riccio, O.; van den Born, M.; Vooijs, M.; Begthel, H.; Cozijnsen, M.; Robine, S.; Winton, D.J.; Radtke, F. & Clevers, H. (2005). Notch/gamma-secretase inhibition turns proliferative cells in intestinal crypts and adenomas into goblet cells. *Nature,* Vol.435, No.7044, (June 2005), pp. 959-963, ISSN 0028-0836

Veeman, M..T.; Axelrod, J.D. & Moon, R.T. A second canon. Functions and mechanisms of beta-catenin-independent Wnt signaling. *Dev. Cell.* Vol.5, No.3, (September 2003), pp. 367-377, ISSN 1534-5807

Vermeulen, L.; De Sousa, E.; Melo, F.; van der Heijden, M.; Cameron, K.; de Jong, J.H.; Borovski, T.; Tuynman, J.B.; Todaro, M.; Merz, C.; Rodermond, H.; Sprick, M.R.; Kemper, K.; Richel, D.J.; Stassi, G. & Medema, J.P. (2010). Wnt activity defines colon cancer stem cells and is regulated by the microenvironment. *Nat Cell Biol,* Vol.12, No.5, (May 2010), pp. 468-476, ISSN 1097-6256

Vermeulen, L.; Sprick, M.R.; Kemper, K.; Stassi, G. & Medema, J.P. (2008). Cancer stem cells--old concepts, new insights. *Cell Death Differ,* Vol.15, No.6, (June 2008), pp. 947-958, ISSN 1350-9047

Vijayaragavan, K., Szabo, E., Bosse, M., Ramos-Mejia, V., Moon, R.T.& Bhatia, M. (2009). Noncanonical Wnt signaling orchestrates early developmental events toward hematopoietic cell fate from human embryonic stem cells. *Cell Stem Cell,* Vol.4, No.3, (March 2009), pp. 248-262, ISSN 19345909.

Vilimas, T.; Mascarenhas, J.; Palomero, T.; Mandal, M.; Buonamici, S.; Meng, F.; Thompson, B.; Spaulding, C.; Macaroun, S.; Alegre, M.L.; Kee, B.; Ferrandom, A.; Miele, L. & Aifantis, I. (2007). Targeting the NF-kappaB signaling pathway in Notch1-induced T-cell leukemia. *Nat Med,* Vol.13, No.1, (January 2007), pp. 70-77, ISSN 1078-8956.

Weng, A.P.; Ferrando, A.A.; Lee, W.; Morris, J.P., 4h; Silverman, L.B.; Sanchez-Irizarry, C; Blacklow, S.C.; Look, A.T. & Aster, J.C. (2004). Activating mutations of NOTCH1 in human T cell acute lymphoblastic leukemia. *Science*, Vol.306, No.5694, (October 2004), pp. 269-271, ISSN 0036-8075.

Westfall, T.A.; Brimeyer, R.; Twedt, J.; Gladon, J.; Olberding, A.; Furutani-Seiki, M. & Slusarski, D.C. (2003) Wnt-5/pipetail functions in vertebrate axis formation as a negative regulator of Wnt/beta-catenin activity, *J. Cell Biol.*, Vol.162, No.5, (September 2003), pp. 889–898, ISSN 0021-9525

Willert, K., Brown, J.D., Danenberg, E., Duncan, AW., Weissman, IL., Reya, T., Yates, JR. 3rd, Nusse, R. (2003). Wnt proteins are lipidmodified and can act as stem cell growth factors. *Nature*, Vol.423, No.6938, (May 2003), pp.448–452, ISSN 00280836.

Wodarz, A. & Nusse, R. (1998) Mechanisms of Wnt signaling in development. *Annu. Rev. Cell Dev. Biol.* Vol.14, (November 1998), pp. 59–88, ISSN 1081-0706

Yamashita YM, Yuan H, Cheng J, Hunt AJ. (2010) Polarity in stem cell division: asymmetric stem cell division in tissue homeostasis, *Cold Spring Harb Perspect Biol*, Vol.2, pp. a001313, ISSN 1943-0264

Yang, Q.; Bermingham, N.A.; Finegold, M.J. & Zoghbi, H.Y. (2001). Requirement of Math1 for secretory cell lineage commitment in the mouse intestine. *Science*, Vol.294, No.5549, (December 2001), pp. 2155-2158, ISSN 0036-8075

Zhao, C., Blum, J., Chen, A., Kwon, HY., Jung, SH., Cook, JM., Lagoo, A. & Reya T. (2007). Loss of beta-catenin impairs the renewal of normal and CML stem cells in vivo. *Cancer Cell*, Vol.12, No.6, (December 2007), pp.528-41, ISSN 15356108.

Zhu, Y., Sun, Z., Han, Q., Liao, L., Wang, J., Bian, C., Li, J., Yan, X., Liu, Y., Shao, C., Zhao, RC.. (2009). Human mesenchymal stem cells inhibit cancer cell proliferation by secreting DKK-1. *Leukemia*, Vol.23, No.5, (May 2009), pp.925-33, ISSN 08876924.

11

Claudins in Normal and Lung Cancer State

V. Morales-Tlalpan[1*], C. Saldaña[2], P. García-Solís[2],
H. L. Hernández-Montiel[2] and H. Barajas-Medina[1]
[1]*Hospital Regional de Alta Especialidad del Bajío,
San Carlos la Roncha, León, Guanajuato,*
[2]*Departamento de Investigación Biomédica. Facultad de Medicina,
Universidad Autónoma de Querétaro, Santiago de Querétaro, Querétaro,*
México

1. Introduction

The epithelial cells form a physical barrier that serves to separate two different environments, and interact with neighboring cells through various kinds of cell-cell communication systems. Among the systems of cell-cell communications have been described 3 types of intercellular junctions: gap-junctions, adherents-junctions and tight-junctions.

The tight junctions are critical for the sealing of cellular sheets and controlling paracellular ion flux. Tight junctions are composed primarily of 3 components: The IgG-like family of junctional adhesion molecules (JAMs), occluding and claudin families. Claudins are the main constituents of tight junctions. The claudin family proteins is composed of approximately 24 transmembrane proteins, all of which are closely related, most of them are well characterized at the level of gene and protein. The claudins are present in variety of normal tissues, hyperplastic conditions, but have also been found in benign neoplams and cancers that exhibit epithelial differentiation. Loss of claudins expression has been reported in various malignant diseases. The differential expression of several members of the family of the claudins in various cancers has been used to confirm the histological identity of certain types of cancer.

The permeability barrier in the terminal airspaces of the lung is due in large part to tight junctions between alveolar epithelial cells, which regulate the flow of molecules between apical and basolateral extracellular compartments. Disruption of the paracellular alveolar permeability barrier is a significant pathological consequence of acute lung injury. Little is known about the expression and localization of claudins in normal bronchial epithelium and lung cancer. So that is in our interest to describe the expression of claudins in normal and lung cancer, also describe the cellular and molecular mechanisms.

2. Tight junctions

The cellular polarity is critical for a variety of cellular functions, such as directed migration, asymmetric cell division and the vectorial transport of molecules. Polarity is studied in

---

* Corresponding Author

epithelial cells where apical and basolateral surface domains with different lipid and protein compositions can be distinguished (Steed, et al., 2010). In vertebrate epithelia, the two membrane domains are separated by tight junctions (TJ), who act as an intramembrane diffusion barrier and also as a paracellular seal that prevents diffusion of molecules across the epithelial cell layer. TJs are structures appearing as discrete sites of fusion between the outer plasma membrane of adjacent cells. The TJ regulates the diffusion of solutes with size and charge selectivity and that it is functionally different in physiologically diverse epithelial cell types. To understand the molecular mechanism controlling TJ structure and function, it is important to determine their molecular composition and organization (Anderson & Cereijido 2001; Steed, et al., 2010).

2.1 Tight junction molecular structure

The molecular components of the TJ have been separated into 3 groups: 1) The integral transmembrane proteins, 2) The peripheral or cytoplasmic and 3) TJ-associated/regulatory proteins. 1) The integral transmembrane proteins are essential for correct assembly of the structure: occludin, claudins and junctional, immunoglobulin superfamily membrane proteins with two extracellular Ig-like domains, including JAM-A, JAM4, coxsackie adenovirus receptor (CAR), and endothelial cell-selective adhesion molecule (ESAM). The integral transmembrane proteins are the critical for correct assembly of the TJ structure and controlling TJ functions via homotypic and heterotypic interactions. 2) The peripheral or cytoplasmic or plaque anchoring proteins: the membrane-associated guanylate kinase (MAGUK) family proteins ZO 1, ZO 2, and ZO 3 bind to the C-terminal cytoplasmic domain of claudins, occludin, tricellulin, and JAM-A. In addition, MAGI-1, MAGI-3, MUPP1, PATJ and ASIP/PAR3 are known to be PDZ domain-containing proteins that directly bind to claudins or other TJ-associated membrane proteins. The plaque anchoring proteins act as a scaffold to bind the raft of TJ molecules together and provide the link to the actin cytoskeleton and the signaling mechanism of the cell. 3) TJ-associated/regulatory proteins – α-catenin, cingulin, paracingulin, etc., (for review see: Blasig, et al., 2006; Furuse 2010; Hamazaki, et al., 2002; Itoh, et al., 1999, Tsukita & Furuse 2000a, 2000b).

2.2 Paracellular transport

Separation of functional compartments is necessary for higher organisms. The structures that separate such compartments, epithelia and endothelia, consist of cell layers with diverse properties according to the organism's actual demands. While such structures prevent uncontrolled diffusion and convection of substances, they also provide selective transport processes via secretion (exocrine and endocrine glands), absorption (intestine), or reabsorption (kidney). Such transport processes are realized via the transcellular pathway involving resorption across the apical membrane, transfer through the cytoplasm, and extrusion at the basolateral membrane. In general, transcellular transport is an energy-dependent process, but it allows the organism to reabsorb substances that are indispensable, even against an existing electrochemical gradient. Moreover, since this pathway is controlled at several steps, it allows fine-tuning according to actual demands. On the other hand, paracellular transport occurs through the intercellular space of adjacent cells. This transport is passive and dependent on an electrochemical gradient. This form of transport allows bulk reabsorption with a minimum of energy expenditure. The key structure of the intercellular space, and thus the major determinant of paracellular transport, is the tight junction (TJ). The ion conductance of tight junctions varies from tissue to tissue and can be experimentally manipulated by expressing or removing specific pores. The specificity of these pores is determined by the claudin composition and, more precisely, by the properties of their extracytoplasmic loops, such as electrostatic interaction sites. The molecular mechanisms that underlie size-selective paracellular diffusion are unclear. However, several studies reported a functional dissociation between transepithelial electrical resistance and size-selective paracellular diffusion upon specific modifications of either junctional components or signaling pathways that affect permeability. Thus, the molecular bases of ion-selective and size-selective permeation seem to be distinct (see Figure1) (Amasheh, et al., 2009; Steed, et al., 2010; Tsukita &Furuse 2000a, 2000b; Will, et al., 2008).

3. Claudins

The tight junctions consist of several components: integral membrane proteins, cytoplasmic proteins and cytoskeletal proteins (Brennan, et al., 2010). To date, a number of integral membrane proteins are associated with TJ, occludin, adhesion molecules, claudin family, etc., (Tsukita & Furuse 2000a, 2000b, Dhawan, et al., 2005; Furuse, 2010). Gradually has been shown that the molecular architecture of these complex is more numerous, the TJ is made up of at least 40 different components (Figure 1) (for review see: Schneeberger & Lynch, 2004; http://www.genome.jp/kegg/pathway/hsa/hsa04530.html). Among the elements that form part of integral membrane proteins, the claudin family (Clds; present active infinitive of *claudeō*, means close), has attracted the attention because of its relatively recent identification, the family includes 24 members in mammals (Furuse 2010, Brennan, et al., 2010), although Tsukita group recently has reported three new genes that code for Cldns 25, 26 and 27 (Mineta, et al., 2011). The Cldns were identified in 1998 by Dr. Tsukita in membrane fractions from chicken liver (enriched with TJ) through sucrose gradients. Among the protein components two bands of 23 kDa were obtained, with similar in size but not identical. Analysis of the amino acid sequence showed that these proteins were structurally related (30% identical at the amino acid sequence), calling claudin 1 and 2, respectively (Furuse et al., 1998). However, earlier reports this had already been described genes with similar sequences (Briehl & Miesfeld, 1991, Katahira, et al., 1997). This information allowed proposes the existence of a large family of proteins. Currently there are over 558 articles that involve claudins studies (updated to April 20, 2011), describing various aspects of molecular, cellular, regulation, operation, including its expression/co-expression, localization in tissues and organs, and their potential involvement in diseases.

Before the discovery of the claudins, it is believed that the tight junctions were composed mainly of occludin. Even thought that occludin and claudin were members of the same family, but the report of the genetic sequence of the Cldns confirmed that these were different from those occludins proteins, and showed no similarity between them (Furuse, et al., 1998a). To date is accepted that the central part proteins responsible for the paracellular barrier are the claudins (Angelow, et al., 2008, Tsukita & Furuse, 2000).

Claudins are found in the tight junction at the interface of the basolateral and apical membranes of polarized epithelial and endothelial cells, and also at paranodes in compact myelin. Transfected claudins are capable of forming tight junction 'strands' or 'fibrils', the freeze-fracture descriptions of a branching and anastomosing network of rows of intramembranous particles characteristic of tight junctions. Claudins are also found in the basolateral membranes, possibly as precursors to the fibrils (Peter & Goodenough 2004).

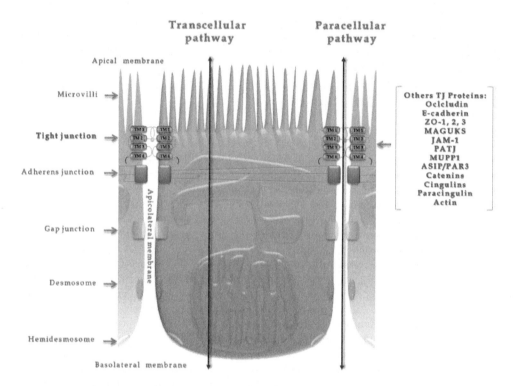

Fig. 1. Schematic diagram of the molecular organization of epithelial.

Cell-cell interactions are mediated by intercellular junctional complexes: gap junctions, adherent junctions, desmosome and thigh junctions, each of which have different had asymmetric distribution in epithelial cells, TJ are located at the apical-basal border and they contribute to maintain cell polarity, regulate the solute and fluid exchange between basolateral and apical domains, and also regulated paracellular permeability. TJ membrane proteins are linked to the cytoskeleton (F-actin) via a complex network of adaptor proteins.

3.1 Claudin evolution

The mechanism by which the family of claudins evolves is unknown; however, the data suggest that this family expanded by gene duplication early in the evolution of chordates (for review sees: Lal-Nag & Morin 2009, Loh, et al., 2004, Kollmar, et al., 2001). When the septate junctions (the corresponding structure of invertebrates), were replaced by tight junctions. In the same way as other groups of genes were extended, the claudins diversified into the body of vertebrates from the chordates, leading to new structures: the skull, pairs of sense organs and appendages (Kollmar, et al., 2001). The search for claudins in the Genebank of *Drosophyla melanogaster* and *Caenorhabditis elegans* showed no similarity to genes previously reported (Venter & Adams, 2001; Kolmar, et al., 2001). However, in *D. melanogaster*, was recently found three genes, which encode for three different proteins that are required in the paracellular transport: Megatrachea (Mega) Sinuous (Sinu) and Kune Kune. Mega, a transmembrane protein homologous to claudins, and show that it acts in septate junctions, this protein has transepithelial barrier function similar to the claudins, and is necessary for normal tracheal cell morphogenesis but not for apico-basal polarity or epithelial integrity (Behr, et al., 2003). The gene sinuous encodes a protein that is molecularly and functionally similar to vertebrate claudins. Sinuous share several characteristic with vertebrate claudins as has all of the amino acids absolutely conserved across vertebrate claudins and has as much sequence similarity to canonical vertebrate claudins as do some of the more divergent vertebrate claudins. Also has functional similarity because it localizes to and is required for the function of paracellular barrier junctions (Wu, et al., 2004). Kune Kune, this protein localizes to septate junctions and is required for junction organization and paracellular barrier function, but not for apical-basal polarity (Nelson, et al., 2010). In *C. elegans* genome database identified four claudin-related, ~20-kDa integral membrane proteins (CLC-1 to -4), which showed sequence similarity to the vertebrate claudins. The expression and distribution of CLC-1 was mainly expressed in the epithelial cells in the pharyngeal region of digestive tubes and colocalized at their intercellular junctions. In CLC-1-deficient worms, the barrier function of the pharyngeal portion of the digestive tubes appeared to be severely in experiments performed with RNA interference. CLC-2 was expressed in seam cells in the hypodermis, and it also appeared to be involved in the hypodermis barrier (Asano, et al., 2003). In addition VAB-9 is a predicted four-pass integral membrane protein that has greatest similarity to BCMP1 (brain cell membrane protein 1, a member of the PMP22/EMP/Claudin family of cell junction proteins) and localizes to the adherents junction domain of *C. elegans* apical junctions. In this nematode *C. elegans* protein VAB-9 regulates adhesion and epidermal morphology (Simske, et al., 2003). In *Danio rerio* (Zebra fish) have been located at least 15 genes for Cldns, some of which have their orthologous in human (Kollmar, et al., 2001), and among non-vertebrate *Halocynthia roretzi* (Sea pineapple) as also found a gene that encodes to claudins (Kollmar, et al., 2001). The presence of these genes suggests that the origin of the claudins may be quite ancient and that a claudin ancestor pre-dates the establishment of the chordates (Kollmar, et al., 2001).

3.2 Claudin structure

The claudins belong to the peripheral myelin protein (PMP22)/ epithelial membrane protein (EMP)/ epithelial membrane protein or membrane protein (MP20)/claudin superfamily of four transmembrane-spanning domains. The 24 mammalian members are 20 to 34 kDa in size (Lal-Nag & Morin, 2009, Peter & Goodenough 2004), and recently others members of the claudin family 25, 26 and 27 were reported (Mineta, et al., 2011). The proteins are predicted, on the basis of hydropathy plots, to have four transmembranal helices (Morita, et al., 1999; Lal-Nag & Morin, 2009), with their NH_2-and COOH-terminal tails extending into the cytoplasm (Lal-Nag & Morin, 2009). Sequence analysis of Cldns has led to classification into two groups: classic claudins (1–10, 14, 15, 17, 19), and non-classic claudins (11–13, 16, 18, 20–24) (Table 2), according to their degree of sequence similarity to conserved structural features at ECL1 for classic claudins (Krause, et al., 2008). The typical claudin protein contains a small intracellular cytoplasmic NH_2-terminal sequence of approximately 4 to 5 residues followed by a huge extracellular loop (EL1) of approximately 60 residues, a short 20-residue intracellular loop, another extracellular loop (EL2) of about 24 residues, and a COOH-terminal cytoplasmic. The size of the COOH-terminal tail is more variable in length; it is typically between 21 and 63 residues. The amino acid sequences of the first and fourth transmembrane domains are highly conserved among Cldns, and the second and third are more diverse. The first loop contains several charged amino acids and, as such, is thought to influence paracellular charge selectivity, and two highly conserved cysteine residues are hypothesized to increase protein stability by the formation of an intramolecular disulfide bond. It is assumed that the first extracellular loop is critical for determining the paracellular tightness and the selective paracellular ion permeability (see Table 2). It has been suggested that the second extracellular loop, can form dimers with Cldns on opposing cell membranes through hydrophobic interactions between conserved aromatic residues and that the second extracellular loop may cause narrowing of the paracellular cleft (Lal-Nag & Morin 2009; Krause, et., al 2008, 2009).

The region that shows the most sequence and size heterogeneity among the claudin proteins is the COOH-terminal tail. It contains a PDZ-domain-binding motif that allows claudins to interact directly with cytoplasmic scaffolding proteins, such as the TJ-associated proteins MUPP1, PATJ, ZO-1, ZO-2 and ZO-3, and MAGUKs (see figure 1 and 2). Furthermore, the COOH-terminal tail upstream of the PDZ-binding motif is required to target the protein to the TJ complex, and also functions as a determinant of protein stability and function. The COOH-terminal tail is the target of various post-translational modifications, such as serine/threonine and tyrosine phosphorylation and palmitoylation, which can significantly alter claudin localization and function. Most cell types express multiple claudins, and the homotypic and heterotypic interactions of claudins from neighboring cells allow strand pairing and account for the TJ properties, although it appears that heterotypic head-to-head interactions between claudins belonging to two different membranes are limited to certain combinations of claudins, and stoichiometry have yet to be determined (Lal-Nag & Morin, 2009; Peter, & Goodenough 2004).

The extracellular domains of claudins in one cell are thought to interact with those in an opposing cell to form a new class of ion channel (see Figure 1). These channels confer ion selectivity to the paracellular pathway between luminal and basolateral extracellular compartments. The permeability properties of the paracellular pathway have the biophysical characteristics of conventional ion channels, including ion selectivity, anomalous mole fraction effects, pH dependence and a diameter of $\sim 6 Å$. Exchanging the first extracellular loop between claudin-2 and claudin-4 changes the Na^+ and Cl^- selectivity of the paracellular pathway in cultured epithelial cells (Peter & Goodenough 2004; Ben-Yosef, et al., 2003).

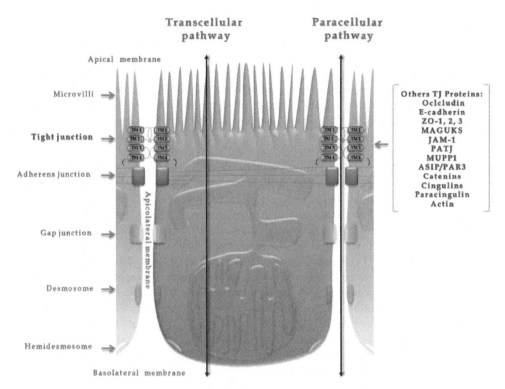

Fig. 2. Schematic representation of structure and molecular interactions of typical claudins. Claudins are proteins with four transmembrane-spanning domains (TM 1, TM 2, TM 3, and TM 4), two extracellular loops (EL1 and EL2) related with paracellular ion selectivity and oligomerization respectively. The NH_2 and COOH terminal are localized in the cytosol, intracellular loops are posttranscriptional modify. COOH terminal contains PDZ-binding domain that is important for signal transduction.

4. Epithelial cells

Epithelia have three basic functions in multicellular organisms, first, cover the outer surface of the body and the cavities and formed a physical barrier that separates two environments. This physical barrier provides protection against mechanical damage, the entry of microorganisms and water losses. Epithelia are also involved in secretion and absorption process. These three functions of epithelia are given primarily by the type of cellular arrangements that characterizes them. Epithelial cells are polarized and form sheets of cells attached to each other through complex mechanisms. Some of these mechanisms of cell-cell interactions will be discussed in this chapter. Thus in an epithelial cell it can distinguish two basic functional components: a) the basal domain and b) the apical domain (Feing & Muthuswamy, 2009). The basal domain participated in cellextracelular matrix (ECM) interactions; in particular, it is in contact with a structure formed by type IV collagen, laminin and proteoglycans called basal lamina (Gumbiner, 1996). The basal lamina allows epithelial cells to be attached to the underlying connective tissue. In specialized epithelial tissue basal domain is also involved in endocrine secretion. The apical domain is located in the opposite direction of the basal. Depending on its functions basal domain is involved in exocrine secretion and absorption. Alternatively, another fundamental property of epithelial tissue is the close cohesion between the cells, which allows the formation adherents selectively permeable layers that are at once very strong mechanical barriers. Cell-cell interactions are mediated by intercellular junctional complexes that consist of gap junctions (GJ), adherent junctions (AJ), desmosomes (Ds) and thigh junctions (TJ), each of which have different functions and properties (Itoh & Bissell et al., 2003; Feing & Muthuswamy, 2009). These junctional complexes had asymmetric distribution in epithelial cells, TJ are located at the apical-basal border and they contribute to maintain cell polarity, regulate the solute and fluid exchange between basolateral and apical domains, and also regulated paracellular permeability (Itoh & Bissell et al., 2003; Feing & Muthuswamy, 2009). TJ are widely explained in section 3 of the present chapter. On the other hand, AJ are basal to TJ and they are considered the primary determinants of cell-cell adhesion. AJ are ubiquitously represented by cadherins, transmembranal Ca^{2+}-dependent receptors, which form complex with catenins, cytoplasmic plaque proteins, and actin cytoskeleton. Cell adhesion regulates the organization of cell patters and architecture of tissues (Gumbiner, 1996). In epithelia, it can be found different cell shapes (flat, cylindrical or cubic) in at least three basic forms of cell arrangement, a) simple epithelia (single sheet of cells), b) stratified (multiple cell sheets) and c) pseudostratified (single sheets of cells with several sizes and shapes that give the appearance of true stratified epithelia).

4.1 Lung epithelial cells

The adult human lung is lined by specialized types of airway epithelia organized in tree-like form with three anatomical and functional units: a) trachea and bronchi (tracheobronchial), b) bronchioles and acinar ducts (bronchiolar), and c) peripheral saccular-alveolar structures (alveoli) (Maeda, et al., 2007). Tracheabronchial and bronchiolar units form conducts that provide inhaled gases to alveoli unit, there; epithelial alveolar cells and capillaries exchange oxygen and carbon dioxide required for respiration (Maeda, et al., 2007). Each of these airway units is composed of distinct types of epithelial cells that are important for maintaining normal lung function, in Table 1 is shown a summary of the main types of intrapulmonary epithelial cells with their respective function.

Trachea and bronchi are characterized by pseudostratified epithelia, whereas bronchioles, acinar ducts and alveoli are mainly characterized by simple cubic epithelia. Epithelial cells throughout of airway show major functions such as protection (Ciliated cells), progenitor cells (Basal and Clara cells), exocrine secretion (Goblet, Clara and Alveolar type II cells), endocrine secretion (Neuroendocrine pulmonary cells), and gas exchange (Alveolar type I cells) (Table 1) (Herzog, et al., 2008; Linnoila, 2006; Maeda, et al., 2007; Reynolds, et al., 2007; Rock, et al., 20010; Rogers, 2007).

Airway unit	Epithelia type	Epithelial cell types	Cell function	Biomarkers
T R A C H E O B R O N C H I A L	Pseudo-stratified	Goblet	Mucus secretion	Mucin 5AC (MUC5AC)
		Basal	Attachment with the basement membrane; Progenitor cells	Transcription factor p63, cytokeratins 5 and 14
		Clara	Mucus secretion; Mucocilliary clearance; Detoxify xenobiotics and oxidant gasses; Progenitor of ciliated cells; Regulation of immune system	Uteroglobin, Surfactant apoproteins A, B and D
		Ciliated	Mucocilliary clearance	Calpastatin, ezrin
B R O N C H I O L A R	Simple columnar-cuboidal	Clara	See above	See above
		Ciliated	See above	See above
		Pulmonary neuroendo-crine cells	Endocrine and paracrine secretion in lung development; Oxygen sensors	Gastrin-releasing peptide, bombesin, calcitonin gene-related peptide, synaptophysin
A L V E O L I	Simple columnar	Alveolar type I cells (squamous cells)	Mediate gas exchange	T1-α, aquaporin-5 (AQP-5)
		Alveolar type II cells	Synthesis, secretion and recycle the lipid and protein component of surfactant Innate immunity	Surfactant apoprotein C and ATP-binding cassette A3 (ABCA3)

Table 1. Characteristics of intrapulmonary epithelia of human. Sources: Herzog, et al., 2008; Linnoila, 2006; Maeda, et al., 2007; Reynolds, et al., 2007; Rock, et al., 2010; Rogers, 2007.

The complex patterns of intrapulmonary epithelial cells, organization, numbers and types of cells, are regulated by several humoral signals and cell-cell interactions. It is know that several physiopathological conditions could modify the lung epithelial cell pattern such as infection, cytokines, inflammatory mediators, pollutants and injury that are associated with common airway diseases, including chronic obstructive pulmonary disease, asthma, cystic fibrosis and cancer (Ballaz & Mulshine, 2003; Maeda, et al., 2007; Rock, et al., 2010).

4.2 Lung cancer

Lung cancer is one of the most important epithelial neoplasias in the world with high incidence and mortality (Jemal, et al., 2011). Currently lung cancer is the most commonly diagnosed cancer, as well as, the leading cause of cancer death in males in worldwide. Among females, it represents the fourth most commonly diagnosed cancer and the second leading cause of cancer death. Even more, lung cancer is the leading cause of cancer-related deaths around the world, accounting for more deaths than those caused by three of the most diagnosed cancers combined (prostate, breast and colorectal cancers). In 2008 estimated lung cancer related deaths in worldwide were 1, 378, 400 whereas estimated related cancer deaths by prostate, breast and colorectal cancers were 1, 004, 900 (Jemal, et al., 2011). Moreover, whereas the five-year survivor over time was improved in prostate, breast and colorectal carcinomas in last 15 years, at 99%, the five-year survivor rate of lung cancer was relatively unchanged at 15% (Borczuck, et al., 2009; Schwartz, et al., 2007). The high mortality of lung cancer could be explained in part by histological heterogeneity and late detection (Borczuck, et al., 2009; Schwartz, et al., 2007). On the other hand, smoking is the most important cause of lung cancer, 80-90% of lung cancer cases are associated with smoking but only 15%of the smokers developed lung cancer and 10% of lung cancers occur in never-smokers (Borczuck, et al., 2009; Schwartz, et al., 2007). Other lung carcinogens are asbestos, arsenic, radon, polycyclic aromatic hydrocarbons and air pollution (Jemal, et al., 2011). The World Health Organization (WHO) reported that the cancer is a leading cause of death worldwide and accounted for 7.6 million deaths (around 13% of all deaths) in 2008. The main types of cancer are: lung (1.4 million deaths), stomach (740 000 deaths), liver (700 000 deaths), colorectal (610 000 deaths) and breast (460 000 deaths). More than 70% of all cancer deaths occurred in low- and middle-income countries. Deaths from cancer worldwide are projected to continue to rise to over 11 million in 2030 (http://www.who.int/mediacentre/factsheets/fs297/en/index.html).

Lung cancer is divided into two histological types, non-small cell lung cancer (NSCLC) and small cell lung cancer (SCLC). NSCLC is the most common lung cancer; it represents between 80-85% of cases and consists in a heterogeneous group of cancers that can divide into three major subtypes: squamous cell carcinoma (SCC), adenocarcinoma (AC) and large-cell carcinoma. This histological heterogeneity is only the reflection of lung cancer biology complexity and it has very important implications in initiation, treatment and prognosis.

Multi-step models of carcinogenesis, genetic and genomic approaches are developed to understand lung carcinogenesis (Borczuck, et al., 2009; Schwartz, et al., 2007; Wistuba, et al., 2002). In this way, an emergent field of lung carcinogenesis is open, the role that play the loss of polarity and dysregulation cell-cell adhesion molecules in initiation and invasion process of cancer cells.

CLDN	Aa	MW	Cl	Transport	Organism	Detection	Disease
1	211	22,744	C	PT	H, C, R, M, D, Cw	Protein RTi_PCR	Up: LC-S Down: LC-A
2	230	24,549	C	PT: Na+, K+, water	H, C, R, M, D,Cw, G	RTi_PCR	
3	220	23,319	C	PB: mono, divalent ions;	H, C, R, M		
4	209	22,077	C	PB: Na+ PT: Cl-	H, C, R, M	RTi_PCR	Up: P-MC
5	218	23,147	C	PB	H, C, R, M	RTi_PCR	Up: LC-A Down: LC-S
6	220	23,292	C		H, C, R, M		
7	211	22,390	C	PB: Na+ PT: Cl-	H, C, R, M	RTi_PCR	
8	225	24,845	C	PBdivalent cations	H, C, R, M	RTi_PCR	
9	217	22,848	C	PT: Na+, K+	H, R, M	RTi_PCR	
10	a: 226 b: 228	24,251 24,488	NC NC		H, C, R, M H, C, R, M		
11	207	21,993	NC		H, C, R, M		
12	244	27,110	NC		H, C, R, M	RTi_PCR	
13					M		
14	239	25,699	C	PB: K+	H, C, R, M	IE	
15	228	24,356	NC		H, C, R, M		
16	305	33,836	NC	PT: Na+, cations	H, C, R, M	RTi_PCR	
17	224	24,603	C		H, C, R, M		
18	a: 261 b: 261	27,856 27,720	NC NC		H, C, R, M H, C, R, M	RTi_PCR	
19	a: 224 b: 211	23,229 22,076	C C	PT: Na+ cations	H, C, R, M H, C, R, M		
20	219	23,515	C		H, C, R, M		
21	229	25,393	NC		H, C, M		
22	220	25,509	NC		H, C, R, M		
23	292	31,915	NC		H, C, R, M		
24	205	22,802	NC		H, C, R, M		
25	276				M		
26	223				M		
27	320				M		

Table 2. Molecular characteristics of claudins. Gene; Aminoacids (Az); Molecular Weight (MW); Classification: Classical (C) and Non-Classical (NC); Paracellular Transport (PT) and Paracellular Barrier (PB); Organism: Human (H), Chimpanzee (C) Rat (R), Mouse (M), Dog (D), Cow (Cw), Chicken (G); Lung expression: Real-time PCR (RT_PCR); Disease type: Adenocarcinoma (LC-A), Lung Cancer (LC-S), Pleura (metastatic adenocarcinoma) (P-MC). Source: Amasheh, et al., 2009; Angelow, et al., 2006; Hewitt, et al., 2006; Hou, et al., 2006; 2007, 2008, 2009; Krause, et al., 2008; Milatz, et al., 2010; Mineta, et al., 2011; Singh, et al., 2010; Wen, et al., 2004; http://www.genecards.org/.

4.3 Claudins and lung cancer

The tight junctions exist in lung epithelium, but knowledge of their development, normal, disease and cancer phases, but exact function and distribution in the developing and adult human lung is incomplete. Epithelial cells often express multiple claudin types, and they show a variable expression profile in different epithelia. Similarly, expression of different claudins varies between different types of epithelial, endothelial and mesothelial tumors Kaarteenaho, et al., 2010). The expression of the different claudins during ontogenesis of human lung might vary since they have distinct expression profiles in normal human lung (Kaarteenaho, et al., 2010). Disruption of the paracellular alveolar permeability barrier is a significant pathologic consequence of acute lung injury. The permeability barrier in terminal airspaces of the lung is due in large part to tight junctions between alveolar epithelial cells, which regulate the flow of molecules between extracellular apical and basolateral compartments (Boitano, et al., 2004). In humans, very little is known about the expression and localization of claudins in normal bronchial epithelium and also in lung cancer. The expression of different claudins was studied in freshly excised human airways using immunfluorescence staining and confocal microscopy bronchi and bronchioles expressed claudins 1, 3, 4, 5, and 7, but not claudins 2, 6, 7, 9, 11, 15, and 16 (Coyne, et al., 2003). Claudins 1, 3, 4, 5, and 7 are expressed in developing human lung from week 12 to week 40 with distinct locations and in divergent quantities. The expression of claudin 1 was restricted to the bronchial epithelium, whereas claudin 3, 4 and 7 were positive also in alveolar epithelium as well as in the bronchial epithelium. All claudins studied are linked to the development of airways, whereas claudin 3, 4, 5 and 7, but not claudin 1, are involved in the development of acinus and the differentiation of alveolar epithelial cells (Kaarteenaho, et al., 2010). In human lung tumors by using cDNA microarray and in large cell carcinomas relatively low levels of claudin-4 and 7 expressions were found as compared with other types of lung cancer, such as adenocarcinoma, squamous cell carcinoma and small cell lung cancer (Garber, et al., 2001). Claudin 1 expression was stronger in squamous cell carcinomas than in adenocarcinomas, whereas claudin 4 and claudin 5 expression was stronger in adenocarcinomas (Jung, et al., 2009). 10 Hydroxycamptothecin (HCPT) elicits strong anti-cancer effects and is less toxic making it widely used in recent clinical trials, HCPT-loaded nanoparticles reduced the expression of cell-cell junction protein claudins, E-cadherin, and ZO-1, and transmission electron microcopy demonstrated a disrupted tight junction ultrastructure (Zhang, et al., 2011). Keratinocyte growth factor (KGF) augments barrier function in primary rat alveolar epithelial cells grown in culture, specifically whether KGF alters tight junction function via claudin expression. KGF significantly increased alveolar epithelial barrier functions in culture as assessed by transepithelial electrical resistance (TER) and paracellular permeability (LaFemina, et al., 2010). Alveolar epithelial cells cultured for 5 days formed high-resistance barriers, which correlated with increased claudin-18 localization to the plasma membrane (Kolval, et al., 2010). Bronchial BEAS-2B cells and SK-LU1 cells respond to tobacco smoke by changing their claudin mRNA synthesis and resulting tight junction permeability changes may thus contribute to tobacco induced carcinogenesis both during initiation and progression (Merikallio, et al., 2011). Zeb1 and twist regulate expression of genes which take part in epitheliomesenchymal transition (EMT). Carcinomas metastatic to the lung showed a significantly higher expression of these transcriptional factors than primary lung tumors, indicating their probable importance in the metastatic process. Zeb1 and twist were inversely associated with several claudins, indicating a role in their down-regulation (Merikalio, et al., 2011).

Despite many questions, recent insights into the molecular structure of tight junctions and claudins are beginning to explain their important physiological differences and contribution to paracellular transport and their importance in several disease and neoplasias, as well as, in healthy tissues.

5. Acknowledgment

This study was supported by PROMEP to Dr. Carlos Saldaña and Dr. Pablo García-Solís. We thank Dr. Gerardo Ortega and Dirección de Planeación, Enseñanza e Investigación (Hospital Regional de Alta Especialidad del Bajío) for their support during the elaboration of this manuscript.

6. References

Amasheh, S.; Milatz, S.; Krug, S.M.; Bergs, M.; Amasheh, M.; Schulzke, J.D. & Fromm, M. (2009). Na+ absorption defends from paracellular back-leakage by claudin-8 upregulation. *Biochem Biophys Res Commun*, Vol. 378, No. 1, (January 2009), pp. 45-50, ISSN: 0006-291X

Amasheh, S., Milatz, S., Krug, S.M., Markov, A.G., Gunzel, D., Amasheh, M. & Fromm, M. (2009). Tight junction proteins as channel formers and barrier builders. *Ann N Y Acad Sci*, Vol. 1165, (May 2009), pp. 211-219, ISSN: 0077-8923

Angelow, S., Kim, K.J. & Yu, A.S. (2006). Claudin-8 modulates paracellular permeability to acidic and basic ions in MDCK II cells. *J Physiol*, Vol. 571, No. Pt 1, (December 2005), pp. 15-26, ISSN: 0022-3751, 1469-7793

Asano, A., Asano, K., Sasaki, H., Furuse, M. & Tsukita, S. (2003). Claudins in Caenorhabditis elegans: their distribution and barrier function in the epithelium. *Curr Biol*, Vol. 13, No. 12, (June 2003), pp. 1042-1046, ISSN: 0960-9822

Ballaz, S. & Mulshine, J.L. (2003). The potential contributions of chronic inflammation to lung carcinogenesis. *Clin Lung Cancer*, Vol. 5, No. 1 (July 2003), pp. 46-62, ISSN: 1525-7304, 938-0690

Behr, M., Riedel, D. & Schuh, R. (2003). The claudin-like megatrachea is essential in septate junctions for the epithelial barrier function in Drosophila. *Dev Cell*, Vol. 5, No. 4, (October 2003), pp. 611-620, ISSN: 1534-5807, 1878-1551

Ben-Yosef, T., Belyantseva, I.A., Saunders, T.L., Hughes, E.D., Kawamoto, K., Van Itallie, C.M., Beyer, L.A., Halsey, K., Gardner, D.J., Wilcox, E.R., Rasmussen, J., Anderson, J.M., Dolan, D.F., Forge, A., Raphael, Y., Camper, S.A. & Friedman, T.B. (2003). Claudin 14 knockout mice, a model for autosomal recessive deafness DFNB29, are deaf due to cochlear hair cell degeneration. *Hum Mol Genet*, Vol. 12, No. 16, (August 2003), pp. 2049-2061, ISSN: 1460-2083, 0964-6906

Blasig, I.E., Winkler, L., Lassowski, B., Mueller, S.L., Zuleger, N., Krause, E., Krause, G., Gast, K., Kolbe, M. & Piontek, J. (2006). On the self-association potential of transmembrane tight junction proteins. *Cell Mol Life Sci*, Vol. 63, No. 4, (February 2006), pp. 505-514, ISSN: 1420-682X, 1420-9071

Borczuk, A.C., Toonkel, R.L. & Powell, C.A. (2009). Genomics of lung cancer. *Proc Am Thorac Soc*, Vol. 6, No. 2, (April 2009), pp. 152-158, ISSN: 1546-3222

Boitano, S., Safdar, Z., Welsh, D.G., Bhattacharya J., & Koval M. (2004). Cell-cell interactions in regulating lung function. *Am J Physiol Lung Cell Mol Physiol, Vol.*287, No. 3, (September 2004), L455 - L459, ISSN: 1040-0605, 1522-1504

Brennan, K., Offiah, G., McSherry, E.A. & Hopkins, A.M. (2010). Tight junctions: a barrier to the initiation and progression of breast cancer? *J Biomed Biotechnol*, Vol. 2010, (November 2009), pp. 460607, ISSN: 1110-7243, 1110-7251

Briehl, M.M. & Miesfeld, R.L. (1991). Isolation and characterization of transcripts induced by androgen withdrawal and apoptotic cell death in the rat ventral prostate. *Mol Endocrinol*, Vol. 5, No. 10, (October 1991), pp. 1381-1388, ISSN: 0888-8809, 1944-9917

Cereijido M & Anderso J.M. (1991). Tight *Junctions* (Kindle Edition), CRC Press, ISBN 0849323835, USA

Coyne, C.B., Gambling, T.M., Boucher, R.C., Carson, J.L. & Johnson L.G. (2003). Role of claudin interactions in airway tight junctional permeability, *Am J Physiol Lung Cell Mol Physiol* Vol, 285, No. 5 (November 2003), pp. 1166–1178, ISSN: 1040-0605, 1522-1504

Dhawan, P., Singh, A.B., Deane, N.G., No, Y., Shiou, S.R., Schmidt, C., Neff, J., Washington, M.K. & Beauchamp, R.D. (2005). Claudin-1 regulates cellular transformation and metastatic behavior in colon cancer. *J Clin Invest*, Vol. 115, No. 7, (July 2005), pp. 1765-1776, ISSN: 0021-9738, 1558-8238

Feigin, M.E. & Muthuswamy, S.K. (2009). Polarity proteins regulate mammalian cell-cell junctions and cancer pathogenesis. *Curr Opin Cell Biol*, Vol. 21, No. 5, (October 2009), pp. 694-700, ISSN: 0955-0674, 1879-0410

Furuse, M., Fujita, K., Hiiragi, T., Fujimoto, K., & Tsukita, S. (1998). Claudin-1 and -2: novel integral membrane proteins localizing at tight junctions with no sequence similarity to occludin. *J Cell Biol*, Vol. 141, No. 7, (June 1998), pp. 1539-1550, ISSN: 0021-9525, 1540-8140

Furuse, M. Molecular basis of the core structure of tight junctions. (2010). *Cold Spring Harb Perspect Biol*, Vol. 2, No. 1, (January 2010), pp. a002907, ISSN: 0014-5793, 1873-3468

Garber, M.E., Troyanskaya, O.G., Schluens, K., Petersen, S., Thaesler, Z., Pacyna-Gengelbach, M., van de Rijn, M., Rosen, G.D., Perou, C.M., Whyte R.I., Altman, R.B., Brown, P.O., Botstein, D., & Petersen, I. (2001) Diversity of gene expression in adenocarcinoma of the lung. *Proc Nat Acad Sci USA*, Vol. 98 No. 24, (November 2001), pp. 13784–13789, ISSN: 0027-8424, 1091-6490

Gumbiner, B.M. (1996). Cell adhesion: the molecular basis of tissue architecture and morphogenesis. *Cell*, Vol. 84, No. 3, (February 1996), pp. 345-357, ISSN: 0092-8674, 1097-4172

Hamazaki, Y., Itoh, M., Sasaki, H., Furuse, M. & Tsukita, S. (2002). Multi-PDZ domain protein 1 (MUPP1) is concentrated at tight junctions through its possible interaction with claudin-1 and junctional adhesion molecule. *J Biol Chem*, Vol. 277, No. 1, (October 2001), pp. 455-461, ISSN: 0021-9258, 1083-351X

Herzog, E.L., Brody, A.R., Colby, T.V., Mason, R. & Williams, M.C. (2008). Knowns and unknowns of the alveolus. *Proc Am Thorac Soc*, Vol. 5, No. 7, (September 2008), pp. 778-782, ISSN: 1546-3222, 1943-5665

Hewitt, K.J., Agarwal, R. & Morin, P.J. (2006). The claudin gene family: expression in normal and neoplastic tissues. *BMC Cancer*, Vol. 6, (July 2006), pp. 186, ISSN:1471-2407

Hou, J., Gomes, A.S., Paul, D.L. & Goodenough, D.A. (2006). Study of claudin function by RNA interference. *J Biol Chem*, Vol. 281, No. 47, (November 2006), pp. 36117-36123, ISSN: 0021-9258, 1083-351X

Hou, J., Shan, Q., Wang, T., Gomes, A.S., Yan, Q., Paul, D.L., Bleich, M. & Goodenough, D.A. (2007). Transgenic RNAi depletion of claudin-16 and the renal handling of magnesium. *J Biol Chem*, Vol. 282, No. 23, (June 2007), pp. 17114-17122, ISSN: 0021-9258, 1083-351X

Hou, J., Renigunta, A., Konrad, M., Gomes, A.S., Schneeberger, E.E., Paul, D.L., Waldegger, S. & Goodenough, D.A. (2008). Claudin-16 and claudin-19 interact and form a cation-selective tight junction complex. *J Clin Invest*, Vol. 118, No. 2, (February 2008), pp. 619-628, ISSN: 0021-9738, 1558-8238

Hou, J., Renigunta, A., Gomes, A.S., Hou, M., Paul, D.L., Waldegger, S. & Goodenough, D.A. (2009). Claudin-16 and claudin-19 interaction is required for their assembly into tight junctions and for renal reabsorption of magnesium. *Proc Natl Acad Sci U S A*, Vol. 106, No. 36, (September 2009), pp. 15350-15355, ISSN: 0027-8424, 1091-6490

Itoh, M., Furuse, M., Morita, K., Kubota, K., Saitou, M. & Tsukita, S. (1999). Direct binding of three tight junction-associated MAGUKs, ZO-1, ZO-2, and ZO-3, with the COOH termini of claudins. *J Cell Biol*, Vol. 147, No. 6, (December 1999), pp. 1351-1363, ISSN: 0021-9525, 1540-8140

Itoh, M. & Bissell, M.J. (2003). The organization of tight junctions in epithelia: implications for mammary gland biology and breast tumorigenesis. *J Mammary Gland Biol Neoplasia*, Vol. 8, No. 4, (October 2003), pp. 449-462, ISSN:1083-3021, 1573-7039

Jemal, A., Bray, F., Center, M.M., Ferlay, J., Ward, E. & Forman, D. (2011) Global cancer statistics. *CA Cancer J Clin*, Vol. 61, No. 2, (March-April 2011), pp. 69-90, ISSN: 0007-9235, 1542-4863

Jung, J.H., Jung, C.K., Choi, H.J., Jun, K.H., Yoo, J., Kang, S.J. & Lee, K.Y. (2009). Diagnostic utility of expression of claudins in non-small cell lung cancer: different expression profiles in squamous cell carcinomas and adenocarcinomas. *Pathol Res Pract*, Vol. 205, No. 6, (February 2009), pp. 409-416, ISSN: 0344-0338, 1618-0631

Kaarteenaho, R., Merikallio, H., Lehtonen, S., Harju, T., & Soini Y. (2010). Divergent expression of claudin -1, -3, -4, -5 and -7 in developing human lung. *Respir Res*, Vol. 17 No. 11 (May 2010), pp.59, ISSN: 1465-9921, 1465-993X

Katahira, J., Inoue, N., Horiguchi, Y., Matsuda, M. & Sugimoto, N. (1997). Molecular cloning and functional characterization of the receptor for Clostridium perfringens enterotoxin. *J Cell Biol*, Vol. 136, No. 6, (March 1997), pp. 1239-1247, ISSN: 0021-9525, 1540-8140

Kollmar, R., Nakamura, S.K., Kappler, J.A. & Hudspeth, A.J. (2001). Expression and phylogeny of claudins in vertebrate primordia. *Proc Natl Acad Sci U S A*, Vol. 98, No. 18, (August 1991), pp. 10196-10201, ISSN: 0027-8424, 1091-6490

Koval, M., Ward, C., Findley, M.K., Roser-Page, S., Helms, M.N., & Roman J. (2010). Extracellular matrix influences alveolar epithelial claudin expression and barrier function. *Am J Respir Cell Mol Biol*, Vol. 42, No. 2, (May 2009), pp. 172-180. ISSN: 1044-1549, 1535-4989

Krause, G., Winkler, L., Mueller, S.L., Haseloff, R.F., Piontek, J. & Blasig, I.E. (2008). Structure and function of claudins. *Biochim Biophys Acta*, Vol. 1778, No. 3, (March 2008), pp. 631-645, ISSN: 0006-3002, 0006-3002

Krause, G, Winkler L, Piehl C, Blasig I, Piontek J and Müller SL. (2009). Structure and function of extracellular claudin domains. *Ann N Y Acad Sci.* Vol 1165, (2009 May), pp. 34-43, ISSN: 0077-8923, 1749-6632

LaFemina, M.J., Rokkam, D., Chandrasena, A., Pan, J., Bajaj, A., Johnson, M., & Frank, J.A. (2010). Keratinocyte growth factor enhances barrier function without altering claudin expression in primary alveolar epithelial cells. *Am J Physiol Lung Cell Mol Physiol,* Vol. 299, No. 6, (September 2010), pp.L724-L734, ISSN: 1040-0605, 1522-1504

Lal-Nag, M. & Morin, P.J. (2009). The claudins. *Genome Biol,* Vol. 10, No. 8, (August 2009), pp. 235, ISSN: 1465-6906, 1465-6914

Linnoila, R.I. (2006). Functional facets of the pulmonary neuroendocrine system. *Lab Invest,* Vol. 86, No. 5, (May 2006), pp. 425-444, ISSN: 0023-6837, 1530-0307

Loh, Y.H., Christoffels, A., Brenner, S., Hunziker, W. & Venkatesh, B. (2004). Extensive expansion of the claudin gene family in the teleost fish, Fugu rubripes. *Genome Res,* Vol. 14, No. 7, (June 2004), pp. 1248-1257, ISSN: 1088-9051, 1549-5469

Maeda, Y., Dave, V. & Whitsett, J.A. (2007). Transcriptional control of lung morphogenesis. *Physiol Rev,* Vol. 87, No. 1, (January 2007), pp. 219-244, ISSN: 0031-9333, 1522-1210

Merikallio, H., Kaarteenaho, R., Pääkkö, P., Lehtonen, S., Hirvikoski, P., Mäkitaro, R., Harju, T., & Soini, Y. (2011). Zeb1 and twist are more commonly expressed in metastatic than primary lung tumours and show inverse associations with claudins. *J Clin Pathol,* Vol. 64, No. 2, (December 2010), pp.136-140. ISSN: 0021-9746, 1472-4146.

Merikallio, H., Kaarteenaho, R., Paakko, P., Lehtonen, S., Hirvikoski, P., Makitaro, R., Harju, T. & Soini, Y. (2011). Impact of smoking on the expression of claudins in lung carcinoma. *Eur J Cancer,* Vol. 47, No. 4, (November 2010), pp. 620-630, ISSN: 0014-2964

Milatz, S., Krug, S.M., Rosenthal, R., Gunzel, D., Muller, D., Schulzke, J.D., Amasheh, S. & Fromm, M. (2010). Claudin-3 acts as a sealing component of the tight junction for ions of either charge and uncharged solutes. *Biochim Biophys Acta,* Vol. 1798, No. 11, (July 2010), pp. 2048-2057, ISSN: 0006-3002

Mineta, K., Yamamoto, Y., Yamazaki, Y., Tanaka, H., Tada, Y., Saito, K., Tamura, A., Igarashi, M., Endo, T., Takeuchi, K. & Tsukita S. (2011). Predicted expansion of the claudin multigene family. *FEBS Lett,* Vol. 585, No. 4, (January 2011), pp. 606-612 ISSN: 0014-5793, 1873-3468

Moldvay, J., Jackel, M., Paska, C., Soltesz, I., Schaff, Z. & Kiss, A. (2007). Distinct claudin expression profile in histologic subtypes of lung cancer. *Lung Cancer,* Vol. 57, No. 2, (April 2007), pp. 159-167, ISSN: 0169-5002, 1872-8332

Morita, K., Furuse, M., Fujimoto, K. & Tsukita, S. (1999). Claudin multigene family encoding four-transmembrane domain protein components of tight junction strands. *Proc Natl Acad Sci U S A,* Vol. 96, No. 2, (January 1999), pp. 511-516, ISSN: 0027-8424, 1091-6490

Nelson, K.S., Furuse, M. & Beitel, G.J. (2010). The Drosophila Claudin Kune-kune is required for septate junction organization and tracheal tube size control. *Genetics,* Vol. 185, No. 3, (April 2010), pp. 831-839, ISSN: 0016-6731, 1943-263

Peter, Y. & Goodenough D. Claudins. *Curr Biol.* Vol.14, No.8, (Apr 2004), pp.R293-R294, ISSN: 0960-9822, 1879-0445

Peter, Y., Comellas, A., Levantini, E., Ingenito, E.P. & Shapiro, S.D. (2009). Epidermal growth factor receptor and claudin-2 participate in A549 permeability and remodeling:

implications for non-small cell lung cancer tumor colonization. *Mol Carcinog*, Vol. 48, No. 6, (June 200), pp. 488-497, ISSN: 0899-1987, 1098-2744

Reynolds, S.D. & Malkinson, A.M. (2010). Clara cell: progenitor for the bronchiolar epithelium. *Int J Biochem Cell Biol*, Vol. 42, No. 1, (January 2010), pp. 1-4, ISSN: 1357-2725, 1878-5875

Rock, J.R., Randell, S.H. & Hogan, B.L. (2010). Airway basal stem cells: a perspective on their roles in epithelial homeostasis and remodeling. *Dis Model Mech*, Vol. 3, No. 9-10, (Sep-Oct 2010), pp. 545-556, ISSN Print: 1754-8403, online: 1754-8411

Rogers, D.F. (2007). Physiology of airway mucus secretion and pathophysiology of hypersecretion. *Respir Care*, Vol. 52, No. 9, (September 2007), pp. 1134-1146; ISSN: 0020-1324

Schneeberger E.E. & Lynch RD. (2004). The tight junction: a multifunctional complex. *Am J Physiol Cell Physiol*, Vol. 286, No. 6, (Jun 2004), pp.C1213-C1228, ISSN: 0363-6143, 1522-1563

Schwartz, A.G., Prysak, G.M., Bock, C.H. & Cote, M.L. (2007). The molecular epidemiology of lung cancer. *Carcinogenesis*, Vol. 28, No. 3, (March 2007), pp. 507-518, ISSN: 0143-3334, 1460-2180

Simske, J.S., Koppen, M., Sims, P., Hodgkin, J., Yonkof, A. & Hardin, J. (2003). The cell junction protein VAB-9 regulates adhesion and epidermal morphology in C. elegans. *Nat Cell Biol*, Vol. 5, No. 7, (July 2003), pp. 619-625, ISSN: 1465-7392, 1476-4679

Singh, A.B., Sharma, A. & Dhawan, P. (2010). Claudin family of proteins and cancer: an overview. *J Oncol*, Vol. 2010, (July 2010), pp. 541957, ISSN: 1687-8450, 1687-8469

Steed, E., Balda, M.S. & Matter, K. (2010). Dynamics and functions of tight junctions. *Trends Cell Biol*, Vol. 20, No. 3, (January 2010), pp. 142-149, ISSN: 0962-8924, 1879-3088

Tsukita, S. & Furuse, M. (2000a). The structure and function of claudins, cell adhesion molecules at tight junctions. *Ann N Y Acad Sci*, Vol. 915, (2000), pp. 129-135, ISSN: 0077-8923, 1749-6632

Tsukita, S. & Furuse, M. (2000b). Pores in the wall: claudins constitute tight junction strands containing aqueous pores. *J Cell Biol*, Vol. 149, No. 1, (April 2000), pp. 13-16, ISSN: 0021-9525, 1540-8140

Tsukita, S., Furuse, M. & Itoh, M. (2001). Multifunctional strands in tight junctions. *Nat Rev Mol Cell Biol*, Vol. 2, No. 4, (April 2001), pp. 285-293, ISSN: 1471-0072, 1471-0080

Venter, J.C., Adams, M.D., Myers, E.W., Li, P.W., Mural, R.J., Sutton, G.G., Smith, H.O., Yandell, M., Evans, C.A., Holt, R.A., Gocayne, J.D., Amanatides, P., Ballew, R.M., Huson, D.H., Wortman, J.R., Zhang, Q., Kodira, C.D., Zheng, X.H., Chen, L., Skupski, M., Subramanian, G., Thomas, P.D., Zhang, J., Gabor Miklos, G.L., Nelson, C., Broder, S., Clark, A.G., Nadeau, J., McKusick, V.A., Zinder, N., Levine, A.J., Roberts, R.J., Simon, M., Slayman, C., Hunkapiller, M., Bolanos, R., Delcher, A., Dew, I., Fasulo, D., Flanigan, M., Florea, L., Halpern, A., Hannenhalli, S., Kravitz, S., Levy, S., Mobarry, C., Reinert, K., Remington, K., Abu-Threideh, J., Beasley, E., Biddick, K., Bonazzi, V., Brandon, R., Cargill, M., Chandramouliswaran, I., Charlab, R., Chaturvedi, K., Deng, Z., Di Francesco, V., Dunn, P., Eilbeck, K., Evangelista, C., Gabrielian, A.E., Gan, W., Ge, W., Gong, F., Gu, Z., Guan, P., Heiman, T.J., Higgins, M.E., Ji, R.R., Ke, Z., Ketchum, K.A., Lai, Z., Lei, Y., Li, Z., Li, J., Liang, Y., Lin, X., Lu, F., Merkulov, G.V., Milshina, N., Moore, H.M., Naik, A.K., Narayan, V.A., Neelam, B., Nusskern, D.,

Rusch, D.B., Salzberg, S., Shao, W., Shue, B., Sun, J., Wang, Z., Wang, A., Wang, X., Wang, J., Wei, M., Wides, R., Xiao, C., Yan, C., Yao, A., Ye, J., Zhan, M., Zhang, W., Zhang, H., Zhao, Q., Zheng, L., Zhong, F., Zhong, W., Zhu, S., Zhao, S., Gilbert, D., Baumhueter, S., Spier, G., Carter, C., Cravchik, A., Woodage, T., Ali, F., An, H., Awe, A., Baldwin, D., Baden, H., Barnstead, M., Barrow, I., Beeson, K., Busam, D., Carver, A., Center, A., Cheng, M. L., Curry, L., Danaher, S., Davenport, L., Desilets, R., Dietz, S., Dodson, K., Doup, L., Ferriera, S., Garg, N., Gluecksmann, A., Hart, B., Haynes, J., Haynes, C., Heiner, C., Hladun, S., Hostin, D., Houck, J., Howland, T., Ibegwam, C., Johnson, J., Kalush, F., Kline, L., Koduru, S., Love, A., Mann, F., May, D., McCawley, S., McIntosh, T., McMullen, I., Moy, M., Moy, L., Murphy, B., Nelson, K., Pfannkoch, C., Pratts, E., Puri, V., Qureshi, H., Reardon, M., Rodriguez, R., Rogers, Y. H., Romblad, D., Ruhfel, B., Scott, R., Sitter, C., Smallwood, M., Stewart, E., Strong, R., Suh, E., Thomas, R., Tint, N. N., Tse, S., Vech, C., Wang, G., Wetter, J., Williams, S., Williams, M., Windsor, S., Winn-Deen, E., Wolfe, K., Zaveri, J., Zaveri, K., Abril, J. F., Guigo, R., Campbell, M. J., Sjolander, K. V., Karlak, B., Kejariwal, A., Mi, H., Lazareva, B., Hatton, T., Narechania, A., Diemer, K., Muruganujan, A., Guo, N., Sato, S., Bafna, V., Istrail, S., Lippert, R., Schwartz, R., Walenz, B., Yooseph, S., Allen, D., Basu, A., Baxendale, J., Blick, L., Caminha, M., Carnes-Stine, J., Caulk, P., Chiang, Y. H., Coyne, M., Dahlke, C., Mays, A., Dombroski, M., Donnelly, M., Ely, D., Esparham, S., Fosler, C., Gire, H., Glanowski, S., Glasser, K., Glodek, A., Gorokhov, M., Graham, K., Gropman, B., Harris, M., Heil, J., Henderson, S., Hoover, J., Jennings, D., Jordan, C., Jordan, J., Kasha, J., Kagan, L., Kraft, C., Levitsky, A., Lewis, M., Liu, X., Lopez, J., Ma, D., Majoros, W., McDaniel, J., Murphy, S., Newman, M., Nguyen, T., Nguyen, N., Nodell, M., Pan, S., Peck, J., Peterson, M., Rowe, W., Sanders, R., Scott, J., Simpson, M., Smith, T., Sprague, A., Stockwell, T., Turner, R., Venter, E., Wang, M., Wen, M., Wu, D., Wu, M., Xia, A., Zandieh, A. & Zhu, X.. (2001). The sequence of the human genome. *Science*, Vol. 291, No. 5507, (February 2001), pp. 1304-1351, ISSN: 0193-4511, 0193-4511

Wen, H., Watry, D.D., Marcondes, M.C. & Fox, H.S. (2004). Selective decrease in paracellular conductance of tight junctions: role of the first extracellular domain of claudin-5. *Mol Cell Biol*, Vol. 24, No. 19, (October 2004), pp. 8408-8417, ISSN: 1471-0072, 1471-0080

Will, C., Fromm, M. & Muller, D. (2008). Claudin tight junction proteins: novel aspects in paracellular transport. *Perit Dial Int*, Vol. 28, No. 6, (November-December 2008), pp. 577-584, ISSN: 0896-8608

Wistuba, II, Mao, L. & Gazdar, A.F. (2002). Smoking molecular damage in bronchial epithelium. *Oncogene*, Vol. 21, No. 48, (October 2002), pp. 7298-7306, ISSN: 0890-6467

Wu, V.M., Schulte, J., Hirschi, A., Tepass, U. & Beitel, G.J. (2004). Sinuous is a Drosophila claudin required for septate junction organization and epithelial tube size control. *J Cell Biol*, Vol. 164, No. 2, (April 2010), pp. 313-323, ISSN: 0021-9525, 1540-8140

Zhang, G., Ding, L., Renegar, R., Wang, X.M., Lu, Q., Huo, S. & Chen, Y.H. (2011). Hydroxycamptothecin-Loaded Fe(3) O(4) Nanoparticles Induce Human Lung Cancer Cell Apoptosis through Caspase-8 Pathway Activation and Disrupt Tight Junctions. *Cancer Sci*, (March 2011), pp. 1216-1222, ISSN: 1347-9032, 1349-7006

http://www.who.int/mediacentre/factsheets/fs297/en/index.html.
http://www.genome.jp/kegg/pathway/hsa/hsa04530.html.
http://www.genecards.org/.

12

The Roles of ESCRT Proteins in Healthy Cells and in Disease

Jasmina Ilievska, Naomi E. Bishop,
Sarah J. Annesley and Paul R. Fisher[*]
*Department of Microbiology,
La Trobe University,
Australia*

1. Introduction

Endocytosis is a process that occurs in all eukaryotes and is an essential mechanism for internalizing membrane proteins and controlling intracellular trafficking (Bishop, 2003). Membrane proteins such as active epidermal growth factor receptors (EGFRs) are endocytosed via clathrin-dependent or independent-pathways and are typically first delivered to the early endosome (Bishop, 1997; Tarrago-Trani & Storrie, 2007) (Figure 1). The early endosomes (or sorting endosomes) have a crucial role in sorting the endocytosed cargo to three alternative destinations: (i) recycling the cargo back to the plasma membrane (receptor sequestration), (ii) transferring the cargo to the *trans* Golgi network (TGN), (iii) transporting the cargo into intraluminal vesicles (ILVs) of maturing endosomes known as multivesicular bodies (MVBs) (reviewed by Gruenberg & Stenmark, 2004; Russel et al., 2006; Piper & Katzmann, 2007). The ultimate consequence of such sorting is the exposure of the ILVs and their contents to lysosomal hydrolases after fusion of the MVB with lysosomes (receptor down-regulation) (reviewed by Sorkin & von Zastrow, 2009; Wegner et al., 2011). MVBs also play an important role in the traffic of lysosomal enzymes from the TGN, and in the secretion of exosomes from cells (Lakkaraju & Rodriguez-Boulan, 2008; Simons & Raposo, 2009, Thery et al., 2009). MVBs functions extend beyond cargo sorting - they also serve as MHC class II compartments for antigen presentation, T-cell secretory granules and melanosomes in specialised cell types (Raiborg et al., 2003).

Efficient sorting at the early endosome and the MVB compartments typically requires mono- or polyubiquitination of cell surface receptors. The molecular machinery that recognises the ubiquitinated cargo at the early endosome and mediates its sorting into MVBs is a set of interacting protein complexes, the endosomal complexes required for transport (ESCRTs'). The ESCRTs' were first identified in yeast and were initially referred to as class E Vps (vacuolar protein sorting) proteins (Raymond et al., 1992). Characterisation of the 18 class E

---

[*] Corresponding Author

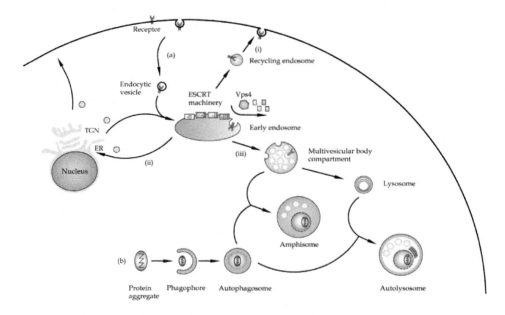

Fig 1. Interrelationships between the endocytic and autophagic pathways.
(a) Receptor-mediated endocytosis involves internalization of plasma membrane cargo into the cell. Endocytosed cargo is first delivered to the early endosome, which also receives cargo from the TGN. From here on, selected cargo can be delivered to three alternative destinations: (i) can be recycled back to the plasma membrane (receptor sequestration), or (ii) sorted into the TGN or (iii) incorporated into ILVs of MVBs. The cargo within the MVB compartment is subsequently transported to the lysosome where the constituents are broken down by lysosomal hydrolases (receptor down-regulation). The MVB biogenesis and the sorting of ubiquitinated cargo is controlled by four ESCRTs', -0, -I, -II -III, and the action of a AAA+ type ATPase Vps4. (b) In contrast to endocytosis, autophagy digests intracellular material by encapsulating damaged organelles or protein aggregates by a phagophore. The resulting autophagosome can fuse directly with the lysosome forming an autolysosome or indirectly via the MVB compartment forming a hybrid organelle, termed an amphisome.

genes revealed that ten of these encode subunits of the ESCRT system, whereas the others encode ESCRT-related proteins or upstream or downstream interactors (reviewed by Slagsvold et al., 2006) (Table 1).

Extensive genetic, biochemical and structural studies using yeast, Drosophila and mammalian model systems have revealed the molecular roles of the ESCRTs'. The ESCRT system consists of four different complexes termed ESCRT-0, -I, -II and –III, and a number of associated proteins such as Vps4 (Babst et al., 2002a, 2002b; Katzmann et al., 2001) (Figure 2). Ubiquitinated endosomal cargo targeted for lysosomal degradation is initially recognised by ESCRT-0. ESCRT-I, -II and -III which are subsequently recruited to the endosomal membrane by protein-protein interactions between the four complexes (reviewed by Roxrud et al., 2010). The ubiquitinated cargo is further concentrated on the endosomal membrane by the action of ESCRT-I and -II, furthermore invaginations that form from the endosomal membrane to become ILVs depend upon ESCRT-III and Vps4 - facilitated membrane abscission (Elia et al., 2011; Babst et al., 2011). The endocytosed contents in the ILVs are ultimately terminated via lysosomal degradation (Figure 1). Following protein sorting into MVBs, the ATPase Vps4 catalyzes the release of the ESCRT machinery from the limiting membrane of the MVB compartment into the cytosol for further rounds of cargo sorting.

The ESCRTs' also have alternative cellular roles beyond lysosomal trafficking. A subset of ESCRTs' have a well-established function in eukaryotic cell abscission (cytokinesis) (Spitzer et al., 2006; Carlton & Martin-Serrano, 2007; Morita et al., 2007), viral budding (Morita & Sundquist, 2004; Fujii et al., 2007) and autophagy (Filimonenko et al., 2007; Lee et al., 2007). Given their importance in fundamental cellular processes, it is not surprising that ESCRT dysfunction is associated with numerous diseases, including neurodegenerative disorders, cancer and infectious diseases. The dynamics and regulation of the ESCRT machinery have been extensively reviewed (Hurley & Emr, 2006; Saksena et al., 2007; Williams & Urbe, 2007; Raiborg & Stanmark, 2009; Hanson et al., 2009; Carlton & Martin-Serrano, 2009; Hurley, 2010; Roxrud et al., 2010; Henne et al., 2011) and will only be mentioned briefly here. This review focuses on understanding the role of the ESCRTs' in disease using model systems, to better understand the mechanisms behind their role in pathogenesis.

2. Evolutionary conservation of ESCRTs'

Comparative genomic and phylogenetic analysis has revealed in great detail the conservation of the molecular machineries involved in cargo sorting and membrane trafficking. The phylogenetic data has shown that most ESCRT genes emerged early during the evolution of eukaryotes (Slater & Bishop, 2006, Field et al., 2007; Leung et al., 2008, Field & Dacks, 2009). However the ESCRT-III complex and Vps4 have been identified in Archaea, suggesting an even earlier, ancestral function for these components (Lindas et al., 2008; Ghazi-Tabatabai et al., 2009; reviewed by Makarova et al., 2010; Samson et al., 2008, 2011). It has even been suggested a similar mechanism may contribute to bacterial outer membrane vesicle production (Kulp & Kuehn, 2011). All of the other ESCRT complexes with the exception of ESCRT-0, are present across all of the eukaryotic lineages. ESCRT-0 appears to be specific to the opisthokonts (metazoa and fungi) and is absent from *Dictyostelium discoideum*, a member of their sister lineage the Amoebozoa, as well as from plants (Winter & Hauser, 2006; Leung et al., 2008; Field & Dacks, 2009). However *D. discoideum* contains instead a minimal, possibly ancestral ESCRT-0 in which DdTom1 interacts with ubiquitin, clathrin and the ESCRT-1 protein Tsg101 (Blanc et al., 2009). MVBs were also recently identified in the basal amoebozoan *Breviata anathema*, strengthening the conclusion that the ESCRTs' are a common feature of this supergroup (Herman et al., 2011). In mammals and plants several *VpsE* genes such as *Vps37*, *Vps4*, *Vps32*, *Mvb12* and *Bro1* have undergone gene duplications. The domain structure of VpsE proteins, especially the domains involved in protein-protein and protein-lipid interactions is well conserved across yeast, metazoa and plants (reviewed by Michelet et al., 2010) (Table 1). Collectively, these data suggests that the fundamental structure and the role of the ESCRTs' is well conserved among many eukaryotic organisms.

ESCRT complex and activity	Yeast Protein names	Metazoan Protein names	Domains/Motifs[1]	Biological function
ESCRT-0 *Clusters ubiquitinated cargo*	Vps27	Hrs	VHS, FYVE, UIM (yeast) DUIM (metazoan), PTAP, GAT, coiled-coil core, clathrin binding	Binds PtdIns3P, ubiquitinated cargo, ESCRT-I and clathrin
	Hse1	STAM1, 2	VHS, UIM, SH3, GAT, coiled-coil core, clathrin binding	Binds ubiquitinated cargo and DUB enzymes
ESCRT-I *Membrane deformation and budding*	Vps23	Tsg101	UEV, Pro-rich linker, stalk, headpiece	Cargo and ESCRT-0 (Vps27) interaction, contains the viral PTAP motif
	Vps28	Vps28	Headpiece, C-terminal	Binds ESCRT-II (Vps36)
	Vps37	Vps37A, B, C, D	Basic helix, head piece	Binds membranes
	Mvb12	MVB12A, B	Stalk, ubiquitin binding domain	Stabilizes ESCRT-I subunits, binds ubiquitin
ESCRT-II *Membrane deformation and budding*	Vps22	EAP30, Snf8	Coiled-coil, WH	Binds membranes
	Vps25	EAP20	PPXY, WH	Binds ESCRT-III (Vps20), cargo
	Vps36	EAP45	GLUE, NZF1, 2 (yeast), WH	Binds membranes, ubiquitin and ESCRT-I (Vps28)
ESCRT-III *Membrane scission*	Vps20	CHMP6	Charged, coiled-coil, MIM	Initiates membrane scission
	Vps32/* (Snf7)	CHMP4A, B, C	Charged, coiled-coil, MIM	Membrane scission, binds Bro1 domains
	Vps24	CHMP3	Charged, coiled-coil, MIM	Completes membrane scission
	Vps2/(Did4)	CHMP2A, B	Charged, coiled-coil, MIM	Recruits Vps4; initiates ESCRT disassembly
Vps4 *Disassembly of ESCRTs*	Vps4	Vps4A,B/(SKD1, 2)	AAA+ ATPase, MIT	ESCRT disassembly and recycling
	Vta1	VTA1/LIP5	MIT, VSL	Positively regulates of Vps4
Other	Vps31/(Bro1)	ALIX/AIP1	Bro1, Proline-rich domain	ESCRT-III interaction by recruiting Snf7, Doa4 recruitment, interacts with apoptosis regulators, contains viral YPXP domains
	Vps60/(Mos10)	CHMP5	Charged, coiled-coil	ESCRT-III like protein, binds Vta1
	Vps46/(Did2)	CHMP1A, B	Charged, coiled-coil	ESCRT-III like protein, recruits Vps4
	Ist1	IST1	MIM1, MIM2	The tandem ESCRTIII domains bind SNF7B and the DUB UBPY (USP8)
	Doa4	UBPY/USP8	Rhod, UBP	Removes ubiquitin

[1]**Domain acronyms:** Bro1, Bro1 domain-containing protein 1; CHMP, charged multivesicular body protein; DID, DOA4-independent degradation protein; DUB, deubiquitylating enzyme; DUIM, double-sided ubiquitin-interacting motif; ESCRT, endosomal sorting complex required for transport; GAT; GLUE, GRAM-like ubiquitin-binding in EAP45; Hrs, hepatocyte growth factor-regulated Tyr kinase substrate; Ist1, increased sodium tolerance protein 1; MIM, MIT-interacting motif; MIT, microtubule-interacting and transport; MVB, multivesicular body; NZF, Npl4-type zinc finger; SH3, SRC homology 3; UBPY, ubiquitin isopeptidase Y; UEV, ubiquitin E2 variant; UIM, ubiquitin-interacting motif; VHS, Vps27, Hrs and STAM; Vps, vacuolar protein sorting; VSL and VTA1; WH2, winged helix 2. * Alternative names are provided in brackets.

Table 1. Components of the ESCRT machinery.
(Table is modified from Hurley & Hanson, 2010, see cited paper for further details on domain/motif structure)

3. Structure and function of ESCRTs' in normal cells

3.1 Composition of the ESCRT complexes

In order to understand the role of the ESCRTs' in disease, a brief overview of the composition of each complex is provided (Figure 2). ESCRT-0, -I and -II are stable heterotetrameric complexes, while ESCRT-III is formed by polymers formed by four core protein subunits.

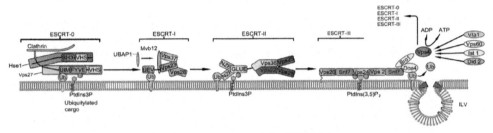

Fig. 2. Composition and molecular interactions of the ESCRTs'.

Interactions between the four ESCRTs' are indicated, as are interactions with ubiquitylated cargo, accessory molecules such as phosphatidylinositol 3-phosphate (PtdIns(3)P), deubiquitylating enzymes (DUBs), Bro1 and the ATPase Vps4. Yeast protein names have been used but the figure above is a composite of data obtained from studies of several model systems containing the ESCRTs'.

3.1.1 ESCRT-0

The ESCRT-0 complex has an early role in MVB biogenesis and in the sorting of ubiquitinated proteins into the MVB pathway. ESCRT-0 binds and clusters ubiquitinated cargo destined for delivery into MVBs, and recruits clathrin and deubiquitinating enzymes (Wollert et al., 2010). The ESCRT-0 complex consists of two subunits, Hrs (Vps27 in yeast) and STAM1/2 (Hse1 in yeast). Hrs contains a FYVE zinc finger domain which binds PtdIns(3)P providing membrane recruitment and endosomal specificity for the ESCRT-0 complex (Mao et al., 2000). Hrs and STAM1/2 bind ubiquitin via their UIM and VHS ubiquitin domains respectively, which are essential for efficient sorting of ubiquitinated proteins (Bishop et al., 2002; Mizuno et al., 2003; Bache et al., 2006). Hrs binds directly to the tumour susceptibility gene-101 product (Tsg101) recruiting ESCRT-I to the endosomal membranes (Bishop et al., 2002) (Figure 2).

3.1.2 ESCRT-I

The ESCRT-I complex along with ESCRT-II is required for further concentrating ubiquitinated cargo on the endosomal membrane and initiating the first stages of membrane invagination (Wollert et al., 2010). Mammalian ESCRT-I is composed of four subunits Tsg101 (Vps23 in yeast), Vps28, Vps37 (four isoforms, A-D) and Mvb12 (two isoforms A/B); the yeast ESCRT-I contains single copies of the four subunits (Chu et al., 2006; Curtiss et al., 2007; Kostelansky et al., 2007; Oestreich et al., 2007) (Figure 2). A novel ESCRT-I component was recently identified in mammalian cells, termed UBAP1. UBAP1 contains a region conserved in Mvb12 and binds Bro1 proteins involved in cytokinesis (Stefani et al., 2011). The ESCRT-I structure is organised as a headpiece core with flexibly connected modules that mediate interactions with other partners such as ESCRT-0, ubiquitin, Alix (Bro1 in yeast) and ESCRT-II. The Tsg101 subunit can also directly bind Vps20, an ESCRT-III component, surpassing both ESCRT-I and –II (Katzmann et al., 2003; Bilodeau et al., 2003; Pornillos et al., 2003).

3.1.3 ESCRT-II

The ESCRT-II complex is recruited to the endosomal membrane by the interaction between the ESCRT-I subunit Vps28 and the ESCRT-II subunit Vps36 (Saksena et al., 2009) (Figure 2). The ESCRT-II complex is a heterotetramer with one copy of Vps22 and Vps36 and two copies of Vps25 (Hierro et al., 2004; Im & Hurley, 2008; Teis et al., 2010). Mammalian Vps36 binds PtdIns(3)P and ubiquitin via the GLUE domain and is important for efficient cargo sorting (Teo et al., 2006). The yeast Vps36 contains a GLUE domain with two NZF insertions. NZF1 binds to ESCRT-I (Gill et al., 2007) and NZF2 binds to ubiquitinated cargo (Alam et al., 2004). The C-terminal domain of Vps25 provides a direct link to ESCRT-III by binding to CHMP6 (Vps20).

3.1.4 ESCRT-III

The ESCRT-III complex plays an important role in membrane scission and is responsible for pinching off the neck of the invagination, forming an ILV (Wollert et al, 2009, Wollert & Hurley, 2010) (Figure 2). Mammalian ESCRT-III consists of multiple subunits, CHMP2 (two isoforms A/B,) (in yeast Vps2), CHMP3 (in yeast Vps24), CHMP4 (four isoforms A-D) (in yeast Snf7), and CHMP6 (in yeast Vps20) (Babst et al, 2002a; Bajorek et al., 2009b). The other ESCRT-III subunits CHMP1 (two isoforms A/B), (in yeast Did2), CHMP5 (in yeast Vps60) and Ist1 are not strictly essential for function and appear to assemble with the rest of the ESCRT-III subunits at a later stage. Did2 and Vps60 recruit and activate Vps4, while Ist1 inhibits Vps4 activity (Nickerson et al., 2006; Dimaano et al., 2008). Vps4 is an AAA-ATPase, which has an important role in catalysing and energizing the dissociation of the ESCRT machinery form the endosomal membrane back to the cytosol, for further rounds of cargo sorting. The ESCRT-III complex does not bind ubiquitin, however it recruits Alix, which plays a key role in the endosomal recruitment of Doa4, a deubiquitinating enzyme (Babst et al., 1997; 1998; Scott et al., 2005; Muziol et al., 2006; Shim et al., 2007; Yu et al., 2008; Teis et al., 2008; Lata et al., 2008; Ghazi-Tabatabai et al., 2009).

3.2 Biological roles of the ESCRTs'

3.2.1 Cytokinesis

In eukaryotes, cytokinesis consists of at least three key steps: (i) assembly of the central spindle, (ii) formation of the cleavage furrow, (iii) and membrane abscission at the midbody (Yang et al., 2008; reviewed by Saksena & Emr, 2009). The membrane scission and the creation of the membrane curvature required in cytokinesis is topologically similar to the curvature needed during MVB sorting and viral budding. Studies have shown that components of ESCRTs' are required for membrane abscission, the final step of cytokinesis. For instance, ESCRT-III is specifically recruited to the midbody to mediate membrane fission and Vps4 is important in the release of ESCRT-III in cytokinesis (Spitzer et al., 2006; Obita et al., 2007; Carlton & Martin-Serrano, 2007). Furthermore, depletion of either Ist1 and Did2 (ESCRT-III and Vps4 human homologues) leads to an arrest in cytokinesis (Agromayor et al., 2009; Bajorek et al., 2009a). Additionally, the ESCRT-I subunit Tsg101 and the ESCRT-III associated protein Alix were found to competitively associate with Cep55 (a multimeric cell division protein essential for late stage cell division) to facilitate recruitment of ESCRT-III and Vps4 for abscission of the two daughter cells (Carlton & Martin-Serrano, 2007; Morita et al., 2007). The role of ESCRT-II in cytokinesis is unclear, although studies conducted by Langelier et al., 2006 indicate that Vps22 of ESCRT-II is located on the centrosomes and is involved in the maturation of these organelles. The mechanisms behind ESCRT mediated scission and their role in microtubule disassembly have been recently reviewed in detail by Henne et al., 2011 and Roxrud et al., 2010 and will not be further discussed in this review.

3.2.2 Autophagy

In the mammalian system there are two pathways that intersect with the lysosome, the MVB pathway as described in the introduction and the autophagy pathway. To date, three autophagy pathways have been described in higher eukaryotes: microautophagy (MA), chaperone-mediated autophagy (CMA) and macroautophagy (Mizushima et al., 2008; Cuervo, 2010). Microautophagy was originally described in yeast, but is not yet well characterised in other eukaryotes (Marzella et al., 1981). In this pathway, the lysosome invaginates and internalizes cytosolic components, which are subsequently degraded in the lumen of the lysosome. Chaperone-mediated autophagy is a more selective autophagy that does not involve vesicle formation but rather a direct translocation of a specific set of proteins across the lysosomal membrane. The cytosolic chaperone hsc70, a major component of the CMA pathway recognises the pentapeptide 'KFERQ' sequence in proteins destined for lysosomal degradation (Sahu et al., 2011). The lysosome-associated protein type 2A (LAMP2A) binds and translocates the KFERQ proteins to the lysosome, through a yet-unclear-mechanism (Orenstein & Cuervo, 2010; reviewed by Shpilka & Elazar, 2011). A recent study has identified a new macroautophagy-like degradation pathway that is distinct from CMA and occurs in lysosomes (Orenstein & Cuervo, 2010). Endosomal microautophagy was shown by Sahu et al., 2011 to occur during MVB formation and requires both ESCRT-I and –III, as well as hsc70 for delivery of KFERQ proteins from the cytosol into MVBs. This study provided fresh insights into the mechanisms of autophagy in mammalian model systems and also extended the role of ESCRTs' to degradation of cytosolic compartments. The role of the ESCRTs' is best characterized in macroautophagy and this will be the focus here.

Macroautophagy (henceforth simply referred to as autophagy) is a bulk degradation pathway responsible for the removal of damaged organelles and for clearance of protein aggregates (reviewed by Mehrpour et al., 2010). The fundamental molecular mechanisms of the autophagy pathway have been extensively studied in yeast, using genetic screening to identify autophagy genes (*atg*) (Klionsky et al., 2003). Subsequent inactivation of *atg* orthologues in higher eukaryotes has shown that the autophagic machinery is highly conserved. The autophagic pathway involves multiple steps: (i) sequestration of cytoplasmic constituents by a double membrane phagophore, resulting in the formation of an autophagosome and (ii) direct fusion of autophagosomes with the lysosome, where the cytoplasmic material is degraded in the resulting autolysosome or alternatively (iii) fusion of the autophagosome with the MVB compartment, forming a hybrid component termed an amphisome, which then fuses with the lysosome (Lawrence & Brown, 1992; Berg et al., 1998; Liou et al., 1997) (Figure 1).

Many age-related neurodegenerative disorders are characterised by an accumulation of ubiquitin-positive aggregates in affected brain regions. Autophagy is necessary for the clearance of these proteins, as aggregates essentially become toxic for postmitotic cells like neurons (reviewed by Eskelinen & Saftig, 2009). Defects in the autophagic pathway are associated with neurodegenerative diseases such as Alzheimer's, Huntington's and Parkinson's diseases. For instance, in Alzheimer's disease (AD) neuronal autophagy is activated in the early stages, however autophagic degradation becomes impaired as the disease progresses (Boland et al., 2008). Similarly in Huntington's disease (HD), active autophagy helps in the clearance of toxic polyglutamine-containing proteins (Ravikumar et al., 2004). In Parkinson's disease (PD) mutant α-synuclein blocks its own degradation via the chaperone-mediated autophagy pathway resulting in a gain-of-function neurotoxicity (Cuervo et al., 2004).

Studies conducted using slime moulds, nematodes, flies and mammals as model systems to study neurodegenerative disease have revealed that the ESCRT machinery plays a role in autophagy. Genetic disruption of ESCRT-I, -II and -III in mammalian and *Drosophila* cells leads to an increase in autophagosomes and toxic protein aggregates increase the severity of HD (Lee et al., 2009). Similarly, in rodent cortical neurons, loss of the CHMP2B subunit leads to an accumulation of autophagosomes (Lee et al,. 2007). Autophagosome and amphisome accumulation was also observed in HeLa cells when Tsg101 and CHMP3/Vps24 were knocked down or CHMP2B was disrupted (Lee et al., 2007). Consistent with the above data, downregulation of Vps4 in HeLa cells resulted in autophagosome accumulation, impaired degradation of autophagy substrates and impaired delivery of endosomal constituents to autophagosomes (Nara et al., 2002). The observed increase in autophagosomes suggests that there is either an enhanced initiation of autophagy in the cell or a decreased autophagic flux. The ESCRT machinery is therefore predicted to be involved in one or more key stages of the autophagic pathway. The possibilities include: (i) ESCRTs' are involved in signalling pathways that induce autophagy, (ii) ESCRTs' are required for phagophore closure or (iii) ESCRTs' are involved in the fusion of autophagosomes with the lysosome and/or the fusion of the autophagosomes with the MVB (reviewed by Rusten & Stenmark, 2009).

To date, little is known about the underlying mechanisms allowing the ESCRTs' to mediate fusion of autophagosomes with the MVB compartment and lysosomes. It has been shown that tethering of lysosomes to endosomes and autophagosomes is mediated by Rab7 (Bucci et al., 2000, Gutierrez et al., 2004; Jager et al., 2004) and the HOPS complex, which brings the membranes in close proximity (Wurmser et al., 2000; Seals et al., 2000; reviewed by Metcalf & Isaacs, 2010). ESCRT proteins interact directly with the HOPS complex which binds Rab7, as determined by a recent study which revealed that mutant CHMP2B (an ESCRT-III subunit) leads to impaired recruitment of Rab7 (Urwin et al., 2010). This suggests that functional ESCRTs' are required either for recruiting the vesicular fusion machinery to the MVB compartment or for delivery of the fusion machinery to lysosomes or autophagosomes. A number of other proteins are also implicated in autophagosome fusion with endosomes/lysosomes including UVRAG, Rubicon and LAMP-2. It is not yet known whether the ESCRT machinery has an effect on these proteins and processes.

3.2.3 Downregulation of receptor-mediated signaling

Receptor tyrosine kinases (RTKs) are growth factor receptors that play a important regulatory roles in controlling cell growth, proliferation, differentiation, survival and metabolism in several tissues and organs (Hunter, 2000; Pawson et al., 2001). Dysfunction of RTKs or mutations in key components of their downstream signaling pathways results in a variety of diseases, such as cancer, diabetes, immune deficiencies and cardiovascular disorders (Blume-Jensen & Hunter, 2001). EGFR is one of the best studied RTKs, and its uncontrolled signaling is associated with the development of a number of human cancers, including mammary carcinomas, squamous carcinomas and glioblastomas (Hunter, 2000; Pawson et al., 2001). The multivesicular body pathway silences RTK signaling via lysosome sequestration and degradation and thus plays an important role in modulating the amplitude and kinetics of amide signaling pathways from activated receptors (Saksena et al., 2007; Hurley & Emr, 2006; Williams & Urbe, 2007). Defects in ESCRT-mediated sorting of these receptors to lysosomal degradation pathways can thus lead to sustained receptor signaling either because of prolonged residence and activity in the endosomal membrane or as a result of increased recycling of the receptors to the plasma membrane.

Drosophila studies have shown that EGFR degradation is impaired and signalling is prolonged by dysfunctional ESCRT-0 (Hrs) (Lloyd et al., 2002), ESCRT-I (Tsg101) (Vaccari & Bilder, 2005) or ESCRT-II (Vps25) (Thompson et al., 2005). In mammals, depleting Tsg101 causes sustained EGFR signaling (Bache et al., 2006), whereas depletion of CHMP3 (ESCRT-III) (Bache et al., 2006) or Eap30 (ESCRT-II) (Malerod et al., 2007) causes delayed EGFR degradation but not sustained signaling (Table 2). Sustained signaling observed in ESCRT-0, -I and -II *Drosophila* mutants and after ESCRT-I depletion in mammals may result from increases in the residence time of receptors in the endosomal membrane and their recycling back to the plasma membrane. Mutations in ESCRT-III subunits do not cause sustained signaling (Bache et al., 2006), possibly because ESCRT–III recruitment occurs after signal termination. This may also explain why ESCRT-III subunits so far have not been implicated in cancer.

The Notch signaling pathway is highly conserved from *Drosophila* to humans and plays a central role in the normal development of many tissues and cell types. It controls various effects on differentiation, survival, and/or proliferation that are highly dependent on signal strength and cellular context. Dysfunction of the Notch signaling pathway leads to many human diseases such as lung and skin cancer (Radtke & Raj, 2003; Allenspach et al., 2002). Studies in *Drosophila* have shown that Notch signaling is terminated via lysosomal degradation suggesting a role for the ESCRT machinery in the regulation of Notch. In *Drosophila*, depletion of Hrs or mutation of Tsg101 or Vps25 leads to an accumulation of the cell-surface receptors Notch, Delta, Thickveins and EGFR (Thompson et al., 2005; Vaccari & Bilder, 2005; Moberg el al., 2005). Notch accumulation stimulates cell proliferation in the eye disc (Chao et al., 2004, Tsai & Sun, 2004) and results in overgrowth phenotypes in surrounding wild-type cells via the JAK/STAT pathway. Furthermore, inactivation of Tsg101 or Vps5 in *Drosophila* results in loss of epithelial cell polarity, which is associated with malignant transformation, suggesting that ESCRT components have a role in organizing the actin and/or microtubule cytoskeleton (Thompson et al., 2005; Vaccari & Bilder, 2005; Moberg el al., 2005; Saksana & Emr, 2009). In summary, there is growing evidence that implicates functional ESCRTs' in suppressing malignant transformation and preventing cancer.

4. The roles of ESCRTs' in disease

4.1 Neurodegenerative diseases

The most direct evidence that ESCRT dysfunction causes neurodegenerative disease comes from the identification of autosomal dominant *CHMP2B* mutations found to cause a rare form of frontotemporal dementia (FTD3) (Skibinski et al., 2005) and amyotrophic lateral sclerosis (ALS) (Parkinson et al., 2006). FTD is the second most common form of early-onset dementia after Alzheimer's disease (Ratnavalli et al., 2002; Harvey et al., 2003) and is characterised by the presence of either tau neurofibrillary tangles or ubiquitin deposits. FTD with the presence of tau or ubiquitin pathology is termed FTLD-U (frontotemporal lobar degeneration with ubiquitin-immunoreactive inclusions) (Neary et al., 2005). Both FTLD-U and ALS are characterised by abnormal accumulation of ubiquitin-positive protein deposits (including TDP-43) that contain p62, tau and α-synuclein-negative neuronal cytoplasmic inclusions (Arai et al., 2006; Neumann et al., 2006). The adapter protein p62 is commonly found in protein inclusions associated with neurodegenerative disease (Talbot & Ansorge, 2006), it binds polyubiquitin (Vadlamudi et al., 1996) and interacts with the autophagic associated protein Atg8/LC3 (Bjorkoy et al., 2005; Pankiv et al., 2007). Collectively, these data implicate p62 as a link between protein accumulation and aggregation with autophagy-mediated clearance (reviewed by Saksena & Emr, 2009). Similarly, ESCRT-depleted cells and cells overexpressing CHMP2 in flies, mice and humans, showed impaired autophagic degradation leading to an accumulation of autophagosomes and protein aggregates containing p62, thereby contributing to the pathogenesis of FTD3. A recent study has shown that deletion of the ESCRT proteins Tsg101 and Vps24 resulted in accumulation of TDP-43, suggesting that impaired MVB function could have a role in TDP-43 aggregate formation in FTLD-U and ALS (Filimonenko et al., 2007). Furthermore, Vps24 was found to be essential in the clearance of expanded polyglutamine aggregates associated with Huntington's disease (Table 2) (Filimonenko et al., 2007). Collectively, these data suggest that efficient autophagic degradation requires functional ESCRTs' and dysfunction of this machinery is associated with neurodegenerative phenotypes and disorders.

Several indirect links also implicate the ESCRTs' in various neurodegenerative disorders, and several ESCRT-interacting proteins are products of genes that are associated with inherited forms of neurodegeneration (reviewed by Stuffers et al., 2009a). For instance, in mice, a null mutation in Mahoganin, an E3 ubiquitin ligase that ubiquitinates Tsg101, causes spongiform neurodegeneration, a recessively transmitted prion-like disease (Kim, et al., 2007; Jiao et al., 2009). Two putative ESCRT-III interacting proteins, spartin and spastin are mutated in spastic paraplegia, an inherited neurodegenerative disease that paralyzes the lower limbs (Reid et al., 2005). The exact mechanism of CHMP4 contribution to this disease remains unclear and requires further investigation. Finally, Niemann-Pick disease type C is an inherited neurodegenerative disorder characterized by a disruption of lipid trafficking and is caused by a mutation in either of the two genes, *npc1* and *npc2* (reviewed by Eskelinen & Saftig, 2009). A dominant-negative mutant of Vps4 was found to cause an accumulation of ubiquitinated

NPC1 (Ohsaki et al., 2006). Together, these data indicate that dysregulation of ESCRT pathways may contribute to a broad spectrum of degenerative diseases.

4.2 Cancer

The first hint that ESCRTs' play a role in cancer came from the identification of *Tsg101* and *Vps37A* as tumour suppressor genes on the basis that they map to chromosomal regions deleted or mutated in cancer (Li & Cohen, 1996; Xu et al., 2003). Genomic deletions and splice variants of *Tsg101* were found in sporadic forms of breast cancer (Li et al., 1997) and other malignancies such as myeloid leukaemia and prostate cancer (Table 2) (Sun et al., 1997; Lin et al., 1998). In addition, Vps37A expression in hepatocellular carcinomas was found to be dramatically reduced or undetected suggesting that Vps37A may be a potential tumour suppressor (Xu et al., 2003). Similar results were observed with CHMP1A, as overexpression of this protein inhibited cell growth and tumour formation in human pancreatic tumor cells (Li et al., 2009).

Mutations that prevent c-Cbl-mediated ubiquitination of EGFRs and thereby inhibit ESCRT-mediated receptor down-regulation are associated with a number of cancers, particularly acute myeloid leukemia. For example, a mutant EGFR lacking only the direct c-Cbl-binding site transduces stronger mitogenic signals when compared to the wild-type receptor (Waterman et al., 2002; Saksena & Emr, 2009). The c-Met RTK (also known as HGFR) regulates invasive growth and is critical for normal development and wound repair. Its overexpression causes uncontrolled proliferation and growth and consequently is associated with a variety of human cancers (Haddad et al., 2001). In part c-Cbl-mediated ubiquitination controls cellular c-Met levels and therefore ubiquitination and functional ESCRTs' are needed to avoid c-Met-related malignant transformation (Peschard et al., 2001).

Collectively, the foregoing studies indicate that the ESCRTs' have a negative regulatory role in growth receptor signaling, however several independent studies have shown that ESCRTs' also have a positive role in growth factor signaling. For instance, Tsg101 was recently found to be overexpressed, rather then reduced in breast, thyroid, ovarian and colon cancer (Ma et al., 2008). Furthermore, depletion of Tsg101 prevented tumorigenicity in several cancer lines (Zhu et al., 2004). To further support ESCRTs' positive role in oncogenic signaling, the ESCRT-0 component Hrs was found to be essential for cell proliferation and tumorigenesis in both HeLa and mouse fibroblast cells (Toyoshima et al., 2007).

A positive regulatory role in growth factor signalling for the ESCRTs' has also been observed in *Drosophila melanogaster* (Vaccari et al., 2005; Thompson et al., 2005; Moberg et al., 2005; Vaccari et al., 2009; Herz et al., 2006; Rodahl et al., 2009). For example, *Tsg101* is essential for normal cell growth and cell survival in the fruit fly and clonal loss of this gene in epithelial cells causes hyperplasia of surrounding tissue despite the mutant cells dying via apoptosis (Moberg et al., 2005; reviewed by Stuffers et al., 2009a). Loss of Vps25 causes a similar effect, whereas loss of Hrs is without effect (Vaccari & Bilder, 2005; Thompson et al., 2005). It is important to note that the proapoptotic signaling pathways Hippo, JNK and Hid are activated in the Vps25 *Drosophila* mutants. Expression of the caspase inhibitor p35 in the Vps25 mutant cells restores cell growth and even results in overgrowth, suggesting that mutations in both the ESCRT pathway and the apoptotic pathway are required for overgrowth. Blocking apoptosis by expressing *Ark* (an essential component of the apoptotic pathway) or *Diap1* (*Drosophila* inhibitor of apoptosis protein 1), again results in overgrowth of the Vps25 mutant tissue. Collectively, these results suggest that the ESCRTs' in *Drosophila* do not act as conventional tumor suppressors.

Overall, the ESCRTs' have been implicated in both positive and negative roles in growth factor receptor signaling and cancer, suggesting that the exact role of the ESCRTs' in tumourigenesis may be cell-type and context-dependent. Alternatively, ESCRT-mediated actions in controlling cell proliferation may reflect diverse endosomal sorting roles on a broad range of molecular targets with many different roles in cellular homeostasis (reviewed by Lobert & Stenmark, 2011). Further research needs to be conducted using different model systems to better understand the complex roles of the ESCRTs' in signaling and cell proliferation. More specifically, future studies need to address whether ESCRTs' act as genuine tumour suppressors in mammals, since at this stage this is still unclear.

4.3 Infectious diseases

4.3.1 Microbial infections

The endocytic and autophagic pathways play an important role in innate immunity. Multiple studies have now shown that these host cell pathways can be manipulated by viruses and microorganisms in order to facilitate infection (von Schwedler et al., 2003; Vieira et al., 2004; Philips et al., 2008; Morita & Sundquist, 2004; Martin-Serrano & Marsh, 2007; McCullough et al., 2008). ESCRTs' play an important role in degenerative endosomal trafficking, so it is not surprising that they are involved in killing many microorganisms. For example, functional ESCRTs' have been shown to restrict mycobacterial growth and infection (Philips et al., 2008). Mycobacteria may invade macrophages and are able to survive and replicate intracellularly due to their ability to prevent fusion of bacteria-containing phagosomes with lysosomes. In both the *Drosophila* model system, and in mammalian macrophages, mutation of ESCRTs' renders cells susceptible to mycobacterial infections. Similarly, overexpression of Vps4 in the host cell results in deficient differentiation and virulence of the intracellular protozoal pathogen *Leishmania major* (Table 2) (Vieira et al., 2004; Philips et al., 2008). Furthermore, autophagosome accumulation was also observed, and both functional endosomal and autophagic pathways are required for optimal *L. major* virulence and infection (Besteiro et al., 2006). The mechanisms by which ESCRTs' mediate resistance to microbial infection have not been defined. It is possible that ESCRTs' are required for the delivery of the pathogen to the lysosome, more specifically having a role in phagosome maturation and fusion between the phagosome and lysosome. Like the involvement of the ESCRTs' in the autophagic pathway these results suggest that the ESCRTs' affect multiple cellular trafficking events. The finding that ESCRT components restrict the growth of intracellular microbial pathogens means that they can now be considered as therapeutic targets for treatment of these infections which cause millions of deaths every year.

In the case of eukaryotic pathogens, the ESCRTs' of the pathogen may also play important roles in virulence. *Candida albicans* causes opportunistic fungal infections and its ESCRT proteins have multiple roles in pathogenesis. The fungal ESCRT components are suggested to contribute to diverse fungal functions including cell signaling, nutrient acquisition and possibly cell wall architecture (Cornet et al., 2005; Wolf et al., 2010). However the role of ESCRTs' in candidiasis is not yet fully understood.

Complex	Component	Dysfunction/disease	Pathogenesis	Model systems
Cancer				
ESCRT-0	Hrs (Vps27)	Tumourigenesis and metastatic potential	Hrs depletion is associated with the upregulation of E-cadherin and reduced β-catenin signalling[1]	Human cancer cells, MEF, mice
ESCRT-0 associated	Hrs (Vps27)	Benign brain tumours (e.g. Schwannomas, meningiomas, ependymomas)	Interaction with neurofibromatosis 2 tumour suppressor protein schwannomin/merlin, regulating STAT signalling[13, 14]	Human cancer cells, rat cells
ESCRT-I	Vps37A	Hepatocellular ca. (HCC) and metastasis	Growth inhibitory protein, suppressing proliferation, transformation and invasion; strongly reduced levels in HCC[2]	Human tissue and cancer cells
	Tsg101 (Vps23)	Ovarian cancer	Up regulation of Tsg101: suppression of p21 expression and posttranslational regulation through MAPK signalling[3, 4]	Human tissue and cancer cells
	Tsg101 (Vps23)	Mammary cancer	Overexpression of Tsg101: increased signalling through MAPK[5]	Human tissue, transgenic mice
	Tsg101 (Vps23)	Papillary thyroid cancer, gastrointestinal stromal tumours	Overexpression of Tsg101 (consequences not known)[6, 7]	Human tissue
ESCRT-I/II	Erupted Tsg101/Vps25	Neoplastic transformation (ovary and imaginal discs), over-proliferation of adjacent WT cells	Enhanced Notch and growth factor signalling in mutant cells[8, 9, 10]	Drosophila
ESCRT-III	CHMP3 (Vps24)	Prostate cancer	CHMP3 induces neuroendocrine cell differentiation[11]	Human cells
	CHMP3 (Vps24)	Non-small cell lung cancer	CHMP3 has a functional role in neuroendocrine cell differentiation[12]	Human cancer cells
ESCRT-III associated	CHMP1A	Ductal pancreatic cancer	Tumour suppressor, regulating tumour growth potentially through p53 signalling pathway[15]	Human cells, mice
Neurodegenerative diseases				
ESCRT-I/III	Tsg101 (Vps23) / CHMP3 (Vps24) / CHMP2B	Neurodegeneration (FTLD-U, ALS, Huntington's disease (HD))	Reduced autophagic degradation, accumulation of Ub-protein aggregates containing TDP-43; reduced clearance of Huntington-positive inclusions[18]	Human cells, mouse cells
ESCRT-I associated	Tsg101 (Vps23)	Spongiform neurodegeneration (hallmark of prion disease)	E3 ubiquitin-protein ligase Mahogunin ubiquitinates Tsg101; depletion of Mahogunin disrupts endosomal trafficking[21]	Human cells, rat tissue
ESCRT-I associated	Tsg101 (Vps23)	Charcot-Marie-Tooth disease (CMT1C)	Interaction with SIMPLE; SIMPLE plays a role in the lysosomal sorting of plasma membrane proteins[22]	
ESCRT-III	CHMP2B (Vps2)	FTLD-U and ALS	Disruption of endosomal trafficking, protein accumulation[16, 17]	Human cells
	CHMP4B (Snf7-2) / CHMP2B	Neurodegeneration (FTLD-U, ALS)	Accumulation of autophagosomes; failure of mutant CHMP2B to dissociate properly leading to dysfunctional ESCRT-III on late endosomes[20]	Drosophila, mice
ESCRT-III associated	CHMP1B	Hereditary spastic	Interaction with spastin; spastin	Monkey cells

Table 2. ESCRT-associated diseases in various model systems (Modified from Stuffers et al., 2009a)

References: [1]Toyoshima et al., 2007; [2]Xu et al., 2003; [3]Young et al., 2007; [4]Young et al., 2007; [5]Oh et al., 2007; [6]Liu et al., 2002; [7]Koon et al., 2004; [8]Moberg et al., 2005; [9]Vaccari & Bilder et al., 2005; [10]Thompson et al., 2005; [11]Wilson et al., 2001; [12]Walker et al., 2006; [13]Gutmann et al., 2001; [14]Scoles et al., 2002; [15]Li et al., 2008; [16]Parkinson et al., 2006; [17]Skibinski et al., 2005; [18]Filimonenko et al., 2007; [19]Rusten et al., 2007; [20]Lee et al., 2007; [21]Kim et al., 2007; [22]Shirk et al., 2005; [23]Reid et al., 2005; [24]Vieira et al., 2004; [25]Cornet et al., 2005; [26]Wolf et al., 2011; [27]Babst et al., 1998; [28]Spitzer et al., 2006; [29]Besteiro et al., 2006; [30]Shiels et al., 2007.

4.3.2 Viral infections

The beneficial role of ESCRTs' in protecting against intracellular bacteria is reversed in viral infections. Many membrane-enveloped viruses hijack the ESCRT machinery to bud out of host cells. Retroviruses (HIV-1), filoviruses (Ebola virus), rhabdoviruses and arenaviruses encode short sequence motifs termed L-domains (late domains) within their structural (Gag) polyproteins that are essential for the release of assembled viruses from the host cells (reviewed by Carlton & Martin-Serrano, 2009; Stuffers et al., 2009b). The P(S/T)AP motif found on the HIV-1 Gag protein for example binds directly to the UEV domain of Tsg101 of ESCRT-I. Even though HIV-1 budding is normally ESCRT-I dependent, if Tsg101 is unavailable, the virus alternatively binds to Alix via the YPxL domain and buds (Stark et al., 2003). Both ESCRT-I and Alix can independently recruit ESCRT-III, which together with Vps4 are required for efficient virus budding. Recent studies have shown that ESCRT-III and Vps4 can be recruited independently of either Tsg101 or Alix by the herpes simplex virus type-1 (Pawliczek & Crump et al., 2009) and the hepatitis C virus (Corless et al., 2010). ESCRT-II was found not to be essential for HIV- budding (Langelier et al., 2006), however ESCRT-II was discovered recently to be essential for release of the avian sarcoma virus (Pincetic et al., 2008). Other viruses such as the rabies virus can indirectly recruit the ESCRTs' by using the PPxY motif to specifically recruit WW-domain-containing E3 ubiquitin ligases of the Nedd4 family (Kikonyogo et al., 2001). Disruption of ESCRT function by RNA interference or dominant-negative Vps4 arrests viral release at the plasma membrane (Garrus et al., 2001; Martin-Serrano & Neil, 2011; Demirov et al., 2002; Strack et al., 2003; reviewed by Carlton & Martin-Serrano, 2009). Collectively, this data confirms that different enveloped viruses require specific proteins for budding and that the ESCRT machinery regulates viral release from the plasma membrane.

5. Conclusions

The ESCRT machinery is ubiquitous in eukaryotes and has been highly conserved in evolution due to its vital functions including endocytosis, cytokinesis and autophagy. Our understanding of the ESCRTs" roles in endocytosis, receptor downregulation, membrane deformation and scission has made great progress over the past few years and the study of various model systems has contributed significantly to this. We know that the ESCRTs', in particular ESCRT-III and Vps4 have an intrinsic budding and scission activity that is focused on the neck of the ILVs and that they are important regulators of cytokinesis (Spitzer et al., 2006; Obita et al., 2007; Carlton & Martin-Serrano, 2007). Model systems have implicated the ESCRTs' in autophagic fusion events and in endosome-lysosome degradation. Impaired function of these pathways causes various neurodegenerative disorders, cancers and is implicated in microbial infections. Genetic disruption of ESCRT-I, -II and –III in mammalian and *Drosophila* systems has been shown to result in an accumulation of autophagosomes and toxic aggregates which accelerates neurodegeneration (Lee et al., 2007). Mutations in the ESCRT-III subunit CHMP2B, have been shown to cause FTD3 (Skibinski et al., 2005) and ALS (Parkinson et al., 2006). Furthermore, the ESCRTs' and their associated proteins are also indirectly implicated in causing spongiform neurodegeneration (Kim, et al., 2007; Jiao et al., 2009), spastic paraplegia (Reid et al., 2005) and Niemann-Pick type C neurodegeneration (Ohsaki et al., 2006). Sustained receptor signaling is a key event in carcinogenesis, and Tsg101 (Li et al., 1997, Sun et al., 1997; Lin et al., 1998), Vps37A (Xu et al., 2003) and CHMP1A (Li et al., 2009) have been identified as potential tumor suppressors. However several other subsequent studies found Tsg101 to play a role in cell cycle control, a conclusion that is in contradiction to the tumor suppressor properties of Tsg101 (Zhu et al., 2004). In *Drosophila* ESCRT-I and -II were found to behave as tumor suppressors (Li & Cohen, 1996; Xu et al., 2003; Li et al., 2008). Tissues expressing mutant ESCRT-I or -II were found to form tumors that are largely attributable to the cell non-automous stimulation of proliferation caused by excessive cytokine production by the mutant cells. This is triggered by overactive Notch signaling from endosomes, signifying that the ESCRT machinery is crucial for silencing Notch signaling and thereby for tumor suppression in flies. It has not yet been clarified whether this is the case in mammals. The ESCRTs' were found to have a beneficial role in innate immunity by restricting microbial growth and infection (von Schwedler et al., 2003; Vieira et al., 2004; Philips et al., 2008; Morita & Sundquist, 2004; Martin-Serrano & Marsh, 2007; McCullough et al., 2008). The ESCRTs' however, are turned against the host in viral infections. Several viruses, such as HIV-1 use the ESCRT components to bud out cells and cause infection (reviewed by Carlton & Martin-Serrano, 2009; Stuffers et al., 2009b). Further dissection of the roles of the ESCRTs' in these events will shed light on the basic mechanism of vesicular traffic and provide new insights into disease pathogenesis and preventative and therapeutic strategies.

6. Acknowledgments

Due to the vast body of primary research that has contributed to our understanding of endocytosis, autophagy and ESCRTs', we have been unable to cite many research papers in the field and we apologise to those authors whose articles are not directly cited. J. Ilievska was the recipient of an Australian postgraduate award scholarship (APA). This work was supported by grants from the Australian Research Council and the Thyne Reid Memorial Trusts. The authors are grateful to F. Nesci for assistance with preparing the Figures and S. Aracic for proofreading the work.

7. References

Agromayor, M., Carlton, J. G., Phelan, J. P., Matthews, D. R., Carlin, L. M., Ameer-Beg, S., Bowers, K. & Martin-Serrano, J.; (2009). Essential role of hIST1 in cytokinesis. *Mol Biol Cell*, 20, 5, (Mar, 2009), 1374-1387.

Alam, S. L., Sun, J., Payne, M., Welch, B. D., Blake, B. K., Davis, D. R., Meyer, H. H., Emr, S. D. & Sundquist, W. I.; (2004). Ubiquitin interactions of NZF zinc fingers. *EMBO J*, 23, 7, (Apr, 2004), 1411-1421.

Allenspach, E. J., Maillard, I., Aster, J. C. & Pear, W. S.; (2002). Notch signaling in cancer. *Cancer Biol Ther*, 1, 5, (Sep, 2002), 466-476.

Arai, T., Hasegawa, M., Akiyama, H., Ikeda, K., Nonaka, T., Mori, H., Mann, D., Tsuchiya, K., Yoshida, M., Hashizume, Y. & Oda, T.; (2006). TDP-43 is a component of ubiquitin-positive tau-negative inclusions in frontotemporal lobar degeneration and amyotrophic lateral sclerosis. *Biochem Biophys Res Commun*, 351, 3, (Dec, 2006), 602-611.

Babst, M., Davies, B. A. & Katzmann, D. J.; (2011). Regulation of Vps4 during MVB sorting and cytokinesis. *Traffic*, 12, 10, (Oct, 2011), 1298-1305.

Babst, M., Sato, T. K., Banta, L. M. & Emr, S. D.; (1997). Endosomal transport function in yeast requires a novel AAA-type ATPase, Vps4. *EMBO J*, 16, 8, (Apr, 1997), 1820-1831.

Babst, M., Wendland, B., Estepa, E. J. & Emr, S. D.; (1998). The Vps4 AAA ATPase regulates membrane association of a Vps protein complex required for normal endosome function. *EMBO J*, 17, 11, (Jun, 1998), 2982-2993.

Babst, M., Katzmann, D. J., Estepa-Sabal, E. J., Meerloo, T. & Emr, S. D.; (2002a). ESCRT-III: an endosome-associated heterooligomeric protein complex required for MVB sorting. *Dev Cell*, 3, 2, (Aug, 2002), 271-282.

Babst, M., Katzmann, D. J., Snyder, W. B., Wendland, B. & Emr, S. D.; (2002b). Endosome-associated complex, ESCRT-II, recruits transport machinery for protein sorting at the multivesicular body. *Dev Cell*, 3, 2, (Aug, 2002), 283-289.

Bache, K. G., Stuffers, S., Malerod, L., Slagsvold, T., Raiborg, C., Lechardeur, D., Walchli, S., Lukacs, G. L., Brech, A. & Stenmark, H.; (2006). The ESCRT-III subunit hVps24 is required for degradation but not silencing of the epidermal growth factor receptor. *Mol Biol Cell*, 17, 6, (Jun, 2006), 2513-2523.

Bajorek, M., Morita, E., Skalicky, J. J., Morham, S. G., Babst, M. & Sundquist, W. I.; (2009a). Biochemical analyses of human IST1 and its function in cytokinesis. *Mol Biol Cell*, 20, 5, (Mar, 2009), 1360-1373.

Bajorek, M., Schubert, H. L., McCullough, J., Langelier, C., Eckert, D. M., Stubblefield, W. M., Uter, N. T., Myszka, D. G., Hill, C. P. & Sundquist, W. I.; (2009b). Structural basis for ESCRT-III protein autoinhibition. *Nat Struct Mol Biol*, 16, 7, (Jul, 2009), 754-762.

Berg, T. O., Fengsrud, M., Stromhaug, P. E., Berg, T. & Seglen, P. O.; (1998). Isolation and characterization of rat liver amphisomes. Evidence for fusion of autophagosomes with both early and late endosomes. *J Biol Chem*, 273, 34, (Aug, 1998), 21883-21892.

Besteiro, S., Williams, R. A., Morrison, L. S., Coombs, G. H. & Mottram, J. C.; (2006). Endosome sorting and autophagy are essential for differentiation and virulence of *Leishmania major*. *J Biol Chem*, 281, 16, (Apr, 2006), 11384-11396.

Bilodeau, P. S., Winistorfer, S. C., Kearney, W. R., Robertson, A. D. & Piper, R. C.; (2003). Vps27-Hse1 and ESCRT-I complexes cooperate to increase efficiency of sorting ubiquitinated proteins at the endosome. *J Cell Biol*, 163, 2, (Oct, 2003), 237-243.

Bishop, N., Horman, A. & Woodman, P.; (2002). Mammalian class E vps proteins recognize ubiquitin and act in the removal of endosomal protein-ubiquitin conjugates. *J Cell Biol*, 157, 1, (Apr, 2002), 91-101.

Bishop, N. E.; (1997). An update on non-clathrin-coated endocytosis. *Rev Med Virol*, 7, 4, (Dec, 1997), 199-209.

Bishop, N. E.; (2003). Dynamics of endosomal sorting. *Int Rev Cytol*, 232, 1-57.

Bjorkoy, G., Lamark, T., Brech, A., Outzen, H., Perander, M., Overvatn, A., Stenmark, H. & Johansen, T.; (2005). p62/SQSTM1 forms protein aggregates degraded by autophagy and has a protective effect on huntingtin-induced cell death. *J Cell Biol*, 171, 4, (Nov, 2005), 603-614.

Blanc, C., Charette, S. J., Mattei, S., Aubry, L., Smith, E. W., Cosson, P. & Letourneur, F.; (2009). *Dictyostelium* Tom1 participates to an ancestral ESCRT-0 complex. *Traffic*, 10, 2, (Feb, 2009), 161-171.

Blume-Jensen, P. & Hunter, T.; (2001). Oncogenic kinase signalling. *Nature*, 411, 6835, (May, 2001), 355-365.

Boland, B., Kumar, A., Lee, S., Platt, F. M., Wegiel, J., Yu, W. H. & Nixon, R. A.; (2008). Autophagy induction and autophagosome clearance in neurons: relationship to autophagic pathology in Alzheimer's disease. *J Neurosci*, 28, 27, (Jul, 2008), 6926-6937.

Bucci, C., Thomsen, P., Nicoziani, P., McCarthy, J. & van Deurs, B.; (2000). Rab7: a key to lysosome biogenesis. *Mol Biol Cell*, 11, 2, (Feb, 2000), 467-480.

Carlton, J. G. & Martin-Serrano, J.; (2007). Parallels between cytokinesis and retroviral budding: a role for the ESCRT machinery. *Science*, 316, 5833, (Jun, 2007), 1908-1912.

Carlton, J. G. & Martin-Serrano, J.; (2009). The ESCRT machinery: new functions in viral and cellular biology. *Biochem Soc Trans*, 37, 1, (Feb, 2009), 195-199.

Chao, J. L., Tsai, Y. C., Chiu, S. J. & Sun, Y. H.; (2004). Localized Notch signal acts through eyg and upd to promote global growth in *Drosophila* eye. *Development*, 131, 16, (Aug, 2004), 3839-3847.

Chu, T., Sun, J., Saksena, S. & Emr, S. D.; (2006). New component of ESCRT-I regulates endosomal sorting complex assembly. *J Cell Biol*, 175, 5, (Dec, 2006), 815-823.

Corless, L., Crump, C. M., Griffin, S. D. & Harris, M.; (2010). Vps4 and the ESCRT-III complex are required for the release of infectious hepatitis C virus particles. *J Gen Virol*, 91, 2, (Feb, 2010), 362-372.

Cornet, M., Bidard, F., Schwarz, P., Da Costa, G., Blanchin-Roland, S., Dromer, F. & Gaillardin, C.; (2005). Deletions of endocytic components VPS28 and VPS32 affect growth at alkaline pH and virulence through both RIM101-dependent and RIM101-independent pathways in *Candida albicans*. *Infect Immun*, 73, 12, (Dec, 2005), 7977-7987.

Cuervo, A. M.; (2004). Autophagy: many paths to the same end. *Mol Cell Biochem*, 263, 1-2, (Aug, 2004), 55-72.

Cuervo, A. M.; (2010). Chaperone-mediated autophagy: selectivity pays off. *Trends Endocrinol Metab*, 21, 3, (Mar, 2010), 142-150.

Curtiss, M., Jones, C. & Babst, M.; (2007). Efficient cargo sorting by ESCRT-I and the subsequent release of ESCRT-I from multivesicular bodies requires the subunit Mvb12. *Mol Biol Cell*, 18, 2, (Feb, 2007), 636-645.

Demirov, D. G., Orenstein, J. M. & Freed, E. O.; (2002). The late domain of human immunodeficiency virus type 1 p6 promotes virus release in a cell type-dependent manner. *J Virol*, 76, 1, (Jan, 2002), 105-117.

Dimaano, C., Jones, C. B., Hanono, A., Curtiss, M. & Babst, M.; (2008). Ist1 regulates Vps4 localization and assembly. *Mol Biol Cell*, 19, 2, (Feb, 2008), 465-474.

Elia, N., Sougrat, R., Spurlin, T. A., Hurley, J. H. & Lippincott-Schwartz, J.; (2011). Dynamics of endosomal sorting complex required for transport (ESCRT) machinery during cytokinesis and its role in abscission. *Proc Natl Acad Sci U S A*, 108, 12, (Mar, 2011), 4846-4851.

Eskelinen, E. L. & Saftig, P.; (2009). Autophagy: a lysosomal degradation pathway with a central role in health and disease. *Biochim Biophys Acta*, 1793, 4, (Apr, 2009), 664-673.

Field, M. C. & Dacks, J. B.; (2009). First and last ancestors: reconstructing evolution of the endomembrane system with ESCRTs', vesicle coat proteins, and nuclear pore complexes. *Curr Opin Cell Biol*, 21, 1, (Feb, 2009), 4-13.

Field, M. C., Gabernet-Castello, C. & Dacks, J. B.; (2007). Reconstructing the evolution of the endocytic system: insights from genomics and molecular cell biology. *Adv Exp Med Biol*, 607, 84-96.

Filimonenko, M., Stuffers, S., Raiborg, C., Yamamoto, A., Malerod, L., Fisher, E. M., Isaacs, A., Brech, A., Stenmark, H. & Simonsen, A.; (2007). Functional multivesicular bodies are required for autophagic clearance of protein aggregates associated with neurodegenerative disease. *J Cell Biol*, 179, 3, (Nov, 2007), 485-500.

Fujii, K., Hurley, J. H. & Freed, E. O.; (2007). Beyond Tsg101: the role of Alix in 'ESCRTing' HIV-1. *Nat Rev Microbiol*, 5, 12, (Dec, 2007), 912-916.

Garrus, J. E., von Schwedler, U. K., Pornillos, O. W., Morham, S. G., Zavitz, K. H., Wang, H. E., Wettstein, D. A., Stray, K. M., Cote, M., Rich, R. L., Myszka, D. G. & Sundquist, W. I.; (2001). Tsg101 and the vacuolar protein sorting pathway are essential for HIV-1 budding. *Cell*, 107, 1, (Oct, 2001), 55-65.

Ghazi-Tabatabai, S., Obita, T., Pobbati, A. V., Perisic, O., Samson, R. Y., Bell, S. D. & Williams, R. L.; (2009). Evolution and assembly of ESCRTs'. *Biochem Soc Trans*, 37, 1, (Feb, 2009), 151-155.

Gill, D. J., Teo, H., Sun, J., Perisic, O., Veprintsev, D. B., Emr, S. D. & Williams, R. L.; (2007). Structural insight into the ESCRT-I/-II link and its role in MVB trafficking. *EMBO J*, 26, 2, (Jan, 2007), 600-612.

Gruenberg, J. & Stenmark, H.; (2004). The biogenesis of multivesicular endosomes. *Nat Rev Mol Cell Biol*, 5, 4, (Apr, 2004), 317-323.

Gutierrez, M. G., Munafo, D. B., Beron, W. & Colombo, M. I.; (2004). Rab7 is required for the normal progression of the autophagic pathway in mammalian cells. *J Cell Sci*, 117, 13, (Jun, 2004), 2687-2697.

Gutmann, D. H., Haipek, C. A., Burke, S. P., Sun, C. X., Scoles, D. R. & Pulst, S. M.; (2001). The NF2 interactor, hepatocyte growth factor-regulated tyrosine kinase substrate (HRS), associates with merlin in the "open" conformation and suppresses cell growth and motility. *Hum Mol Genet*, 10, 8, (Apr, 2001), 825-834.

Haddad, R., Lipson, K. E. & Webb, C. P.; (2001). Hepatocyte growth factor expression in human cancer and therapy with specific inhibitors. *Anticancer Res*, 21, 6B, (Dec, 2001), 4243-4252.

Hanson, P. I., Shim, S. & Merrill, S. A.; (2009). Cell biology of the ESCRT machinery. *Curr Opin Cell Biol*, 21, 4, (Aug, 2009), 568-574.

Harvey, R. J., Skelton-Robinson, M. & Rossor, M. N.; (2003). The prevalence and causes of dementia in people under the age of 65 years. *J Neurol Neurosurg Psychiatry*, 74, 9, (Sep, 2003), 1206-1209.

Henne, W. M., Buchkovich, N. J. & Emr, S. D.; (2011). The ESCRT pathway. *Dev Cell*, 21, 1, (Jul, 2011), 77-91.

Herman, E. K., Walker, G., van der Giezen, M. & Dacks, J. B.; (2011). Multivesicular bodies in the enigmatic amoeboflagellate *Breviata anathema* and the evolution of ESCRT 0. *J Cell Sci*, 124, 4, (Feb, 2011), 613-621.

Herz, H. M., Chen, Z., Scherr, H., Lackey, M., Bolduc, C. & Bergmann, A.; (2006). Vps25 mosaics display non-autonomous cell survival and overgrowth, and autonomous apoptosis. *Development*, 133, 10, (May, 2006), 1871-1880.

Hierro, A., Sun, J., Rusnak, A. S., Kim, J., Prag, G., Emr, S. D. & Hurley, J. H.; (2004). Structure of the ESCRT-II endosomal trafficking complex. *Nature*, 431, 7005, (Sep, 2004), 221-225.

Hong, L., Ning, X., Shi, Y., Shen, H., Zhang, Y., Lan, M., Liang, S., Wang, J. & Fan, D.; (2004). Reversal of multidrug resistance of gastric cancer cells by down-regulation of ZNRD1 with ZNRD1 siRNA. *Br J Biomed Sci*, 61, 4, 206-210.

Hunter, T.; (2000). Signaling-2000 and beyond. *Cell*, 100, 1, (Jan, 2000), 113-127.

Hurley, J. H. & Emr, S. D.; (2006). The ESCRT complexes: structure and mechanism of a membrane-trafficking network. *Annu Rev Biophys Biomol Struct*, 35, (Nov, 2006) 277-298.

Hurley, J. H., Boura, E., Carlson, L. A. & Rozycki, B.; (2010). Membrane budding. *Cell*, 143, 6, (Dec, 2010), 875-887.

Hurley, J. H & Hanson, P. I (2010). Membrane budding & scission by the ESCRT machinery : its all in the neck. *Nat Rev Mol Cell Biol*, 11, 8, 556-566.

Im, Y. J. & Hurley, J. H.; (2008). Integrated structural model and membrane targeting mechanism of the human ESCRT-II complex. *Dev Cell*, 14, 6, (Jun, 2008), 902-913.

Jager, S., Bucci, C., Tanida, I., Ueno, T., Kominami, E., Saftig, P. & Eskelinen, E. L.; (2004). Role for Rab7 in maturation of late autophagic vacuoles. *J Cell Sci*, 117, 20, (Sep, 2004), 4837-4848.

Jiao, J., Sun, K., Walker, W. P., Bagher, P., Cota, C. D. & Gunn, T. M.; (2009). Abnormal regulation of TSG101 in mice with spongiform neurodegeneration. *Biochim Biophys Acta*, 1792, 10, (Oct, 2009), 1027-1035.

Katzmann, D. J., Babst, M. & Emr, S. D.; (2001). Ubiquitin-dependent sorting into the multivesicular body pathway requires the function of a conserved endosomal protein sorting complex ESCRT-I. *Cell*, 106, 2, (Jul, 2001), 145-155.

Katzmann, D. J., Stefan, C. J., Babst, M. & Emr, S. D.; (2003). Vps27 recruits ESCRT machinery to endosomes during MVB sorting. *J Cell Biol*, 162, 3, (Aug, 2003), 413-423.

Kikonyogo, A., Bouamr, F., Vana, M. L., Xiang, Y., Aiyar, A., Carter, C. & Leis, J.; (2001). Proteins related to the Nedd4 family of ubiquitin protein ligases interact with the L domain of Rous sarcoma virus and are required for gag budding from cells. *Proc Natl Acad Sci U S A*, 98, 20, (Sep, 2001), 11199-11204.

Kim, B. Y., Olzmann, J. A., Barsh, G. S., Chin, L. S. & Li, L.; (2007). Spongiform neurodegeneration-associated E3 ligase Mahogunin ubiquitylates TSG101 and regulates endosomal trafficking. *Mol Biol Cell*, 18, 4, (Apr, 2007), 1129-1142.

Klionsky, D. J., Cregg, J. M., Dunn, W. A., Jr., Emr, S. D., Sakai, Y., Sandoval, I. V., Sibirny, A., Subramani, S., Thumm, M., Veenhuis, M. & Ohsumi, Y.; (2003). A unified nomenclature for yeast autophagy-related genes. *Dev Cell*, 5, 4, (Oct, 2003), 539-545.

Koon, N., Schneider-Stock, R., Sarlomo-Rikala, M., Lasota, J., Smolkin, M., Petroni, G., Zaika, A., Boltze, C., Meyer, F., Andersson, L., Knuutila, S., Miettinen, M. & El-Rifai, W.; (2004). Molecular targets for tumour progression in gastrointestinal stromal tumours. *Gut*, 53, 2, (Feb, 2004), 235-240.

Kostelansky, M. S., Schluter, C., Tam, Y. Y., Lee, S., Ghirlando, R., Beach, B., Conibear, E. & Hurley, J. H.; (2007). Molecular architecture and functional model of the complete yeast ESCRT-I heterotetramer. *Cell*, 129, 3, (May, 2007), 485-498.

Kulp, A. & Kuehn, M. J.; (2011). The recognition of {beta}-strand motifs by RseB is required for {sigma}E activity in *Escherichia coli*. *J Bacteriol*, 193, 22, (Sep, 2011), 6179-6186.

Lakkaraju, A. & Rodriguez-Boulan, E.; (2008). Itinerant exosomes: emerging roles in cell and tissue polarity. *Trends Cell Biol*, 18, 5, (May, 2008), 199-209.

Langelier, C., von Schwedler, U. K., Fisher, R. D., De Domenico, I., White, P. L., Hill, C. P., Kaplan, J., Ward, D. & Sundquist, W. I.; (2006). Human ESCRT-II complex and its role in human immunodeficiency virus type 1 release. *J Virol*, 80, 19, (Oct, 2006), 9465-9480.

Lata, S., Schoehn, G., Jain, A., Pires, R., Piehler, J., Gottlinger, H. G. & Weissenhorn, W.; (2008). Helical structures of ESCRT-III are disassembled by VPS4. *Science*, 321, 5894, (Sep, 2008), 1354-1357.

Lawrence, B. P. & Brown, W. J.; (1992). Autophagic vacuoles rapidly fuse with pre-existing lysosomes in cultured hepatocytes. *J Cell Sci*, 102, 3, (Jul, 1992), 515-526.

Lee, J. A., Liu, L. & Gao, F. B.; (2009). Autophagy defects contribute to neurodegeneration induced by dysfunctional ESCRT-III. *Autophagy*, 5, 7, (Oct, 2009), 1070-1072.

Lee, J. A., Beigneux, A., Ahmad, S. T., Young, S. G. & Gao, F. B.; (2007). ESCRT-III dysfunction causes autophagosome accumulation and neurodegeneration. *Curr Biol*, 17, 18, (Sep, 2007), 1561-1567.

Leung, K. F., Dacks, J. B. & Field, M. C.; (2008). Evolution of the multivesicular body ESCRT machinery; retention across the eukaryotic lineage. *Traffic*, 9, 10, (Sep, 2008), 1698-1716.

Li, J., Belogortseva, N., Porter, D. & Park, M.; (2008). Chmp1A functions as a novel tumor suppressor gene in human embryonic kidney and ductal pancreatic tumor cells. *Cell Cycle*, 7, 18, (Sep, 2008), 2886-2893.

Li, J., Orr, B., White, K., Belogortseva, N., Niles, R., Boskovic, G., Nguyen, H., Dykes, A. & Park, M.; (2009). Chmp 1A is a mediator of the anti-proliferative effects of all-trans retinoic acid in human pancreatic cancer cells. *Mol Cancer*, 8, 7, (Sept, 2008), 2886-2893.

Li, L. & Cohen, S. N.; (1996). Tsg101: a novel tumor susceptibility gene isolated by controlled homozygous functional knockout of allelic loci in mammalian cells. *Cell*, 85, 3, (May, 1996), 319-329.

Li, L., Li, X., Francke, U. & Cohen, S. N.; (1997). The TSG101 tumor susceptibility gene is located in chromosome 11 band p15 and is mutated in human breast cancer. *Cell*, 88, 1, (Jan, 1997), 143-154.

Lin, P. M., Liu, T. C., Chang, J. G., Chen, T. P. & Lin, S. F.; (1998). Aberrant TSG101 transcripts in acute myeloid leukaemia. *Br J Haematol*, 102, 3, (Aug, 1998), 753-758.

Lindas, A. C., Karlsson, E. A., Lindgren, M. T., Ettema, T. J. & Bernander, R.; (2008). A unique cell division machinery in the Archaea. *Proc Natl Acad Sci U S A*, 105, 48, (Dec, 2008), 18942-18946.

Liou, W., Geuze, H. J., Geelen, M. J. & Slot, J. W.; (1997). The autophagic and endocytic pathways converge at the nascent autophagic vacuoles. *J Cell Biol*, 136, 1, (Jan, 1997), 61-70.

Liu, R. T., Huang, C. C., You, H. L., Chou, F. F., Hu, C. C., Chao, F. P., Chen, C. M. & Cheng, J. T.; (2002). Overexpression of tumor susceptibility gene TSG101 in human papillary thyroid carcinomas. *Oncogene*, 21, 31, (Jul, 2002), 4830-4837.

Lloyd, T. E., Atkinson, R., Wu, M. N., Zhou, Y., Pennetta, G. & Bellen, H. J.; (2002) Hrs regulates endosome membrane invagination and tyrosine kinase receptor signaling in *Drosophila*. *Cell* 108, 2, (Jan, 2002), 261–269.

Lobert, V. H. & Stenmark, H.; (2011). Cell polarity and migration: emerging role for the endosomal sorting machinery. *Physiology*, 26, 3, (Jun, 2011), 171-180.

Ma, X. R., Edmund Sim, U. H., Pauline, B., Patricia, L. & Rahman, J.; (2008). Overexpression of WNT2 and TSG101 genes in colorectal carcinoma. *Trop Biomed*, 25, 1, (Apr, 2008), 46-57.

Makarova, K. S., Yutin, N., Bell, S. D. & Koonin, E. V.; (2010). Evolution of diverse cell division and vesicle formation systems in Archaea. *Nat Rev Microbiol*, 8, 10, (Oct, 2010), 731-741.

Malerod, L., Stuffers, S., Brech, A. & Stenmark, H.; (2007). Vps22/EAP30 in ESCRT-II mediates endosomal sorting of growth factor and chemokine receptors destined for lysosomal degradation. *Traffic* 8, 11, (Nov, 2007) 1617–1629.

Mao, Y., Nickitenko, A., Duan, X., Lloyd, T. E., Wu, M. N., Bellen, H. & Quiocho, F. A.; (2000). Crystal structure of the VHS and FYVE tandem domains of Hrs, a protein involved in membrane trafficking and signal transduction. *Cell*, 100, 4, (Feb, 2000), 447-456.

Martin-Serrano, J. & Marsh, M.; (2007). ALIX catches HIV. *Cell Host Microbe*, 1, 1, (Mar, 2007), 5-7.

Martin-Serrano, J. & Neil, S. J.; (2011). Host factors involved in retroviral budding and release. *Nat Rev Microbiol*, 9, 7, (Jul, 2011), 519-531.

Marzella, L., Ahlberg, J. & Glaumann, H.; (1981). Autophagy, heterophagy, microautophagy and crinophagy as the means for intracellular degradation. *Virchows Arch B Cell Pathol Incl Mol Pathol*, 36, 2-3, 219-234.

McCullough, J., Fisher, R. D., Whitby, F. G., Sundquist, W. I. & Hill, C. P.; (2008). ALIX-CHMP4 interactions in the human ESCRT pathway. *Proc Natl Acad Sci U S A*, 105, 22, (Jun, 2008), 7687-7691.

Mehrpour, M., Esclatine, A., Beau, I. & Codogno, P.; (2010). Overview of macroautophagy regulation in mammalian cells. *Cell Res*, 20, 7, (Jul, 2010), 748-762.

Metcalf, D. & Isaacs, A. M.; (2010). The role of ESCRT proteins in fusion events involving lysosomes, endosomes and autophagosomes. *Biochem Soc Trans*, 38, 6, (Dec, 2010), 1469-1473.

Michelet, X., Djeddi, A. & Legouis, R.; (2010). Developmental and cellular functions of the ESCRT machinery in pluricellular organisms. *Biol Cell*, 102, 3, (Mar, 2010), 191-202.

Mizuno, E., Kawahata, K., Kato, M., Kitamura, N. & Komada, M.; (2003). STAM proteins bind ubiquitinated proteins on the early endosome via the VHS domain and ubiquitin-interacting motif. *Mol Biol Cell*, 14, 9, (Sep, 2003), 3675-3689.

Mizushima, N., Levine, B., Cuervo, A. M. & Klionsky, D. J.; (2008). Autophagy fights disease through cellular self-digestion. *Nature*, 451, 7182, (Feb, 2008), 1069-1075.

Moberg, K. H., Schelble, S., Burdick, S. K. & Hariharan, I. K.; (2005). Mutations in erupted, the *Drosophila* ortholog of mammalian tumor susceptibility gene 101, elicit non-cell-autonomous overgrowth. *Dev Cell*, 9, 5, (Nov, 2005), 699-710.

Morita, E. & Sundquist, W. I.; (2004). Retrovirus budding. *Annu Rev Cell Dev Biol*, 20, (Jun, 2004), 395-425.

Morita, E., Sandrin, V., Chung, H. Y., Morham, S. G., Gygi, S. P., Rodesch, C. K. & Sundquist, W. I.; (2007). Human ESCRT and ALIX proteins interact with proteins of the midbody and function in cytokinesis. *EMBO J*, 26, 19, (Oct, 2007), 4215-4227.

Muziol, T., Pineda-Molina, E., Ravelli, R. B., Zamborlini, A., Usami, Y., Gottlinger, H. & Weissenhorn, W.; (2006). Structural basis for budding by the ESCRT-III factor CHMP3. *Dev Cell*, 10, 6, (Jun, 2006), 821-830.

Nara, A., Mizushima, N., Yamamoto, A., Kabeya, Y., Ohsumi, Y. & Yoshimori, T.; (2002). SKD1 AAA ATPase-dependent endosomal transport is involved in autolysosome formation. *Cell Struct Funct*, 27, 1, (Feb, 2002), 29-37.

Neary, D., Snowden, J. & Mann, D.; (2005). Frontotemporal dementia. *Lancet Neurol*, 4, 11, (Nov, 2005), 771-780.

Neumann, M., Sampathu, D. M., Kwong, L. K., Truax, A. C., Micsenyi, M. C., Chou, T. T., Bruce, J., Schuck, T., Grossman, M., Clark, C. M., McCluskey, L. F., Miller, B. L., Masliah, E., Mackenzie, I. R., Feldman, H., Feiden, W., Kretzschmar, H. A., Trojanowski, J. Q. & Lee, V. M.; (2006). Ubiquitinated TDP-43 in frontotemporal lobar degeneration and amyotrophic lateral sclerosis. *Science*, 314, 5796, (Oct, 2006), 130-133.

Nickerson, D. P., West, M. & Odorizzi, G.; (2006). Did2 coordinates Vps4-mediated dissociation of ESCRT-III from endosomes. *J Cell Biol*, 175, 5, (Dec, 2006), 715-720.

Obita, T., Saksena, S., Ghazi-Tabatabai, S., Gill, D. J., Perisic, O., Emr, S. D. & Williams, R. L.; (2007). Structural basis for selective recognition of ESCRT-III by the AAA ATPase Vps4. *Nature*, 449, 7163, (Oct, 2007), 735-739.

Oestreich, A. J., Davies, B. A., Payne, J. A. & Katzmann, D. J.; (2007). Mvb12 is a novel member of ESCRT-I involved in cargo selection by the multivesicular body pathway. *Mol Biol Cell*, 18, 2, (Feb, 2007), 646-657.

Oh, K. B., Stanton, M. J., West, W. W., Todd, G. L. & Wagner, K. U.; (2007). Tsg101 is upregulated in a subset of invasive human breast cancers and its targeted overexpression in transgenic mice reveals weak oncogenic properties for mammary cancer initiation. *Oncogene*, 26, 40, (Aug, 2007), 5950-5959.

Ohsaki, Y., Sugimoto, Y., Suzuki, M., Hosokawa, H., Yoshimori, T., Davies, J. P., Ioannou, Y. A., Vanier, M. T., Ohno, K. & Ninomiya, H.; (2006). Cholesterol depletion facilitates ubiquitylation of NPC1 and its association with SKD1/Vps4. *J Cell Sci*, 119, 3, (Jul, 2006), 2643-2653.

Orenstein, S. J. & Cuervo, A. M.; (2010). Chaperone-mediated autophagy: molecular mechanisms and physiological relevance. *Semin Cell Dev Biol*, 21, 7, (Sep, 2010), 719-726.

Pankiv, S., Clausen, T. H., Lamark, T., Brech, A., Bruun, J. A., Outzen, H., Overvatn, A., Bjorkoy, G. & Johansen, T.; (2007). p62/SQSTM1 binds directly to Atg8/LC3 to facilitate degradation of ubiquitinated protein aggregates by autophagy. *J Biol Chem*, 282, 33, (Aug, 2007), 24131-24145.

Parkinson, N., Ince, P. G., Smith, M. O., Highley, R., Skibinski, G., Andersen, P. M., Morrison, K. E., Pall, H. S., Hardiman, O., Collinge, J., Shaw, P. J. & Fisher, E. M.; (2006). ALS phenotypes with mutations in CHMP2B (charged multivesicular body protein 2B). *Neurology*, 67, 6, (Sep, 2006), 1074-1077.

Pawliczek, T. & Crump, C. M.; (2009). Herpes simplex virus type 1 production requires a functional ESCRT-III complex but is independent of TSG101 and ALIX expression. *J Virol*, 83, 21, (Nov, 2009), 11254-11264.

Pawson, T., Gish, G. D. & Nash, P.; (2001). SH2 domains, interaction modules and cellular wiring. *Trends Cell Biol*, 11, 12, (Dec, 2001), 504-511.

Peschard, P., Fournier, T. M., Lamorte, L., Naujokas, M. A., Band, H., Langdon, W. Y. & Park, M.; (2001). Mutation of the c-Cbl TKB domain binding site on the Met receptor tyrosine kinase converts it into a transforming protein. *Mol Cell*, 8, 5, (Nov, 2001), 995-1004.

Philips, J. A., Porto, M. C., Wang, H., Rubin, E. J. & Perrimon, N.; (2008). ESCRT factors restrict mycobacterial growth, *Proc Natl Acad Sci U S A*, 105, 8, (Feb, 2008), 3070-3075.

Pincetic, A., Medina, G., Carter, C. & Leis, J.; (2008). Avian sarcoma virus and human immunodeficiency virus, type 1 use different subsets of ESCRT proteins to facilitate the budding process. *J Biol Chem*, 283, 44, (Oct, 2008), 29822-29830.

Piper, R. C. & Katzmann, D. J.; (2007). Biogenesis and function of multivesicular bodies. *Annu Rev Cell Dev Biol*, 23, 519-547.

Pornillos, O., Higginson, D. S., Stray, K. M., Fisher, R. D., Garrus, J. E., Payne, M., He, G. P., Wang, H. E., Morham, S. G. & Sundquist, W. I.; (2003). HIV Gag mimics the Tsg101-recruiting activity of the human Hrs protein. *J Cell Biol*, 162, 3, (Aug, 2003), 425-434.

Radtke, F. & Raj, K.; (2003) The role of Notch in tumorigenesis: oncogene or tumour suppressor? *Nat Rev Cancer*, 3, 10, (Oct, 2003), 756-767.

Raiborg, C. & Stenmark, H.; (2009). The ESCRT machinery in endosomal sorting of ubiquitylated membrane proteins. *Nature*, 458, 7237, (Mar, 2009), 445-452.

Raiborg, C., Rusten, T. E. & Stenmark, H.; (2003). Protein sorting into multivesicular endosomes. *Curr Opin Cell Biol*, 15, 4, (Aug, 2003), 446-455.

Ratnavalli, E., Brayne, C., Dawson, K. & Hodges, J. R.; (2002). The prevalence of frontotemporal dementia. *Neurology*, 58, 11, (Jun, 2002), 1615-1621.

Ravikumar, B., Vacher, C., Berger, Z., Davies, J. E., Luo, S., Oroz, L. G., Scaravilli, F., Easton, D. F., Duden, R., O'Kane, C. J. & Rubinsztein, D. C.; (2004). Inhibition of mTOR induces autophagy and reduces toxicity of polyglutamine expansions in fly and mouse models of Huntington disease. *Nat Genet*, 36, 6, (Jun, 2004), 585-595.

Raymond, C. K., Howald-Stevenson, I., Vater, C. A. & Stevens, T. H.; (1992). Morphological classification of the yeast vacuolar protein sorting mutants: evidence for a prevacuolar compartment in class E vps mutants. *Mol Biol Cell*, 3, 12, (Dec, 1992), 1389-1402.

Reid, E., Connell, J., Edwards, T. L., Duley, S., Brown, S. E. & Sanderson, C. M.; (2005). The hereditary spastic paraplegia protein spastin interacts with the ESCRT-III complex-associated endosomal protein CHMP1B. *Hum Mol Genet*, 14, 1, (Jan, 2005), 19-38.

Rodahl, L. M., Haglund, K., Sem-Jacobsen, C., Wendler, F., Vincent, J. P., Lindmo, K., Rusten, T. E. & Stenmark, H.; (2009). Disruption of Vps4 and JNK function in *Drosophila* causes tumour growth. *PLoS One*, 4, 2, (2009).

Roxrud, I., Stenmark, H. & Malerod, L.; (2010). ESCRT & Co. *Biol Cell*, 102, 5, (May, 2010), 293-318.

Russell, M. R., Nickerson, D. P. & Odorizzi, G.; (2006). Molecular mechanisms of late endosome morphology, identity and sorting. *Curr Opin Cell Biol*, 18, 4, (Aug, 2006), 422-428.

Rusten, T. E. & Stenmark, H.; (2009). How do ESCRT proteins control autophagy? *J Cell Sci*, 122, 13, (Jul, 2009), 2179-2183.

Rusten, T. E., Vaccari, T., Lindmo, K., Rodahl, L. M., Nezis, I. P., Sem-Jacobsen, C., Wendler, F., Vincent, J. P., Brech, A., Bilder, D. & Stenmark, H.; (2007). ESCRTs' and Fab1 regulate distinct steps of autophagy. *Curr Biol*, 17, 20, (Oct, 2007), 1817-1825.

Sahu, R., Kaushik, S., Clement, C. C., Cannizzo, E. S., Scharf, B., Follenzi, A., Potolicchio, I., Nieves, E., Cuervo, A. M. & Santambrogio, L.; (2011). Microautophagy of cytosolic proteins by late endosomes. *Dev Cell*, 20, 1, (Jan, 2011), 131-139.

Saksena, S. & Emr, S. D.; (2009). ESCRTs' and human disease. *Biochem Soc Trans*, 37, 1, (Feb, 2009), 167-172.

Saksena, S., Sun, J., Chu, T. & Emr, S. D.; (2007). ESCRTing proteins in the endocytic pathway. *Trends Biochem Sci*, 32, 12, (Dec, 2007), 561-573.

Saksena, S., Wahlman, J., Teis, D., Johnson, A. E. & Emr, S. D.; (2009). Functional reconstitution of ESCRT-III assembly and disassembly. *Cell*, 136, 1, (Jan, 2009), 97-109.

Samson, R. Y., Obita, T., Freund, S. M., Williams, R. L. & Bell, S. D.; (2008). A role for the ESCRT system in cell division in archaea. *Science*, 322, 5908, (Dec, 2008), 1710-1713.

Samson, R. Y., Obita, T., Hodgson, B., Shaw, M. K., Chong, P. L., Williams, R. L. & Bell, S. D.; (2011). Molecular and structural basis of ESCRT-III recruitment to membranes during archaeal cell division. *Mol Cell*, 41, 2, (Jan, 2011), 186-196.

Scoles, D. R., Nguyen, V. D., Qin, Y., Sun, C. X., Morrison, H., Gutmann, D. H. & Pulst, S. M.; (2002). Neurofibromatosis 2 (NF2) tumor suppressor schwannomin and its interacting protein HRS regulate STAT signaling. *Hum Mol Genet*, 11, 25, (Dec, 2002), 3179-3189.

Scott, A., Chung, H. Y., Gonciarz-Swiatek, M., Hill, G. C., Whitby, F. G., Gaspar, J., Holton, J. M., Viswanathan, R., Ghaffarian, S., Hill, C. P. & Sundquist, W. I.; (2005). Structural and mechanistic studies of VPS4 proteins. *EMBO J*, 24, 20, (Oct, 2005), 3658-3669.

Seals, D. F., Eitzen, G., Margolis, N., Wickner, W. T. & Price, A.; (2000). A Ypt/Rab effector complex containing the Sec1 homolog Vps33p is required for homotypic vacuole fusion. *Proc Natl Acad Sci U S A*, 97, 17, (Aug, 2000), 9402-9407.

Shiels, A., Bennett, T. M., Knopf, H. L., Yamada, K., Yoshiura, K., Niikawa, N., Shim, S. & Hanson, P. I.; (2007). CHMP4B, a novel gene for autosomal dominant cataracts linked to chromosome 20q. *Am J Hum Genet*, 81, 3, (Sep, 2007), 596-606.

Shim, S., Kimpler, L. A. & Hanson, P. I.; (2007). Structure/function analysis of four core ESCRT-III proteins reveals common regulatory role for extreme C-terminal domain. *Traffic*, 8, 8, (Aug, 2007), 1068-1079.

Shirk, A. J., Anderson, S. K., Hashemi, S. H., Chance, P. F. & Bennett, C. L.; (2005). SIMPLE interacts with NEDD4 and TSG101: evidence for a role in lysosomal sorting and implications for Charcot-Marie-Tooth disease. *J Neurosci Res*, 82, 1, (Oct, 2005), 43-50.

Shpilka, T. & Elazar, Z.; (2011). Shedding light on mammalian microautophagy. *Dev Cell*, 20, 1, (Jan, 2011), 1-2.

Simons, M. & Raposo, G.; (2009). Exosomes-vesicular carriers for intercellular communication. *Curr Opin Cell Biol*, 21, 4, (Aug, 2009), 575-581.

Skibinski, G., Parkinson, N. J., Brown, J. M., Chakrabarti, L., Lloyd, S. L., Hummerich, H., Nielsen, J. E., Hodges, J. R., Spillantini, M. G., Thusgaard, T., Brandner, S., Brun, A., Rossor, M. N., Gade, A., Johannsen, P., Sorensen, S. A., Gydesen, S., Fisher, E. M. & Collinge, J.; (2005). Mutations in the endosomal ESCRT-III complex subunit CHMP2B in frontotemporal dementia. *Nat Genet*, 37, 8, (Aug, 2005), 806-808.

Slagsvold, T., Pattni, K., Malerod, L. & Stenmark, H.; (2006). Endosomal and non-endosomal functions of ESCRT proteins. *Trends Cell Biol*, 16, 6, (Jun, 2006), 317-326.

Slater, R. & Bishop, N. E.; (2006). Genetic structure and evolution of the Vps25 family, a yeast ESCRT-II component. *BMC Evol Biol*, 6, (Aug, 2006), 59.

Sorkin, A. & von Zastrow, M.; (2009). Endocytosis and signalling: intertwining molecular networks. *Nat Rev Mol Cell Biol*, 10, 9, (Sep, 2009), 609-622.

Spitzer, C., Schellmann, S., Sabovljevic, A., Shahriari, M., Keshavaiah, C., Bechtold, N., Herzog, M., Muller, S., Hanisch, F. G. & Hulskamp, M.; (2006). The *Arabidopsis* elch mutant reveals functions of an ESCRT component in cytokinesis. *Development*, 133, 23, (Dec, 2006), 4679-4689.

Stark, P., Bodemer, W., Hannig, H., Luboshitz, J., Shaklai, M. & Shohat, B.; (2003). Human T lymphotropic virus type 1 in a seronegative B chronic lymphocytic leukemia patient. *Med Microbiol Immunol*, 192, 4, (Nov, 2003), 205-209.

Stefani, F., Zhang, L., Taylor, S., Donovan, J., Rollinson, S., Doyotte, A., Brownhill, K., Bennion, J., Pickering-Brown, S. & Woodman, P.; (2011). UBAP1 is a component of an endosome-specific ESCRT-I complex that is essential for MVB sorting. *Curr Biol*, 21, 14, (Jul, 2011), 1245-1250.

Strack, B., Calistri, A., Craig, S., Popova, E. & Gottlinger, H. G.; (2003). AIP1/ALIX is a binding partner for HIV-1 p6 and EIAV p9 functioning in virus budding. *Cell*, 114, 6, (Sep, 2003), 689-699.

Stuffers, S., Brech, A. & Stenmark, H.; (2009a). ESCRT proteins in physiology and disease. *Exp Cell Res*, 315, 9, (May, 2009), 1619-1626.

Stuffers, S., Sem Wegner, C., Stenmark, H. & Brech, A.; (2009b). Multivesicular endosome biogenesis in the absence of ESCRTs'. *Traffic*, 10, 7, (Jul, 2009), 925-937.
Sun, Z., Pan, J., Bubley, G. & Balk, S. P.; (1997). Frequent abnormalities of TSG101 transcripts in human prostate cancer. *Oncogene*, 15, 25, (Dec, 1997), 3121-3125.
Talbot, K. & Ansorge, O.; (2006). Recent advances in the genetics of amyotrophic lateral sclerosis and frontotemporal dementia: common pathways in neurodegenerative disease. *Hum Mol Genet*, 15, 2, (Oct, 2006), 182-187.
Tarrago-Trani, M. T. & Storrie, B.; (2007). Alternate routes for drug delivery to the cell interior: pathways to the Golgi apparatus and endoplasmic reticulum. *Adv Drug Deliv Rev*, 59, 8, (Aug, 2007), 782-797.
Teis, D., Saksena, S. & Emr, S. D.; (2008). Ordered assembly of the ESCRT-III complex on endosomes is required to sequester cargo during MVB formation. *Dev Cell*, 15, 4, (Oct, 2008), 578-589.
Teis, D., Saksena, S., Judson, B. L. & Emr, S. D.; (2010). ESCRT-II coordinates the assembly of ESCRT-III filaments for cargo sorting and multivesicular body vesicle formation. *EMBO J*, 29, 5, (Mar, 2010), 871-883.
Teo, H., Gill, D. J., Sun, J., Perisic, O., Veprintsev, D. B., Vallis, Y., Emr, S. D. & Williams, R. L.; (2006). ESCRT-I core and ESCRT-II GLUE domain structures reveal role for GLUE in linking to ESCRT-I and membranes. *Cell*, 125, 1, (Apr, 2006), 99-111.
Thery, C., Ostrowski, M. & Segura, E.; (2009). Membrane vesicles as conveyors of immune responses. *Nat Rev Immunol*, 9, 8, (Aug, 2009), 581-593.
Thompson, B. J., Mathieu, J., Sung, H. H., Loeser, E., Rorth, P. & Cohen, S. M.; (2005). Tumor suppressor properties of the ESCRT-II complex component Vps25 in *Drosophila*. *Dev Cell*, 9, 5, (Nov, 2005), 711-720.
Toyoshima, M., Tanaka, N., Aoki, J., Tanaka, Y., Murata, K., Kyuuma, M., Kobayashi, H., Ishii, N., Yaegashi, N. & Sugamura, K.; (2007). Inhibition of tumor growth and metastasis by depletion of vesicular sorting protein Hrs: its regulatory role on E-cadherin and beta-catenin. *Cancer Res*, 67, 11, (Jun, 2007), 5162-5171.
Tsai, Y. C. & Sun, Y. H.; (2004). Long-range effect of upd, a ligand for JAK/STAT pathway, on cell cycle in *Drosophila* eye development. *Genesis*, 39, 2, (Jun, 2004), 141-153.
Urwin, H., Authier, A., Nielsen, J. E., Metcalf, D., Powell, C., Froud, K., Malcolm, D. S., Holm, I., Johannsen, P., Brown, J., Fisher, E. M., van der Zee, J., Bruyland, M., Van Broeckhoven, C., Collinge, J., Brandner, S., Futter, C. & Isaacs, A. M.; (2010). Disruption of endocytic trafficking in frontotemporal dementia with CHMP2B mutations. *Hum Mol Genet*, 19, 11, (Jun, 2010), 2228-2238.
Vaccari, T. & Bilder, D.; (2005). The *Drosophila* tumor suppressor Vps25 prevents nonautonomous overproliferation by regulating notch trafficking. *Dev Cell*, 9, 5, (Nov, 2005), 687-698.
Vaccari, T., Rusten, T. E., Menut, L., Nezis, I. P., Brech, A., Stenmark, H. & Bilder, D.; (2009). Comparative analysis of ESCRT-I, ESCRT-II and ESCRT-III function in *Drosophila* by efficient isolation of ESCRT mutants. *J Cell Sci*, 122, 14, (Jul, 2009), 2413-2423.
Vadlamudi, R. K., Joung, I., Strominger, J. L. & Shin, J.; (1996). p62, a phosphotyrosine-independent ligand of the SH2 domain of p56lck, belongs to a new class of ubiquitin-binding proteins. *J Biol Chem*, 271, 34, (Aug, 1996), 20235-20237.

Vieira, O. V., Harrison, R. E., Scott, C. C., Stenmark, H., Alexander, D., Liu, J., Gruenberg, J., Schreiber, A. D. & Grinstein, S.; (2004). Acquisition of Hrs, an essential component of phagosomal maturation, is impaired by mycobacteria. *Mol Cell Biol*, 24, 10, (May, 2004), 4593-4604.

von Schwedler, U. K., Stuchell, M., Muller, B., Ward, D. M., Chung, H. Y., Morita, E., Wang, H. E., Davis, T., He, G. P., Cimbora, D. M., Scott, A., Krausslich, H. G., Kaplan, J., Morham, S. G. & Sundquist, W. I.; (2003). The protein network of HIV budding. *Cell*, 114, 6, (Sep, 2003), 701-713.

Walker, G. E., Antoniono, R. J., Ross, H. J., Paisley, T. E. & Oh, Y.; (2006). Neuroendocrine-like differentiation of non-small cell lung carcinoma cells: regulation by cAMP and the interaction of mac25/IGFBP-rP1 and 25.1. *Oncogene*, 25, 13, (Mar, 2006), 1943-1954.

Waterman, H., Katz, M., Rubin, C., Shtiegman, K., Lavi, S., Elson, A., Jovin, T. & Yarden, Y.; (2002). A mutant EGF-receptor defective in ubiquitylation and endocytosis unveils a role for Grb2 in negative signaling. *EMBO J*, 21, 3, (Feb, 2002), 303-313.

Wegner, C. S., Rodahl, L. M. & Stenmark, H.; (2011). ESCRT proteins and cell signalling. *Traffic*, 12, 10, (Oct, 2011), 1291-1297.

Williams, R. L. & Urbe, S.; (2007). The emerging shape of the ESCRT machinery. *Nat Rev Mol Cell Biol*, 8, 5, (May, 2007), 355-368.

Wilson, E. M., Oh, Y., Hwa, V. & Rosenfeld, R. G.; (2001). Interaction of IGF-binding protein-related protein 1 with a novel protein, neuroendocrine differentiation factor, results in neuroendocrine differentiation of prostate cancer cells. *J Clin Endocrinol Metab*, 86, 9, (Sep, 2001), 4504-4511.

Winter, V. & Hauser, M. T.; (2006). Exploring the ESCRTing machinery in eukaryotes. *Trends Plant Sci*, 11, 3, (Mar, 2006), 115-123.

Wolf, J. M., Johnson, D. J., Chmielewski, D. & Davis, D. A.; (2010). The *Candida albicans* ESCRT pathway makes Rim101-dependent and -independent contributes to pathogenesis. *Eukaryot Cell*, 9, 8, (Jun, 2010), 1203-1215.

Wollert, T. & Hurley, J. H.; (2010). Molecular mechanism of multivesicular body biogenesis by ESCRT complexes. *Nature*, 464, 7290, (Apr, 2010), 864-869.

Wollert, T., Wunder, C., Lippincott-Schwartz, J. & Hurley, J. H.; (2009). Membrane scission by the ESCRT-III complex, *Nature*, 458, 7235, (Mar, 2009), 172-177.

Wurmser, A. E., Sato, T. K. & Emr, S. D.; (2000). New component of the vacuolar class C-Vps complex couples nucleotide exchange on the Ypt7 GTPase to SNARE-dependent docking and fusion. *J Cell Biol*, 151, 3, (Oct, 2000), 551-562.

Xu, Z., Liang, L., Wang, H., Li, T. & Zhao, M.; (2003). HCRP1, a novel gene that is downregulated in hepatocellular carcinoma, encodes a growth-inhibitory protein. *Biochem Biophys Res Commun*, 311, 4, (Nov, 2003), 1057-1066.

Yang, D., Rismanchi, N., Renvoise, B., Lippincott-Schwartz, J., Blackstone, C. & Hurley, J. H.; (2008). Structural basis for midbody targeting of spastin by the ESCRT-III protein CHMP1B. *Nat Struct Mol Biol*, 15, 12, (Dec, 2008), 1278-1286.

Young, T. W., Mei, F. C., Rosen, D. G., Yang, G., Li, N., Liu, J. & Cheng, X.; (2007a). Up-regulation of tumor susceptibility gene 101 protein in ovarian carcinomas revealed by proteomics analyses. *Mol Cell Proteomics*, 6, 2, (Feb, 2007), 294-304.

Young, T. W., Rosen, D. G., Mei, F. C., Li, N., Liu, J., Wang, X. F. & Cheng, X.; (2007b). Upregulation of tumor susceptibility gene 101 conveys poor prognosis through suppression of p21 expression in ovarian cancer. *Clin Cancer Res*, 13, 13, (Jul, 2007), 3848-3854.

Yu, Z., Gonciarz, M. D., Sundquist, W. I., Hill, C. P. & Jensen, G. J.; (2008). Cryo-EM structure of dodecameric Vps4p and its 2:1 complex with Vta1p. *J Mol Biol*, 377, 2, (Mar, 2008), 364-377.

Zhu, G., Gilchrist, R., Borley, N., Chng, H. W., Morgan, M., Marshall, J. F., Camplejohn, R. S., Muir, G. H. & Hart, I. R.; (2004). Reduction of TSG101 protein has a negative impact on tumor cell growth. *Int J Cancer*, 109, 4, (Apr, 2004), 541-547.

13

Autologous Grafts of Mesenchymal Stem Cells – Between Dream and Reality

Frédéric Torossian, Aurelie Bisson, Laurent Drouot,
Olivier Boyer and Marek Lamacz
*Institute for Research and Innovation in Biomedicine,
Faculty of Medicine & Pharmacy, University of Rouen,
France*

1. Introduction

During the last decade, the characterization of Adult Stem Cells (ASC) incited extraordinary infatuation for the development of autologous cellular therapy. The number of directed cellular differentiation essays of hematopoietic and mesenchymal stem cells demonstrated, against the classical rules of embryology, unsuspected capacities to generate *ex vivo* practically all cellular types. Nevertheless, the difficulties of revealing these spectacular capacities during clinical applications of tissue reparation suggest a random evolution of cell cultures. The major influence on this reality may play fragmentary knowledge of transductional mechanisms controlling the cellular fate and above all, the quality of isolated cells. This last condition seems to represent the one of essential technical barriers. In fact, the process of cellular isolation principally residing in the employment of cellular adherence and/or magnetic field able to retain cells marked by tagged antibodies. Unfortunately, the proteins recognized by antibodies are not expressed by sole stem cells but also by committed progenitors. In fact, the bone marrow precursors represent very heterogenic population of mononuclear cells whose affiliation seemed to be recently questioned. For instance, the antibody against CD34 protein is employed for isolation of Hematopoietic Stem Cells (HSC) while the absence of CD45 (universal hematopoietic cell marker) is recognized as sufficient to qualify Mesenchymal Stem Cells (MSC). Consequently, if previous observations revealed promising potential of bone marrow stem cells to be used for development of cellular therapy, their utilization must be preceded by detailed studies of their biology with particular focus on specific markers and transductional pathways permitting a high purity of isolation and control of differentiation protocols. The proposed chapter is based on our recent work indicating the heterogeneity of mesenchymal stem cells isolated from rabbit bone marrow which, placed in the context of recent studies, allows to propose a novel hierarchic organization of bone marrow cells. As *ex vivo* differentiation of stem cells would be dependent on transductional mechanisms we also propose to discuss how pharmacological modulation of activity of molecular target implicated in calcium homeostasis may influence cellular differentiation.

2. Heterogeneity of MSC

2.1 Introduction

Besides tissues having properties of self-renewal such bone marrow, the liver represents in Man the sole internal organ endowed with a spectacular capacity of regeneration illustrated already by the ancient myth of Prometheus. Interestingly, this process intervenes only after physical damage of the hepatic parenchyma which, destabilizing the entirety of the extracellular matrix, highlights the crucial role of epigenetic modulation on the proliferation and cellular differentiation (Michalopoulos & DeFrances, 1997; for review). Even if this natural phenomenon does not seem to be reproduced in internal organs, the recent isolation of adult multipotent, dormant within the various organs and tissues, stem cells seemed open the way towards a Regenerating Cellular Therapy. Indeed, the bibliographical data indicate a great plasticity of stem cells and in particular those taking from bone marrow like hematopoietic stem cells (HSC) and mesenchymal stem cells (MSC). Certain reports, already conclusive in the rat and the mouse, indicate a possibility of directed in situ and in vitro differentiation of stem cells and open exciting therapeutic prospects for tissue and various organs repair, without exposing the host to the failure of an allogeneic transplant rejection. Thus, contrary to embryonic stem cells (ESC) whose clinical application is still not unanimously accepted, ASC initially appeared as an ideal solution to prepare various autologous graft. Nevertheless, this dream about the imminent clinical application of ASC to cure a number of diseases as diabetes, cystic fibrosis, myocardial infarction and many others physiopathological states appeared more difficult to accomplish than initially expected. With the perspective of recent dynamic works it is conceivable to think that this is just a problem of a better knowledge of their diversity as well on the fundamental level as from the point of view of their therapeutic use.

This observed *ex vivo* pluripotency of HSC and MSC was a cause of noted infatuation. Surprisingly, HSC known to date for their capacity of renewal of blood morphotical elements, were also able differentiated toward skeletal muscle (Ferrari, 1998), cardiac (Orlic et al., 2001), nervous (Mezey et al., 2000), liver (Lagasse et al., 2000) or epithelial cells (Krause et al., 2001). Despite the notable example of post infarct myocardium wall repair, these works did not open the way to the routine clinical application (Agbulut et al., 2004). Troublesome, these results appeared as not reproducible and the parabiosis experiments between exposed to radiation and green fluorescent protein (GFP)-transgenic mousses did not demonstrate this supposed regenerating power of HSC (Wagers et al., 2002).

The *ex vivo* experiments carried out with MSC reporting relatively similar observations. For a long time, MSC were considered as having potential of differentiation limited to mesenchymal family cells as osteoblasts, chondrocytes, adipocytes or muscle precursors (Ashton et al., 1980). More recently, MSC revealed *in vitro* abilities to generate cells distinguished also by ecto- and endodermal features (Reyes et al., 2002; Woodbury et al., 2000; Sato et al., 2005). However, this pluripotentiality was objected by certain unsettled findings. At first, Hardeman et al. (1986) showed formation of cellular hybrids like myofibroblast which was forming by a fusion of fibroblasts with myoblastes able to conserve muscular character (Hardeman et al., 1986). Recently, the fusion of neuronal stem cells with ESC yielded cells expressing both characters (Ying et al., 2002). Since labeling technique with DNA coding enzymes or fluorescent proteins, these observations question the reality of observed differentiation *in situ*. On the other hand, the possibility of phenotype modification of gene expression according to culture conditions could contribute to observed *in vitro* differentiation (Discher et al., 2009). It is also plausible that these divergent observations reflect a greate heterogeneity of MSC characterized by a certain ability of multipotency revealed during *ex vivo* manipulation where large majority of them represent committed progenitor cells rather than really pluripotent stem cells. Then, the random results of directed differentiation may be explained by imperfect approach of cell isolation. In fact, majority of protocols used is based rather on adhesion capacity of MSC than on specificity of a membrane marker not yet identified.

At the first time, on the basis of morphological differences, the heterogeneity of MSC was brought up by Colter et al. (2000) which proposed three types of MSC. Two first, named RS-1 and RS-2, characterized by a little size and absence or presence of granulations, were considered as self-renewal cells. The third type, distinguished by apparently bigger size, seemed corresponded to already partially differentiated (CSMm) cells (Colter et al., 2000). Thus, the authors hypothesized that RS-1 and RS-2 cells were progenitors of CSMm but since more quiescent state, RS-1 population appeared as precursor of RS-2 that differentiated to CSMm cells. In the proposed schema, RS-2 population would have had the capacity to maintain equilibrium of CSMm production by the ability to reprogramming towards the ground RS-1 state (Colter et al., 2000). In reality, the ulterior antigenic study of these three cellular populations that matching these morphological differences revealed yet more important cellular heterogeneity than initially supposed (Colter et al., 2001).

2.2 Evidence of rabbit MSC heterogeneity

The above data indicated the necessity to explore this proposed heterogeneity of MSC on the molecular level with particular insight into differences between clonal colonies which seems to be essential in elaboration of final approach of directed differentiation. Thus, we carried out the study having for objective the molecular characterization of colonies proliferating from individual CD45- mononuclear cells isolated from bone marrow of rabbit. This model was chosen for relative facility to obtain a biological material. This advantage being unfortunately associated with limited knowledge of rabbit genome, we have employed the Differential-Display Reverse Transcription-Polymerase Chain Reaction (DDRT-PCR) technique to analyze expressed respective mRNAs (Sturtevant, 2000). This approach resides in use of several non-specific primers able to hybridize with certain extracted mRNA during low temperature reaction (for more details see original paper of Sturtevant, 2000). In this way, the comparison of obtained patterns of mRNA in analyzed colonies showed differences in genes expression. The colonies were cultured separately after isolation of each clone proliferated from one cell on the surface delimited by cylinder (Figure 1A). After the harvest, the mRNA extract of each colony was analyzed with DDRT-PCR approach. In the figure 1B, we present the DDRT-PCR patterns of amplicons obtained after analyze of 14 colonies with couple 1 of DDRT-PCR primers purchased from Seegen (Seoul, Korea). Thus, these patterns, despite a certain similitude, vary by five differentially expressed mRNAs marked by the arrows. These genes correspond to proteins implicated in different cellular functions as follows: 1 - TBC1D7- cellular growth and proliferation; 2 - Filamine - cell migration; 3 - Cystatine 10 - chondrogenesis; 4- LUC7-like - inhibition of myogenic differentiation; 5 - MTHFR - inhibition of intracellular methylation. Their expression seem to be convergent with expression of OCT-4 gene (Figure 1C) considering as a marker of non differentiated cellular state (Tondreau et al., 2005).

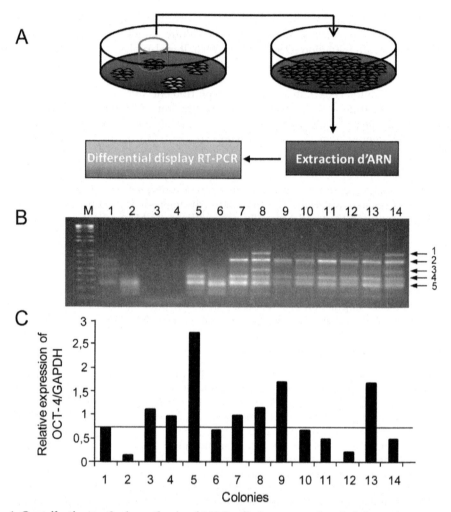

Fig. 1. **Contribution to the hypothesis of MSC cells heterogeneity.** A. Schematic representation of MSC clonal colonies development. After delimitation of one cell in the cylinder space and its proliferation, young colony is displaced in Petri dish where continues to proliferate. Just before confluence, cells are harvested for extract of RNAs. B. Differential Display Reverse Transcription-Polymerase Chain Reaction (DDRT-PCR) gel containing the amplicons obtained from 14 colonies extracts. The arrows indicate sequenced bands corresponding to TBC1D7, Filamine, Cystatine 10, LUC7-like, MTHFR gens. See the text for more details. C. Histograms representing the expression of OCT-4 gen. The line visualizes the mean level of relative OCT-4 expression.

It was interesting to observe, that in the medium LIF-free (Leukemia Induced Factor is employed in view to preserve non differentiated state during cell proliferation) the colonies were able spontaneously differentiated to muscle precursor cells with unequal capacities (Figure 2).

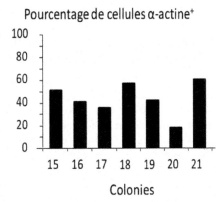

Fig. 2. **Relative spontaneous expression of α-actin gene in studied colonies.** The presence of actin was revealed by indirect immunocytochemistry where the presence of specific antibody was detected with horseradish peroxydase system.

All these results clearly indicate that mononuclear CD45- cells, still considered as MSC, form a heterogenic population characterized by different non differentiated and committed state. Our results did not determine the definitive number of cellular clones but suggest that currently practiced isolation of MSC may conducted toward random results of directed differentiation.

2.3 Toward a new hierarchy of bone marrow stem cells

Our conclusion seems to be strongly supported by the similar results obtained by microarray analysis of human MSC (Mareddy et al., 2009; Tormin et al., 2009). In addition, common distribution of certain membranes markers as CD44, CD73, CD90 or CD105 indicates that use of relative antibodies cannot be considered as discriminative tool for cell isolation. Thus, our results raised the question concerning hierarchy of organisation of bone marrow stem cells and place the observations previously published by groups of Verfaillie and Ratajczak at the special place. These key studies made mention of the very special cells named multipotent adult progenitor cells (MAPC) and very small embryonic-like cells (VSEL) respectively (Jiang et al., 2002; Kucia et al., 2006). Even if these results cannot be reproduced by other laboratories, MAPC possessing similar morphology to MSC are able generate mature cells characterizing by ecto-, meso- and endodermal features. This pluripotentiality, attesting their immature character, allows thinking that MAPC may be direct precursor of MSC as well as HSC (Jiang et al., 2002). Conceivably, this hypothesis may explain the random results of directed *ex vivo* MSC differentiation.

In contrast, VSEL cells are a very small, morphologically similar to embryo cells which being probably attracted by the chemical gradient of SDF-1, colonize bone marrow during embryogenesis (Kucia et al., 2006). Amazingly, the grafts of VSEL in irradiated mice indicated that their weak number seems to be responsible for acceleration of senescence process which suggesting their participation in internal organs and tissues regeneration (Kucia et al., 2008). Convergently, the increased number of VSEL, expressing myogenic Nkx2.5 protein, detected in general circulation in patients suffering from cardiac ischemia suggests the possibility of their implication in cardiovascular repair (Wojakowski et al., 2009). In the case of definitive clinical confirmation, this observation may open extremely promising horizons of cellular therapy. Nevertheless, identification of these new cellular populations not responds to the question concerning the origin of MSC.

In this context, our results support hypothesis that MSC cannot be considered as pluripotent stem cells having the potential to generate all cells naturally deriving from tree embryonic layers. In the Figure 3, we propose to attribute this role to MAPC and VSEL cells. There are two possibilities, either we observe a coexistence of three cellular populations or existence an ontogenetic hierarchy. In the first situation, each cellular type possesses the variable potency of differentiation: i) VSEL that of committed precursor, ii) MAPC would be pluripotent and iii) MCS just mesodermal. In the second situation, VSEL would be direct precursor of MAPC generating MSC among other. In this way, the apparition of all cellular populations in bone marrow would reflect an ontogenic hierarchy formed during embryogenesis where initial number of VSEL cells determines a capacity of hypothetic organ repair during adult life. It appears that however the reality may be, the strategy of preparation of graft from MSC should be revised taking into account the recent clinical trials lacking therapeutic effects as published recently (Menasche, 2011). In fact, the clinical use of MSC is actually recognize for their immunomodulatory effects known in diminution of graft reject or graft versus host disease (Ringden et al., 2006; Ucceli et al., 2007; Le Blanc et al., 2008). In this way, two novel axes of fundamental research seem to be profiled: **i)** definite establishment of hierarchy of bone marrow cells in regard of MAPC and VSEL cells, **ii)** exploration of intracellular pathways in view to determine cellular fate during directed ex vivo differentiation.

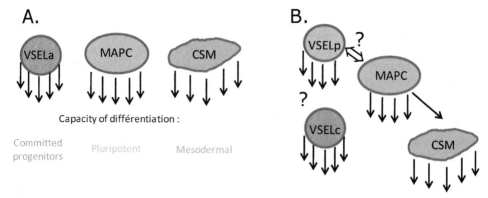

Fig. 3. **Schematic representation of bone marrow stem cells hierarchy.** A. Are we in the presence of three types of CSA which evolved separately according to the physiological regulation? B) Can one envisage a hierarchic ontogenic organization where VSEL cells would be pluripotent stem cells generating all bone marrow cells?

3. Calcium signaling and fate of MSC

3.1 Introduction

The random results of *ex vivo* differentiation of MSC during preparation of autologous grafts indicate that promising potential may be revealed by the microenvironment of cellular culture. This hypothesis seems to open a new area for proceedings of directed differentiation which may be based on modulation of activity of molecular targets of MSC. Given the dependence of phenotypical gene expression, self renewing, migration or proteolytical enzyme secretion on increase of cytosolic calcium concentration $[Ca^{2+}]_c$, pharmacological modulation of calcium signalling would represent a key to control cellular fate. The various membrane channels and ionic transporters having potential to modulate $[Ca^{2+}]_c$ appears then as appropriate molecular targets. Recent reports indicate that MSC express several ionic membrane channels generating sodium (Na), calcium (Ca) and potassium (K) inward and outward currents characterized by molecular biology and patch-clamp approaches (Li et al., 2005; Li et al., 2006; Deng et al., 2006; Kawano et al., 2003; Kawano et al., 2002; Heubach et al., 2004). Nevertheless, the capacity of these channels to modulate $[Ca^{2+}]_c$ in MSC was not yet evaluated. Some data indicate that in human MSC, the $[Ca^{2+}]_c$ may change upon cyclic oscillatory variations via a mechanism implicating inositol trisphosphate receptors (IP$_3$Rs), store operating channels (SOCs), L-type voltage dependent calcium channels (L-VDCaCs) as well as Na^+-Ca^{2+} exchangers (NCX) (Kawano et al., 2003; Kawano et al., 2002). We have recently shown that MSC express also several genes coding the proteins of transient receptor potential cation channel (TRPC1/2/4/6) family (Torossian et al., 2010) possessing a major role in cell proliferation as already documented in cancers (El Boustany et al., 2008). Over it, dependence of immature cell proliferation or myoblasts fusion (Lory et al., 2006) on activity of voltage dependent T-type calcium channel reinforces idea that pharmacological modulation of calcium signalling could reveal potential to improve efficiency of protocols employed in directed differentiation of adult stem cells.

In the present study, using functional and molecular biology approaches, we pursued two major objectives: *i)* evaluation of efficiency of membrane voltage dependent ionic channels (VDCaC, VDNaC, VDKC) and transporters (Na/K-dependent ATPase and NCX) to modulate calcium homeostasis on the basis of kinetics of $[Ca^{2+}]_c$ variations occasioned by selective activators and blockers, *ii)* demonstration that inactivation of chosen targets such T- or L- type VDCaC and TRPC1 reduced cellular proliferation and that high concentration of nifedipine activated neuroglial differentiation.

3.2 Efficiency of molecular targets to modulate calcium homeostasis

Figure N°4 illustrates that equilibrium state in single MSC is disturbed by modifications of the extracellular medium or by the presence of selective pharmacological agents which changing Ca^{2+}, Na^+ or K^+ gradients induce the $[Ca^{2+}]_c$ variations with different kinetics. The gathered histograms representing the areas under curves (AUC) were obtained from individual profiles whose averages are expressed in the Figures 4 and 5. The highest calcium mobilization was observed in the presence of depolarizing solution of KCl as well as 2-diazo-4,6-dinitrophenol (DDNP) or bepridil, well known respective blockers of BK_{Ca} channels and NCX. Even if each product activated this increase by different mechanism, the obtained AUCs were very similar and corresponded to 5.5, 5.25 and 5.1 µM/L (Fig. 4). The depolarizing solution of KCl imposing membrane potential to value inferior to -30 mV activates low threshold VDCaCs as "L"- and/or "N"-type channels. Similar effect obtained with DDNP (Fig. 5D) revealed a high capacity of the BK_{Ca} channel inactivation to membrane depolarization subsequent to cytosolic K^+ accumulation. The action of bepridil eliciting the reverse mode of NCX action is responsible for calcium influx which considered its proximity with endoplasmic reticulum, induces intracellular calcium mobilization (Niggli et al., 1991). In the same way, figure 4 shows also that the action of Na/K-dependent ATPase having the capacity to modify sodium gradient induced cytosolic calcium increase by recruitment of NCX (Hilgemann et al., 1992). In contrast, the effect of other depolarizers as $CaCl_2$ solution, tetraethylammonium (TEA) or veratridine (Fig 5B, 4C and Fig 6) appeared as less efficient.

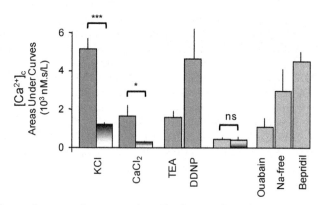

Fig. 4. **Mean values of area under curves (AUC) of cytosolic calcium mobilization in rabbit mesenchymal (MSC) stem cells.** KCl (25 mM) in the absence (n= 10) or in the presence of 10 µM nifedipine (▣ ; n= 15), $CaCl_2$ (10 mM) in the absence (n= 14) or in the presence of 10 µM nifedipine (▣ ; n= 17), TEA (3 mM; n= 14), DDNP (10 µM; n= 6), veratridine alone (100 µM; n= 29) or in the presence of 1 µM TTX (▣ ; n= 22), ouabain (10 µM; n= 6), Na-free (n= 11) and bepridil (100 µM; n= 17) were injected in the vicinity of MSC (***, $p<0.001$; **, $p<0.01$; *, $p<0.1$). The mean values (±SEM) of areas under curves were numerically integrated from individual microfluorimetric (Indo-1) recordings by trapeze method using Excel programme. Each recording represents the same number of points acquired every 250 ms by PC-assisted system developed by Notocord Systems (Paris, France).

The type of VDCaCs was determined with nifedipine, a dihydropyridine derived L- type channel blocker and by RT-PCR experiments. The significantly reduced, but never totally abolished stimulatory effect of KCl and $CaCl_2$ solutions on calcium mobilization in the presence of the blocker indicating the involvement of both nifedipine-sensitive and insensitive VDCaCs. RT-PCR convergent experiments, carried out with Cav1.2, Cav2.2 and Cav3.3 specific primers, confirmed expression of L-, N- and T-type of VDCaCs in MSC (Fig. 4, boxes). L-type channels (Cav1.2 subunits) were already reported in human and rat MSC (Li et al., 2005; Li et al., 2006). In contrast, the existence of T channels in MSC are a matter of debate since contradictory reports concluding to the absence of Cav3.1 and Cav3.2 subunits (Li et al., 2005; Heubach et al., 2004) or to the presence of Cav3.2 whose functionality was however not determined (Kawano et al. 2002). The importance of expression of T channel in MSC is illustrated by observations in ESC where the sustained increase in $[Ca^{2+}]_c$ is responsible for cell proliferation (Lory et al., 2006) or fusion of differentiated myoblasts (Bijlenga et al., 2000). The expression of N-type VDCaCs in MSC is not surprising because its functionality in differentiating cells evolves through an expression pattern (Arnhold et al., 2000). Thus, during neuronal differentiation of ESC, transitory high expression of N-type channel in initially apolar phenotype matched with cellular migration whereas its reappearance in differentiated neuron coincided, similarly to mature cells (Yokoyama et al., 2005), with synaptogenesis and modification of the exocytose level (Jones et al., 1997). As MSC are known for their secretory and migratory activities, similar functionality may be expected. Consequently, calcium fluxes in MSC can be modified by opening of three types of VDCaCs which filling up the different cytosolic microdomains with calcium can separately control gene expression, cellular proliferation and migration or exocytosis (Lory et al., 2006; Yokoyama et al., 2005; Yang et al., 2006; Yoo et al., 2007).

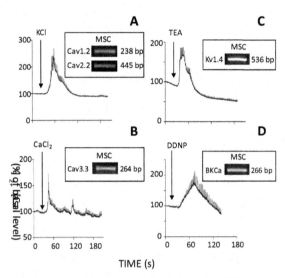

Fig. 5. **Effects of modulation of calcium and potassium channels on [Ca^{2+}]$_c$ in rabbit mesenchymal stem cells (MSC).** The arrows indicate the pressure-ejected administration of depolarizing solution KCl (25 mM) (A), CaCl$_2$ (10 mM) (B) and voltage or calcium dependent potassium channels blockers as TEA (3 mM) (C) and DDNP (10 µM) (D) in the vicinity of the cells. The right-placed boxes represent RT-PCR obtained amplicons of mRNA coding L, N, T-type voltage-dependent calcium channels (Cav1.2, Cav2.2 and Cav3.3 subunits), voltage dependent (Kv1.4 subunit) and calcium dependent (BK$_{Ca}$) potassium channels. The curves represent a mean from 10 (A), 14 (B), 14 (C) and 6 (D) individual cell recordings. The spontaneous level of [Ca^{2+}]$_c$ (100% basal level) was calculated for each experiment as the mean concentration during 30 s preceding the administration of ionic solutions or potassium channel blockers.

Interestingly, the activation of T-type VDCaC *in vivo* appears to be directly dependent on a potassium gradient demonstrating the crucial role of K$^+$ channels in the evolution of stem cell fate. This astute mechanism is based on cooperation between three types of ionic channels. Briefly, VDKC or/and CaDKC provoke transitory membrane hyperpolarization conducting to depolarizing potassium influx through delayed-rectifier potassium channel responsible for T-type VDCaC activation and [Ca2+]c increase. Such membrane hyperpolarization, detected in rat MSC (Deng et al., 2007) during progress of cell cycle from G(1) to S phase, seems to be dependet on the balance of expression between KCa3.1 and delayed-rectifier (Kv1.2/Kv2.1) subunits which since their down-regulation with the specific RNAi appeared crucial for cell proliferation. Thus, the cooperation between IK_{Ca} and KDR channels that generate hyperpolarizing efflux and subsequently delaying influx of K^+ may vary membrane polarity near the threshold value of T-type VDCaC activation.

Amazingly, unlike the mechanism described above, we found pharmacological way to obtain *ex vivo* a similar effect on [Ca2+]c increase in MSC. Using functional and RT-PCR experiments, we observed that the blockage of Kv1.4 and BK_{Ca} channels (fig.4C, D) by TEA (51%; 32 of 63 cells) and DDNP (46%; 39 of 84 cells) induced $[Ca^{2+}]_c$-increase after VDCaC activation due to intracellular membrane depolarization triggered by cytosolic K^+ accumulation. Further studies are needed to show whether such blockage of BK_{Ca} channel would stabilize cell proliferation and immaturity.

The pharmacological activation of VDNaC represents another way to augment [Ca2+]c. Similarly to excitable cells like neurons or cardiocytes, the opening of VDCaCs in MSC results also from progressive membrane depolarization initiated by low threshold T-type VDCaC and/or VDNaC. In our experiments, veratridine (non-selective opener of VDNaCs) (Yang et al., 2006) started Na-induced depolarization which reaching activation threshold of VDCaCs was responsible for increase of $[Ca^{2+}]_c$ (84% given 54 of 64 cells). Not significant reduction of this effect by TTX, a VDNaCs blocker (t=0.38; 79% given 55 of 69 cells) (Fig. 5A) and identification of mRNA encoding Nav1.9 subunit (Fig 5A) indicated the expression of TTX-resistant VDNaCs in MSC. Noticeably, the type of VDNaC expression in MSC appears to be controversial. Using identical primer as Deng et al. (2006), we were unable to confirm their observation on expression of Nav1.1 subunit in rabbit MSC but we found relative transcript in extracts from rabbit nervous system which suggests non-expression of this subunit in our cultures. Divergent findings on the expression of VDNaC may also be noted in human MSC. While Heubach et al. (2004) failed to identify both TTX-resistant and TTX-sensitive channels, Li et al. (2005) detected a functional TTX-sensitive inward current. These discrepancies may result from the different experimental protocols used. In our study, mononuclear cells were separated with CD45 antibody instead of their capacity to adhesion already reported (Li et al., 2005). Moreover, our mRNA samples were obtained at the final stage of the first passage contrary to the 4th or even the 8th as previously described (Li et al., 2005; Deng et al., 2006). As expression of sodium channel unit *in vivo* changes throughout cellular maturation (Benn et al., 2001), these observed *in vitro* differences reveal modulation of gene expression by microenvironment. Nevertheless, the weak kinetics of calcium mobilization induced by veratridine seems indicate that VDNaCs did not appear as interesting target to modulate a fate of MSC.

On the contrary, NCX having capacity to exchange cytosolic/extracellular Ca^{2+} for Na^+ in normal or reverse mode (Niggli et al., 1991) within chemical gradient of both ionic populations, appears as powerful $[Ca^{2+}]_c$ enhancer in MSC. As shown in Fig.6, Na-free medium (38% given 17 of 45 cells), bepridil (44% given 31 of 70 cells) or ouabain (62% given 23 of 37 cells) led to transient increases in $[Ca^{2+}]_c$. The RT-PCR-detected expression of genes coding NCX and Na^+/K^+-ATPases (Figs. 6B, C) matched our functional observations. Similarly to other cellular models (Hilgemann et al., 1992), cytosolic overloading with sodium after ouabain-induced inactivation of Na^+/K^+-ATPase triggered a $[Ca^{2+}]_c$ increase resulting from exchange of sodium for calcium during reverse mode action of NCX (Niggli et al., 1991). In human MSC, the NCX seems to take part in the induction of calcium

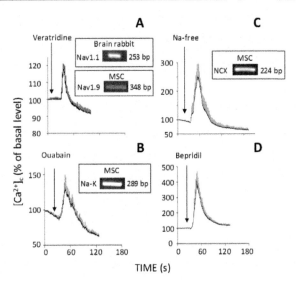

Fig. 6. **Effects of modulation of voltage dependent sodium channels, ATPase Na+/K+ dependent and sodium-calcium exchanger activities on [Ca^{2+}]$_c$ in rabbit mesenchymal stem cells (MSC).** The arrows indicate the pressure-ejected administration of Veratridine (100 μM) (A), Ouabain (10 μM) (B), Na+-free medium (C) and Bepridil (100 μM) in the vicinity of MSC. The right-placed boxes represent the RT-PCR obtained amplicons of mRNA coding voltage-dependant sodium channel (Nav1.9 subunit) in MSC, Nav1.1 being detected only in brain rabbit extract, ATPase Na-K dependend (B) and Na+-Ca^{2+} exchanger (NCX) (C). The curves represent a mean from 22 (A), 6 (B), 11 (C) and 17 (D) individual cell recordings. The spontaneous level of [Ca^{2+}]$_c$ (100% basal level) was calculated for each experiment as the mean concentration during 30 s preceding the administration of ionic solutions or potassium channel blockers.

oscillations (Kawano et al., 2003). In the present work, its activation induced a transient increase in [Ca^{2+}]$_c$ followed by a slow basal calcium level recovery. It is like during early stage of cardiomyocyte differentiation of mouse ESC, where without modifying transient calcium variations, NCX enhanced the basal level of [Ca^{2+}]$_c$ (Fu et al., 2006). This may indicate the crucial role of NCX in the stabilization of higher basal [Ca^{2+}]$_c$ in immature cells where its activity may be improved by direct intracellular phosphorylation or by increase of Na-gradient during opening of VDNaC or Na+/K+-ATPase inhibition.

3.3 Effects of VDCaCs inactivation on MSC cell culture

Taken account of highest capacity to modify calcium homeostasis, VDCaCs was chosen as more appropriate target to evaluate pharmacological modulation of MSC fate. Another choice is related to TRPC1 protein which being largely expressed by rabbit MSC (Torossian et al., 2010) is known as one of essential factors managing calcium distribution during cancer cell proliferation (El Boustany et al., 2008; El Hiani et al., 2009).

Then, the blockage of L- and T-type VDCaC pointed their implication in the control of cellular proliferation and differentiation. Mibefradil and nifedipine induced a dose-dependent decrease in cell numbers corresponding to 25 and 15 % of cells respectively at 10 µM concentrations attaining very significant inhibition (65 and 50 %; p<0.005) when treated with 30 µM doses (Fig. 7). Similarly, inducing 45% inhibition of MSC proliferation, the specific siRNA demonstrated a major role of TRPC1 protein in this process.

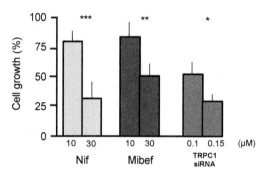

Fig. 7. **Relative mesenchymal stem cells (MSC) proliferation in the presence of calcium channel blockers.** MSC were cultured during 48 hours in the medium containing nifedipine, mibefradil (10 and 30 µM, both) or siRNA of TRPC1 (0.1 or 0.15 µM). The results represent the means (±SEM) from four independent experiments expressed as a percentage of proliferating cells. Each culture contained initially 20000 MSC and after 24h incubation period in expansion medium, the blockers at respective concentrations were administered in the plates. After 48h period of incubation the cells were fixed in acetic alcohol, stained with crystal Violet and extracted with acetic acid after drying. The optical density of extractions was evaluated using spectrophotometric measurement at 570 nm and compared to the standard range to obtain the number of cells. The relative effect of the drugs on cell proliferation was evaluated in comparison to non-treated cells (***, p<0.001; **, p<0.01; *, p<0.1).

Noticeably, the presence of a higher concentration of nifedipine (100 µM) induced apparition of two types attached irregularly shaped cells. The first type, representing about 85%, was characterized by expression of Glial Fibrillary Acidic Protein (GFAP) (Fig. 8B, D, F) whereas the second remained GFAP negative. This result may be particularly relevant in comparison to the control LIF-free culture where cells showed varied morphology and ability to spontaneously differentiate into myogenic precursor cells since relative to α-smooth muscle actin staining (Fig. 8A)

For the first time, we show that blockade of L-type channels in MSC may generate neural precursor cells already shown for their GFAP staining (Imura et al., 2003). We observed two kinds of GFAP+ cells corresponding to a low number of neural-like cells accompanied the large majority of staining cells displaying astrocyte-like morphology. The absence of GFAP staining in the LIF-free expansion medium and the disappearance of myogenic character after treatment with nifedipine, fully support the idea that pharmacological modulation of calcium homeostasis would reinforce strategy for directed differentiation of stem cells. These observations suggest that the reduction of higher and persistent $[Ca^{2+}]_c$ appears like a turning point between proliferation and differentiation where favouring proliferation, the persistent calcium level avoids differentiation.

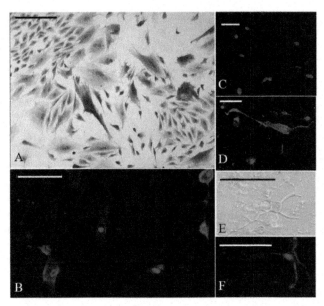

Fig. 8. **Spontaneous myogenic and nifedipine induced GFAP+ cells derived from mesenchymal stem cells (MSC).** (A) MSC stained with α-smooth muscle actin cultured in the LIF-free expansion medium. (B) Expression of GFAP in MSC cultured in the LIF-free expansion medium supplemented with 100 μM nifedipine. (C) Absence of GFAP expression in MSC cultured in the LIF-free expansion medium in the absence of nifedipine and counter staining with nuclear dye Hoechst. (D) Neural-like GFAP+ **cells.** (E,F) Morphological aspects of astrocyte-like cultured MSC in the LIF-free expansion medium supplemented with 100 μM nifedipine (E, viable cell; F, GFAP+ fixed cell). Scale bars, 100 μm.

The results obtained during prolonged exposition of MSC on both anticalcics seem to corroborate this hypothesis. As shown in the Fig. 9, the 10 μM doses were able introduce morphological modifications indicating the initiation of differentiation process. Cells growing in the presence of mibefradil (Fig. 9B) seem to display a more elongated and spindle shape while nifedipine favoured formation of cell extensions (Fig. 8C). Theirs action coincided with apparition of numerous vacuoles apparently more large and swollen in the presence of nifedipine (Fig. 8B, C). Since negative staining with oil red O, hematoxylin-eosine, toluidine blue or periodic acid-Schiff (data not shown) these vacuoles did not contain lipids, glycoproteins nor mucopolysaccharides. Such formation, attributable to an intensification of autophagy process (Mizushima & Levy, 2010), was transiently observed during erythrocyte or lymphocyte differentiation (Kundu et al., 2008; Mortensen et al., 2010) and appeared crucial to adipogenesis (Baerga et al., 2009). According to information recently reported in human U-251 glioblastoma cells (Johnson et al., 2006) or maturating foetal hepatocytes (Matsunga et al., 2008), the mechanism of this process may be explain by not well understood dependency of initial stage of differentiation upon Ca-dependent PI3-kinase activity. Interestingly, the siRNA-inactivation of TRPC1 expression did not modify cell morphology suggesting that unlike T and L type channels this protein is not implicated in MSC differentiation.

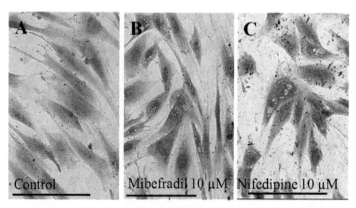

Fig. 9. **Morphological modifications of mesenchymal stem cells (MSC) induced by calcium channel blockers.** Culture of MSC in the LIF-free expansion medium in the absence (A) and in the presence of mibefradil, "T-type" (B) or nifedipine, "L-type" calcium channel blockers (C). Both blockers were employed at 10 µM concentration. Scale bar, 100 µm.

3.4 Conclusion

Taken together, our data demonstrated that pharmacological modulation of ionic carriers' activity, as particularly T- and L-type VDCaC or TRPC1 protein, may reinforce the strategies employed *ex vivo* for directed differentiation of stem cells. Further studies should demonstrate whether pharmacologically induced modulation of $[Ca^{2+}]_c$ in stem cells would maintain their immaturity or begin their differentiation.

4. General discussion and perspectives

The heterogeneity of MSC and their various commitments, as discussed above, perfectly explain why clinical application appears restraint their supposed pluripotentiality on immunological and mesenchymal capacity. In fact, not expressing II class MHC molecule, MSC are therefore not antigen-presenting cells and would be ignored by the host's immune system (Tse et al., 2003; Krampera et al., 2003). By their constitutive secretory activity (Caplan, 2009), MSC have capacity to create microenvironment favourable to combat graft-versus-host-disease (Koc et al., 2000) as well as attenuate inflammatory bowel symptoms in Crohn disease grafted patients (Caplan, 2009). Their aptitude for differentiate into osteoblasts was exploited in clinical trial for the treatment of *osteogenesis imprfecta* patients (Horowitz et al., 1999; 2002). One of very interesting work representing the regeneration of surgically amputated meniscus in goat by knee injection of MSC with hyaluronan delivery vehicle, provides perspective in the treatment of arthritis (Murphy et al., 2003). In contrast, use of MSC in view of cardiac post infracted reparation which seems to provide therapeutic improvement appeared to be not exerted by cardiomyocyte differentiation (Caplan, 2009). Convergently, the results of clinical trials realized with autologous MSC, HSC and mononuclear bone marrow cells (MNC) in about 1600 (Menasche, 2011) patients suffering from acute myocardial infarction, refractory angina or chronic heart failure did not give expected benefits indicating that heterogenic MNC of bone marrow, while remaining immunologically neutral, appear to be not therapeutically reliable to repair other than hard tissues like bone or cartilage. Nevertheless, the existence of great variability in the functionality of MSC retrieved from patients indicating that pluripotent differentiation would be ascribed to more immature cells which are able generating MSC. Our study shows clearly that MSC should be considered as heterogeneous and composed by lineage-committed cells that may be multipotent but certainly non pluripotent cells. In addition, parallelism between decrease in MSC number with age (Lennon et al., 1996) and acceleration of the senescence process in mouse grafted with a low number of VSEL cells (Kucia et al., 2008) strongly suggests that this role may be ascribed to VSEL cells which would represented this pluripotent cellular population. In this way, the unequal number of VSEL cells in isolated samples may explain the random results of *ex vivo* differentiation. It is therefore conceivable that specific isolation of this cell population represents the first problem to resolve. In fact, the number of data suggest that bone marrow may be considered as reserve of pluripotent cells but this property cannot be attributed to MSC.

Our study of calcium signaling raises a second problem of directed differentiation representing by epigenetic reprogramming of gene expression which in an unpredictable manner would change the cellular fate. This conclusion is supported by divergence concerning the expression of VDCaCs and VDNaC in MSC. This inconvenience could be avoided in the cultures composed of a homogenous population of stem cells able reproducing stable microenvironment. Microenvironmental stability appears then as one of the more important conditions allowing prediction of cellular evolution and an objective comparison of the effects occasioned by experimentally introduced modifications. Our results indicated that pharmacological modulation of calcium homeostasis may influence cellular behavior seem open the perspectives for research of experimental protocols having potential to control the cell proliferation and differentiation.

Taken together, it can be concluded that in view to realize a dream about autologous regenerative grafts it would be necessary to direct the basic research toward two major objectives: i) to find the strategy to facilely isolate pluripotent stem cells from the bone marrow and ii) to perfect protocols allowing control the evolution of cellular cultures.

5. References

Agbulut, O.; Vandervelde, S., Al Attar, N., Larghero, J., Ghostine , S., Leobon, B., Robidel, E., Borsani, P., Le Lorc'h, M., Bissery, A., Chomienne, C., Bruneval, P., Marolleau, J.P., Vilquin, J.T., Hagege, A., Samuel, J.L. & Menasche, P. (2004). Comparison of human skeletal myoblasts and bone marrow-derived CD133+ progenitors for the repair of infarcted myocardium. *J Am Coll Cardiol*, 44, pp. (458-63).

Arnhold, S.; Andressen, C, Angelov, D.N., et al. (2000). Embryonic stem-cell derived neurones express a maturation dependent pattern of voltage-gated calcium channels and calcium-binding proteins. *Int J Dev Neurosci*, 18, pp. (201-212).

Ashton, B.A.; Allen, T.D., Howlett, C.R., Eaglesom, C.C., Hattori, A. & Owen, M. (1980). Formation of bone and cartilage by marrow stromal cells in diffusion chambers in vivo. *Clin Orthop Relat Res*, 151, pp. (294-307).

Baerga, R.; Zhang, Y., Chen, P.H., Goldman, S. & Jin, S. (2009). Targeted deletion of autophagy-related 5 (atg5) impairs adipogenesis in a cellular model and in mice. *Autophagy*, 5, pp. (1118-1130).

Benn, S.C.; Costigan, M., Tate, S., Fitzgerald, M. & Woolf, C.J. (2001). Developmental expression of the TTX-resistant voltage-gated sodium channels Nav1.8 (SNS) and Nav1.9 (SNS2) in primary sensory neurons. *J Neurosci,* 21, pp. (6077-6085).

Bijlenga, P; Liu, J.H., Espinos, E., et al. (2000) T-type alpha 1H Ca2+ channels are involved in Ca2+ signaling during terminal differentiation (fusion) of human myoblasts. *Proc Natl Acad Sci USA,* 97, pp. (7627-7632).

Caplan, A. (2009). Why are MSCs therapeutics? New data: new insight. *J Pathol,* 217, pp. (318-324).

Colter, D.C.; Class, R., DiGirolamo, C.M. & Prockop, D.J. (2000). Rapid expansion of recycling stem cells in cultures of plastic-adherent cells from human bone marrow. *Proc Natl Acad Sci USA,* 97, pp. (3213-8).

Colter, D.C.; Sekiya, I. & Prockop, D.J. (2001). Identification of a subpopulation of rapidly self-renewing and multipotential adult stem cells in colonies of human marrow stromal cells. *Proc Natl Acad Sci USA,* 98, pp. (7841-5).

Deng, X.L.; Sun, H.Y., Lau, C.P. & Li, G.R. (2006). Properties of ion channels in rabbit mesenchymal stem cells from bone marrow. *Biochem Biophys Res Commun,* 348, pp. (301-309).

Deng, X.L.; Lau, C.P., Lai, K., Cheung, K.F., Lau, G.K. & Li, G.R. (2007). Cell cycle-dependent expression of potassium channels and cell proliferation in rat mesenchymal stem cells from bone marrow. *Cell Prolif,* 40, pp. (656-670).

Discher, D.E.; Mooney, D.J. & Zandstra P.W. (2009). Growth factors, matrices, and forces combine and control stem cells. *Science,* 324, pp. (1673-7).

El Boustany, C.; Bidaux, G., Enfissi, A., Delcourt, P., Prevarskaya, N. & Capiod, T. (2008). Capacitative calcium entry and transient receptor potential canonical 6 expression control human hepatoma cell proliferation. *Hepatology,* 47, pp. (2068-2077).

El Hiani, Y.; Ahidouch, A., Lehen'kyi, V., et al. (2009). Extracellular signal-regulated kinases 1 and 2 and TRPC1 channels are required for calcium-sensing receptor-stimulated MCF-7 breast cancer cell proliferation. *Cell Physiol Biochem,* 23, pp. (335-346).

Ferrari, G. ; Cusella-De Angelis, G., Coletta, M., Paolucci, E., Stornaiuolo, A., Cossu, G. & Mavilio, F. (1998). Muscle regeneration by bone marrow-derived myogenic progenitors. *Science,* 279, pp. (1528-30).

Fu, J.D; Yu, H.M., Wang, R., Liang, J. & Yang, H.T. (2006). Developmental regulation of intracellular calcium transients during cardiomyocyte differentiation of mouse embryonic stem cells. *Acta Pharmacol Sin,* 27, pp. (901-910).

Hardeman, E.C.; Chiu, C.P., Minty, A. & Blau, H.M. (1986). The pattern of actin expression in human fibroblast x mouse muscle heterokaryons suggests that human muscle regulatory factors are produced. *Cell,* 47, pp. (123-30).

Heubach, J.F.; Graf, E.M., Leutheuser, J., et al. (2004). Electrophysiological properties of human mesenchymal stem cells. *J Physiol,* 554, pp. (659-672).

Hilgemann, D.W; Matsuoka, S, Nagel, G.A. & Collins, A. (1992). Steady-state and dynamic properties of cardiac sodium-calcium exchange. Sodium-dependent inactivation. *J Gen Physiol,* 100, pp. (905-932).

Horwitz, E.M.; Prockop, D.J., Fitzpatrick, L.A., Koo, W.W.K., Gordon, P.L., Neel, M., et al. (1999). Transplantability and therapeutic effects of bone marrow-derived mesenchymal cells in children with osteogenesis imperfecta. *Nat Med,* 5, pp. (309-313).

Horwitz, E.M.; Gordon, P.L., Koo, W.K., Marx, J.C., Neel, M.D., McNall, R.Y., et al. (2002). Isolated allogeneic bone marrow-derived mesenchymal cells engraft and stimulate growth in children with osteogenesis imperfecta: implications for cell therapy of bone. *Proc Natl Acad Sci USA,* 99, pp. (8932-8937).

Imura, T.; Kornblum, H.I. & Sofroniew, M.V. (2003). The predominant neural stem cell isolated from postnatal and adult forebrain but not early embryonic forebrain expresses GFAP. *J Neurosci,* 23, pp. (2824-2832).

Jiang, Y.; Jahagirdar, B.N., Reinhardt, R.L., Schwartz, R.E., Keene, C.D., Ortiz-Gonzalez, X.R., Reyes, M., Lenvik, T., Lund, T., Blackstad, M., Du, J., Aldrich, S., Lisberg, A., Low, W.C., Largaespada, D.A. & Verfaillie, C.M. (2002). Pluripotency of mesenchymal stem cells derived from adult marrow. *Nature,* 418, pp. (41-9).

Johnson, E.E.; Overmeyer, J.H., Gunning, W.T. & Maltese, W.A. (2006). Gene silencing reveals a specific function of hVps34 phosphatidylinositol 3-kinase in late versus early endosomes. *J Cell Sc,* 119, pp. (1219-1232).

Jones, O.T.; Bernstein, G.M., Jones, E.J., et al. (1997). N-Type calcium channels in the developing rat hippocampus: subunit, complex, and regional expression. *J Neurosci,* 17, pp. (6152-6164).

Kawano, S.; Otsu, K., Shoji, S., Yamagata, K. & Hiraoka, M. (2003). Ca(2+) oscillations regulated by Na(+)-Ca(2+) exchanger and plasma membrane Ca(2+) pump induce fluctuations of membrane currents and potentials in human mesenchymal stem cells. *Cell Calcium,* 34, pp. (145-156).

Kawano, S.; Shoji, S., Ichinose, S., Yamagata, K., Tagami, M. & Hiraoka M. (2002). Characterization of Ca(2+) signaling pathways in human mesenchymal stem cells. *Cell Calcium,* 32, pp. (165-174).

Koc, O.N.; Gerson, S.L., Cooper, B.W., Dyhouse, S.M., Haynesworth, S.E., Caplan, A.I., et al. (2000). Rapid hematopoietic recovery after co-infusion of autologous blood stem cells and culture expanded marrow mesenchymal stem cells in advanced breast cancer patients receiving high dose chemotherapy. *J Clin Oncol,* 18, pp. (307-316).

Krampera, M.; Glennie, S., Dyson, J., Scott, D., Laylor, R., Simpson, E., et al. (2003). Bone marrow mesenchymal stem cells inhibit the response of naive and memory antigen-specific T cells to their cognate peptide. *Blood,* 101, pp. (3722).

Kucia, M.; Reca, R., Campbell, F.R., Zuba-Surma, E., Majka, M., Ratajczak, J. & Ratajczak, M.Z. (2006). A population of very small embryonic-like (VSEL) CXCR4(+)SSEA-1(+)Oct-4+ stem cells identified in adult bone marrow. *Leukemia,* 20, pp. (857-69).

Kucia, M.J.; Wysoczynski, M., Wu, W., Zuba-Surma, E.K., Ratajczak, J. & Ratajczak, M.Z. (2008). Evidence that very small embryonic-like stem cells are mobilized into peripheral blood. *Stem Cells,* 26, pp. (2083-92).

Kundu, M.; Lindsten, T., Yang, C.Y., et al. (2008). Ulk1 plays a critical role in the autophagic clearance of mitochondria and ribosomes during reticulocyte maturation. *Blood,* 112, pp. (1493-1502).

Mortensen, M.; Ferguson, D.J., Edelmann, M., et al. (2010). Loss of autophagy in erythroid cells leads to defective removal of mitochondria and severe anemia in vivo. *Proc Natl Acad Sci USA*, 107, pp. (832-837).

Krause, D.S.; Theise, N.D., Collector, M.I., Henegariu, O., Hwang, S., Gardner, R., Neutzel, S. & Sharkis, S.J. (2001). Multi-organ, multi-lineage engraftment by a single bone marrow-derived stem cell. *Cell,*105, pp. (369-77).

Lagasse, E.; Connors, H., Al-Dhalimy, M., Reitsma, M., Dohse, M., Osborne, L., Wang, X., Finegold, M., Weissman, I.L. & Grompe M. (2000). Purified hematopoietic stem cells can differentiate into hepatocytes in vivo. *Nat Med*, 6, pp. (1229-34).

Le Blanc, K.; Frassoni, F., Ball, L., Locatelli, F., Roelofs, H., Lewis, I., Lanino, E., Sundberg, B., Bernardo, M.E., Remberger, M., Dini, G., Egeler, R.M, Bacigalupo, A., Fibbe, W. & Ringden, O. (2008). Mesenchymal stem cells for treatment of steroid-resistant, severe, acute graft-versus-host disease: a phase II study. *Lancet*, 371, pp. (1579-86).

Lennon, D.P.; Haynesworth, S.E., Bruder, S.P., Jaiswall, N. & Caplan A.I. (1996). Human and animal mesenchymal progenitor cells from bone marrow: identification of serum for optimal selection and proliferation. *In vitro Cell Dev Biol*, 32, pp. (602-611).

Li, G.R.; Sun, H., Deng, X. & Lau, C.P. (2005) Characterization of ionic currents in human mesenchymal stem cells from bone marrow. *Stem Cells*, 23, pp. (371-382).

Li, G.R.; Deng, X.L., Sun, H., Chung, S.S., Tse, H.F. & Lau, C.P. (2006) Ion channels in mesenchymal stem cells from rat bone marrow. *Stem Cells*, 24, pp. (1519-1528).

Lory, P.; Bidaud, I. & Chemin, J. (2006). T-type calcium channels in differentiation and proliferation. *Cell Calcium*, 40, pp. (135-146).

Mareddy, S.; Broadbent, J., Crawford, R. & Xiao, Y. (2009). Proteomic profiling of distinct clonal populations of bone marrow mesenchymal stem cells. *J Cell Biochem*, 106, pp. (776-86).

Matsunaga, T.; Toba, M., Teramoto, T., Mizuya, M., Aikawa, K. & Ohmori, S. (2008). Formation of large vacuoles induced by cooperative effects of oncostatin M and dexamethasone in human fetal liver cells. *Med Mol Morphol*, 41, pp. (53-58).

Menasche, P. (2011). Cardiac cell therapy: lessons from clinical trials. *J Mol Cell Cardiol*, 50, pp. (258-265).

Mezey, E.; Chandross, K.J., Harta, G., Maki, R.A. & McKercher, S.R. (2000). Turning blood into brain: cells bearing neuronal antigens generated in vivo from bone marrow. *Science* 290, pp. (1779-82).

Michalopoulos, G.K. & DeFrances, M.C. (1997). Liver regeneration. *Science*, 276, pp. (60-66).

Mizushima N & Levine B. (2010). Autophagy in mammalian development and differentiation. *Nat Cell Biol*, 12, pp. (823-830).

Murphy, J.; Fink, D., Hunsiker, E. & Barry, F. (2003). Stem cell therapyin a caprine model of osteoarthritis. *Arthrit Rheum*, 48, pp. (3464–3474).

Niggli, E. & Lederer, W.J. (1991. Molecular operations of the sodium-calcium exchanger revealed by conformation currents. *Nature*, 349, pp. (621-624).

Orlic, D.; Kajstura, J., Chimenti, S., Jakoniuk, I., Anderson, S.M., Li, B., Pickel, J., McKay, R., Nadal-Ginard, B., Bodine, D.M., Leri, A. & Anversa P. (2001). Bone marrow cells regenerate infarcted myocardium. *Nature*, 410, pp. (701-5).

Reyes, M.; Lund, T., Lenvik, T., Aguiar, D., Koodie, L. & Verfaillie, C.M. (2001). Purification and ex vivo expansion of postnatal human marrow mesodermal progenitor cells. *Blood*, 98, pp. (2615-2625).
Reyes, M.; Dudek, A., Jahagirdar, B., Koodie, L., Marker, P.H. & Verfaillie C.M. (2002). Origin of endothelial progenitors in human postnatal bone marrow. *J Clin Invest*, 109, pp. (337-46).
Ringden, O.; Uzunel, M., Rasmusson, I., Remberger, M., Sundberg, B., Lonnies, H., Marschall, H.U., Dlugosz, A., Szakos, A., Hassan, Z., Omazic, B., Aschan, J., Barkholt, L. & Le Blanc, K. (2006). Mesenchymal stem cells for treatment of therapy-resistant graft-versus-host disease. *Transplantation*, 81, pp. (1390-7).
Sato, Y.; Araki, H., Kato, J., Nakamura, K, Kawano, Y., Kobune, M., Sato, T., Miyanishi, K., Takayama T, Takahashi M, Takimoto R, Iyama S, Matsunaga T, Ohtani S, Matsuura Hamada, H. & Niitsu, Y. (2005). Human mesenchymal stem cells xenografted directly to rat liver are differentiated into human hepatocytes without fusion. *Blood*, 106, pp. (756-63).
Sturtevant, J. (2000). Applications of differential-display reverse transcription-PCR to molecular pathogenesis and medical mycology. *Clin Microbiol Rev*, 13, pp. (408-27).
Tondreau, T.; Meuleman, N., Delforge, A., et al. (2005). Mesenchymal stem cells derived from CD133-positive cells in mobilized peripheral blood and cord blood: proliferation, Oct4 expression, and plasticity. *Stem Cells*, 23, pp. (1105-1112).
Tormin, A.; Brune, J.C., Olsson, E., Valcich, J., Neuman, U., Olofsson, T., Jacobsen, S.E. & Scheding, S. (2009). Characterization of bone marrow-derived mesenchymal stromal cells (MSC) based on gene expression profiling of functionally defined MSC subsets. *Cytotherapy*, 11, pp. (114-28).
Torossian, F.; Bisson, A., Vannier, J.P., Boyer, O. & Lamacz, M. (2010). TRPC expression in mesenchymal stem cells. *Cell Mol Biol Lett*, 15, pp. (600-610).
Tse, W.T.; Pendleton, J.D., Beyer, W.M., Egalka, M.C. & Guinan, E.C. (2003). Suppression of allogeneic T cell proliferation by human marrow stromal cells: implications in transplantation. *Transplantation*, 75, pp. (389).
Uccelli, A.; Pistoia, V. & Moretta L. (2007). Mesenchymal stem cells: a new strategy for immunosuppression? *Trends Immunol*, 28, pp. (219-26).
Wagers, A.J.; Sherwood, R.I., Christensen, J.L. & Weissman, I.L. (2002). Little evidence for developmental plasticity of adult hematopoietic stem cells. *Science*, 297, pp. (2256-9).
Wojakowski, W.; Tendera, M., Kucia, M., Zuba-Surma, E., Paczkowska, E., Ciosek, J., Halasa, M., Krol, M., Kazmierski, M., Buszman, P., Ochala, A., Ratajczak, J., Machalinski, B. & Ratajczak, M.Z. (2009). Mobilization of bone marrow-derived Oct-4+ SSEA-4+ very small embryonic-like stem cells in patients with acute myocardial infarction. *J Am Coll Cardiol*, 53, pp. (1-9).
Woodbury, D.; Schwarz, E.J., Prockop, D.J. & Black, I.B. (2000). Adult rat and human bone marrow stromal cells differentiate into neurons. *J Neurosci Res*, 61, pp. (364-70).
Yang, S.N. & Berggren, P.O. (2006). The role of voltage-gated calcium channels in pancreatic beta-cell physiology and pathophysiology. *Endocr Rev*, 27, pp. (621-676).

Ying, Q.L.; Nichols, J., Evans, E.P. & Smith A.G. (2002). Changing potency by spontaneous fusion. *Nature,* 416, pp. (545-8).
Yokoyama, C.T.; Myers, S.J., Fu, J., Mockus, S.M., Scheuer, T. & Catterall, W.A. (2005). Mechanism of SNARE protein binding and regulation of Cav2 channels by phosphorylation of the synaptic protein interaction site. *Mol Cell Neurosci,* 28, pp. (1-17).
Yoo, J.G. & Smith, L.C. (2007). Extracellular calcium induces activation of Ca(2+)/calmodulin-dependent protein kinase II and mediates spontaneous activation in rat oocytes. *Biochem Biophys Res Commun,* 359, pp. (854-859).

Section 3

New Methods in Cell Biology

14

Salivary Glands: A Powerful Experimental System to Study Cell Biology in Live Animals by Intravital Microscopy

Monika Sramkova, Natalie Porat-Shliom, Andrius Masedunkas, Timothy Wigand, Panomwat Amornphimoltham and Roberto Weigert
Intracellular Membrane Trafficking Unit, Oral and Pharyngeal Cancer Branch, National Institutes of Dental and Craniofacial Research, National Institutes of Health, Bethesda, USA

1. Introduction

Mammalian cell biology has been studied primarily by using *in vitro* models. Among them, cell cultures are the most extensively used since they make possible to study in great detail the molecular machineries regulating the biological process of interest. Indeed, cell cultures offer several advantages such as, being amenable to both pharmacological and genetic manipulations, reproducibility, and relatively low costs. However, their major limitation is that the architecture and physiology of cells *in vitro* may differ considerably from the *in vivo* environment. This reflects the fact that cells in a living organism i) have a three-dimensional architecture, ii) interact with other cell populations, iii) are surrounded by an extracellular matrix with a specific and unique composition, and iv) receive a number of cues from the vasculature and from the nervous system that are essential for maintaining their functions and differentiation state (Cukierman et al., 2001; Ghajar and Bissell, 2008; Xu et al., 2009). In the last two decades, cell biology has greatly benefited from major technological advances in light microscopy that have enabled imaging virtually any cellular process at different levels of resolution. The development of genetically-encoded fluorescently tagged proteins (Chalfie et al., 1994) has triggered the development of novel technologies such as FRAP, FLIM, FRET, BRET, photo-activation, photo-switching and photo-conversion (Diaspro, 2002; Lippincott-Schwartz, 2011a, b), and the realization of more sophisticated microscopes, which have significantly improved the limits of light microscopy in terms of both temporal (spinning disk, resonant scanners) and spatial (PALM, STORM, STED) resolution (Lippincott-Schwartz, 2011a, b). However, the application of these very powerful technologies has been primarily restricted to *in vitro* systems. One of the major breakthroughs in light microscopy is the realization of instruments based on non-linear emissions (Denk et al., 1990; Mertz, 2004; Zipfel et al., 2003b), which has opened the door to the development of intravital microscopy (IVM). IVM encompasses a series of light microscopy-based techniques aimed at studying several physiological processes in live animals (Amornphimoltham et al., 2011; Fukumura et al., 2010; Weigert et al., 2010). In particular, two-photon microscopy (TPM) has been instrumental in developing fields such as neuroscience, immunology and tumor biology For example,, TPM has made possible imaging the behavior of single neuronal populations in the brain of live animals leading to fundamental discoveries in neuronal plasticity and neurotransmission, thus increasing our understanding of pathological conditions such as the Alzheimer's disease or ischemia-induced damages (Serrano-Pozo et al., 2011; Svoboda and Yasuda, 2006; Zhang and Murphy, 2007). In immunology, TPM has been instrumental in analyzing the interactions among the cells of the immune system during the immune response and has provided valuable information on host-pathogen interactions (Cahalan and Parker, 2008; Germain et al., 2005; Miller et al., 2002; Textor et al., 2011). Finally, the ability to image tumors *in situ* during cell growth and invasion, and to monitor the tumor microenvironment has provided with formidable tools to unravel several key mechanisms regulating tumor progression, thus leading to the design and test of novel therapeutic approaches (Andresen et al., 2009; Fukumura et al., 2010; Fukumura and Jain, 2008; Orth et al., 2011). The first attempt to image submicron structures in a live animal has been in the brain, where long term imaging of dendritic spines has been accomplished (Pan and Gan, 2008), whereas the first attempts to image the internalization of fluorescently labeled molecules into highly dynamics sub-cellular structures, such as the endosomes, has been performed in the kidney of live rats and mice (Dunn et al., 2002; Dunn et al., 2003; Sandoval et al., 2004; Sandoval and Molitoris, 2008). However, the motion artifacts due the heartbeat and the respiration of the animal have precluded a detailed analysis of the dynamics of these events. Recently, we have developed an experimental system that has enabled us to follow the dynamics of endosomes and secretory granules in the salivary glands (SGs) of live rodents by using IVM (Masedunskas et al., 2011; Masedunskas and Weigert, 2008; Sramkova et al., 2009). In this chapter, we will review some of the most recent applications of IVM, aimed at studying various aspects of cell biology in live rodents, and will highlight the fact that the salivary glands (SGs) represent a perfect model organ for these studies since they offer unique advantages: first, they can be easily externalized and positioned to completely eliminate the motion artifacts due to the respiration and the heartbeat (Masedunskas and Weigert, 2008), and second they can be easily manipulated both pharmacologically and genetically providing thus with the opportunity to dissect and unravel molecular machineries (Masedunskas and Weigert, 2008, Sramkova et al., 2009). Our goal is to persuade the readers that this approach has a wide range of applicability in different areas of the biomedical field and has the potential to address several fundamental biological questions.

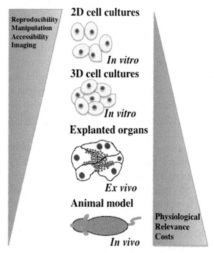

Fig. 1. Comparison among the various experimental systems utilized to study cell biology in mammalian system

Increased complexity in the architecture of the experimental model systems utilized to study cell biology: cell cultures grown on bi-dimensional surfaces (i.e. plastic or glass), cell cultures grown in three dimensions (i.e. purified components of the extracellular matrix), explanted organs, and live animals.

2. Basic principles of intravital microscopy

Biologists have been always fascinated by the possibility to observe biological process in live organisms. One of the major challenges in IVM is to expose the tissue of interest taking care of minimizing damages and maintaining its functionality during the observation period. To this aim, appropriate surgical techniques have been developed since the early days of IVM. The first intravital studies were performed in the early 30's, although they were limited to the examination of the vasculature and its cellular components by using bright field illumination (Beck and Berg, 1931). Advances in optical methods and particularly the development of fluorescence light microscopy, have increased the level of resolution, thus extending the number of biological processes that can be observed *in vivo* (Amornphimoltham et al., 2011; Weigert et al., 2010). Fluorescence light microscopy is based on the generation of contrast by the excitation of the energy levels of molecules (referred as fluorophores) that are either naturally present in the tissue of interest or are administered exogenously. The excitation is achieved by illuminating the specimen with a light source such as a mercury lamp or a laser. The emission can be either directly proportional to the excitation (linear) or exhibit a more complex dependence (non-linear). In the last two decades, microscopes based on e various non-linear processes have been developed, making possible to perform deep tissue imaging (Denk et al., 1990; Mertz, 2004; Zipfel et al., 2003b). Below, we will briefly describe and compare some of the linear (confocal microscopy, CM) and non-linear (multi-photon and harmonic generation) techniques that are commonly used to perform IVM.

2.1 Confocal microscopy

In CM, the excitation of the fluorophore is achieved by using single photons with wavelengths ranging from ultraviolet (UV) to visible light (Fig.2). In order to gather the signal coming from the focal plane and to avoid off-focus emissions that reduce the spatial resolution, the emitted light is forced to pass through a pinhole. This allows to modify the thickness of the sampled area providing an easy way to balance resolution and signal intensity. Confocal microscopes are widespread tools and have been extensively used for IVM (Guan et al., 2009; Masedunskas et al., 2011). However, CM has some limitations. First, UV and visible light are scattered by biological specimen, thus limiting the imaging to the first 50-60 μm below the surface of the specimen and making CM the optimal choice for cell cultures and optically transparent tissues. Second, long term illumination with UV and visible light may lead to photobleaching and phototoxicity, limiting the use of CM to short term imaging as documented by several reports of radiation-induced cellular damage or impairment in tissue development (Dela Cruz et al., 2010). However, when tissues are homogeneous and biological processes are not dependent on the depth, CM can be successfully used providing a better spatial resolution than other techniques (Masedunskas et al., 2011).

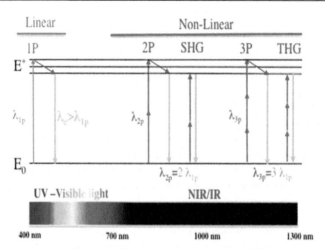

Fig. 2. Comparison between linear and non-linear modalities of fluorescence light microscopy.

Single photon excitation: the energy gap between the ground (E_0) and the excited (E^*) state in the fluorophore is filled by a single photon in the UV or visible range. Following some energy dissipation, a single photon is emitted at a higher wavelength (red-shift). Two- and three-photon excitation: the same energy gap is filled by two or three photons respectively, which have half or a third of the energy required for single photon excitation (NIR or IR light). Second and third harmonic generation (SHG and THG): two or three photons interact with the fluorophore and recombine generating a photon with half or a third of the wavelength of the incident ones.

2.2 Multiphoton microscopy (MPM)

Multiphoton emission is based on the fact that a fluorophore can be excited by the almost simultaneous absorption (within femto or atto seconds) of two or three photons that have a half or a third of the energy required to fill the energy gap in the fluorophore (Fig. 2). This requires the use of infrared (IR) light which has a lower intrinsic scattering in biological specimen when compared to UV or visible light. The non-linear nature of multi-photon excitation and the low probability for a multiphoton transition to occur, require that a high number of photons are focused in a restricted volume (1 fl–1 µm³). This is achieved with pulsed lasers, such as the tunable titanium:sapphire laser, which generates high power beams (in the order of 2-4 W), that are focused in the focal point with high numerical aperture lenses (McMullen and Zipfel, 2010). This implies that all the emitted light generated from the focal point can be utilized to generate the image without the need for a pinhole. This simplifies the geometry of multiphoton microscopes, which require detectors with high sensitivity placed as close as possible to the specimen. Another implication of the fact that photons are absorbed in a confined volume is that photobleaching and phototoxicity are reduced, extending the duration of the experimental observations without any tissue damage, and enabling the realization of long term longitudinal studies that are fundamental in fields such as tumor biology. In terms of depth, MPM enables to extend the range of observation when compared with CM. For example, by using high numerical aperture objectives, subcellular structures can be resolved up to a depth of 100-150 μm. Lowering the level of resolution and using lenses with longer working distances cellular structures can be routinely resolved at a depth of 300-500 μm. Furthermore, some tissues, such as the brain, exhibit lower light scattering enabling imaging up to a 1 mm depth. Recently, alternative approaches based on the use of either longer excitation wavelengths through the use of optical parametric oscillators (OPO) or regenerative amplifiers, have extended the limits of imaging depth (Andresen et al., 2009; Theer et al., 2003). Two final advantages of multiphoton excitation are: first, the fact that several endogenous molecules can be easily excited providing a contrast that provides numerous information on tissue and cell architecture (Campagnola and Loew, 2003; Dela Cruz et al., 2010; Weigert et al., 2010; Zipfel et al., 2003a), and second, that due to their broad multiphoton absorption spectra, multiple fluorophores can be excited simultaneously using a single excitation wavelength. This avoids the use of multiple lasers, thus reducing further the risk of photodamage.

2.3 Second and third harmonic generation (SHG and THG)

SHG and THG do not involve energy absorption since the incident photons are scattered and recombined into a single photon in a process without energy loss (Campagnola and Loew, 2003; Schenke-Layland et al., 2008; Zoumi et al., 2002). Molecules that generate second harmonic signals such as, collagen, microtubules, and muscle myosin are usually assembled in highly ordered and repeated structures with non-centrosymmetric symmetries, whereas third harmonic signals are typically generated at the interface between optically heterogeneous biological materials (Campagnola and Loew, 2003; Debarre et al., 2006; Gualda et al., 2008). SHG has been extensively utilized to study the properties of the extracellular matrix under both physiological and pathological conditions, and shows also an incredible potential for diagnostic purposes. THG has been used to image lipid bodies in small organisms, to study early embryogenesis dynamics in zebrafish, and to study the process of demyelination in models for neurodegenerative disorders. SHG and THG have the advantage of being nontoxic since no energy is absorbed by the specimen during imaging (Fig. 2), and they can be combined with MPM providing with the opportunity to perform multimodal imaging (Campagnola and Loew, 2003; Chen et al., 2009; Debarre et al., 2006; Farrar et al., 2011; Radosevich et al., 2008).

3. The salivary glands as a versatile model to perform intravital microscopy and to manipulate cellular pathways

3.1 Architecture and physiology of the salivary glands

Salivary glands (SGs) are major exocrine glands responsible for the production and secretion of saliva into the oral cavity (Gorr et al., 2005; Melvin et al., 2005). In mammals there are two kinds of SGs: the major and the minor glands. The major SGs include: parotid glands, which secrete primarily enzymes involved in digestion (e.g. amylase), submandibular glands, which secrete enzymes required to defend the oral cavity from pathogens (e.g. peroxidases, kallikrein), and sublingual glands, which secrete molecules required to protect the oral cavity (e.g. mucins). Saliva is a mixture of water, proteins, and electrolytes that is primarily released from the acini, the main secretory units of the SGs, into the acinar canaliculi and from there discharged into the ductal system (Fig. 3). Acini are formed by polarized acinar cells, with the apical plasma membrane (APM) facing the lumen of the acinar canaliculi, and the basolateral membrane facing the basement membrane and the stroma. The secretion of water and proteins is under the control of G protein-coupled receptors: muscarinic stimulation is the primary signal regulating water secretion, whereas protein secretion is regulated by either the beta-adrenergic (submandibular and parotid) or the muscarinic receptors (sublingual) (Gorr et al., 2005; Melvin et al., 2005). The ductal system is also formed by polarized cells and its main function is to modify the electrolyte composition of the primary saliva and convey it into the oral cavity. In rodents, a subpopulation of the ductal cells, the granular convoluted tubules, secrete large amount of growth factors (such as EGF and NGF) that are stored in large secretory granules (Peter et al., 1995).

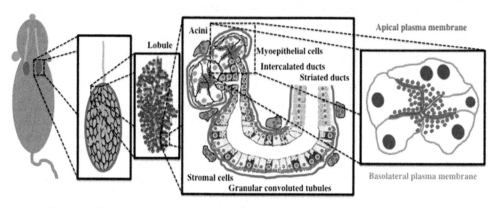

Fig. 3. Diagram of rodent submandibular salivary glands.

The SGs are formed by interconnected lobules, which contain both acini and ductal structures. Acini are formed by polarized epithelial cells, which contain secretory granules that fuse with the apical plasma membrane and release their content in the acinar canaliculi. The canaliculi merge in larger ducts, the intercalated ducts, than enlarge forming the granular convoluted tubules and later the striated ducts. The striated ducts merge with the major excretory duct

3.2 Delivery of molecules, drugs and gene transduction in the salivary glands

SGs are ideal organs to perform IVM for various reasons. First, in rodents the glands are located in the neck area, where the motion artifacts due to the heartbeat and the respiration are significantly reduced (Masedunskas and Weigert, 2008). Second, the SGs can be exposed with relatively minor surgical procedures, which do not involve the exposure of the body cavity, which may effect the overall health status of the animal. Finally, the epithelium of the SGs can be easily accessed from the oral cavity by introducing fine polyethylene tubings into the major excretory ducts (called Wharton's duct in the submandibular glands and Stensen's duct in the parotid glands) that can be utilized to selectively deliver various molecules into the ductal system (Masedunskas et al., 2011; Masedunskas and Weigert, 2008; Sramkova et al., 2009) (Fig. 4A). We have shown that fluorescent dyes can be delivered into the ductal system through injection or by gravity diffusion, and utilized to study endocytosis, exocytosis or various aspects of water secretion (Fig. 4B) (Masedunskas and Weigert, 2008; Sramkova et al., 2009, Masedunskas et al., 2011). The same route has been utilized to selectively deliver drugs to the SGs. This approach offers two advantages: 1) to specifically target the SGs avoiding the side effects due to systemic injections, and second, to precisely control the doses of the drugs administered (Masedunskas and Weigert, 2008).

SGs have been widely used as a target organ for the viral-mediated expression and gene delivery of various transgenes both in live animals and in humans (Baum et al., 2010; Cotrim and Baum, 2008). Indeed, these organs have the potential to be utilized for gene therapy to correct various diseases including Sjogren's syndrome and protein deficiencies (Baum et al., 2004; Voutetakis et al., 2004). Notably, for viral-mediated gene therapy in humans, the SGs offer several advantages with respect to other organs: i) the encapsulation of the SG tissue prevents the dissemination of the virus in the rest of the body (Baum et al., 2004; Voutetakis et al., 2004), ii) in case of potential health issues the SGs can be removed since they are not essential for life, iii) the differentiation of the cells provides a relatively stable cell populations for non-integrating vectors, and iv) duacrine (both exocrine and endocrine) protein secretion allows to direct the expressed molecules into either the saliva or the blood stream (Baum et al., 2004). Numerous studies have shown successful gene transfer into both rat and mouse submandibular glands using viral-based approaches, which offer the advantage of a more robust expression of the transgenes (Andresen et al., 2009; Baum and Tran, 2006; Delporte et al., 1996; Honigman et al., 2001; Mastrangeli et al., 1994; Morita et al., 2011; Palaniyandi et al., 2011; Perez et al., 2011; Samuni et al., 2008; Wang et al., 2000; Zheng et al., 2009). However, non viral-mediated approaches have also been utilized, although limited to a small percentage of the cells in the parenchyma (Goldfine et al., 1997; Honigman et al., 2001; Niedzinski et al., 2003a; Niedzinski et al., 2003b; Passineau et al., 2010; Sramkova et al., 2009). Furthermore, the majority of the studies on rodent SGs were focused on submandibular glands and only few studies were performed in parotid glands. Recent studies demonstrated efficient gene transfer into rat parotid glands, as shown by the effective delivery of human erythropoietin and human parathyroid hormone (Adriaansen et al., 2010; Kagami et al., 1998; Mastrangeli et al., 1994; Zheng et al., 2009). The rationale behind developing strategies to deliver transgene into parotid glands is their use in humans as main target for clinical applications (Zheng et al., 2011).

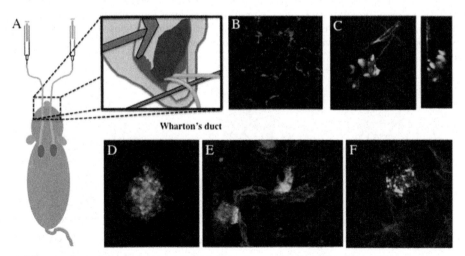

Fig. 4. Non viral-mediated gene transfer in the submandibular SGs of live rats.

A. Fine polyethylene cannulae are introduced in the oral cavity in the Wharton's duct of live rats. B- 10 kDa Texas Red-dextran is injected in the SGs and the ductal system is highlighted. C-F. Plasmid DNAs encoding for the fluorescent protein pVenus (B), GFP-ribonuclases (D), Aquaporin 5-GFP (E), and GFP-Clathrin (F) were injected in the submandibular glands. After 16 hrs, the glands were exposed and imaged by two-photon IVM (C and D) or excised, fixed, labeled with rhodamine-phalloidin, and imaged by CM (E and F). C. Cluster of pVenus expressing-cells (green) localized below the surface of the glands as shown by SHG, which reveals the collagen fibers (cyan). Excitation 930 nm. D. GFP-ribonuclease is localized in intracellular vesicles. Excitation 930 nm. E,F. GFP-Acquaporin is expressed in large ducts (E) and GFP-clathrin in acini (F) as revealed by labeling for the actin cytoskeleton (red).

We have utilized plasmid DNA in live rats and shown that the transgenes under the appropriate conditions can be targeted to specific subpopulations of the SGs (Sramkova et al., 2009). The main advantage in using naked DNA vs. viral-based vectors is the possibility to screen very rapidly for multiple genes without dealing with the time-consuming steps of designing, cloning and preparing the viral particles. We have injected plasmid DNA encoding for various fluorescent proteins into the Wharton's duct of rat submandibular glands, and after 16 hours we have observed their expression in the SGs epithelium (Fig 4C-F). Specifically, we have found that when plasmid DNA is injected alone, the reporter molecule is expressed in approximately 0.05% of the cells of the parenchyma, which we have identified as intercalated ducts. The addition of empty replication-defective adeno-viral (rAd5) particles increases the level of transduction up to 0.5-2% of the cells and notably, the fluorescent reporter is expressed primarily in the large striated and granular ducts and to a lesser extent in the acinar cells (Fig. 4C, 4E and 5). In both instances, the expression of the reporter molecule is transient and lasts for 72 hrs, a window of time sufficient to be utilized for IVM. Furthermore, since our goal is to transduce genes primarily into acinar structures, we sought to find a more specific way to target these cells. We reasoned that plasmid DNA might be internalized by the acinar cells via the endocytic pathways and for this reason we stimulated compensatory endocytosis by activating the beta-adrenergic receptors during plasmid DNA injection (see below). Notably, under these conditions, the reporter molecule is expressed in 1% of the cells of the parenchyma and primarily in the acinar cells (Fig. 4D, 4F and 5) (Sramkova et al., 2009). It is important to emphasize, that although the efficiency of gene expression is low, the absolute number of cells that can be imaged by IVM is still very high. This implies that viral-based approaches may still have to be utilized whenever a different readout, such as a biochemical assay, needs to be used.

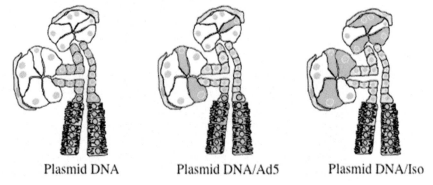

Fig. 5. Naked DNA is targeted to different cell populations of the salivary glands.

When naked DNA is injected in the absence of any other manipulation the transgene is expressed in the intercalated ducts. When naked DNA is pre-mixed with empty adenoviral particles, the transgene is expressed primarily in the large ducts but also in some acinar cells. When naked DNA is injected and compensatory endocytosis is elicited by stimulation of the beta-adrenergic receptor (sub-cutaneous injections of isoproterenol) the transgene is expressed in acinar cells.

This approach can be utilized to express any fluorescently tagged protein, enabling the expansion of the repertoire of compartments that can be visualized in a short period of time (Fig. 4C-F). Moreover, this strategy has provided us with a powerful tool to interfere with specific molecular machineries by introducing molecules acting as positive or negative regulators of the specific processes of interest. Finally, this approach can be used to genetically modify the target tissue by silencing certain genes. Small interfering RNA (siRNA) has been successfully delivered in live rats. siRNA targeting the cystic fibrosis transmembrane conductance regulator (CFTR) were injected intraductallly into rat submandibular glands resulting in the effective silencing of CFTR (Ishibashi et al., 2008; Ishibashi et al., 2006). Notably, in order to complement and confirm the results with the siRNA, a specific inhibitor of CFTR (CFTR$_{inh-}$ 172) and suramin, a non-specific P2 receptor antagonist, were also injected, further highlighting the power of this approach. Although the efficiency in siRNA delivery is low, novel approached has been introduced to overcome this issue. For example, silencing of GAPDH in rat parotid glands was performed in combination with microbubble-enhanced sonoporation, improving the efficiency of siRNA transfer by 10-50% (Sakai et al., 2009).

4. Imaging membrane trafficking and the actin cytoskeleton in salivary glands by intravital microscopy

Membrane traffic is an important field in cell biology that studies the processes and the machineries involved in the transport of various molecules among different compartments within the cell. Transport steps are mediated by membranous containers, termed "transport intermediates", which are very heterogeneous in size, shape, contents and modality of transport. Their biogenesis, trafficking, delivery to the target compartments, and dynamic behavior are dictated by the architecture of the cells and by the organization of the cytoskeletal elements (e.g. microtubules and microfilaments). Most of the data on the dynamics of the transports steps have been derived from cell culture models. As, previously discussed, the architecture of the cells in a living organism differs considerably from the architecture of cells in culture. Although IVM offers a very powerful opportunity to study membrane trafficking in physiological conditions, the challenges in controlling the motion artifacts have discouraged several investigators from pursuing this approach and only few labs have invested in high resolution imaging of live animals. For example, submicron structures were imaged dynamically in the brain of live mice, where structural changes in the architecture of dendritic spines were observed under conditions such as epileptic seizures (Mizrahi et al., 2004; Pan and Gan, 2008; Svoboda and Yasuda, 2006). In kidney, various subcellular processes were analyzed, such as endocytosis of selected molecules (Dunn et al., 2002; Dunn et al., 2003; Molitoris and Sandoval, 2006; Sandoval et al., 2004; Sandoval and Molitoris, 2008), exocytosis of renin (Toma et al., 2006), and mitochondrial function (Hall et al., 2009). Recently, mitochondrial dynamics and lipid bodies have been analyzed in live animals (Debarre et al., 2006; Roberts et al., 2008; Zhong et al., 2008).

Another area where imaging membrane trafficking in vivo has provided novel information is tumor biology. Very recently, nuclear dynamics and mitotis were observed using in murine xenograft model of human cancer and compared to cells in culture (Orth et al., 2011) with profound implications for drug development and cancer therapy (Amornphimoltham et al., 2011). Furthermore, using QD conjugated either to an anti-HER2 antibody or to EGF-conjugated nanotubes the delivery and the uptake of these molecules by tumor cells was analyzed (Bhirde et al., 2009; Tada et al., 2007).

Here, we review our work using the SGs, which represent a robust model to study several aspects of membrane trafficking, particularly because the motion artifacts can be easily reduced using various strategies described in detail elsewhere (Masedunskas et al. 2011b). Although the SGs are exocrine glands, which represent a perfect model system to study exocytosis, they are also a powerful model to study endocytic processes, such as receptor-mediated endocytosis, which occur at the basolateral plasma membrane of the epithelium, compensatory endocytosis that is triggered upon exocytosis at APM, fluid phase endocytosis in stromal cells, polarized trafficking of plasma membrane proteins in the epithelium, and mitochondrial dynamics.

4.1 Endocytosis

The endosomal system is utilized as a transport route, to shuffle proteins, lipids and membranes to and from the cell surface, and towards other sub-cellular organelles. Endocytosis occurs in every cell and is involved in several processes such as nutrient uptake, cell adhesion, migration, cytokinesis, polarity and signaling (Maxfield and McGraw, 2004; Mellman, 1996). Endocytosis and recycling mediate the removal and retrieval of membrane components from the cell surface and these tightly coupled processes are highly regulated. Notably, endocytic pathways are very diversified in terms of molecular machinery, as shown by the fact that multiple endocytic routes have been described (Conner and Schmid, 2003; Doherty and McMahon, 2009; Grant and Donaldson, 2009; Mayor and Pagano, 2007). Much of our understanding of the endosomal system is derived from studies performed in cell cultures and few studies have been performed in live organisms such as rodents. The first attempt to image endocytosis in vivo was realized in the kidney of live rats and mice, where fluorescently labeled dextrans of different molecular weight were injected systemically (Dunn et al., 2002). Similar studies were performed imaging the receptor-mediated endocytosis of the antibiotic gentamicin and the internalization of the folate-receptor (Dunn et al., 2002; Sandoval et al., 2004). Although the diffusion of the injected probes along the tubular system in the kidney was imaged, their internalization could be followed only for short period of times due to the high levels of the motion artifacts. In this respect, the SGs provide with a more controlled experimental system. Indeed, we injected fluorescently-labeled molecules in the tail artery and tracked the dynamics of the endosomal compartments for over 60 minutes (Masedunskas and Weigert, 2008) (Fig. 6A). To further distinguish among the different endocytic sub-compartments, Texas red-dextran (TXR-D) was injected systemically and allowed to accumulate into the lysosomes (Fig. 6B). After 24 hours, Alexa-488 dextran (488-D) was injected and imaged in time-lapse mode. 488-D was first internalized into small vesicles, and then delivered to early endosomes that grew over time due to homotypic fusion events. Later, 488-D was delivered to late endosomes and lysosomes and the process was imaged providing novel insight on the dynamics of the endo-lysosomal system (Fig. 6B) (Masedunskas and Weigert, 2008). In a separate study lysosomal fusion events were captured at a higher resolution, almost comparable to that achieved in cell culture (Weigert et al., 2010).

Interestingly, Cytochalasin D and Latrunculin A, two actin-disrupting agents significantly reduced the uptake of fluorescent dextran in SGs, suggesting for a role of actin during internalization (Fig. 6C and 6D) (Masedunskas and Weigert, 2008). The requirement for actin during endocytosis has been demonstrated in yeast but has been controversial in mammalian cells (Galletta et al., 2010). The conflicting results could be due to differences in the organization of the actin cytoskeleton or to a different organization of the endocytic pathways. Additional work is required to address these fundamental questions and particularly in defining the endocytic routes *in vivo* and their reciprocal relationship.

Another important issue is the fact that different cell populations within the same organ exhibit different rate of internalization. For example we found that in the SGs, fibroblasts and dendritic cells internalized molecules at a much faster rates than the acinar or the ductal cells in the parenchyma (Fig. 6E). This may reflect the presence of barriers such as the basement membrane and the tight junctions that controls the delivery of molecules from the blood stream.

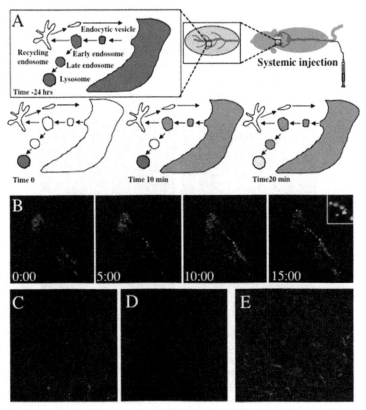

Fig. 6. Endocytosis of systemically-injected fluorescently-labeled probes in the salivary glands of live rats

A. Diagram of the experimental design to study endocytosis in SGs. The fluorescent probes are injected into the tail artery, reach the SGs through the circulation, and diffuse out of the vasculature from the fenestrated capillaries. B. TXR-D (red) is accumulated into the lysosomes after 24 hrs from the injection (time 0:00). Alexa 488-dextran is first internalized into early endosomes (time 5:00 and 10:00) and later reach the lysosome (time 15:00 and inset). C,D. TXR-D was injected and the SGs were imaged after 20 min. C. Control glands D. SGs treated with latrunculin A. E. Lower magnification of the SGs after the injection of TXR-D. The probe is accumulated in the stroma, internalized in stromal cells, and was excluded from the acini that are revealed by two-photon-stimulated intrinsic emission (Masedunskas and Weigert, 2008; Weigert et al., 2010).

Endocytosis from the APM of the SGs was also analyzed by either injection or slow gravity-mediated infusion of small molecular weight dextrans though the Wharton's duct. Under resting conditions most of the probes underwent a low but detectable level of endocytosis in both the ducts and the acinar cells. However, upon stimulation of protein but not water secretion, the probes were primarily internalized into the acini, most likely by the activation of a process known as compensatory endocytosis (Masedunskas et al., 2011; Sramkova et al., 2009). Interestingly, this process did not involve any of the currently characterized endocytic processes and studies are undergoing to further elucidate this it s machinery. Finally, in order to study receptor-mediated endocytosis at the basolateral plasma membrane we have transfected in the acinar cells of live rats, transferrin receptor, as a model for constitutive endocytosis and beta2-adrenergic receptor, as a model for agonist-induced endocytosis. The ectopically expressed receptors are properly targeted to the acinar cells and at the plasma membrane as predicted (Fig. 7C and 7D).

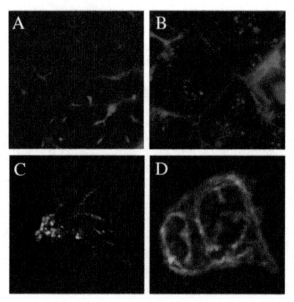

Fig. 7. Endocytosis from the apical and the basolateral plasma membrane of the salivary glands

A,B. 10 kDa TXR-D was injected into the Wharton's duct to fill the acinar canaliculi. After 20 minutes, minimal internalization of the probe was observed (A). Five minutes after the SC injection of isoproterenol smaller endocytic vesicles formed from the APM (B). C,D. GFP-Transferrin receptor (C) and YFP-beta2-adrenergic receptor were transfected in the acinar cells of live rats and imaged by using IVM. Maximal projections of Z-stacks are shown. Transferrin receptor is primarily localized in intracellular vesicles, whereas beta2-adrenergic receptor is primarily localized at the basolateral plasma membrane.

4.2 Exocytosis

SGs are a well-established model for exocrine secretion. Proteins destined to secretion are synthesized in the endoplasmic reticulum and transported through the Golgi apparatus to the trans-Golgi network (TGN) where they are packed in large vesicles, secretory granules (SCGs), which are released into the cytoplasm, and transported to the cell periphery. Here, upon stimulation of the appropriate G protein-coupled receptor (GPCR), the SCGs fuse with the APM, releasing their content into the lumen of the canaliculi. Although, exocytosis in SGs has been extensively studied in ex-vivo models (Castle et al., 2002; Castle, 1998; Gorr et al., 2005), very little is known about the molecular mechanisms regulating this process. Several studies have reported contradicting findings about the stimuli triggering exocytosis, the modalities of fusion, and the requirement for the actin cytoskeleton in this process (Eitzen, 2003; Nashida et al., 2004; Segawa et al., 1998; Segawa and Riva, 1996; Segawa et al., 1991; Sokac and Bement, 2006; Warner et al., 2008). This variability probably reflects the different experimental conditions utilized to isolate and culture *ex-vivo* the acinar cells. To overcome this issue, we have utilized IVM and studied the dynamics of the SCGs in live animals. To this aim, we used a series of transgenic mouse models expressing selected fluorescently labeled molecules, combined with the ability to transduce genes, and selectively deliver molecules and pharmacological agents, as described above and elsewhere (Masedunskas et al., 2011b). This approach has provided novel and valuable information on the structure and the physiology of the acinar cells (Masedunskas et al., 2011). We estimated that in resting conditions, the major SGs contain approximately 2500-3000 granules per acinus, most of them accumulated in the sub apical area of the PM. Our analysis on the effect of various agonists of GPCRs has revealed three major differences between in vivo and ex-vivo models: 1) the stimulation of the beta-adrenergic but not the muscarinic receptors, enhances the mobility of the secretory granules promoting their docking and subsequent fusion at the APM; 2) muscarinic receptors do not play any synergistic role with the adrenergic receptor during exocytosis; and 3) the maximal rate of fusion of the secretory granules in live animal (10-15 granules/cell/min) is 3-4 times faster than previously reported for *ex-vivo* systems. Furthermore, by using another mouse model, which expresses the Tomato fluorescent protein fused with a di-palmitoylated peptide (m-Tomato), a well-established marker for the plasma membrane, we discovered that the secretory granules after fusing with the plasma membrane completely collapse within 30-40 seconds (Masedunskas et al., 2011). This result underscores another major difference between *in vivo* and *ex-vivo* models, in which compound exocytosis (i.e. the sequential fusion of strings of SCGs), has been described as the primary modality of fusion (Warner et al., 2008). Notably, we also observed that the granules in close proximity of the APM recruit a series of cytosolic proteins including actin, suggesting a role for the cytoskeleton during granule exocytosis. To address this issue we have transduced the salivary glands of live rats with the small peptide Lifeact fused with GFP, a novel tool to label dynamically F-actin (Riedl et al., 2008). We determined that F-actin filaments are polymerized onto the surface of the granules only after fusion has occurred, and persisted until their complete collapse. The impairment of the dynamics of the actin cytoskeleton, using pharmacological agents such as cytochalasin D (cyto D) or latrunculin A (lat A), did not affect the fusion of the secretory granules with the APM, but it blocked substantially their collapse leading to the accumulation of fused granules which often expanded in size. Finally, we found that myosin IIa and IIb, two actin-based motor proteins are recruited on the fused secretory granules and that their motor activity is required to drive the gradual collapse of the granules. These results suggest that the acto-myosin complex provides a contractile scaffold around the secretory granules that facilitates the completion of the fusion at the APM and preventing an aberrant influx of membrane inside the cell (Masedunskas et al., 2011). This novel approach provided new insights into the molecular mechanisms of exocytosis in SGs, captured the exocytosis process dynamically and established important tools to study this process. This approach can be extended to study exocytosis in other exocrine glands such as pancreas, lacrimal glands, and mammary glands

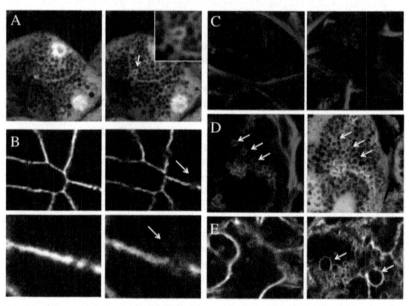

Fig. 8. Regulated exocytosis in the salivary glands of live rodents.

A. SGs of mice expressing cytoplasmic GFP. GFP is excluded from the large SCGs that appear as dark vesicles in the cytoplasm. Upon stimulation with isoproterenol, SCGs fuse with the APM. Fusing SCGs are characterized by an accumulation of GFP on the limiting membranes of the SCGs (arrow), as previously described (Masedunskas et al., 2011). B. SGs in the m-Tomato mice. The m-Tomato probe labels all the cellular membranes. Upon stimulation with isoproterenol, the m-Tomato diffuses into the membrane of the SCGs (arrows). C. The SGs of a live rat were labeled with rhodamine-phalloidin to reveal the actin cytoskeleton. The left panel shows the enrichment of actin at the APM in control conditions. The right panel shows the recruitment of actin around the SCGs upon stimulation with isoproterenol. D. A GFP mouse was stimulated with isoproterenol and labeled with rhodamine-phalloidin. Arrows point to the SCGs at the plasma membrane that are enriched in GFP and actin. E. The effect on exocytosis of actin-disrupting agents. The SGs of a live m-Tomato mouse were treated with latrunculin A and stimulated with isoproterenol (right panel). SCGs fail to collapse at the plasma membrane and increase in size forming large vacuolar structures.

4.3 Actin cytoskeleton

The actin cytoskeleton plays a fundamental role in many cellular events. We have recently shown a role for actin in endocytosis and a novel role for actin and the actin motor protein myosin II in exocytosis in the SGs of live animals (Masedunskas et al., 2011; Masedunskas and Weigert, 2008). Notably, by using IVM we revealed a novel function for the actomyosin complex that was not completely appreciate in cell cultures. Specifically, we found that actin serves: 1) as a barrier preventing the unwanted homotypic fusion between the SCGs, 2) as a scaffold to prevent the hydrostatic pressure generated by fluid secretion to disrupt the exocytic events, and 3) as a platform to generate a contractile scaffold that facilitate the collapse of the SCGs with the apical plasma membrane (Masedunskas et al., 2011). We took an advantage of the methodology reviewed here to pharmacologically disrupt the actin cytoskeleton and fluorescently tag proteins for labeling the cytoskeleton *in vivo*. Specifically, we have used GFP-lifeact as a tool to follow the dynamics of F-actin (Riedl et al., 2008). This tool is more effective and less toxic than GFP-actin (Fig. 9A and 9C). However, one of the drawbacks of our transfection system is that the expression of the protein of interest is limited to one or two cells per acinus, limiting the possibility to study the behavior of the actin cytoskeleton in groups of cell. Recently, a transgenic mouse expressing GFP-life act has been generated (Riedl et al., 2010). These mice are superior to the GFP-actin transgenic mice in terms of viability, level of protein expression and cellular toxicity and represent a formidable tool to study several aspect of the involvement of actin in various cellular processes. These mice are nicely complemented by other transgenic mice expressing the GFP-tagged versions of the actin motor proteins myosin IIa and IIb (Bao et al., 2007).

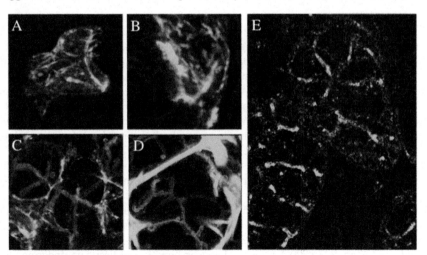

Fig. 9. Imaging the actin cytoskeleton in the salivary glands of live animals.

A, B. GFP-actin or GFP-lifeact were transfected in the SGs of live rats. Both molecules were expressed in acinar cells and exhibited filamentous cortical localization. However, the precise localization of the two probes with respect to the apical or the basolateral pole cannot be assessed in a single cell. C, D. Transgenic mice expressing GFP-actin (C) or GFP-lifeact (D). The acinar structure in the salivary glands show the typical enrichment at the apical plasma membrane. E. Transgenic mouse expressing GFP-myosin IIb. In the acini of the SGs, GFP-myosin IIb is localized at the apical plasma membrane as previously described for the endogenous myosins (Masedunskas et al., 2011).

5. Conclusions

We have provided several examples from our recent work that the combination of IVM as imaging technique, and the SGs as a model organ is a very versatile tool that can be successfully used to address several biological questions under physiological conditions. The possibility to either express or down regulate proteins in an acute fashion, combined with the plethora of available transgenic and knockdown mouse models, offer a unique set of opportunities to study processes in live animals at a molecular level. Furthermore, the ability to image at a subcellular level in a live animal has opened the possibility to study several aspects of cell biology in a dynamic fashion. We have utilized this approach to study membrane trafficking and the dynamics of the actin cytoskeleton, which represent only a fraction of the fields that can be studied *in vivo*. We envision that soon IVM will be extended to study other processes, such as cell cycle, signal transduction, mitochondrial dynamics and metabolisms to name a few.

6. Acknowledgments

This research was supported by the Intramural Research Program of the NIH, National Institute of Dental and Craniofacial Research.

7. References

Adriaansen, J., Zheng, C., Perez, P., and Baum, B.J. (2010). Production and sorting of transgenic, modified human parathyroid hormone in vivo in rat salivary glands. Biochem Biophys Res Commun *391*, 768-772.

Amornphimoltham, P., Masedunskas, A., and Weigert, R. (2011). Intravital microscopy as a tool to study drug delivery in preclinical studies. Adv Drug Deliv Rev *63*, 119-128.

Andresen, V., Alexander, S., Heupel, W.M., Hirschberg, M., Hoffman, R.M., and Friedl, P. (2009). Infrared multiphoton microscopy: subcellular-resolved deep tissue imaging. Curr Opin Biotechnol *20*, 54-62.

Bao, J., Ma, X., Liu, C., and Adelstein, R.S. (2007). Replacement of nonmuscle myosin II-B with II-A rescues brain but not cardiac defects in mice. J Biol Chem *282*, 22102-22111.

Baum, B.J., Adriaansen, J., Cotrim, A.P., Goldsmith, C.M., Perez, P., Qi, S., Rowzee, A.M., and Zheng, C. (2010). Gene therapy of salivary diseases. Methods Mol Biol *666*, 3-20.

Baum, B.J., and Tran, S.D. (2006). Synergy between genetic and tissue engineering: creating an artificial salivary gland. Periodontol 2000 *41*, 218-223.

Baum, B.J., Voutetakis, A., and Wang, J. (2004). Salivary glands: novel target sites for gene therapeutics. Trends Mol Med *10*, 585-590.

Beck, J.S., and Berg, B.N. (1931). The Circulatory Pattern in the Islands of Langerhans. Am J Pathol 7, 31-36 31.

Beerling, E., Ritsma, L., Vrisekoop, N., Derksen, P.W., and van Rheenen, J. (2011). Intravital microscopy: new insights into metastasis of tumors. J Cell Sci 124, 299-310.

Bhirde, A.A., Patel, V., Gavard, J., Zhang, G., Sousa, A.A., Masedunskas, A., Leapman, R.D., Weigert, R., Gutkind, J.S., and Rusling, J.F. (2009). Targeted killing of cancer cells in vivo and in vitro with EGF-directed carbon nanotube-based drug delivery. ACS Nano 3, 307-316.

Cahalan, M.D., and Parker, I. (2008). Choreography of cell motility and interaction dynamics imaged by two-photon microscopy in lymphoid organs. Annu Rev Immunol 26, 585-626.

Campagnola, P.J., and Loew, L.M. (2003). Second-harmonic imaging microscopy for visualizing biomolecular arrays in cells, tissues and organisms. Nat Biotechnol 21, 1356-1360.

Castle, A.M., Huang, A.Y., and Castle, J.D. (2002). The minor regulated pathway, a rapid component of salivary secretion, may provide docking/fusion sites for granule exocytosis at the apical surface of acinar cells. J Cell Sci 115, 2963-2973.

Castle, J.D. (1998). Protein secretion by rat parotid acinar cells. Pathways and regulation. Ann N Y Acad Sci 842, 115-124.

Chalfie, M., Tu, Y., Euskirchen, G., Ward, W.W., and Prasher, D.C. (1994). Green fluorescent protein as a marker for gene expression. Science 263, 802-805.

Chen, H., Wang, H., Slipchenko, M.N., Jung, Y., Shi, Y., Zhu, J., Buhman, K.K., and Cheng, J.X. (2009). A multimodal platform for nonlinear optical microscopy and microspectroscopy. Opt Express 17, 1282-1290.

Conner, S.D., and Schmid, S.L. (2003). Regulated portals of entry into the cell. Nature 422, 37-44.

Cotrim, A.P., and Baum, B.J. (2008). Gene therapy: some history, applications, problems, and prospects. Toxicol Pathol 36, 97-103.

Cukierman, E., Pankov, R., Stevens, D.R., and Yamada, K.M. (2001). Taking cell-matrix adhesions to the third dimension. Science 294, 1708-1712.

Debarre, D., Supatto, W., Pena, A.M., Fabre, A., Tordjmann, T., Combettes, L., Schanne-Klein, M.C., and Beaurepaire, E. (2006). Imaging lipid bodies in cells and tissues using third-harmonic generation microscopy. Nat Methods 3, 47-53.

Dela Cruz, J.M., McMullen, J.D., Williams, R.M., and Zipfel, W.R. (2010). Feasibility of using multiphoton excited tissue autofluorescence for in vivo human histopathology. Biomed Opt Express 1, 1320-1330.

Delporte, C., O'Connell, B.C., He, X., Ambudkar, I.S., Agre, P., and Baum, B.J. (1996). Adenovirus-mediated expression of aquaporin-5 in epithelial cells. J Biol Chem 271, 22070-22075.

Denk, W., Strickler, J.H., and Webb, W.W. (1990). Two-photon laser scanning fluorescence microscopy. Science 248, 73-76.

Diaspro (2002). Confocal and Two-Photon Microscopy. Foundations, Applications, and Advances.

Doherty, G.J., and McMahon, H.T. (2009). Mechanisms of endocytosis. Annu Rev Biochem 78, 857-902.

Dunn, K.W., Sandoval, R.M., Kelly, K.J., Dagher, P.C., Tanner, G.A., Atkinson, S.J., Bacallao, R.L., and Molitoris, B.A. (2002). Functional studies of the kidney of living animals using multicolor two-photon microscopy. Am J Physiol Cell Physiol *283*, C905-916.

Dunn, K.W., Sandoval, R.M., and Molitoris, B.A. (2003). Intravital imaging of the kidney using multiparameter multiphoton microscopy. Nephron Exp Nephrol *94*, e7-11.

Eitzen, G. (2003). Actin remodeling to facilitate membrane fusion. Biochim Biophys Acta *1641*, 175-181.

Farrar, M.J., Wise, F.W., Fetcho, J.R., and Schaffer, C.B. (2011). In vivo imaging of myelin in the vertebrate central nervous system using third harmonic generation microscopy. Biophys J *100*, 1362-1371.

Fukumura, D., Duda, D.G., Munn, L.L., and Jain, R.K. (2010). Tumor microvasculature and microenvironment: novel insights through intravital imaging in pre-clinical models. Microcirculation *17*, 206-225.

Fukumura, D., and Jain, R.K. (2008). Imaging angiogenesis and the microenvironment. APMIS *116*, 695-715.

Galletta, B.J., Mooren, O.L., and Cooper, J.A. (2010). Actin dynamics and endocytosis in yeast and mammals. Curr Opin Biotechnol *21*, 604-610.

Germain, R.N., Castellino, F., Chieppa, M., Egen, J.G., Huang, A.Y., Koo, L.Y., and Qi, H. (2005). An extended vision for dynamic high-resolution intravital immune imaging. Semin Immunol *17*, 431-441.

Ghajar, C.M., and Bissell, M.J. (2008). Extracellular matrix control of mammary gland morphogenesis and tumorigenesis: insights from imaging. Histochem Cell Biol *130*, 1105-1118.

Goldfine, I.D., German, M.S., Tseng, H.C., Wang, J., Bolaffi, J.L., Chen, J.W., Olson, D.C., and Rothman, S.S. (1997). The endocrine secretion of human insulin and growth hormone by exocrine glands of the gastrointestinal tract. Nat Biotechnol *15*, 1378-1382.

Gorr, S.U., Venkatesh, S.G., and Darling, D.S. (2005). Parotid secretory granules: crossroads of secretory pathways and protein storage. J Dent Res *84*, 500-509.

Grant, B.D., and Donaldson, J.G. (2009). Pathways and mechanisms of endocytic recycling. Nat Rev Mol Cell Biol *10*, 597-608.

Gualda, E.J., Filippidis, G., Voglis, G., Mari, M., Fotakis, C., and Tavernarakis, N. (2008). In vivo imaging of cellular structures in Caenorhabditis elegans by combined TPEF, SHG and THG microscopy. J Microsc *229*, 141-150.

Guan, Y., Worrell, R.T., Pritts, T.A., and Montrose, M.H. (2009). Intestinal ischemia-reperfusion injury: reversible and irreversible damage imaged in vivo. Am J Physiol Gastrointest Liver Physiol *297*, G187-196.

Hall, A.M., Unwin, R.J., Parker, N., and Duchen, M.R. (2009). Multiphoton imaging reveals differences in mitochondrial function between nephron segments. J Am Soc Nephrol *20*, 1293-1302.

Honigman, A., Zeira, E., Ohana, P., Abramovitz, R., Tavor, E., Bar, I., Zilberman, Y., Rabinovsky, R., Gazit, D., Joseph, A., *et al.* (2001). Imaging transgene expression in live animals. Mol Ther *4*, 239-249.

Ishibashi, K., Okamura, K., and Yamazaki, J. (2008). Involvement of apical P2Y2 receptor-regulated CFTR activity in muscarinic stimulation of Cl(-) reabsorption in rat submandibular gland. Am J Physiol Regul Integr Comp Physiol *294*, R1729-1736.

Ishibashi, K., Yamazaki, J., Okamura, K., Teng, Y., Kitamura, K., and Abe, K. (2006). Roles of CLCA and CFTR in electrolyte re-absorption from rat saliva. J Dent Res *85*, 1101-1105.

Kagami, H., Atkinson, J.C., Michalek, S.M., Handelman, B., Yu, S., Baum, B.J., and O'Connell, B. (1998). Repetitive adenovirus administration to the parotid gland: role of immunological barriers and induction of oral tolerance. Hum Gene Ther *9*, 305-313.

Lippincott-Schwartz, J. (2011a). Bridging structure and process in developmental biology through new imaging technologies. Dev Cell *21*, 5-10.

Lippincott-Schwartz, J. (2011b). Emerging in vivo analyses of cell function using fluorescence imaging (*). Annu Rev Biochem *80*, 327-332.

Masedunskas, A., Sramkova, M., Parente, L. and Weigert, R. (2011b). Intravital microscopy to image membrane trafficking in live rats. Method in Mol Biol. (in press)

Masedunskas, A., Sramkova, M., Parente, L., Sales, K.U., Amornphimoltham, P., Bugge, T.H., and Weigert, R. (2011). Role for the actomyosin complex in regulated exocytosis revealed by intravital microscopy. Proc Natl Acad Sci U S A *108*, 13552-13557.

Masedunskas, A., and Weigert, R. (2008). Intravital two-photon microscopy for studying the uptake and trafficking of fluorescently conjugated molecules in live rodents. Traffic *9*, 1801-1810.

Mastrangeli, A., O'Connell, B., Aladib, W., Fox, P.C., Baum, B.J., and Crystal, R.G. (1994). Direct in vivo adenovirus-mediated gene transfer to salivary glands. Am J Physiol *266*, G1146-1155.

Maxfield, F.R., and McGraw, T.E. (2004). Endocytic recycling. Nat Rev Mol Cell Biol *5*, 121-132.

Mayor, S., and Pagano, R.E. (2007). Pathways of clathrin-independent endocytosis. Nat Rev Mol Cell Biol *8*, 603-612.

McMullen, J.D., and Zipfel, W.R. (2010). A multiphoton objective design with incorporated beam splitter for enhanced fluorescence collection. Opt Express *18*, 5390-5398.

Mellman, I. (1996). Endocytosis and molecular sorting. Annu Rev Cell Dev Biol *12*, 575-625.

Melvin, J.E., Yule, D., Shuttleworth, T., and Begenisich, T. (2005). Regulation of fluid and electrolyte secretion in salivary gland acinar cells. Annu Rev Physiol *67*, 445-469.

Mertz, J. (2004). Nonlinear microscopy: new techniques and applications. Curr Opin Neurobiol *14*, 610-616.

Miller, M.J., Wei, S.H., Parker, I., and Cahalan, M.D. (2002). Two-photon imaging of lymphocyte motility and antigen response in intact lymph node. Science *296*, 1869-1873.

Mizrahi, A., Crowley, J.C., Shtoyerman, E., and Katz, L.C. (2004). High-resolution in vivo imaging of hippocampal dendrites and spines. J Neurosci *24*, 3147-3151.

Molitoris, B.A., and Sandoval, R.M. (2006). Pharmacophotonics: utilizing multi-photon microscopy to quantify drug delivery and intracellular trafficking in the kidney. Adv Drug Deliv Rev *58*, 809-823.

Morita, T., Tanimura, A., Shitara, A., Suzuki, Y., Nezu, A., Takuma, T., and Tojyo, Y. (2011). Expression of functional Stim1-mKO1 in rat submandibular acinar cells by retrograde ductal injection of an adenoviral vector. Arch Oral Biol.

Nashida, T., Yoshie, S., Imai, A., and Shimomura, H. (2004). Presence of cytoskeleton proteins in parotid glands and their roles during secretion. Arch Oral Biol 49, 975-982.
Niedzinski, E.J., Chen, Y.J., Olson, D.C., Parker, E.A., Park, H., Udove, J.A., Scollay, R., McMahon, B.M., and Bennett, M.J. (2003a). Enhanced systemic transgene expression after nonviral salivary gland transfection using a novel endonuclease inhibitor/DNA formulation. Gene Ther 10, 2133-2138.
Niedzinski, E.J., Olson, D.C., Chen, Y.J., Udove, J.A., Nantz, M.H., Tseng, H.C., Bolaffi, J.L., and Bennett, M.J. (2003b). Zinc enhancement of nonviral salivary gland transfection. Mol Ther 7, 396-400.
Orth, J.D., Kohler, R.H., Foijer, F., Sorger, P.K., Weissleder, R., and Mitchison, T.J. (2011). Analysis of mitosis and antimitotic drug responses in tumors by in vivo microscopy and single-cell pharmacodynamics. Cancer Res 71, 4608-4616.
Palaniyandi, S., Odaka, Y., Green, W., Abreo, F., Caldito, G., De Benedetti, A., and Sunavala-Dossabhoy, G. (2011). Adenoviral delivery of Tousled kinase for the protection of salivary glands against ionizing radiation damage. Gene Ther 18, 275-282.
Pan, F., and Gan, W.B. (2008). Two-photon imaging of dendritic spine development in the mouse cortex. Dev Neurobiol 68, 771-778.
Passineau, M.J., Zourelias, L., Machen, L., Edwards, P.C., and Benza, R.L. (2010). Ultrasound-assisted non-viral gene transfer to the salivary glands. Gene Ther 17, 1318-1324.
Perez, P., Adriaansen, J., Goldsmith, C.M., Zheng, C., and Baum, B.J. (2011). Transgenic alpha-1-antitrypsin secreted into the bloodstream from salivary glands is biologically active. Oral Dis 17, 476-483.
Peter, B., Van Waarde, M.A., Vissink, A., s-Gravenmade, E.J., and Konings, A.W. (1995). Degranulation of rat salivary glands following treatment with receptor-selective agonists. Clin Exp Pharmacol Physiol 22, 330-336.
Radosevich, A.J., Bouchard, M.B., Burgess, S.A., Chen, B.R., and Hillman, E.M. (2008). Hyperspectral in vivo two-photon microscopy of intrinsic contrast. Opt Lett 33, 2164-2166.
Riedl, J., Crevenna, A.H., Kessenbrock, K., Yu, J.H., Neukirchen, D., Bista, M., Bradke, F., Jenne, D., Holak, T.A., Werb, Z., et al. (2008). Lifeact: a versatile marker to visualize F-actin. Nat Methods 5, 605-607.
Riedl, J., Flynn, K.C., Raducanu, A., Gartner, F., Beck, G., Bosl, M., Bradke, F., Massberg, S., Aszodi, A., Sixt, M., et al. (2010). Lifeact mice for studying F-actin dynamics. Nat Methods 7, 168-169.
Roberts, M.S., Roberts, M.J., Robertson, T.A., Sanchez, W., Thorling, C., Zou, Y., Zhao, X., Becker, W., and Zvyagin, A.V. (2008). In vitro and in vivo imaging of xenobiotic transport in human skin and in the rat liver. J Biophotonics 1, 478-493.
Sakai, T., Kawaguchi, M., and Kosuge, Y. (2009). siRNA-mediated gene silencing in the salivary gland using in vivo microbubble-enhanced sonoporation. Oral Dis 15, 505-511.
Samuni, Y., Zheng, C., Cawley, N.X., Cotrim, A.P., Loh, Y.P., and Baum, B.J. (2008). Sorting of growth hormone-erythropoietin fusion proteins in rat salivary glands. Biochem Biophys Res Commun 373, 136-139.

Sandoval, R.M., Kennedy, M.D., Low, P.S., and Molitoris, B.A. (2004). Uptake and trafficking of fluorescent conjugates of folic acid in intact kidney determined using intravital two-photon microscopy. Am J Physiol Cell Physiol *287*, C517-526.

Sandoval, R.M., and Molitoris, B.A. (2008). Quantifying endocytosis in vivo using intravital two-photon microscopy. Methods Mol Biol *440*, 389-402.

Schenke-Layland, K., Xie, J., Angelis, E., Starcher, B., Wu, K., Riemann, I., MacLellan, W.R., and Hamm-Alvarez, S.F. (2008). Increased degradation of extracellular matrix structures of lacrimal glands implicated in the pathogenesis of Sjogren's syndrome. Matrix Biol *27*, 53-66.

Segawa, A., Loffredo, F., Puxeddu, R., Yamashina, S., Testa Riva, F., and Riva, A. (1998). Exocytosis in human salivary glands visualized by high-resolution scanning electron microscopy. Cell Tissue Res *291*, 325-336.

Segawa, A., and Riva, A. (1996). Dynamics of salivary secretion studied by confocal laser and scanning electron microscopy. Eur J Morphol *34*, 215-219.

Segawa, A., Terakawa, S., Yamashina, S., and Hopkins, C.R. (1991). Exocytosis in living salivary glands: direct visualization by video-enhanced microscopy and confocal laser microscopy. Eur J Cell Biol *54*, 322-330.

Serrano-Pozo, A., Mielke, M.L., Gomez-Isla, T., Betensky, R.A., Growdon, J.H., Frosch, M.P., and Hyman, B.T. (2011). Reactive Glia not only Associates with Plaques but also Parallels Tangles in Alzheimer's Disease. Am J Pathol *179*, 1373-1384.

Sokac, A.M., and Bement, W.M. (2006). Kiss-and-coat and compartment mixing: coupling exocytosis to signal generation and local actin assembly. Mol Biol Cell *17*, 1495-1502.

Sramkova, M., Masedunskas, A., Parente, L., Molinolo, A., and Weigert, R. (2009). Expression of plasmid DNA in the salivary gland epithelium: novel approaches to study dynamic cellular processes in live animals. Am J Physiol Cell Physiol *297*, C1347-1357.

Svoboda, K., and Yasuda, R. (2006). Principles of two-photon excitation microscopy and its applications to neuroscience. Neuron *50*, 823-839.

Tada, H., Higuchi, H., Wanatabe, T.M., and Ohuchi, N. (2007). In vivo real-time tracking of single quantum dots conjugated with monoclonal anti-HER2 antibody in tumors of mice. Cancer Res *67*, 1138-1144.

Textor, J., Peixoto, A., Henrickson, S.E., Sinn, M., von Andrian, U.H., and Westermann, J. (2011). Defining the quantitative limits of intravital two-photon lymphocyte tracking. Proc Natl Acad Sci U S A *108*, 12401-12406.

Theer, P., Hasan, M.T., and Denk, W. (2003). Two-photon imaging to a depth of 1000 microm in living brains by use of a Ti:Al2O3 regenerative amplifier. Opt Lett *28*, 1022-1024.

Toma, I., Kang, J.J., and Peti-Peterdi, J. (2006). Imaging renin content and release in the living kidney. Nephron Physiol *103*, p71-74.

Voutetakis, A., Kok, M.R., Zheng, C., Bossis, I., Wang, J., Cotrim, A.P., Marracino, N., Goldsmith, C.M., Chiorini, J.A., Loh, Y.P., *et al.* (2004). Reengineered salivary glands are stable endogenous bioreactors for systemic gene therapeutics. Proc Natl Acad Sci U S A *101*, 3053-3058.

Wang, S., Baum, B.J., Yamano, S., Mankani, M.H., Sun, D., Jonsson, M., Davis, C., Graham, F.L., Gauldie, J., and Atkinson, J.C. (2000). Adenoviral-mediated gene transfer to mouse salivary glands. J Dent Res 79, 701-708.

Warner, J.D., Peters, C.G., Saunders, R., Won, J.H., Betzenhauser, M.J., Gunning, W.T., 3rd, Yule, D.I., and Giovannucci, D.R. (2008). Visualizing form and function in organotypic slices of the adult mouse parotid gland. Am J Physiol Gastrointest Liver Physiol 295, G629-640.

Weigert, R., Sramkova, M., Parente, L., Amornphimoltham, P., and Masedunskas, A. (2010). Intravital microscopy: a novel tool to study cell biology in living animals. Histochem Cell Biol 133, 481-491.

Xu, R., Boudreau, A., and Bissell, M.J. (2009). Tissue architecture and function: dynamic reciprocity via extra- and intra-cellular matrices. Cancer Metastasis Rev 28, 167-176.

Zhang, S., and Murphy, T.H. (2007). Imaging the impact of cortical microcirculation on synaptic structure and sensory-evoked hemodynamic responses in vivo. PLoS Biol 5, e119.

Zheng, C., Cotrim, A.P., Sunshine, A.N., Sugito, T., Liu, L., Sowers, A., Mitchell, J.B., and Baum, B.J. (2009). Prevention of radiation-induced oral mucositis after adenoviral vector-mediated transfer of the keratinocyte growth factor cDNA to mouse submandibular glands. Clin Cancer Res 15, 4641-4648.

Zheng, C., Shinomiya, T., Goldsmith, C.M., Di Pasquale, G., and Baum, B.J. (2011). Convenient and reproducible in vivo gene transfer to mouse parotid glands. Oral Dis 17, 77-82.

Zhong, Z., Ramshesh, V.K., Rehman, H., Currin, R.T., Sridharan, V., Theruvath, T.P., Kim, I., Wright, G.L., and Lemasters, J.J. (2008). Activation of the oxygen-sensing signal cascade prevents mitochondrial injury after mouse liver ischemia-reperfusion. Am J Physiol Gastrointest Liver Physiol 295, G823-832.

Zipfel, W.R., Williams, R.M., Christie, R., Nikitin, A.Y., Hyman, B.T., and Webb, W.W. (2003a). Live tissue intrinsic emission microscopy using multiphoton-excited native fluorescence and second harmonic generation. Proc Natl Acad Sci U S A 100, 7075-7080.

Zipfel, W.R., Williams, R.M., and Webb, W.W. (2003b). Nonlinear magic: multiphoton microscopy in the biosciences. Nat Biotechnol 21, 1369-1377.

Zoumi, A., Yeh, A., and Tromberg, B.J. (2002). Imaging cells and extracellular matrix in vivo by using second-harmonic generation and two-photon excited fluorescence. Proc Natl Acad Sci U S A 99, 11014-11019.

15

Regeneration and Recycling of Supports for Biological Macromolecules Purification

Marcello Tagliavia[1] and Aldo Nicosia[2]
[1]BioNat Italia S.r.l., Palermo,
[2]IAMC-CNR, U.O.S. Capo Granitola, Torretta Granitola,
Italy

1. Introduction

Great evolution and improvement in biological molecules purification have been achieved in the last 20 years, giving advantages in both product quality and yield, and speed of purification methods. Purity levels, once poorly suitable for large scale production, as well as requiring long and very expensive procedures, are today achievable using simple procedures. For example, till few decades ago, the only way to obtain ultrapure nucleic acids was the ultracentrifugation on cesium chloride gradient; today, the same -if not higher- purity is easily achieved using solid-phase anion-exchange separations. The purification of recombinant proteins has also been dramatically improved by using more precise affinity techniques.

However, despite their routine use, these procedures are often quite expensive, making it very convenient the possibility of using the same purification devices several times, instead of wasting them after one use only.

The possibility of recycling purification matrices has to be considered desiderable and convenient not only in the laboratory research field, where most purifications are anyway performed on small-medium scale, but also in large scale production. Several attempts to reuse purification systems have been made in the last years, showing how many critical points have to be considered.

In fact, most of the previous procedures tested, especially on DNA purification columns, failed to fully decontaminate them (Chang *et al.*,1999; Fogel and McNally, 2000; Kim *et al.*, 2000), resulting in a substantial carry-over contamination, because of the remaining of substantial amounts of material into the matrix after elution (Esser *et al.*, 2005), so that the main challenge in every regeneration procedure is the complete removal of any detectable trace of the previously purified molecules, to avoid the presence of contaminating molecules in downstream applications. Moreover, the use of very sensitive analysis systems (like PCR) reduced dramatically the threshold of acceptable contamination levels. Only in recent years reliable decontamination methods have been proposed, as it will be discussed in the chapter.

Regeneration procedures of columns used in nucleic acids purification are based on nucleic acids hydrolysis, and take advantage of DNA and RNA chemical properties, so that different protocols may be needed for DNA and RNA efficient removal, depending on the decontamination procedure used. To this aim, the proper knowledge of the basis of binding and elution (including the incomplete release of sample molecules), chemical properties and tolerance of both binding matrices and biomolecules to different reagents is needed, to ensure the proper management and improvement of decontamination procedures without impairing the matrix performances.

1.1 DNA columns

DNA purification strategies using columns take advantage of the chemical nature of the molecule, an highly negatively charged polyanion. Silica, glassfiber and silica-based anion exchange supports are among the most used DNA purification systems, ensuring rapid procedures with good yields and quality without any organic extraction. Such columns are commercially available for the purification of either small molecules (PCR fragments, plasmids, etc.) and genomic DNA, suitable for most applications. Anion-exchange matrices are usually based on polysaccharidic or mineral (silica) derivatized supports, where the active chemical group on resin surface is usually the DEAE (DiEthyl-AminoEthyl). The density of DEAE groups, much higher in silica based matrices (as found in Qiagen resins, Fig. 1A), seems to strongly affect not only the column capacity, but also the binding properties and the selective release of different molecules, each in specific conditions.

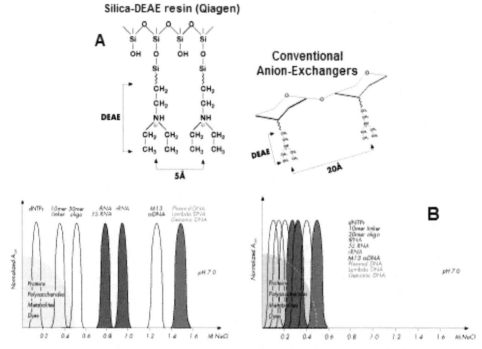

Fig. 1. Structure of anion-exchange DEAE resins (A) and elution profiles (B) (from Qiagen, Purification Technologies)

Such devices have a very broad range of separation, and allow the highly selective binding of nucleic acids, due to the negatively charged phosphate backbone, to the positively charged DEAE groups. Different column suppliers propose various purification conditions, using either changes of both pH and ionic strength or constant pH (usually at neutral values) and different saline concentration to achieve binding (sometimes in the presence of some amount of ethanol or isopropanol) and elution. These parameters are also tuned to sequentially purify DNA and RNA, or nucleic acid molecules of different sizes (Fig. 1B)

The DNA purification by means of silica gel, proposed first in batch and later in column, has been the most largely used method to recover DNA from both dissolved gel bands and cell lysates, giving high quality DNA without time consuming procedures.

Silica and glassfiber matrices bind DNA because of its negative charge. However, the interactions between silica and nucleic acids occur in different ways depending on chemical conditions.

Binding properties of silica surface depend on its hydratation status (silica-gel) and on the pH value.

In particular, acidic or neutral pH combined with high ionic strength (Fig. 2) allows the silica surface to be positively charged and the DNA to tightly bind silica particles. Binding occurs in the presence of high concentrations of chaotropic salts (usually guanidinium hydrochloride or thiocyanate, sometimes sodium perchlorate). These chemicals alter the hydratation status of macromolecules and facilitate silica-nucleic acid interactions. Under these conditions, DNA remains selectively bound to the matrix, while other molecules (RNA, proteins, polysaccharides and other biological molecules) flow through. Such interaction is still strong under low ionic strength in the presence of high alcohol concentrations, so that this condition is used to perform one or more washes, allowing any trace of salts or soluble contaminants to flush out. Finally, DNA is recovered by elution in pure water or very low salt buffer (frequently Tris-HCl 10 mM, pH 7.5-8.5). In most columns, the maximum DNA release occurs above pH 8.

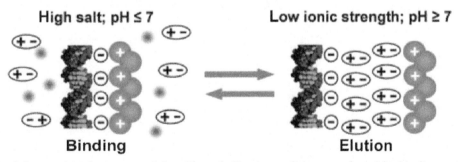

Fig. 2. Interactions between nucleic acids and silica in conditions employed for binding and elution (from Esser et al., 2006)

1.2 RNA columns

Devices for fast total RNA purification are based on the same principles described above. In these columns, a guanidinium salt is always used to protect RNA from degradation, besides promoting its binding to the resin.

Columns can be employed which allow the purification of DNA or RNA, or both (silica-DEAE resins in particular), simply using different buffers. Some silica-based columns, like silica gel itself, show an enhanced RNA binding capacity when the chaotropic, high salt binding buffer is supplemented with ethanol and/or 2-propanol; in these conditions, DNA binds the silica with low efficiency. Due to the different supports and/or binding and elution conditions, molecules shorter than 200 nucleotides usually fail to bind or are lost during washes, although recently columns allowing the recovery of the whole RNA population, including molecules few tens of nucleotides long, became commercially available.

Affinity columns are largely used to purify the poly-A^+ fraction of eukaryotic mRNA. Such purifications can be performed either in batch or in columns, although the mechanism is the same. The resin employed consists of polymeric, hydrophilic beads (often agarose) whose surface is coated of covalently linked oligo-dT. The interaction with target RNAs occur because of the complementarity between the mRNA poly-A tail and the resin-linked oligo-dT, leaving other molecules unbound. The following elution allows the recovery of the poly-A^+ fraction (Fig. 3).

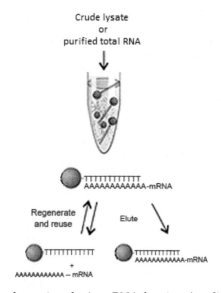

Fig. 3. Purification of the eukaryotic poly-A^+ mRNA fraction. (modified from Invitrogen)

All the systems described above are largely used in most molecular biology laboratories and in many biotechnological companies for both research applications and large scale productions, as they offer the possibility to get large quantities of high quality products using short and simple procedures.

2. Limits

The major disadvantage of the purification systems and devices described above is the cost, as they can only be used once because of the substantial amount of DNA which remains into the matrix after elution.

In fact, the nucleic acids recovery has been estimated to be not more than 90-95% of the input. The remaining part is lost during purification (because of binding failure or leakage during washes) or remains inside the matrix. The incomplete elution may have various explanations:

1. some molecules are not released from the resin;
2. a small volume of eluent is always retained by the resin, leaving free molecules inside;
3. some molecules might be included into unsolubilized protein particles;
4. some molecules might be associated with cellular fragments, especially of bacterial origin.

Whatever the cause, after the elution (even if sequential rounds are performed) the resin contains relatively large amounts of sample (Fig. 4), making it impossible to reuse the same matrix for further purifications, especially of different samples, because of the high risk of cross-contamination and reduction in the column binding capacity due to trapped particles (especially when a crude lysate had passed trough the column).

All these conditions require particular attention for any attempt of recycling purification columns, so that only methods whose reliability has been proven should be employed.

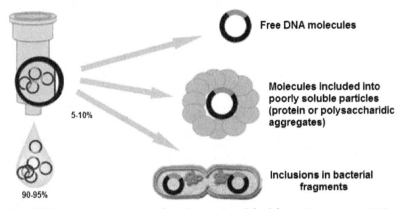

Fig. 4. Residual DNA into the column after elution (modified from Esser *et al.*, 2006).

3. Regeneration

Whereas working with a large number of DNA samples (or large volumes) could represent a problem because of the columns or, in general, purification matrices cost, the possibility of recycling them becomes attractive. The main challenge in every regeneration procedure is the complete elimination of any detectable DNA trace.

In the last years several attempts have been made to set fast and safe procedures ensuring a true complete decontamination without impairing the resin binding properties.

Thus, chemical procedures are needed that could ensure the complete hydrolysis or functional/chemical inactivation of nucleic acids. Moreover, the treatment should be able to remove, at least partially, particles (mainly composed of proteins) trapped into the matrices.

The methods proposed earlier did not avoid carry-over contamination (Chang *et al.*, 1999; Fogel and McNally, 2000; Kim *et al.*, 2000).

The first reliable method has been proposed (Esser *et al.*, 2005) and became commercially available as a kit (patented in USA in 2009).

It is based on the hydrolysis (single and double strand breaks) of nucleic acids in the presence of ferric salts in a reducing, buffered acidic environment. Other biomolecules, including lipids and proteins are damaged by the treatment, too. This procedure is claimed to be effective within minutes or hours, ensuring the complete decontamination (assessed by PCR assay) of columns used to purify plasmids or PCR products. Moreover, the composition of the two active solutions included in the kit is claimed to be safe and not hazardous. In fact, the first solution presumably consist of ferric chloride, citric acid, ascorbic acid (as reducing agent), detergents and phosphate buffer at mild acidic pH.

The ability of iron salts (and other transition elements salts, like those of copper, zinc, cobalt, etc.) to damage DNA and RNA is well known since a long time, so that Fe^{2+}-mediated DNA hydrolysis has been used in earlier studies on chromatin structure (Hertzberg and Dervan, 1982) and even today for the footprinting of DNA-protein complexes (Swapan and Tullius, 2008).

Several reactions have been hypothesized to occur, and the so called Fenton's reaction is the most widely recognized and used up today in several fields where fast and efficient oxydative degradation of organic compounds is required.

Fenton's reaction (see below) consists in the iron(II) salt-dependent decomposition of hydrogen peroxide (hypothesized to occur *via* an oxoiron(IV) intermediate), which generates the highly reactive hydroxyl radical. When a reducing agent is added, it leads to a cycle which greatly enhance the damage to biomolecules.

$$Fe^{2+} + H_2O_2 \rightarrow Fe^{3+} + OH^{\cdot} + OH^-$$

This reaction is believed to occur together with the Haber-Weiss reaction, which triggers the following cycle:

$$H_2O_2 + OH^{\cdot} \rightarrow H_2O + O_2^- + H^+$$

$$H_2O_2 + O_2^- \rightarrow O_2 + OH^- + OH^{\cdot}$$

Although the exact sequence of these reactions and the identity of the reactive species involved in the various conditions (i.e. presence or absence of chelating agents and/or reducing agents) is still controversal (for a discussion, see Barbusiński, 2009), it's widely accepted that the formation of the hydroxyl radical is determinant for the subsequent DNA damage.

Besides the reactions reported above, the presence of chelated Fe(III) instead of Fe(II), together with reducing agents, can lead to strong DNA damage resulting in multiple strand breaks. For example, it has been reported that Fe(III)-nitrilotriacetate (NTA) in the presence of either H_2O_2 (able to act both as reductant and oxidizing agent (Buettner and Jurkiewica, 1996), ascorbate (which in certain conditions can act as pro-oxidizing) or cysteine, produced DNA single and double strand breaks as a function of reductant concentration, *via* a mechanism involving the reduction of Fe(III) to Fe(II) and the formation of H_2O_2. The latter, in turn, enters in the Fenton/Haber-Weiss reactions, where the presence of a reducing agent supports the "iron redox cycle". In all these cases, H_2O_2 seems to be a common intermediate (in fact, catalase activity is able to block these events, leading to the reduction of DNA damage), while the $OH^{\cdot}$ hydroxyl radical is the reactive species which attacks DNA (Toyokuni and Sagripanti, 1992). Moreover, the auto-oxidation of ascorbate in the presence of Fe(II) ions, chelating agents and phosphate buffer, with the concurrent formation of hydroxyl radicals (OH˙) has been reported (Prabhu and Krishnamurthy, 1993). Figure 5 shows the chemical reaction resulting in DNA (or RNA) multiple breaks, although other reactions may occur simultaneously, which result in direct bases damage caused by the OH˙ radical.

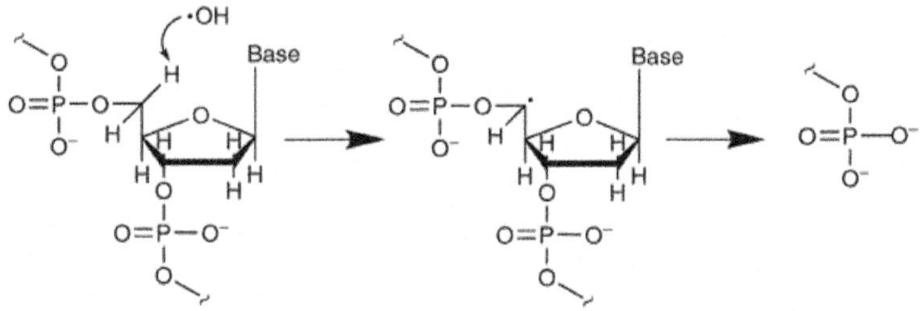

Fig. 5. DNA break by exposure to hydroxyl radicals (modified from Swapan and Tullius, 2008)

Thus, this method makes use of relatively non-toxic and environmental friendly chemicals, allowing the virtually complete removal of small nucleic acid molecules from purification columns. A disadvantage is that the solutions are guaranteed for 18 months only, because reagents undergo chemical modifications making them ineffective. The supplier of the recycling kit claims that resins can undergo about 20 regeneration cycles. Unfortunately, no data are available on the possibility of efficiently decontaminate devices used for genomic DNA purification, whose features are expected to make it more resistant to chemical treatmets, thus leading to the possibility of DNA carry-over.

The first report of a simple and efficient home-made decontamination method has been published in 2008. It achieves the nucleic acids removal by acidic hydrolysis, but its major limits were the need of very long incubation times (Siddappa et al., 2007), and the secure efficacy on low molecular weight nucleic acids only.

The principle used in that method was the DNA and RNA degradation by a treatment with strong acids. In particular, nucleic acids are known to be susceptible to acid catalyzed hydrolysis, which involves the cleavage of the N-glycosidic bond of purine nucleosides. As shown in Fig. 6, the reaction results in the formation of an AP-site (apurinic or abasic sites), which causes DNA or RNA break. In fact, the hydrolysis of the N-glycosidic bond unmasks the latent aldehyde functionality at the C1' position, rendering the 3'-phosphate group susceptible to β-elimination (1), which results in strand break. Moreover, such products are highly sensitive to further alkaline hydrolysis (Fig. 6), so that depurinate molecules can easily undergo fragmentation following exposure to bases (2). These are, for example, reactions used for the controlled, partial DNA hydrolysis in molecular biology protocols, where the short exposure to relatively low concentrations of HCl lead to DNA fragmentation.

However, when the acid concentration is high and/or the exposure time is extended, the depurination extent is so high that, after alkaline treatment, very short DNA fragments or even nucleotides are obtained.

The column regeneration method proposed by Siddappa et al. (2007) consisted in a 24 hours incubation of used columns in a HCl solution, followed by several washes. Data reported showed that no detectable nucleic acids were still present in the column, whose binding capacity were claimed to be mantained.

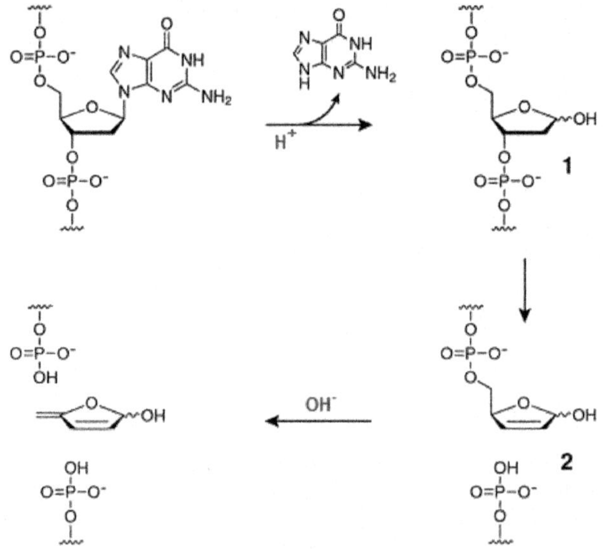

Fig. 6. DNA strand break resulting from H⁺ catalyzed depurination and subsequent β-elimination at the AP site. Hydrolysis results in release of the purinic nucleotide and formation of an AP site (1). The α- and β-hemiacetals are in equilibrium with the open chain aldehyde, which is susceptible to β-elimination that results in cleavage of the adjacent 3′ phosphoester (2). This product in turn undergoes cleavage of the 5′ phosphoester under alkaline conditions. (modified from Sheppard et al., 2000).

The lack of evidences about the efficacy of the method even on columns contaminated by genomic DNA and the time-expensiveness of the procedure prompted further tests to improve the procedure.

In fact, if silica columns are used to purify small molecules, contaminating DNA can be virtually completely eliminated by commercial kits (Esser et al., 2005) or using the procedure reported in Siddappa et al. (2007), as they make any trace of the previous sample undetectable. However, the efficacy of both methods in eliminating genomic DNA remains uncertain.

The fastest and most effective home-made procedure available up today for the decontamination of silica-based columns consist in an improvement of that described above, as it's also based on DNA depurination and hydrolysis, and addresses the main limits of previously proposed protocols.

Silica-bound DNA could be expected to be efficiently depurinated and removed by treatments with strong acids even after short exposures. However, after such a regeneration procedure small amounts of amplifiable DNA are actually still detectable.

Such failure might be hypothesized to be due to an incomplete permeation of the acidic solution into the silica matrix, where the nucleic acid might be still bound to silica or trapped because of its high molecular weight (Esser et al., 2005). Moreover, any molecules included into aggregates might be somewhat resistant to chemical treatments. All these conditions might allow variable amounts of DNA to escape the depurinating agent, resulting in residual amplifiable traces, making it necessary a very long incubation in HCl solutions.

These limitations have been overcome by the procedure described by Tagliavia et al. (2009) and reported below. It can be completed in about 45' (instead of more than 24 hours), and allows not only to regenerate silica columns contaminated by DNA of any size, but also to save time.

The method consists in sequential alkaline and acidic treatments which denature and depurinate, respectively, any DNA still present into the column (depurination rate in denatured DNA is higher than in native DNA (Lindahl and Nyberg, 1972). A further alkaline treatment hydrolyzes long depurinated DNA molecules reducing them into very small fragments (Siddappa et al., 2007). These chemical treatments are performed in the presence of a non-ionic detergent at low concentration, which seems to enhance their action. In fact, given the structure of the column resins, the detergent is supposed to allow a more even permeation of the solutions employed in the treatment, as it modifies their surface tension. Moreover the tensioactive (which is important to be non-ionic to reduce any dependence of its action on pH and ionic strength), along with the initial alkaline treatment, helps dissolving aggregates, making trapped molecules more exposed to NaOH and HCl.

The efficacy of the method has been demonstrated both by assays using radiolabeled DNA and by PCR, using columns contaminated by large amounts of either genomic DNA or short PCR products.

The protocol steps are briefly reported in box 1.

Box 1. DNA silica column regeneration protocol

1. Load the silica columns with a 1 N NaOH/0.1% Triton X-100^{TM} solution
2. Incubate 5 minutes at room temperature
3. Spin for 30 seconds
4. Load the silica columns with a 1 N HCl/0.1% Triton X-100^{TM} solution
5. Let a little amount of the solution flow through by gravity
6. If dropping tends to empty the column, put it into a tube containing the same solution
7. Incubate at room temperature for 30 minutes
8. Spin for 30 seconds
9. Load the silica columns with a 1 N NaOH/0.1% Triton X-100^{TM} solution
10. Incubate at room temperature for 5 minutes
11. Spin for 30 seconds
12. Load the column with sterile ddH$_2$O
13. Spin 30 seconds

Note: it's essential to treat not only the resin, but any surface even potentially contaminated by DNA.

The use of the regeneration systems described above is safer on silica-based columns, but not on those supports consisting of polysaccharidic compounds, as they might be hydrolysed or their structure impaired by chemicals employed. Alternative methods, some of which based on radical-driven nucleic acids degradation different from that described above, are under investigation.

Regeneration methods, besides their first application in reusing purification supports, might become of wider use, even for the pre-treatment of new columns before first use. There are many commercially available kits that rely on DNA binding columns to extract and purify DNA from tissues or cultured cells, and a recent paper (Erlwein et al., 2011) reported that, in independent tests, some DNA purification columns from different kits were contaminated with DNA of diverse provenance, including human and murine DNA. Although further investigations are needed, the need of a preliminary columns decontamination step should be considered, at least for particular experiments or analyses have to be carried out.

3.1 RNA columns

Total RNA purification is carried out using columns working exactly like those employed in DNA purification. As discussed earlier, different conditions used for binding and/or elution allow the selective recovery of RNA only.

The same problems described for DNA columns occur in RNA columns, too. However, RNA is well known to be very sensitive to a variety of conditions and chemicals, but treatments are needed that ensure not only the complete degradation of any residual RNA, but also the maintainance of the columns RNase-free state.

The commercial system based on the earlier discussed iron-mediated degradation is effective, but a home-made, simple and inexpensive method is available (Nicosia et al., 2010).

In fact, the methods described in two previous reports (Siddappa et al., 2008; Tagliavia et al., 2009) are time expensive or include steps not required for RNA hydrolysis, so that a faster and more efficient protocol has been set up. It is based on the RNA high sensitivity to alkali, omitting acidic treatments. Indeed, the exposure of RNA to high pH is able to completely hydrolyse RNA, since it is directly cleavable by the OH- due to the presence of the 2'-OH group in the molecule (Fig. 7).

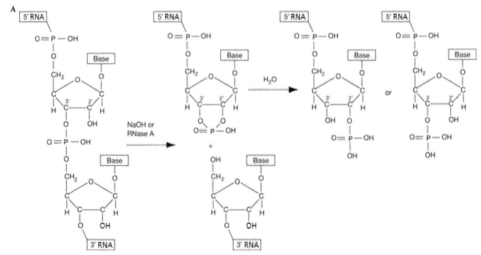

Fig. 7. Alkali catalyzed RNA hydrolysis. The 2'-OH group, present in RNA only, makes it OH- sensitive. Besides a 5'-OH end, a cyclic 2',3'-P intermediate is released, which in turn produces a 3'-P or 2'-P end (modified from Vengrova and Dalgaard, 2005).

Thus, a strong base like NaOH is employed in the presence of low concentrations of a non-ionic surfactant, whose role has been earlier discussed. Treatments are performed using prewarmed solutions, so as to allow the reduction of both alkali concentration and exposure time. Indeed, it should be remembered that silica does not tolerate high alkali concentrations, as it forms silicates, resulting in matrix destruction and loss of binding properties. This is the reason why time of exposure to NaOH, its concentration and the temperature, as described in Nicosia et al. (2010), are crucial for the successful decontamination without impairing the columns integrity and efficieny, making it possible to reuse them several times.

The regeneration protocol is briefly reported below.

Box 2. RNA silica column regeneration protocol

1. Fill the silica column with a prewarmed (75°C) solution containing 0.2 N NaOH and 0.1% (v/v) Triton X-100™
2. Incubate 5 minutes at room temperature
3. Spin for 1 minute
4. Repeat step 1, incubate for 10'
5. Spin for 1 minute
6. Add a volume of a 50 mM sodium acetate/acetic acid buffer (pH 4) solution
7. Incubate at room temperature for 1 minute
8. Spin for 1 minute
9. Load the column with sterile ddH$_2$O
10. Spin for 1 minute

A different strategy is used to purify the poly-A$^+$ fraction of eukaryotic mRNAs, aiming to exclude the most abundant RNA classes like rRNAs, where oligo-dT covalently linked on the surface of polysaccharidic beads or similar solid supports are employed, as described earlier.

Many suppliers indicate, in instruction of such kits, that oligo-dT supports may be reused, and provide regeneration protocols always based on RNA hydrolysis by NaOH treatments, which will destroy any RNA traces, leaving unmodified the DNA component (oligo-dT).

4. Protein purification resins

4.1 IMAC

The use of immobilized-metal affinity chromatography (IMAC) for protein purification was firstly described and showed by Porath et al. (1975). Initially developed for purification of native proteins with an intrinsic affinity to metal ions, IMAC shows numerous application fields spanning from chromatographic purification of metallo and phosphorylated proteins, antibodies and recombinant His-tagged proteins. IMAC is also used in proteomics approaches where fractions of the cellular protein pool are enriched and analyzed differentially (phosphoproteome and metalloproteome).

IMAC is a chromatography method that can simply be scaled up linearly from milliliter to liter volumes (Block et al., 2008; Hochuli et al., 1988; Kaslow and Shiloach, 1994; Schäfer et al., 2000) and Ni-NTA Superflow columns are in use for biopharmaceutical production processes.

It is based on the known affinity of transition metal ions such as Zn^{2+}, Cu^{2+}, Ni^{2+}, and Co^{2+} to certain amino acid in aqueous solutions (Hearon, 1948). Amino acids as histidine, cysteine, tryptophan, tyrosine, or phenylalanine, working as electron donors on the surface of proteins, are able to reversibly bind transition metal ions that have been immobilized by a chelating group covalently bound to a solid support. Histidine represents the preferential choice in protein purification using IMAC since it binds selectively immobilized metal ions even in presence of free metal ions excess (Hutchens and Yip, 1990b); additionally, copper and nickel ions have the greatest affinity for histidine.

Great improvement in development of IMAC chromatographic procedures was achieved by the introduction of DNA engineering techniques allowing the construction of fusion proteins in which specific affinity tags as 6xHis tag are added to the N-terminal or C-terminal protein sequence; the use of these strategies simplifies purification of the recombinant fusion proteins (Hochuli et al., 1988). Moreover the identification or invention of chelating agent able to be both covalently bound to a support and interact with transitional metal ions contributed to the definition of IMAC for high-quality protein purification.

The chelating group that has been first used for IMAC proteins purification is iminodiacetic acid (IDA) (Porath *et al.*, 1975). IDA was charged with metal ions such as Zn^{2+}, Cu^{2+}, or Ni^{2+}, and then used to purify a variety of different proteins and peptides (Sulkowski, 1985)

The tridentate IDA group binds to three sites within the coordination sphere of divalent metal ions such as copper, nickel, zinc, and cobalt (Fig. 8). When copper ions (coordination number of 4) are bound to IDA, only one site remains available for interaction with proteins (Hochuli *et al.*, 1987). For nickel ions (coordination number of 6) bound to IDA, three valencies are available for imidazole ring interaction while it is unclear whether the third is sterically able to participate in the interaction binding to proteins. Thus Cu^{2+}-IDA complexes are stable on the column but have lower capacity for protein binding. Conversely, Ni^{2+}-IDA complexes bind proteins more avidly, but Ni^{2+}-protein complexes are more likely to dissociate from the solid support.

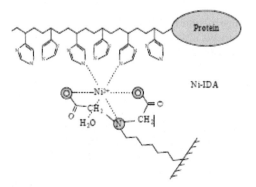

Fig. 8. Model of the interaction between residues in the His tag and the metal ion in tridentate (IDA) IMAC ligand.

The development of a new metal-chelating adsorbent, nitrilotriacetic acid (NTA), has provided a convenient and inexpensive tool for purification of proteins containing histidine residues (Hochuli *et al.*, 1987). The NTA chelating agent coordinates Ni^{2+} with four valencies (tetradentate, coordination number 4) leaving two valencies available for binding to electron donor groups (i.e., histidine) on the surface of proteins (Fig. 9).

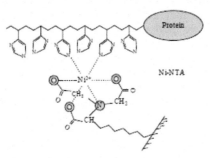

Fig. 9. Model of the interaction between residues in the His tag and the metal ion in tetradentate (NTA) IMAC ligand.

The coordination number plays an important role regarding to the quality of the purified protein fraction but not in protein yield. IDA has only 3 metal-chelating sites and cannot tightly bind metal ions, a relative weak binding leads to ion leaching after loading with strongly chelating proteins or during washing steps. This results in impure products, and metal-ion contamination of isolated proteins; meanwhile protein recovery is usually similar between the two chelating agent. Thus the advantage of NTA over IDA is that the divalent ion is bound by four rather than three of its coordination sites. This minimizes leaching of the metal from the solid support and allows for more stringent purification conditions (Hochuli, 1989).

The NTA also binds Cu^{2+} ions with high affinity, but this occupies all of the coordination sites, rendering the resulting complex ineffective for IMAC. Another tetradentate ligand is a chelating agent commercially known as Talon resin, consisting in carboxymethyl aspartate (CM-Asp), available as cobalt-charged (Chaga et al., 1999).

The lowest metal leaching is obtained using N,N,N'-tris(carboxymethyl)ethylenediamine (TED), a pentadentate ligand (Fig.10). Because TED coordinates ions extremely tightly, such chelators represent a valid alternative expecially if low metal ion contamination is needed; nevertheless only one coordination site is avaiable for His tag binding and protein recovery is substantially lower than IDA or NTA

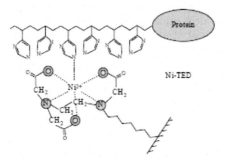

Fig. 10. Model of the interaction between residues in the His tag and the metal ion in pentadentate (TED) IMAC ligand.

The choice of the metal ion immobilized on the IMAC ligand depends on the application. Whereas trivalent cations such as Al^{3+}, Ga^{3+}, and Fe^{3+}(Andersson and Porath, 1986; Muszynska et al., 1986; Posewitz and Tempst, 1999) or tetravalent Zr^{4+} usually immobilized to IDA (Zhou et al., 2006) are preferred for phosphoproteins and phosphopeptides capturing, divalent Cu^{2+}, Ni^{2+}, Zn^{2+}, and Co^{2+} ions are preferentially used for purification of His-tagged proteins. Combinations of a tetradentate ligand that ensure strong immobilization, and a metal ion that leaves two coordination sites available free for imidazole interaction (Ni^{2+} and Co^{2+}) allow similar recovery yield and eluted proteins quality. Immobilized copper or nickel ions bind native proteins with a K_d of 1×10^{-5} M and 1.7×10^{-4} M, respectively (Hutchens and Yip, 1990a). The K_d value is reduced for protein produced, using recombinant DNA technology, as chimeric constructs with an epitope containing six or more histidine residues. Addition of six histidines to the protein results only in 0.84 kDa protein mass excess whereas other fusion protein systems utilize much larger affinity groups that must be often removed to allow normal protein function (e.g., glutathione-S-transferase, protein A, Maltose Binding Protein). Furthermore the lack of His-tag immunogenic activity allows injection into animals for antibody production without tag removal. Addition of a His-tag results in an enhanced affinity for Ni^{2+}-NTA complex binding due to K_d value of 10^{-13} M at pH 8.0 even in the presence of detergent, ethanol, 2 M KCl (Hoffmann and Roeder, 1991), 6 M guanidine hydrochloride (Hochuli et al., 1988), or 8 M urea (Stüber et al., 1990) allowing protein purification under both native and denaturing conditions, as well as both oxidizing and reducing conditions providing a stringent environment avoiding host strain proteins co-purification (Jungbauer et al., 2004). Nevertheless proteins intrinsically expressing chelating amino acids, such as histidine on their surface, are able to interact with an IMAC support and, although usually with lower affinity than a His-tagged protein, co-purify. In E. coli, proteins observed to copurify with His-tagged target proteins, especially in native conditions, can be classified into four groups (Bolanos-Garcia and Davies, 2006):

1. proteins with natural metal-binding motifs,
2. proteins displaying histidine clusters or stretches on their surfaces,
3. proteins interacting directly or not with heterologously expressed His-tagged proteins,
4. proteins showing affinity to IMAC support such agarose or sepharose based supports.

Furthermore, some copurifying proteins seem to have a binding preference for Co^{2+} over Ni^{2+} (or other ions) and others vice versa. Several options have been developed in order to reduce the contaminating amount of copurified quote or avoiding their adsorption to the matrix, including additional purification steps, adjusting the His-tagged protein to resin ratio, to using an engineered host strain that does not express certain proteins, using an alternative support, tag cleavage followed by reverse chromatography and reduction of non specific binding by including imidazole in the lysis and washing buffer.

Since there is an higher potential of binding background contaminants under native conditions than under denaturing conditions, low concentrations of imidazole in lysis and wash buffers (10-20 mM) could be used. The imidazole ring is part of histidine structure and it's responsible for Ni-NTA interaction (Fig. 11).

At low imidazole concentrations, non specific binding is prevented, while 6xHis-tagged proteins, because of the K_d value derived, still bind strongly to the Ni-NTA matrix allowing greater purity in fewer steps.

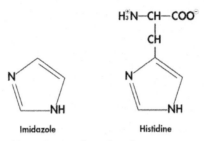

Fig. 11. Chemical structures of histidine and imidazole.

Binding of tagged proteins to Ni-NTA resin is not conformation-dependent and is relatively not affected, within a certain concentration range, by most detergents and denaturants, so Triton X-100 and Tween 20 (up to 2%), or high salt concentrations (up to 2 M NaCl) can be used, resulting in nonspecific binding reduction without affecting specific interaction.

As previously described, purification of tagged proteins under native conditions is often associated with copurification of coupled proteins such as enzyme subunits and binding proteins present in the expressing host (Le Grice and Grueninger-Leitch, 1990; Flachmann and Kühlbrandt 1996). Purification in denaturing condition is performed in presence of strong chaotropic agents such as 6 M GuHCl or 8 M Urea. Under these conditions the 6xHis tag on the protein surface is fully exposed so that binding to the Ni-NTA matrix will improve, and the efficiency of the purification procedure will be maximized by reducing the potential of non specific binding. The histidine tail binds to the Ni^{2+}-NTA resin via the imidazole ring of the histidine residues. At pH ≥7.0, the imidazole side chain is deprotonated, leading to a net negative charge interacting with Ni^{2+}-NTA; at pH 5.97 (corresponding to imidazole pK_a), 50% of the histidines are protonated; finally, within pH values ≤4.5, almost all of the histidines are protonated and unable to interact with Ni^{2+}-NTA. Thus, there are, generally, three different methods for His-tagged proteins recovery after washing steps based on chemical and cinetical counterpart features that can be used for both native or denaturing purifications.

A "competition derived approach" based on Ni^{2+}-NTA affinity for imidazole, working as competitor, increasing imidazole concentrations results in protein displacement from the support at constant pH. Under these conditions the 6xHis-tagged protein can no longer bind to the nickel ions and will dissociate from the Ni-NTA resin.

An alternative procedure uses buffers of decreasing pH to elute the histidine tail ensuring efficient recovery from Ni^{2+}-NTA (Hochuli et al., 1988). Disadvantages are that the pH must be maintained accurately at all temperatures and that some proteins may not be able to withstand the extreme pH change required for protein elution.

An optional method is based on the stripping ability of certain reagents such as EDTA or EGTA in chelation of nickel ions and their removal from the NTA groups. This results in the 6xHis-tagged protein elution as a protein–metal complex. NTA resins, so stripped, appear white in color because they have lost their nickel ions and must be recharged if additional purification steps have to been performed.

Whereas all elution methods (imidazole, pH, and EDTA) are equally effective, imidazole is recommended under native conditions, when the protein would be damaged by a pH reduction or when the presence of metal ions in the eluate needs to be avoided

4.2 Cleaning and regeneration of Ni-NTA resins

The suitability of IMAC for industrial production purposes has been largely demonstrated and it can be expected that IMAC-based procedures will acquire increasing application because of its robustness and relatively low requirements for individual optimization. In contrast to these facilities it's noteworthy the production of a large amount of discarted materials consisting in metal-chelating groups, IMAC supports such as agarose and sepharose ones and, above all, considerable metal transition amounts to be disposed.

In order to reduce the environmental impact of such wastes, several IMAC commercially manufacturers have introduced and developed protocols allowing to reuse the same resin after regeneration and equilibration step cycles. Regeneration methods, enabling the flush out of any contaminating materials from previously purified samples, can be divided into 2 different classes:

1. CIP (cleaning-in-place) protocols;
2. Stripping and recharging.

A simple and effective cleaning procedure for Ni-NTA resins used to purify proteins from different samples is represented by the incubation of such resins with a non-flammable, bacteriostat 0.5M NaOH solution for 30 min in 15 column volumes (Schäfer *et al.*, 2000) allowing denaturation and desorption of unspecifically resin-attached proteins.

Box 3. Ni-NTA agarose regeneration protocol

1. Wash the column with 2 volumes of Regeneration Buffer (6 M GuHCl, 0.2 M acetic acid).
2. Wash the column with 5 volumes of H_2O.
3. Wash the column with 3 volumes of 2% SDS.
4. Wash the column with 1 volume of 25% EtOH.
5. Wash the column with 1 volume of 50% EtOH.
6. Wash the column with 1 volume of 75% EtOH.
7. Wash the column with 5 volumes of 100% EtOH.
8. Wash the column with 1 volume of 75% EtOH.
9. Wash the column with 1 volume of 50% EtOH.
10. Wash the column with 1 volume of 25% EtOH.
11. Wash the column with 1 volume of H_2O.
12. Wash the column with 5 volumes of 100 mM EDTA, pH 8.0.
13. Wash the column with H_2O.
14. Recharge the column with 2 volumes of 100 mM $NiSO_4$.
15. Wash the column with 2 volumes of H_2O.
16. Wash the column with 2 volumes of Regeneration Buffer.
17. Equilibrate with 2 volumes of a suitable buffer

Resins stored for long terms in up to 1 M NaOH do not show any significant effect on metal-leaching rates corresponding to 1 ppm under any conditions without compromising its performance.

For repeated reuse of a Ni-NTA column, the CIP procedures had to be followed by a reequilibration step. Furthermore for long-term storage, resin may be kept in 30% (v/v) ethanol to inhibit microbial growth. No significant changes of metal-ion leaching were observed during five CIP runs, moreover the binding capacities for 6xHis-tagged protein of Ni-NTA resins remained unchanged from run 1 to run 5 (Schäfer et al., 2000).

Due to the high chelating strength and the resulting low metal-leaching rate of all Ni-NTA IMAC resins, stripping is not required even after repeated reuse or long-term storage. However, reduction in binding capacity or resin damages for example, by repeated purification of samples containing chelating agents, could happens. In this cases Ni-NTA may be stripped and recharged with nickel or a different metal ion using combination of chelating steps (EDTA treatments) ensuring a Ni^{2+} free medium, followed by nickel salts incubation. Metal chloride and sulfate salts, (e.g. 0.1 M $NiSO_4$) are commonly used. Here we report (box 3)a stripping and recharging protocol based on Qiagen instruction for relative Ni-NTA agarose resins

4.3 IMAC for industrial-scale protein production and Ni^{2+} environmental impact

IMAC for production of proteins in industrial scale, has not been used until quite recently due to worries regarding allergenic effects of nickel leaching from an IMAC matrix. During protein purification 1ml or resins is usually used for each 30-40 mg recombinant proteins. Several data describing nickel leaching from resins show that nickel concentrations in the peak elution fractions is below 1 ppm under all conditions, including denaturant or native conditions. More specifically even after several purification steps followed by CIP, the level of nickel contamination in the peak elution fractions is comprised between 0.3 and 0.6 ppm for native and denaturing conditions, respectively (Schäfer et al., 2005). The discarded cations are released as liquid or dry waste into the environment where it's just present under many forms.

Nickel, occurs naturally in the earth's crust, in various forms such as nickel sulphides and oxides, its sources arise from earth's molten core where it is trapped and unusable to volcanic eruptions, soils, ocean floors, and ocean water (Stimola, 2007).

Such divalent cation is used not only in metallurgic industries to make stainless steel but also in other application fields such as in coinage in various forms of 'costume' or 'fashion' jewellery. The different forms of nickel include elemental nickel (Ni), nickel oxide (NiO), nickel chloride ($NiCl_2$), nickel sulphate ($NiSO_4$), nickel carbonate ($NiCO_3$), nickel monosulfide (NiS), and nickel subsulfide (Ni_3S_2) (ATSDR, 2005).

Human exposure to nickel is associated with drinking water, food, or smoking tobacco containing nickel or direct contact with nickel-containing products, such as jewelry, stainless steel and coins. The average concentration of nickel in different categories of soil span from 4 to 80 ppm, but this number has increased significantly (up to 9,000 ppm) around nickel producing industries (ATSDR, 2005). Skin contact is the usual source of contamination from the ground unless for children who are more likely to ingest soil particles. Foods such as tea, coffee, chocolate, cabbage, spinach and potatoes contain high levels of nickel, making these foods a major source of exposure. The average amount of nickel introduced is 70 micrograms of nickel per day.

This rapid analysis suggests nickel concentrations typically observed in protein preparations obtained from tetradentate IMAC resins are low and content in expected daily doses of protein used such as biopharmaceutical will be far below the typical daily intake of nickel.

4.4 Amylose affinity chromatography

The expression and purification of recombinant proteins compared to native ones represent an efficient system to product any protein. As previously described for IMAC tag, recombinant DNA techniques allow the construction of fusion proteins in which specific affinity tags are added to the protein sequence of interest, facilitating the recombinant fusion proteins purification by the use of affinity chromatography methods.

Maltose-binding protein (MBP) is one of the older and more popular fusion partners used for recombinant proteins production in bacterial cells; it's coded by the *malE* gene of *Escherichia coli* as part of maltose/maltodextrin system (Nikaido, 1994). MBP, despite the molecular weight (42.5 kDa) is considered one of the best choises to solve problems related to heterologous protein expression since it acts as protein production and solubilisation enhancer by mechanisms far to be completely understood (Randall *et al.*, 1998; Nomine *et al.*, 2001; Sachdev and Chirgwin, 1998). Several commercial plasmid DNA vectors have been constructed allowing expression of a cloned protein or peptide by fusing it to MBP (Guan *et al.*, 1988; Bedouelle and Duplay, 1988; Maina *et al.*, 1988). The isolation and purification of recombinant proteins MBP fused can be performed using an easy affinity column procedure amylose based resins dependeding on MBP affinity for maltose packaged in the amylose resins (K_d value of MBP for maltose is 3.5 µM) (Kellerman and Ferenci, 1982). A crude cell extract, in absence of detergent or chaotropic agents, is prepared and passed over a column containing an agarose resin derivatized with amylose, a polysaccharide consisting of maltose subunits.

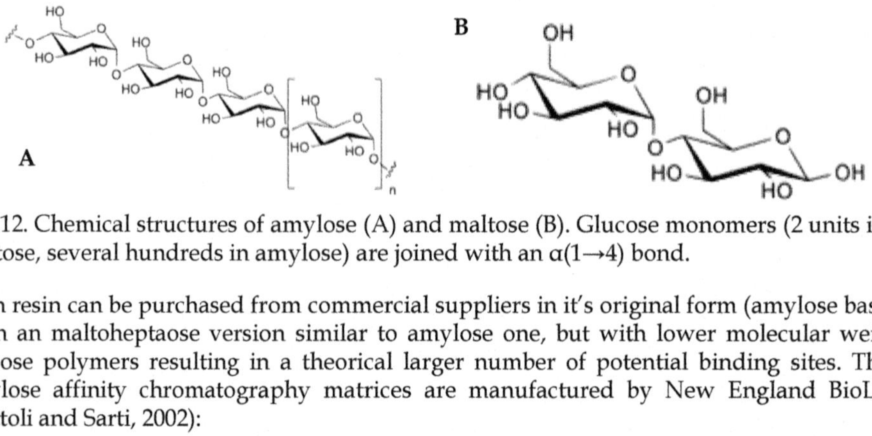

Fig. 12. Chemical structures of amylose (A) and maltose (B). Glucose monomers (2 units in maltose, several hundreds in amylose) are joined with an α(1→4) bond.

Such resin can be purchased from commercial suppliers in it's original form (amylose based) or in an maltoheptaose version similar to amylose one, but with lower molecular weight glucose polymers resulting in a theorical larger number of potential binding sites. Three amylose affinity chromatography matrices are manufactured by New England BioLabs (Cattoli and Sarti, 2002):

1. Amylose magnetic beads;
2. Amylose agarose resin;
3. High flow support matrix.

Amylose magnetic beads have a binding capacity up to 10 µg/mg (supplied as a 10 mg/ml suspension). Amylose agarose has a binding capacity of 3 mg/mL for MBP and 6 mg/ml for an MBP-β-galactosidase protein. The typical flow velocity of the amylose resin is 1 ml/min in a 2.5 cm x 10 cm column, and the matrix can withstand small manifold vacuums (universally known as "piglet"). The amylose matrix can suffer from flow restrictions. So that total protein loading should be ≤2.5 mg/ml. Amylose high flow has a binding capacity of approximately 7 mg/ml for an MBP-paramyosin protein. The exact chemical nature of the matrix is not described but has a pressure limit of 0.5 MPa (75 psi), a maximum flow velocity of 300 cm/h, and recommended velocities are below 60 cm/h being 10-25 ml/min (for Ø1.6-cm and Ø2.5-cm columns respectively).

Alternatively, home-made amylose-agarose resin can be prepared following procedures described by Lee et al. (1990). Pratically, sepharose beads are washed with water and then incubated with 1M sodium carbonate pH 11 allowing to react in presence of vinyl sulfonic acid. Activated resin is derivatized by mixing, in 1 M sodium carbonate pH 11 environment, with an amylose solution. The resulting matrix can be freshly used or in 20% ethanol stored.

In contrast with an IMAC conformation-independent binding of tagged proteins to Ni-NTA resin, MBP's affinity to amylose and maltose depends on hydrogen bonds patterns derived from the three-dimensional structure of the protein; agents interfering with hydrogen bonds or the protein structure interfere with binding as well. For these reasons protein purification of tagged proteins can be performed under native conditions only, (Tris-HCl, MOPS, HEPES, and phosphate, buffers at pH values between 6.5 and 8.5) in presence or absence of optional additives as 1 mM sodium azide, 10 mM β-mercaptoethanol or 1 mM DTT. Such reducing agents can be added to mantein reduced cysteins avoiding non specific disulphide bridges formation resulting in tedious aggregations. Moreover higher ionic strength does not adversely affect MBP binding to amylose, so that 1M NaCl can be used to reduce non specific protein binding to resin.

Despite MBP's affinity of some fusions to amylose is dramatically reduced in presence of nonionic detergents (0.2% Triton X-100 or 0.25% Tween 20) resulting in<5% binding, other fusions are unaffected. Binding is efficient in the presence of 5% ethanol or acetonitrile, as well as in 10% glycerol. 0.1% SDS completely eliminates binding.

Furthermore low levels of residual detergents, especially from regeneration solutions, (see below) can still remain; removal of detergent and mixed micelles can be achieved using dilute methanol-containing solutions

After several washing steps, protein elution and recovery is performed in a "competition derived approach" based on MBP affinity for maltose. Maltose working as competitor at 10 mM concentration, results in protein displacement from amylose at constant pH value.

Because the presence of substantial amounts of amylases in the crude extracts interferes with binding, by cutting the fusion off the column or by releasing maltose that elutes the fusion from the column, the amylose resin "half-life" depends on incubation time with trace amounts of contaminant. Manufacturers instructions and recommendations explain (e.g. NEB): "Under normal conditions defined as 15 ml of amylose agarose matrix processing, 1 liter of LB media supplemented with 0.2% glucose (producing 40 mg MBP fusion protein); a matrix binding capacity reduction of 1-3% after each purification step is reported. It is stated that such a column may be used up to 5 times before a decrease in yield is detectable (5-15% lost binding capacity), and up to 10 times to achieve an evident reduction (10-30% lost binding capacity)".

Column reuse and regeneration can be performed according to New England BioLabs following sequence of washes in water, saline buffer (20 mM Tris-HCl, 200 mM NaCl and 1 mM EDTA), and 0.1% SDS, or by a very short treatment using 0.1 N NaOH followed by a neutralization step.

Alternatively Pattenden et al. (2008) proposed a regeneration procedure based on sequential amylose resin treatments with two different regeneration solutions:

1. Regeneration 1: 50 mM HEPES, 4 M Urea, 0.5% w/v SDS pH 7.4.
2. Regeneration 2: 50 mM HEPES, 150 mM $(NH_4)_2SO_4$, 2 mM EDTA, 2 mM EGTA pH 7.4.

Regenerated resin can be stored in 20% ethanol at 4 °C

5. References

[1] Agency for Toxic Substances and Disease Registry (ATSDR), (2005). Toxicological Profile for Nickel (Update). Atlanta, GA: U.S. Department of Public Health and Human Services, Public Health Service. Available at: http:/www.atsdr.cdc.gov

[2] Andersson, L. and Porath, J. (1986). Isolation of phosphoproteins by immobilized metal (Fe^{3+}) affinity chromatography. Anal. Biochem.154:250-254.

[3] Baneyx, F., Mujacic, M. (2004).Recombinant protein folding and misfolding in *Escherichia coli*. Nat Biotechnol. Nov; 22(11):1399-408.

[4] Barbusiński, K. (2009). Fenton reaction-controversy concerning the chemistry. Ecological Chemistry and Engineering. Vol. 16 (3), pp. 347-358.

[5] Bedouelle, H. and Duplay, P. (1988). Production in *Escherichia coli* and one-step purification of bifunctional hybrid proteins which bind maltose. Eur. J. Biochem. 171:541-549.

[6] Block, H., Kubicek, J., Labahn, J., Roth, U., and Schäfer, F. (2008). Production and comprehensive quality control of recombinant human Interleukin-1b: A case study for a process development strategy. Protein Expr. Purif. 27, 244–254.

[7] Bolanos-Garcia, V. M., and Davies, O. R. (2006). Structural analysis and classification of native proteins from E. coli commonly co-purified by immobilized metal affinity chromatography. Biochim. Biophys. Acta 1760, 1304–1313.

[8] Buettner, G.R., Jurkiewica, B. A. (1996). Catalytic metals, ascorbate and free radicals: combinations to avoid. Radiat. Res. 145:532-541)

[9] Cattoli, F., and Sarti, G. C. (2002). Separation of MBP fusion proteins through affinity membranes, Biotechnol. Prog. 18, 94.

[10] Chaga, G., Hopp, J., and Nelson, P. (1999). Immobilized metal ion affinity chromatography on Co^{2+}-carboxymethylaspartate–agarose Superflow, as demonstrated by one-step purification of lactate dehydrogenase from chicken breast muscle. Biotechnol. Appl. Biochem. 29, 19–24.

[11] Chang, V.W., Wu, R., and Ho, Y.S. (1999). Recycling of anion-exchange resins for plasmid DNA purification. Biotechniques 26(6):1056

[12] Duplay, P., Bedouelle, H., Fowler, A., Zabin, I., Saurin, W., and Hofnung, M. (1984). Sequences of the *malE* gene and of its product, the maltosebinding protein of *Escherichia coli* K12. J. Biol. Chem. 259:10606-10613.

[13] Erlwein, O., Robinson, M.J., Dustan, S., Weber, J., Kaye, S., McClure, M.O. (2011). DNA Extraction Columns Contaminated with Murine Sequences. PLoS ONE 6(8): e23484. doi:10.1371/journal.pone.0023484

[14] Esser, K.H., Marx,W. H., and Lisowsky, T. (2005).Nucleic acid-free matrix: Regeneration of DNA binding columns. Biotechniques 39(2):270-271

[15] Esser, K.H., Marx, W. H., and Lisowsky, T. (2006). maxXbond: first regeneration system for DNA binding silica matrices. Nat. Met. Jan, i-ii, DOI:0.1038/NMETH845

[16] Flachmann, R., and Kühlbrandt, W. (1996). Crystallization and assembly defect of recombinant antenna complexes produced in transgenic tobacco plants. Proc. Nat. Acad. Sci. USA 93, 14966–14971.

[17] Fogel, B.L., and McNally, M.T. (2000). Trace contamination following reuse of anion-exchange DNA purification resins. Biotechniques 28(2):299-302

[18] Guan, C., Li, P., Riggs, P.D., and Inouye, H. (1987). Vectors that facilitate the expression and purification of foreign peptides in Escherichia coli by fusion to maltose-binding protein. Gene 67:21- 30.
[19] Hearon, J. (1948). The configuration of cobaltihistidine and oxy-bis (cobalthistidine). J. Natl. Cancer Inst. 9, 1–11.
[20] Hertzberg, R.P., and Dervan, P.B. (1982) Cleavage of double-helical DNA by (Methidiumpropyl- EDTA) iron(II). J. Am. Chem. Soc. 104: 313-315.
[21] Hochuli, E. (1990). Purification of recombinant proteins with metal chelate adsorbent. In Genetic Engineering, Principles and Practice, Vol. 12 (J. Setlow, ed.) pp. 87-98. Plenum, New York.
[22] Hochuli, E., Dobeli, H., and Schacher, A. (1987). New metal chelate adsorbent selective for proteins and peptides containing neighbouring histidine residues. J. Chromatogr. 411:177-184.
[23] Hochuli, E., Bannwarth, W., Dobeli, H., Gentz, R., and Stüber, D. (1988). Genetic approach to facilitate purification of recombinant proteins with a novel metal chelate adsorbent. Bio/Technology 6:1321-1325
[24] Hochuli, E. (1989) Genetically designed affinity chromatography using a novel metal chelate absorbent, Biologically Active Molecules 217–239.
[25] Hoffmann, A. and Roeder, R. (1991). Purification of His-tagged proteins in nondenaturing conditions suggests a convenient method for protein interaction studies. Nucl. Acids Res. 19:6337-6338
[26] Hutchens, T.W. and Yip, T.-T. (1990a). Protein interactions with immobilized transition metal ions: Quantitative evaluations of variations in affinity and binding capacity. Anal. Biochem. 191:160-168.
[27] Hutchens, T.W. and Yip, T.-T. (1990b). Differential interaction of peptides and protein surface structures with free metal ions and surface-immobilized metal ions. J. Chromatogr. 500:531-542
[28] Jungbauer, A., Kaar, W., and Schlegl, R. (2004). Folding and refolding of proteins in chromatographic beds. Curr. Opin. Biotechnol. 15, 487–494.
[29] Kaslow, D. C., and Shiloach, J. (1994). Production, purification and immunogenicity of a Malaria transmission-blocking vaccine candidate: TBV25H expressed in yeast and purified using Ni-NTA Agarose. Biotechnology 12, 494–499.
[30] Kellerman, O.K. and Ferenci, T. (1982). Maltose binding protein from E. coli. Methods Enzymol. 90:459-463.
[31] Kim, A.I., Hebert, S.P. and Denny, C.T. (2000). Cross-contamination limits the use of recycled anion exchange resins for preparing plasmid DNA. Biotechniques 28(2):298
[32] Lee, R.T., Ichikawa, Y., Allen, H.J,, Lee, Y.C. (1990). Binding characteristics of galactoside-binding lectin (galaptin) from human spleen.J Biol Chem. May 15;265(14):7864-71.
[33] Le Grice, S.F.J. and Grueninger-Leitch, F. (1990). Rapid purification of homodimer HIV-I reverse transcriptase by metal chelate affinity chromatography. Eur. J. Biochem. 187, 307–314.
[34] Lindahl, T. and Nyberg, B. (1972) Rate of depurination of native deoxyribonucleic acid. Biochemistry, 11 (19), pp 3610–3618
[35] Maina, C.V., Riggs, P.D., Grandea, A.G. III, Slatko, B.E., Moran, L.S., Tagliamonte, J.A., McReynolds, L.A., and Guan, C. (1988). A vector to express and purify foreign proteins in Escherichia coli by fusion to, and separation from, maltose binding protein. Gene 74:365-373.

[36] Muszynska, G., Dobrowolska, G., Medin, A., Ekman, P., and Porath, J.O. (1992). Model studies on iron(III) affinity chromatography. II. Interaction of immobilized iron (III) ions with phosphorylated amino acids, peptides and proteins. J. Chromatogr. 604:19-28.
[37] Nicosia, A., Tagliavia, M., Costa, S. (2010). Regeneration and recycling of total RNA purification silica columns. Biomed. Chrom. 24(12):1263-4
[38] Nikaido H. (1994). Maltose transport system of Escherichia coli: an ABC-type transporter. FEBS Lett. Jun 6;346(1):55-8.
[39] Nominé, Y., Ristriani, T., Laurent, C., Lefèvre, J.F., Weiss, E., Travé, G. (2001). A strategy for optimizing the monodispersity of fusion proteins: application to purification of recombinant HPV E6 oncoprotein. Protein Eng. Apr; 14(4):297-305.
[40] Pattenden, L.K., Thomas, W.G. (2008) Amylose affinity chromatography of maltose-binding protein: purification by both native and novel matrix-assisted dialysis refolding methods. Methods Mol Biol.;421:169-89.
[41] Prabhu, H.R. and Krishnamurthy, S. (1993). Ascorbate-dependent formation of hydroxyl radicals in the presence of iron chelates. Indian J. Biochem. Biophys. 30(5):289-92.
[42] Porath, J., Carlsson, J., Olsson, I., and Belfrage, G. (1975). Metal chelate affinity chromatography, a new approach to protein fractionation. Nature 258:598-599
[43] Posewitz, M. C., and Tempst, P. (1999). Immobilized gallium (III) affinity chromatography of phosphopeptides. Anal. Chem. 71, 2883–2892.
[44] Randall, L.L., Topping, T.B., Smith, V.F., Diamond, D.L., Hardy, S.J. (1998). SecB: a chaperone from *Escherichia coli*. Methods Enzymol.; 290:444-59.
[45] Sachdev, D., and Chirgwin, J.M., (1998). Solubility of proteins isolated from inclusion bodies is enhanced by fusion to maltose-binding protein or thioredoxin. Protein Expr. Purif. 12, 122.
[46] Schäfer, F., Blümer, J., Römer, U., and Steinert, K. (2000). Ni-NTA for large-scale processes — systematic investigation of separation characteristics, storage and CIP conditions, and leaching. QIAGEN. QIAGEN News 4, 11–15.
[47] Sheppard, T.L., Ordoukhanian, P. and Joyce G.F. (2000). A DNA enzyme with N-glycosylase activity. PNAS 97(14): 7802-7807)
[48] Siddappa, N.B., Avinash, A., Venkatramanan, M. and Ranga, U. (2007). Regeneration of commercial nucleic acid extraction columns without the risk of carry-over contamination. Biotechniques 42:186-192
[49] Stimola, A. (2007). Understanding the elements of the periodic table. 1. New York: The Rosen Publishing Group;
[50] Stüber, D., Matile, H., and Garotta, G. (1990). System for high-level production in *Escherichia coli* and rapid purification of recombinant proteins: Application to epitope mapping, preparation of antibodies, and structure-function analysis. *Immunol. Methods* 4:121-152.
[51] Sulkowski, E. (1985). Purification of proteins by IMAC. Trends Biotechnol. 3, 1–7.
[52] Swapan, S.J. and Tullius T.D. (2008). Footprinting protein-DNA complexes using the hydroxyl radical. Nature Protocols 3, 1092 - 1100
[53] Tagliavia, M., Nicosia, A., Gianguzza, F. (2009). Complete decontamination and regeneration of DNA purification silica columns. Anal. Biochem. 1;385(1):182-3
[54] Toyokuni, S. and Sagripanti J.L. (1992). Iron-mediated DNA damage: Sensitive detection of DNA strand breakage catalyzed by iron. J. Inorg. Biochem. 47(1):241-248
[55] Zhou, H., Xu, S., Ye, M., Feng, S., Pan, C., Jiang, X., Li, X., Han, G., Fu, Y., and Zou, H. (2006). Zirconium phosphonate-modified porous silicon for highly specific capture of phosphopeptide and MALDI-TOF MS analysis. J. Proteome Res. 5, 2431–2437.

16

Evaluation of Mitochondrial DNA Dynamics Using Fluorescence Correlation Analysis

Yasutomo Nomura
Department of Systems Life Engineering, Maebashi Institute of Technology, Japan

1. Introduction

Mitochondria are the sites of oxidative phosphorylation and generate ATP when electron is transferred from respiratory substrates to oxygen by a series of redox reaction in which respiratory enzymes pump protons across the mitochondrial inner membrane from the matrix space[1]. In isolated mitochondria as well as in intact cells, respiration frequently produces reactive oxygen species (ROS). Especially, ROS increased if respiration is perturbed, e.g., ischemia-reperfusion injury [2]. ROS can attack almost all biomolecules unspecifically. In case of DNA, ROS causes single- and double –strand breaks, and base damage[3]. Nevertheless mitochondria metabolize them only partially. Due to the defense system, mitochondrial DNA (mtDNA, ~17 kbp) that mitochondria contain independently of nucleus is particularly vulnerable because it is partially associated with the inner mitochondrial membrane as shown in Fig.(**1A**) [4]. Moreover, mtDNA repair system is weaker than that of nucleus[5]. Since mtDNA codes a part of respiratory enzymes[6], the damages would be harmful to mitochondrial function. On the other hand, because mtDNA forms a complex with proteins and is not naked, a concept that mtDNA have a resistance against ROS is also favored[7]. Therefore, one may need to reconsider mtDNA damage using a newly developed methodology which is able to detect symptoms failed to be found in the previous studies. Among symptoms, the changes in mtDNA dynamics are noticed.

2. mtDNA nucleoid

As shown in Fig.(**1B**), mtDNA molecules are usually clustered within mitochondria as protein-DNA complexes called nucleoid[8]. Cells contain tens to hundreds of nucleoid dependent on the species, growth conditions, differentiation, developmental stage and so on. Each nucleoid contains several mtDNA copies. Among nucleoid proteins reported previously, main proteins were (i) transcription factor A of mitochondria, TFAM[9], (ii) mitochondrial single-stranded DNA binding protein, mtSSB[10], (iii) mtDNA helicase, Twinkle[11], and (iv) mtDNA polymerase, POLG[8]. These proteins would participate in the maintenance of mtDNA. When a cell divides during cell cycle, daughter cells need to receive mtDNA. However, since mtDNA damage probably affects the interaction with nucleoid proteins, transmission of damaged mtDNA may differ from that of intact mtDNA as discussed in heteroplasmy[15]. Indeed, in contrast to neutral polymorphisms, severe mtDNA mutations responsible for diseases in a heteroplasmic state, almost never returned

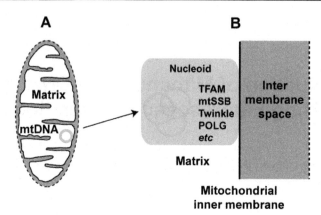

Fig. 1. Structural organization of mitochondrial DNA, mtDNA. (A) Mitochondria have own genetic materials within the matrix. (B) mtDNA nucleoid structure. Several mtDNA molecules within a nucleoid associate to mitochondrial inner membrane through the complex formed with nucleoid proteins such as TFAM, mtSSB, Twinkle, POLG and so on (see text).

to homoplasmy[16]. Large deletions (~5 kb) of mtDNA were very rarely transmitted to the offspring[17]. It was reported that deleterious heteroplasmic mtDNA mutation occurred in early stage of development of primary oocytes from a woman carrying a mtDNA mutation responsible for MELAS, mitochondrial encephalomyopathy with lactic acidosis and stroke-like episodes[18].

Moreover, mitochondria itself are dynamic organelles in post mitotic state as well [12]. When mitochondria move along cytoskeletal tracks, each mitochondrion encounters and undergoes fusion[13]. Consequently, mitochondrial networks spread within an entire cell in some cases. On the other hand, each mitochondrion yields two or more shorter mitochondria when fission occurs[14]. Therefore, in addition to mtDNA damage, mutations in nuclear-encoded genes that play an important role in mitochondria dynamics would also cause changes in mtDNA dynamics. For example, heterozygous mutations in optic atrophy gene 1 product, OPA1, caused autosomal dominant optic atrophy, the most common heritable form of optic neuropathy[19]. Mutations in gene encoding Twinkle or POLG cause the autosomal dominant progressive external opthalmoplegia[11, 20]. Although the organization of nucleoid proteins in fixed cells was revealed by immunocytochemistry, effects of the organization on mtDNA dynamics remained fully unclear. Methods by which mtDNA dynamics can be evaluated would permit to diagnose biopsy samples from patients suspected of having these diseases. Moreover, when we search candidates for compounds that modulate mtDNA dynamics by high throughput screening, image correlation method is one of the useful methods.

3. Direct measurement of mtDNA dynamics

Time-lapse fluorescence microscopy was successfully used to study mtDNA dynamics. When nucleoids were observed in cells that expressed GFP-Twinkle, the average displacement velocity of GFP-Twinkle spots was 0.01 μm/s [21]. However, in this method, because only a few mtDNAs selected within a cell were analyzed, it is probably hard to determine mtDNA dynamics in a whole cell. In order to analyze mtDNA dynamics in wide area of cytoplasm, image correlation spectroscopy (ICS) are proper because fluorescent particles, namely mtDNAs, are not selected. ICS is an imaging analog of fluorescence correlation spectroscopy (FCS)[22]. Therefore, prior to ICS, principles of FCS are described. In FCS, fluorescent molecules entering a tiny detection area generated by confocal optics emit photons and those exiting the area due to Brownian motion cease to emit them[23, 24]. Based on the fluctuations of fluorescence intensity, the motion of fluorescent molecules is evaluated as diffusion constant. As shown in Fig.(2), the fluctuation signal is dependent on molecular weight and the number of molecules.

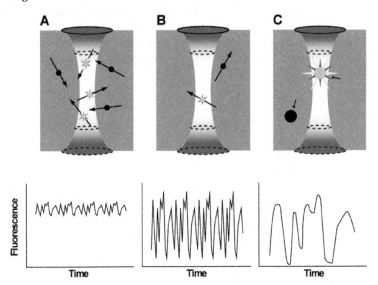

Fig. 2. Effect of molecule number and molecular weight on fluorescence fluctuation. (A) When fluorescent molecules enter and exit tiny detection area (confocal volume element) due to Brownian motion, fluorescence intensity fluctuates. (B) When the fluorescent molecules decrease, the relative fluctuation of fluorescence intensity against the average value increases. (C) When the fluorescent molecules become larger, the fluctuation become slower.

Since fluorescence intensity fluctuates with only a few fluorescent molecules diffusing in and out of the volume element, the intensity at time t, $I(t)$, changes into $I(t+\tau)$, τ seconds later. The normalized autocorrelation function commonly used is calculated from the random fluctuation of fluorescence intensity:

$$G(\tau) = \frac{\langle I(t)I(t+\tau)\rangle}{\langle I(t)\rangle^2} \quad (1)$$

To evaluate the experimentally obtained autocorrelation function, the following analytical expression has been derived [25]:

$$G(\tau) = 1 + \frac{1}{N}\left(1 + \frac{\tau}{\tau_d}\right)^{-1}\left(1 + \frac{s^2\tau}{\tau_d}\right)^{-\frac{1}{2}} \qquad (2)$$

In fact, the equation indicates simple diffusion properties, and then represents the time-dependent correlation function for translational diffusion based on fluorescence fluctuation due to Brownian motion of three dimensions, where N is the average number of molecules in the volume element. τ_d is the diffusion time that the molecules take to traverse the detection area in the radial direction. s is the ratio of the axial half-axis to the lateral half-width of the detection area and it can be previously obtained with an authentic material such as rhodamine 6G. When Eq. (2) is fitted to the experimentally obtained autocorrelation function, τ_d and N can be obtained. Although Eq. (2) represents a one-component model for the autocorrelation function, depending on the application, practically, a two-component [26-28] or multicomponent model [29, 30], or even analytical expression for the cross-correlation function [31, 32] is also adopted.

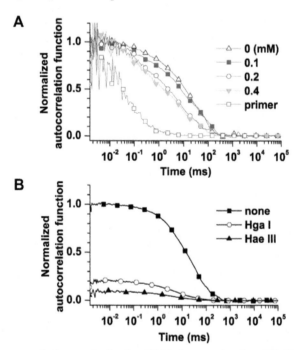

Fig. 3. Fluorescence correlation analysis of mtDNA damage *in vitro*. (A) Changes in normalized autocorrelation functions of long PCR products (~17 kbp) for mtDNA isolated from the cells exposed to H_2O_2 (0 ~ 0.4 mM). For comparison, normalized autocorrelation function of primer is also shown. A decrease in fraction of slow-moving components (long PCR products) shifted normalized autocorrelation function to the left hand side. (B) Effect of restriction digestion on the normalized autocorrelation function of long PCR products for mtDNA. An increase in fluorescent molecules due to the fragmentation resulted in the decrease in amplitude of autocorrelation function.

In order to estimate the vulnerability of mtDNA to oxidative stress, using FCS, the complete mtDNA genome isolated from the cells exposed to H_2O_2 was amplified by long PCR and the product (~17 kbp) was fluorescently labeled with an intercalating dye, YOYO-1 [34]. As shown in Fig.(**3A**), normalized autocorrelation function (normalized ($G(\tau)$-1)) of long PCR for mtDNA product was shifted to the left with the increment of H_2O_2 concentration. When the data were analyzed by a 2-component model, a decrease in the slow component due to mtDNA damage was revealed. In further study, we quantified size distribution of restriction fragments in long PCR product for mtDNA with Hga I and Hae III [30] (Fig.(**3B**)), which indicated changes in molecular number due to fragmentation. Using a multi-component model which was considered as a fragment length-weighted correlation function, we calculated the correlation amplitude expected theoretically and compared it to that measured by FCS (refer [30]). Since these were coincident well, the amplitude measured by FCS would be a very useful index for primary screening for alterations in the entire mitochondrial genome using restriction enzymes that have several polymorphic restriction sites.

4. Image correlation spectroscopy

In ICS which is an imaging analog of FCS, the raw data for image correlation analyses is an image series which is recorded as a function of space and time. The images are usually obtained from a confocal laser scanning microscope (LSM), two-photon LSM or evanescent wave imaging[35]. A generalized spatiotemporal correlation function is defined as:

$$r(\xi,\eta,\tau) = \frac{\langle \delta I(x,y,t)\delta I(x+\xi,y+\eta,t+\tau)\rangle}{\langle I(x,y,t)\rangle^2} \quad (3)$$

where a fluctuation in fluorescence, $\delta I(x, y, t)$, is given by:

$$\delta I(x,y,t) = I(x,y,t) - \langle I(x,y,t)\rangle \quad (4)$$

where $I(x,y,t)$ is the intensity at pixel (x, y) in the image recorded at time t, and $<I(x,y,t)>$ is the average intensity of that image at time t. Every image acquired on a LSM is a convolution of point spread function (PSF) for the microscope with the point-source emission from the fluorophores due to diffraction [36]. This convolution causes the signal from a point-emitter to be spread over a number of pixels. Correlation of fluctuations arising from fluorescent particles within the microscope PSF also confers some critical limitations on ICS approaches. In ICS introduced here, the spatial correlation function is firstly computed and then number of particle is obtained. Next, using the value, when the temporal correlation function is fitted to an analytical model derived from diffusion theory, diffusion coefficient is calculated.

With spatial ICS, a spatial autocorrelation function is calculated from the intensities recorded in the pixels of individual images (Fig.(**4A**)) As shown in the colored surface of Fig.(**4B**), the spatial autocorrelation function of the image is given by Eq. 3 when $\tau = 0$:

$$r(\xi,\eta,0) = \frac{\langle \delta I(x,y,t)\delta I(x+\xi,y+\eta,t)\rangle}{\langle I(x,y,t)\rangle^2} \quad (5)$$

where the angular brackets denote spatial averaging over the image, and ξ and η are spatial lag variables corresponding to pixel shifts of the image relative to itself in the x and y directions. The correlation function is then fitted to a 2D Gaussian using a nonlinear least squares algorithm (the grey mesh in Fig.(4B)):

$$r(\xi,\eta,0) = g(0,0,0) \cdot \exp\left[-\frac{\xi^2 + \eta^2}{\omega_0^2}\right] + g_\infty \qquad (6)$$

where $g(0,0,0)$ is the zero-lags amplitude, and g_∞ is the long-spatial lag offset to account for an incomplete decay of the correlation function. Fitted parameters are $g(0,0,0)$ and g_∞. In Eq.(6), a Gaussian function is used because the laser beam acts as the spatial correlator and has a Gaussian intensity profile. The zero-lags amplitude of the correlation function is inversely proportional to the number of independent fluorescent particles per beam area. The beam radius of the microscope PSF (ω_0) can be determined using methods such as imaging of fluorescent microspheres with diameter less than diffraction limit [36]. Because the size of ω_0 is wavelength-dependent, the PSF should be measured at excitation wavelength same as the ICS experiment.

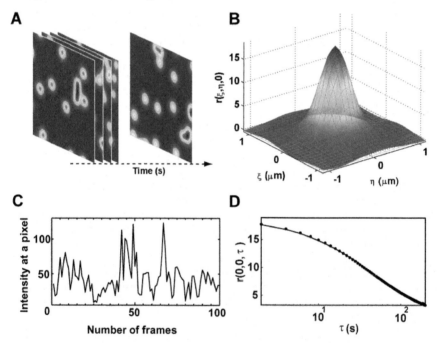

Fig. 4. Image correlation analysis of computer-generated simulations in the case of $\tau_d = 0.001$ $\mu m^2/s$, 0.1 particles/μm^2 and $\omega_0 = 0.4$ μm. (A) Temporal image series with 5 μm × 5 μm / 2 s. (B) The raw correlation function is denoted by the colored surface, and the fitted 2D Gaussian function is denoted by the grey mesh. (C) Intensity at a pixel fluctuates with number of frames separated 2 s interval. (D) Temporal image correlation function derived from temporal image series (A).

Next, as shown in Figs.(**4C** and **D**), temporal autocorrelation function of an image series as a function of time lag τ is obtained from Eq. 3 when ξ and $\eta = 0$:

$$r(0,0,\tau) = \frac{\langle \delta I(x,y,t)\delta I(x,y,t+t)\rangle}{\langle I(x,y,t)\rangle^2} \quad (7)$$

where the angular brackets denote spatial and temporal averaging. Experimentally, τ values are determined by the time between subsequent images in the image series. Depending on the microscope system used, sampling time of image acquisition is usually between 0.03 and 10 s. Here, because it can be assumed that mtDNA in cytoplasm of cells attached strongly on a culture dish behaves as 2D diffusion, the correlation function $r(0,0,\tau)$ was fitted to a simple one component model which was diffusing freely:

$$r(0,0,\tau) = \frac{g(0,0,0)}{\left(1+\dfrac{\tau}{\tau_d}\right)} + g_\infty \quad (8)$$

where $g(0,0,0)$ is the zero-lags amplitude dependent on number of fluorescent particles, and g_∞ is the long-time offset. For confocal excitation, the characteristic diffusion time, τ_d is related to the diffusion coefficient, D by:

$$D = \frac{\langle\omega_0\rangle^2}{4\tau_d} \quad (9)$$

where ω_0 is e^{-2} radius of the focused beam of the microscope.

5. mtDNA dynamics

Prior to analysis of mtDNA dynamics in living cells, mtDNA localization was determined by cross-correlation analysis of dual-labeled images with specific dyes for mtDNA (PicoGreen [37], PG) and mitochondria (MitoTracker Deep Red, MT) as shown in Fig.(**5A**). The cross-correlation function was calculated by shifting the red image over a distance ξ in the x-direction with respect the green image with $-4 \leq \xi \leq 4\mu m$ [38]. For each value of ξ, Pearson's correlation coefficient $r_p(\xi)$ was calculated according to:

$$r_p(\xi) = \frac{\sum_{(x,y)}\left[I_g(x,y)-\langle I_g\rangle\right]\left[I_r(x+\xi,y)-\langle I_r\rangle\right]}{\sqrt{\sum_{(x,y)}\left[I_g(x,y)-\langle I_g\rangle\right]^2}\sqrt{\sum_{(x,y)}\left[I_r(x,y)-\langle I_r\rangle\right]^2}} \quad (10)$$

where $I_g(x,y)$ and $I_r(x,y)$ are the intensity of green and red channel at pixel (x,y), and $<I_g>$ and $<I_r>$ are average intensity, respectively. As shown in Fig.(**5B**), the cross-correlation function was obtained by plotting $r_p(\xi)$ against ξ. In principle, a cross-correlation function can be determined for shifts in any direction of the x, y, z-space, but shifts in the x,y-plane are

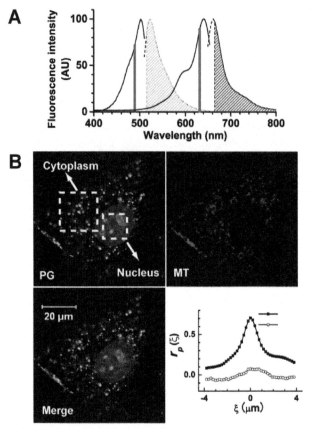

Fig. 5. Localization of mtDNA within mitochondria. (**A**) fluorescence spectra of suspension of cells dual-stained with PicoGreen (ex. 488 nm) for mtDNA and MitoTracker Deep Red (ex. 633 nm) for mitochondria. A solid line shows each excitation spectrum. Using LSM, the hatched area of emission spectra (broken line) was observed. (**B**) Confocal images of mtDNA (PG) and mitochondria (MT) are merged. In right panel, cross-correlation analysis of a dual-labeled image is shown in cytoplasmic (filled squares) and nuclear area (open circles).

preferred because of the limited z-resolution of LSM. For simplicity, cross-correlation function for shifts in the x-direction was analyzed here. As shown in Fig.(5B), in contrast to nucleus, PG signals in cytoplasm was partially localized within mitochondria, as confirmed by $r_p(0)$. Therefore, PG signal would show mtDNA.

As shown in Fig.(6), temporal autocorrelation function of mtDNA in living cells could be fitted well using Eq.8. Average of diffusion coefficient of mtDNA was 9.4×10^{-3} μm^2/s (25 cells), was stable among 0.2~2 s of sampling time, and was comparable to that of mitochondria (8.2×10^{-3} μm^2/s) although graphical data are not shown. Therefore, using sequential frames acquired by a LSM, ICS allowed evaluating both dynamics simultaneously. Diffusion coefficient of mtDNA in single living cells was two digits smaller than that *in vitro* [34], and was comparable to that of mitochondria. This suggests that mtDNA would be bound to an internal structure of mitochondria matrix.

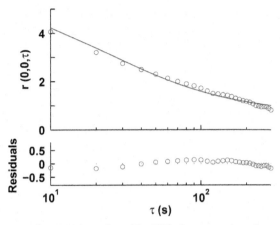

Fig. 6. Typical dynamics of mtDNA analyzed by ICS. In upper panel, temporal image correlation function of raw data (open circles) and fitted analytical model (solid line) are shown. Lower panel is the residuals.

6. Concluding remarks

Because diffusion coefficient depends on molecular weight and/or molecular interaction, diffusion coefficient of mtDNA would allow detecting large deletion of ~5 kb and abnormal transmission. In contrast to recent methods studying mitochondrial genetics such as its RNA expression (e.g., [39]), the present technique is useful for quantifying mtDNA dynamics in single living cells.

7. Acknowledgments

This study was financially supported in part by Grant-in-Aid for Scientific Research (C) (2) (17500299) and Knowledge Cluster Initiative (Hakodate Marine Bio-Industrial Cluster) from the Ministry of Education, Science, Sports and Culture of Japan, and Research for Promoting Technological Seeds (02-297) from Japan Science and Technology Agency.

8. References

[1] Santos, J.H.; Hunakova, L.; Chen, Y.; Bortner, C., and Van Houten, B. (2003) *J Biol Chem*, 278(3): 1728.
[2] Das, D.K. and Maulik, N. (2003) *Arch Biochem Biophys*, 420(2): 305.
[3] Demple, B. and Harrison, L. (1994) *Annu Rev Biochem*, 63: 915.
[4] Iborra, F.J.; Kimura, H., and Cook, P.R. (2004) *BMC Biol*, 2: 9.
[5] Yakes, F.M. and Van Houten, B. (1997) *Proc Natl Acad Sci U S A*, 94(2): 514.
[6] Wallace, D.C. (1999) *Science*, 283(5407): 1482.
[7] Spelbrink, J.N. (2010) *IUBMB Life*, 62(1): 19.
[8] Bogenhagen, D.F.; Rousseau, D., and Burke, S. (2008) *J Biol Chem*, 283(6): 3665.
[9] Alam, T.I.; Kanki, T.; Muta, T.; Ukaji, K.; Abe, Y.; Nakayama, H.; Takio, K.; Hamasaki, N., and Kang, D. (2003) *Nucleic Acids Res*, 31(6): 1640.
[10] Maier, D.; Farr, C.L.; Poeck, B.; Alahari, A.; Vogel, M.; Fischer, S.; Kaguni, L.S., and Schneuwly, S. (2001) *Mol Biol Cell*, 12(4): 821.

[11] Spelbrink, J.N.; Li, F.Y.; Tiranti, V.; Nikali, K.; Yuan, Q.P.; Tariq, M.; Wanrooij, S.; Garrido, N.; Comi, G.; Morandi, L.; Santoro, L.; Toscano, A.; Fabrizi, G.M.; Somer, H.; Croxen, R.; Beeson, D.; Poulton, J.; Suomalainen, A.; Jacobs, H.T.; Zeviani, M., and Larsson, C. (2001) *Nat Genet*, 28(3): 223.
[12] Liesa, M.; Palacin, M., and Zorzano, A. (2009) *Physiol Rev*, 89(3): 799.
[13] Detmer, S.A. and Chan, D.C. (2007) *Nat Rev Mol Cell Biol*, 8(11): 870.
[14] Berman, S.B.; Pineda, F.J., and Hardwick, J.M. (2008) *Cell Death Differ*, 15(7): 1147.
[15] Malka, F.; Lombes, A., and Rojo, M. (2006) *Biochim Biophys Acta*, 1763(5-6): 463.
[16] Chinnery, P.F.; Thorburn, D.R.; Samuels, D.C.; White, S.L.; Dahl, H.M.; Turnbull, D.M.; Lightowlers, R.N., and Howell, N. (2000) *Trends Genet*, 16(11): 500.
[17] Chinnery, P.F.; DiMauro, S.; Shanske, S.; Schon, E.A.; Zeviani, M.; Mariotti, C.; Carrara, F.; Lombes, A.; Laforet, P.; Ogier, H.; Jaksch, M.; Lochmuller, H.; Horvath, R.; Deschauer, M.; Thorburn, D.R.; Bindoff, L.A.; Poulton, J.; Taylor, R.W.; Matthews, J.N., and Turnbull, D.M. (2004) *Lancet*, 364(9434): 592.
[18] Brown, D.T.; Samuels, D.C.; Michael, E.M.; Turnbull, D.M., and Chinnery, P.F. (2001) *Am J Hum Genet*, 68(2): 533.
[19] Alexander, C.; Votruba, M.; Pesch, U.E.; Thiselton, D.L.; Mayer, S.; Moore, A.; Rodriguez, M.; Kellner, U.; Leo-Kottler, B.; Auburger, G.; Bhattacharya, S.S., and Wissinger, B. (2000) *Nat Genet*, 26(2): 211.
[20] Van Goethem, G.; Dermaut, B.; Lofgren, A.; Martin, J.J., and Van Broeckhoven, C. (2001) *Nat Genet*, 28(3): 211.
[21] Garrido, N.; Griparic, L.; Jokitalo, E.; Wartiovaara, J.; van der Bliek, A.M., and Spelbrink, J.N. (2003) *Mol Biol Cell*, 14(4): 1583.
[22] Wiseman, P.W.; Brown, C.M.; Webb, D.J.; Hebert, B.; Johnson, N.L.; Squier, J.A.; Ellisman, M.H., and Horwitz, A.F. (2004) *J Cell Sci*, 117(Pt 23): 5521.
[23] Nomura, Y.; Nakamura, T.; Feng, Z., and Kinjo, M. (2007) *Curr Pharm Biotechnol*, 8(5): 286.
[24] Rigler, R.; Mets, U.; Widengren, J., and Kask, P. (1993) *Eur. Biopys. J.* , 22: 166.
[25] Maiti, S.; Haupts, U., and Webb, W.W. (1997) *Proc Natl Acad Sci U S A*, 94(22): 11753.
[26] Kinjo, M. and Rigler, R. (1995) *Nucleic Acids Res*, 23(10): 1795.
[27] Saito, K.; Ito, E.; Takakuwa, Y.; Tamura, M., and Kinjo, M. (2003) *FEBS Lett*, 541(1-3): 126.
[28] Walter, N.G.; Schwille, P., and Eigen, M. (1996) *Proc Natl Acad Sci U S A*, 93(23): 12805.
[29] Kinjo, M.; Nishimura, G.; Koyama, T.; Mets, and Rigler, R. (1998) *Anal Biochem*,
[30] Nomura, Y.; Fuchigami, H.; Kii, H.; Feng, Z.; Nakamura, T., and Kinjo, M. (2006) *Exp Mol Pathol*, 80(3): 275.
[31] Eigen, M. and Rigler, R. (1994) *Proc Natl Acad Sci U S A*, 91(13): 5740.
[32] Saito, K.; Wada, I.; Tamura, M., and Kinjo, M. (2004) *Biochem Biophys Res Commun*, 324(2): 849.
[33] Driggers, W.J.; LeDoux, S.P., and Wilson, G.L. (1993) *J Biol Chem*, 268(29): 22042.
[34] Nomura, Y.; Fuchigami, H.; Kii, H.; Feng, Z.; Nakamura, T., and Kinjo, M. (2006) *Anal Biochem*, 350(2): 196.
[35] Kolin, D.L. and Wiseman, P.W. (2007) *Cell Biochem Biophys*, 49(3): 141.
[36] Yoo, H.; Song, I., and Gweon, D.G. (2006) *J Microsc*, 221(Pt 3): 172.
[37] Ashley, N.; Harris, D., and Poulton, J. (2005) *Exp Cell Res*, 303(2): 432.
[38] van Steensel, B.; van Binnendijk, E.P.; Hornsby, C.D.; van der Voort, H.T.; Krozowski, Z.S.; de Kloet, E.R., and van Driel, R. (1996) *J Cell Sci*, 109 (Pt 4): 787.
[39] Ozawa, T.; Natori, Y.; Sato, M., and Umezawa, Y. (2007) *Nat Methods*, 4(5): 413.

Permissions

The contributors of this book come from diverse backgrounds, making this book a truly international effort. This book will bring forth new frontiers with its revolutionizing research information and detailed analysis of the nascent developments around the world.

We would like to thank Prof. Dr. Stevo Najman, for lending his expertise to make the book truly unique. He has played a crucial role in the development of this book. Without his invaluable contribution this book wouldn't have been possible. He has made vital efforts to compile up to date information on the varied aspects of this subject to make this book a valuable addition to the collection of many professionals and students.

This book was conceptualized with the vision of imparting up-to-date information and advanced data in this field. To ensure the same, a matchless editorial board was set up. Every individual on the board went through rigorous rounds of assessment to prove their worth. After which they invested a large part of their time researching and compiling the most relevant data for our readers. Conferences and sessions were held from time to time between the editorial board and the contributing authors to present the data in the most comprehensible form. The editorial team has worked tirelessly to provide valuable and valid information to help people across the globe.

Every chapter published in this book has been scrutinized by our experts. Their significance has been extensively debated. The topics covered herein carry significant findings which will fuel the growth of the discipline. They may even be implemented as practical applications or may be referred to as a beginning point for another development. Chapters in this book were first published by InTech; hereby published with permission under the Creative Commons Attribution License or equivalent.

The editorial board has been involved in producing this book since its inception. They have spent rigorous hours researching and exploring the diverse topics which have resulted in the successful publishing of this book. They have passed on their knowledge of decades through this book. To expedite this challenging task, the publisher supported the team at every step. A small team of assistant editors was also appointed to further simplify the editing procedure and attain best results for the readers.

Our editorial team has been hand-picked from every corner of the world. Their multi-ethnicity adds dynamic inputs to the discussions which result in innovative outcomes. These outcomes are then further discussed with the researchers and contributors who give their valuable feedback and opinion regarding the same. The feedback is then collaborated with the researches and they are edited in a comprehensive manner to aid the understanding of the subject.

Apart from the editorial board, the designing team has also invested a significant amount of their time in understanding the subject and creating the most relevant covers. They scrutinized every image to scout for the most suitable representation of the subject and create an appropriate cover for the book.

The publishing team has been involved in this book since its early stages. They were actively engaged in every process, be it collecting the data, connecting with the contributors or procuring relevant information. The team has been an ardent support to the editorial, designing and production team. Their endless efforts to recruit the best for this project, has resulted in the accomplishment of this book. They are a veteran in the field of academics and their pool of knowledge is as vast as their experience in printing. Their expertise and guidance has proved useful at every step. Their uncompromising quality standards have made this book an exceptional effort. Their encouragement from time to time has been an inspiration for everyone.

The publisher and the editorial board hope that this book will prove to be a valuable piece of knowledge for researchers, students, practitioners and scholars across the globe.

List of Contributors

Yong Zhong Xu, Cynthia Kanagaratham and Danuta Radzioch
McGill University, Canada

Sally-Anne Stephenson, Inga Mertens-Walker and Adrian Herington
Queensland University of Technology, Australia

Hassan Bousbaa
Centro de Investigação em Ciências da Saúde (CICS), Instituto Superior de Ciências da Saúde Norte/CESPU, Portugal
Centro de Química Medicinal da Universidade do Porto (CEQUIMED-UP), Portugal

Rita M. Reis
Centro de Investigação em Ciências da Saúde (CICS), Instituto Superior de Ciências da Saúde Norte/CESPU, Portugal

Andrea Leibfried
EMBL Heidelberg, Germany

Yohanns Bellaïche
Institut Curie Paris, France

Olaf Bossinger
Institute of Molecular and Cellular Anatomy (MOCA), RWTH Aachen University, Aachen, Germany

Michael Hoffmann
Department of General Pediatrics, University Children's Hospital, Heinrich-Heine-University Düsseldorf, Düsseldorf, Germany

Nuri Faruk Aykan
Istanbul University, Institute of Oncology, Turkey

Andrey I. Shukalyuk
Institute of Biomaterials and Biomedical Engineering, University of Toronto, Toronto, Ontario, Canada

Valeria V. Isaeva
A. V. Zhirmunsky Institute of Marine Biology of The Far Eastern Branch of The Russian Academy of Sciences, Vladivostok and A. N. Severtsov Institute of Ecology and Evolution of The Russian Academy of Science, Moscow, Russia

Maria N. Garnovskaya
Department of Medicine (Nephrology Division), Medical University of South Carolina, Charleston, USA

David Alejandro Silva, Elizabeth Richey and Hongmin Qin
Department of Biology, Texas A&M University, College Station, USA

Elena Lazzeri, Anna Peired, Lara Ballerini and Laura Lasagni
University of Florence, Italy

V. Morales-Tlalpan and H. Barajas-Medina
Hospital Regional de Alta Especialidad del Bajío, San Carlos la Roncha, León, Guanajuato, Mexico

C. Saldaña, P. García-Solís and H. L. Hernández-Montiel
Departamento de Investigación Biomédica. Facultad de Medicina, Universidad Autónoma de Querétaro, Santiago de Querétaro, Querétaro, México

Jasmina Ilievska, Naomi E. Bishop, Sarah J. Annesley and Paul R. Fisher
Department of Microbiology, La Trobe University, Australia

Frédéric Torossian, Aurelie Bisson, Laurent Drouot, Olivier Boyer and Marek Lamacz
Institute for Research and Innovation in Biomedicine, Faculty of Medicine & Pharmacy, University of Rouen, France

Monika Sramkova, Natalie Porat-Shliom, Andrius Masedunkas, Timothy Wigand, Panomwat Amornphimoltham and Roberto Weigert
Intracellular Membrane Trafficking Unit, Oral and Pharyngeal Cancer Branch, National Institutes of Dental and Craniofacial Research, National Institutes of Health, Bethesda, USA

Marcello Tagliavia
BioNat Italia S.r.l., Palermo, Italy

Aldo Nicosia
IAMC-CNR, U.O.S. Capo Granitola, Torretta Granitola, Italy

Yasutomo Nomura
Department of Systems Life Engineering, Maebashi Institute of Technology, Japan

CPSIA information can be obtained
at www.ICGtesting.com
Printed in the USA
BVOW10*1954300416
446123BV00035B/3/P